BIOPHARMACEUTICALS,
AN INDUSTRIAL PERSPECTIVE

BIOPHARMACEUTICALS, AN INDUSTRIAL PERSPECTIVE

Edited by

Gary Walsh
*University of Limerick,
Limerick, Ireland*

and

Brendan Murphy
*Limerick Institute of Technology,
Limerick, Ireland*

KLUWER ACADEMIC PUBLISHERS
DORDRECHT / BOSTON / LONDON

Library of Congress Cataloging-in-Publication Data

ISBN 0-7923-5746-9

Published by Kluwer Academic Publishers,
P.O. Box 17, 3300 AA Dordrecht, The Netherlands.

Sold and distributed in North, Central and South America
by Kluwer Academic Publishers,
101 Philip Drive, Norwell, MA 02061, U.S.A.

In all other countries, sold and distributed
by Kluwer Academic Publishers,
P.O. Box 322, 3300 AH Dordrecht, The Netherlands.

Printed on acid-free paper

All Rights Reserved
© 1999 Kluwer Academic Publishers
No part of the material protected by this copyright notice may be reproduced or
utilized in any form or by any means, electronic or mechanical,
including photocopying, recording or by any information storage and
retrieval system, without written permission from the copyright owner.

Printed in the Netherlands.

Contents

Contributors	v
Acknowledgements	vii
Preface	ix
Biopharmaceuticals, an overview GARY WALSH	1
Abciximab: The First Platelet Glycoprotein IIb/IIIa Receptor Antagonist ROBERT E. JORDAN, MARIAN T. NAKADA, HARLAN F. WEISMAN	35
Recombinant Coagulation Factor IX (BeneFix®) JOHN EDWARDS, NEIL KIRBY	73
Biopharmaceutical Drug Development: A Case History MARYANN FOOTE, AND THOMAS BOONE	109
Follitropin beta (Puregon) HENK J. OUT	125
Insulin Lispro (Humalog) RONALD E. CHANCE, N. BRADLY GLAZER AND KATHLEEN L. WISHNER	149

Interferon beta-1b - the first long-term effective treatment of relapsing-remitting and secondary progressive multiple sclerosis (MS) R. Horowski, J.-F. Kapp, M. Steinmayr, St. Stuerzebecher	173
Reteplase, a recombinant plasminogen activator Michael Waller And Ulrich Kohnert	185
Stabilisation of biopharmaceutical products and finished product formulations Maninder S. Hora And Bao-Lu Chen	217
Patent Law for Biopharmaceuticals R. Stephen Crespi	249
The development of new medicines: an overview John C. Stinson	269
The EMEA and regulatory control of (bio)pharmaceuticals within the European Union Gary Walsh	289
Biopharmaceutical Validation: an overview Stephen Slater	311
Validation of Biopharmaceutical Chromatography Systems K.F. Williams (*) And C.J.A. Davis (**)	337
Validation of Water for Injections (WFI) for Biopharmaceutical Manufacture Paschal Baker And Wael Allan	363
Information retrieval and the biopharmaceutical industry: an introductory overview Patricia O'Donnell	389
Information technology and the internet as a resource of biopharmaceutical information J.P. Jenuth, D. Fieldhouse, J.C.-M. Yu	405
Marketing Issues for the (Bio)pharmaceutical sector Scott Spinka	421

Contents

Viral mediated gene therapy　　　　443
　　BRENDAN MURPHY

Pharmaceutical gene medicines for non-viral gene therapy　　　　471
　　A. ROLLAND, S. SULLIVAN, K. PETRAK

Index　　　　505

Contributors

Paschal Baker and Wael Allan, Raytheon Engineers & Constructors UK, Validation and GMP Compliance Group;
Ronald E. Chance, N. Bradly Glazer and Kathleen L. Wishner, Eli Lilly and Company, Inidanapolis, USA
R. Stephen Crespi, European Patent Attorney, West Sussex, UK;
John Edwards, Neil Kirby, Genetics Institute Inc., 87 Cambridge Park Drive, Cambridge, Mass.;
Maryann Foote and Thomas Boone, Amgen Inc., USA;
Maninder S. Hora and Bao-Lu Chen, Dept. of Formulation Development, Chiron Corporation, 4560 Horton Street, Emeryville, Ca 94608, USA;
R. Horowski, J.-F. Kapp, M. Steinmayr, St. Stuerzebecher, Schering AG, SBU Therapeutics, D-13342, Berlin, Germany;
J.P. Jenuth, D. Fieldhouse, J.C.-M. Yu, Base4, Bioinformatics Inc.;
Robert E. Jordan, Marian T. Nakada, Harlan F. Weisman, Centocor Inc., Malvern, Pensylvania, USA;
Brendan Murphy, University of Limerick, Ireland;
Patricia O'Donnell, University of Limerick, Ireland;
Henk J. Out, N.V. Organon, PO Box 20, 5340, BH OSS, The Netherlands;
A. Rolland, S. Sullivan, K. Petrak, Gene Medicine Inc., 8301 New Trails Drive, The Woodlands, Texas, USA
Stephen Slater, Raytherm Engineers and Constructors;
Scott Spinka, CareMerica Inc., 16508 Kingspointe, Lake Lane, Chesterfield, Mo., USA;
Dr. John C. Stinson, Leo Laboratories Ltd., Crumlin, Dublin 12, Ireland;

Dr Michael Waller and Dr Ulrich Kohnert, Boehringer Mannheim Therapeutics, Mannheim and Penzberg, Germany;

K.F. Williams, Validation Technologies (Europe) Ltd., Sutton Place, 49 Stoney St., Nottingham, NG1 1LX, UK and C.J.A. Davis, Tanvec Ltd., Alexandra Court, Carrs Road, Cheadle, SK8 2JY, UK;

Dr. Gary Walsh, University of Limerick

Acknowledgements

The editors wish to thank the authors of individual chapters for providing such excellent contributions, and for their cooperation during the post writing phase of the publication process. A special word of thanks to Sandy Lawson, for her professionalism and efficiency in reformating the chapters to comply with publication requirements. Finally, thank you to Janet Hoffman and her colleagues at Kluwer for all their help.

Preface

The beginning of the modern biotech era can be traced to the mid-1970s, with the development of recombinant DNA technology and hybridoma technology. Thus far, the most prominent applied impact of these technologies has been the successful development of biotech-derived therapeutic agents - the biopharmaceuticals. This class of pharmaceutical product has rapidly become established. The first such product, Humulin (recombinant human insulin, Eli Lilly) was approved in the USA in 1982. Today there are in excess of 50 biopharmaceutical products approved for medical use, with almost another 400 undergoing clinical trials. While all the biopharmaceutical products approved to date are protein-based, nucleic acid-derived products are likely to gain regulatory approval within the next decade.

Given the undoubted scientific and commercial prominence of this sector, relatively few books detailing biopharmaceutical products or issues of practical relevance to the biopharmaceutical industry have been published thus far. This book aims to complement the previously published texts which focus upon this area. The initial chapters are largely concerned with specific biopharmaceutical products, which have, in the main, gained regulatory approval in the relatively recent past. Subsequent chapters focus upon various issues of practical relevance to the biopharmaceutical industry, such as product stabilization, patenting and regulatory issues. The final two chapters focus upon gene therapy, a therapeutic approach currently at the cutting edge of pharmaceutical research and development.

The book, whose contributors are largely drawn from industry, is primarily aimed at an industrial audience. However, it should also prove a useful reference source to research and educational personnel with a direct interest in this field.

In conclusion, the editors wish to thank all those who have contributed to the successful completion of this book. Chief amongst these are the various chapter authors (and their employers), as well as Kluwer Academic Publishers, whose professionalism was much in evidence at all stages of the publication process. A special word of thanks is reserved for Sandy Lawson, whose patience and word processing skills yet again proved to be second to none.

Gary Walsh

Brendan Murphy

Limerick

September 1998

Chapter 1

Biopharmaceuticals, an overview

Dr. Gary Walsh
Lecturer, Industrial Biochemistry Programme, University of Limerick, Ireland

Key words: Bipharmaceuticals, drug, therapeutic agents, blood products, cytokines, gene therapy

Abstract: The modern pharmaceutical industry is barely 100 years old. Amongst the most recent product types developed are the biopharmaceuticals; therapeutic substances produced by modern biotechnological techniques. Thus far, in excess of 50 such substances have gained regulatory approval for medical use. All are proteins produced by recombinant DNA technology or (in the case of monoclonal antibodies) by hybridoma technology.

Biopharmaceuticals approved to date include blood factors, anticoagulants and thrombolytic agents, therapeutic enzymes, hormones and haemopoietic growth factors. Also approved are a number of interferons and an interleukin. Recombinant vaccines and several monoclonal antibody based products are also now on the market.

In addition to these, in excess of 350 potential biopharmaceutical products are currently under evaluation in clinical trials. Prominent amongst these is a new sub-class of biopharmaceutical - nucleic acid. Nucleic acid based products find application in the emerging therapeutic techniques of gene therapy and anti-sense technology. These techniques will likely provide medical practitioners with an additional powerful tool with which to treat conditions such as genetic diseases, cancer and infectious diseases.

The biopharmaceutical sector will continue to grow strongly for the foreseeable future. Its current global market value of $7-$8 billion is likely to triple within the next 5-6 years. This sector, born less than 20 years ago, is quickly reaching maturity.

1. DEVELOPMENT OF THE PHARMACEUTICAL INDUSTRY

The modern pharmaceutical industry is a premier global industry, both commercially and technologically. It employs several hundreds of thousands of people, and its annual global sales value exceeds $200 billion. Currently, there are over 10,000 pharmaceutical companies in existance, manufacturing over 5,000 different medicinal products. The majority of the 100 or so large multinational companies in this sector originated from Europe and the USA, with many of the remainder having been founded in Japan.

At the turn of the century, there were only 4 drugs available which had been scientifically proven to be effective in treating their target indications:

- Digitalis, which consisted of extracts of foxglove, was shown to stimulate heart muscle and, hence, proved effective in treating various heart conditions. The active ingredients were subsequently shown to be two cardiac glycosides: digoxin and digitoxin.

- Quinine, an alkaloid obtained from the bark and roots of the fever tree (Cinchona species), was found to be effective in treating Malaria.

- Pecacuanha, obtained from the bark and roots of the plant species, Cephaelis, was effective in treating dystentry. (The active ingredients of this preparation turned out to be a mixture of alkaloids).

- Mercury, which was used to treat syphilis.

From such modest beginnings, the pharmaceutical industry has grown rapidly.

Most medicines now available can be categorized into one of four groups, depending upon their method of manufacture. The majority of medicinal substances are relatively low molecular weight organic compounds manufactured by direct chemical synthesis. Others (e.g. taxol and semi-synthetic antibiotics) are obtained by semi-synthesis, while a smaller, but important, group of drugs are obtained by direct extraction from their native biological source (Table 1). The fourth group are 'products of biotechnology' or 'biopharmaceuticals'. By and large, these are protein-based therapeutic agents (Table 2). However, several nucleic acid-based biopharmaceuticals are likely to gain regulatory approval within the next few years.

Table 1. Some pharmaceuticals which may be obtained by direct extraction from biological source material. Note that, in some cases, recombinant versions of the same product are also available

Substance	Medical application
Blood products (e.g. clotting factors)	Treatment of blood disorders such as haemophilia A or B
Vaccines	Vaccination against various diseases
Antibodies	Passive immunization against various diseases
Insulin	Treatment of diabetes mellitus
Enzymes	Used as thrombolytic agents, digestive aids, debriding agents (i.e. cleansing of wounds), etc.
Antibiotics	Treatment of various infectious conditions
Plant extractives (e.g alkaloids)	Various, including pain relief

Table 2. Most biopharmaceuticals approved or in clinical trials are proteins. Functionally, they may be classified as belonging to one or other of the families of proteins listed below

Blood clotting factors	Monoclonal antibodies
Colony stimulating factors	Neurotrophic factors
Enzymes	Polypeptide anticoagulants
Growth factors	Polypeptide hormones
Interferons	Thrombolytic agents
Interleukins	Vaccines

1.1 The birth of the biopharmaceutical industry

Over the years, advances in biomedical research has identified various biomolecules synthesized naturally by the body whose therapeutic potential was obvious. Early examples include insulin and various blood clotting factors. More recently discovered examples include interferons, interleukins and other cytokines which regulate aspects of immunity, inflammation and other processes of central importance to maintaining a healthy state.

As the majority of these substances were complex macromolecules (predominantly proteins), their direct chemical synthesis proved to be technically challenging/impossible and economically unattractive. Some (e.g. blood products and various hormones) are produced naturally in quantities sufficient to facilitate their direct extraction from biological source material in medically useful quantities. In many cases, however, (e.g. most cytokines), these biomolecules are produced in exceedingly low concentrations in the body. This made their isolation difficult and routine large scale production impossible.

In addition to such problems of source availability, extraction from natural sources carried with it the possibility of accidental transmission of disease.

Well publicized examples include the accidental transmission of HIV and other blood borne viruses via infected blood products and the transmission of Creutzfeldt-Jacob disease via human growth hormone extracted from the pituitaries of deceased human donors.

The development in the 1970s of the twin technologies of genetic engineering and hybridoma technology largely overcame these problems of source availability and accidental transmission of disease. Genetic engineering essentially facilitates the production of limitless quantities of any protein of interest, while hybridoma technology allows production of limitless quantities of a chosen monoclonal antibody.

These biotechnological innovations, along with an increasing understanding of the molecular mechanisms underlining both health and disease, rendered possible the development of a new generation of biotech.-derived drugs - the biopharmaceuticals. By the late 1970s, hundreds of start-up biotechnological companies had been formed to develop such products. Most such ventures were founded in the USA, mainly by academics and technical experts in the biotech. arena. These companies were largely financed by speculative monies. While they boasted significant technical expertise, most of these companies lacked practical experience in the drug development process. In the earlier years, most of the established large pharmaceutical companies failed to appreciate the potential of biotechnology as a means to produce drugs and, consequently, were slow to invest in this technology. As its medical potential became apparent, many of these companies did diversify into this area. While some initiated biotech. efforts in house, most either acquired small established biopharmaceutical firms, or entered strategic alliances with them. An example of the latter was the alliance formed between Genentech and Eli Lilly with regard to the development and marketing of recombinant human insulin.

Many of the original biopharmaceutical companies set up in the late 1970s and 1980s no longer exist. In addition to mergers, acquisitions and alliances, many were forced out of business due to lack of capital, or disappointing clinical trial results. However, a number of the early start-up companies have successfully developed products and are now well established within the biopharmaceutical sector. Major examples include Genentech and Amgen. A list of pharmaceutical companies who now manufacture and/or market biopharmaceutical products is provided in Table 3.

Table 3. (Bio)pharmaceutical companies which manufacture and/or market biopharmaceutical products which have gained regulatory approval in the USA and/or the EU. (Note: several of these companies have a presence in both regions)

Company	Company	Company
Amgen (CA, USA)	Cytogen (NJ, USA)	Novo-Nordisk (NJ, USA)
Bayer Corp. (CT, USA)	Eli Lilly (IN, USA)	N.V. Organon (The Netherlands)
Baxter Healthcare (MA, USA)	Galenus Mannheim (Germany)	Ortho-biotech (NJ, USA)
Behringwerke A.G. (Germany)	Genentech (CA, USA)	Ortho McNeil Pharmaceuticals (NJ, USA)
Berlex Labs (NJ, USA)	Genetics Institute (MA, USA)	Pharmacia & Upjohn (ML, USA)
Biogen (MA, USA)	Genzyme (MA, USA)	Schering Plough (NJ, USA)
Bio-Technology General (NJ, USA)	Hoechst AG (Germany)	Serono Labs (MA, USA)
Boehringer-Mannheim (Germany)	Hoechst Marion Roussel (MO, USA)	SmithKline Beecham (PA, USA)
Boehringer-Ingelheim (Germany)	Hoffman La Roche (NJ, USA)	Sorin biomedica diagnostica (Italy)
Centocor (PA, USA)	Immunex (WA, USA)	
Chiron (CA, USA)	Immunomedics (NJ, USA)	
Ciba Europharm (UK)	Interferon Sciences (NJ, USA)	
CISbio (France)	Merck (NJ, USA)	

1.2 Biopharmaceuticals; market value

From a zero starting point in the edarly 1980s, the world-wide sales value of biopharmaceuticals reached US$5 billion by 1993. (The first biopharmaceutical to gain marketing authorization was recombinant human Insulin in 1982). By 1997, the global market value had surpassed the $7 billion mark. By 2003, this figure is projected to be in the region of $35 billion, which will represent some 15% of the total global pharmaceutical market (1-3). Biopharmaceuticals are amongst the most expensive of therapeutic agents. The annual cost of erythropoietin, for example, per patient per year is in the region of $4000-$6000, while that of human growth hormone can be $12,000-$18,000. In monetary terms, erythropoietin is the single largest selling biopharmaceutical product, and was the first such product to surpass an annual sales value of $1 billion. The estimated sales value of some notable biopharmaceutical products is presented in Table 4.

Table 4. Some major biopharmaceuticals currently on the market. The value of each product quoted represents its estimated annual global sales value

Biopharmaceutical	Indication	Year first approved	Value ($ million)
α-Interferon	Cancer, Viral infection	1986	1,000
β-Interferon	Multiple sclerosis, Viral infection	1993	35
γ-Interferon	Chronic granulomatous disease	1990	45
Erythropoietin	Anaemia	1989	1,800
Factor VIII	Haemophilia	1993	445
Granulocyte-colony stimulating factor	Neutropenia	1991	870
Human growth hormone	Growth deficiency	1985	660
Insulin	Diabetes mellitus	1982	1,000
Interleukin 2	Cancer	1992	50
OKT 3 Monoclonal antibody	Kidney transplant rejection	1986	160
Tissue plasminogen activator	Cardiovascular disease	1987	120

2. SOURCES AND MANUFACTURE OF BIOPHARMACEUTICAL PRODUCTS

The vast majority of biopharmaceutical products currently on the market are produced by recombinant DNA technology in either *E. coli* or Chinese Hamster ovary (CHO) cell lines (4). Most monoclonal antibody based products are predictably still produced by hybridoma technology, although the technical methodology now exists to facilitate production of antigen-binding antibody fragments by recombinant means (5, 6).

E. coli represents a popular recombinant expression system for a number of reasons (7, 8). In addition to its ease of culture and rapid growth rates, *E. coli* has long served as the model system of the prokaryotic geneticist. Its genetic characteristics are thus exceedingly well-characterized and reliable standard protocols for its genetic manipulation have been developed. Appropriate fermentation technology is well established, and high expression levels of recombinant proteins are generally attained. *E. coli*, however, does display some disadvantages as a recombinant production system. Recombinant proteins generally accumulate intracellularly, complicating downstream processing and (often more critically) *E. coli* lacks the ability to glycosylate proteins (or carry out any other post-translational modifications).

Many proteins of therapeutic interest are naturally glycosylated and lack of the carbohydrate component can, potentially, adversely affect its biological activity, solubility, or *in vivo* half-life.

Recombinant proteins may be expressed in a number of other microbial systems which do contain the enzymatic activities to facilitate post-translational processing. Various proteins have been expressed, both in yeast (particularly *Saccharomyces cerevisiae*) and fungi (especially various *Aspergilli*) (9-11). While such microorganisms are capable of glycosylating recombinant therapeutic proteins, the pattern of glycosylation usually differs to that associated with such proteins when expressed naturally in the human body. Such microbial expression systems exhibit a number of characteristic advantages and disadvantages in terms of recombinant protein production. Thus far, however, few recombinant biopharmaceuticals developed are produced in either yeast or fungal systems. Two approved biopharmaceuticals are produced in *Saccharomyces cerevisiae*: Refludan (recombinant hirudin, an anticoagulant marketed by Behringwerke AG) and recombinant hepatitis B surface antigen, incorporated into various combination vaccines by SmithKline Beecham.

More recently, a number of recombinant therapeutic proteins produced in various animal cell lines have gained marketing approval. Chinese hamster ovary (CHO) cells have become popular recombinant production systems, as have baby hamster kidney (BHK) cell lines (12). Patterns of glycosylation associated with recombinant glycoprotein biopharmaceuticals produced in such systems resemble most closely the native glycosylation pattern when the protein is produced naturally in the body.

The production of recombinant therapeutic proteins in the milk of transgenic animals has also gained much publicity over the last few years (13, 14). A variety of therapeutically significant proteins, including tissue plasminogen activator, α_1-antitrypsin, interleukin 2 and factor IX have been produced in this matter (15, 16). It is likely that therapeutic proteins produced in such systems will gain regulatory approval within the next few years.

2.1 Upstream processing

After its initial construction, the recombinant producer cell line is thoroughly characterized and its genetic stability verified. The cell line is then used to construct a 'master' and 'working' cell bank system (4). Initial stages of upstream processing invariably involves lab-scale culture of the contents of a single vial from the working cell bank. This, in turn, is used to innoculate a larger volume of media which (after cell growth) is, in turn, used to innoculate the production scale bioreactor. The scale of fermentation depends upon the level of production required, but generally production scale bioreactors would vary in capacity from one thousand litres to several tens of thousands of litres.

2.2 Downstream processing

All biopharmaceutical products must be exhaustively purified in order to remove virtually all contaminants from the product stream. Such contaminants include proteins (related or unrelated to the protein product), DNA, pyrogens, viral particles and microorganisms.

Downstream processing is initiated by recovery of the crude protein product from the fermentation media (if produced extracellularly) or cell paste (if produced intracellularly). The crude product is then usually concentrated (often by ultrafiltration, but ammonium sulphate precipitation or ion exchange chromatography may also be used). It is next subjected to high resolution chromatographic purification (17, 18). Generally, at least three different chromatographic steps (e.g. ion-exchange, gel filtration, hydrophobic interaction chromatography or affinity chromatography) are employed, yielding a product which is 98-99% pure.

While chromatographic fractionation is designed to remove contaminant proteins from the protein of interest, several chromatographic steps are also quite effective in removing additional potential contaminants from the product stream. Gel filtration chromatography, for example, is usually quite effective in removing any contaminant viruses.

After chromatography, excipients are added (19) and the product potency is adjusted by dilution/concentration as necessary. As therapeutic proteins are heat labile, product sterilization is by filtration and this is followed by aseptic filling into final product containers. Although some products may be marketed in liquid format, most are freeze dried (20, 21). Freeze dried products generally are more stable, exhibiting a longer shelf life than analogous liquid formulations. An example of a generalized biopharmaceutical production procedure is provided in Figure 1.

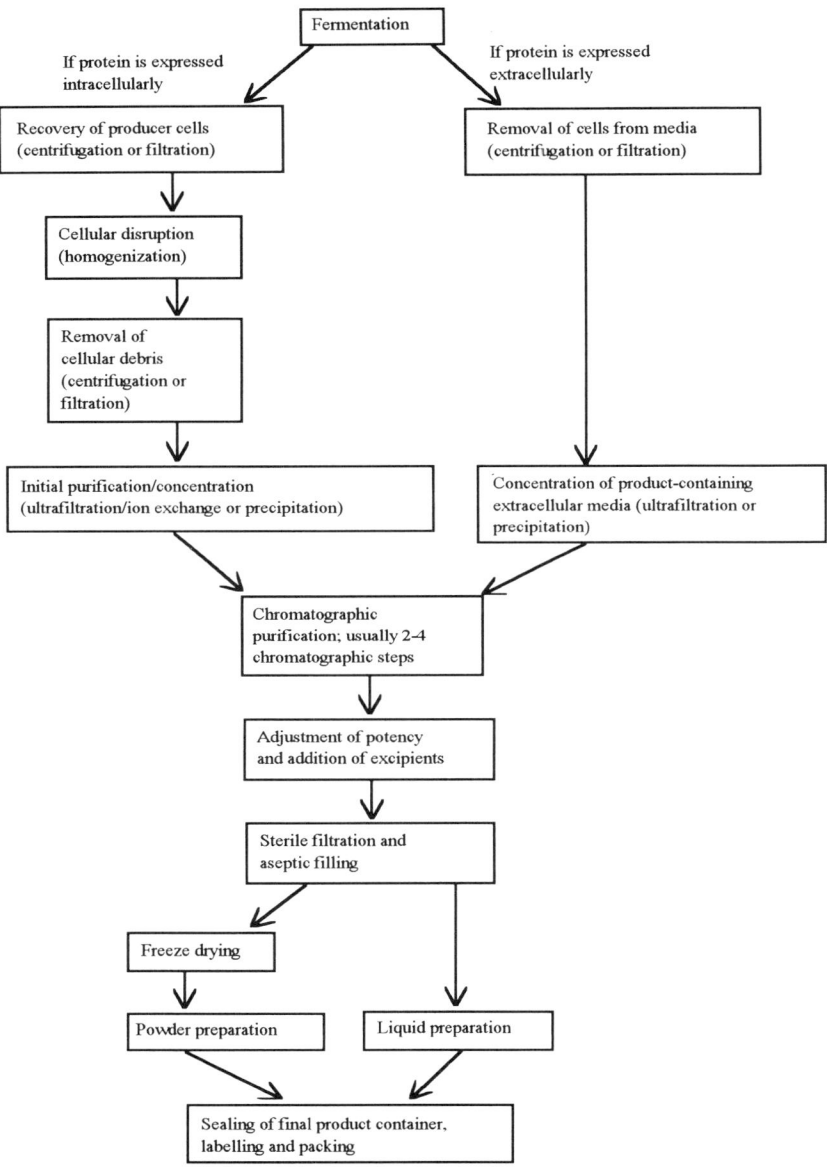

Figure 1. Overview of a generalized downstream processing procedure employed to produce a finished-product (protein) biopharmaceutical. Quality control also plays a prominent role in downstream processing. QC personnel collect product samples during/after each stage of processing. These samples are analysed to ensure that various in-process specifications are met. In this way, the production process is tightly controlled at each stage. (Reproduced from 'Biopharmaceuticals: Biochemistry and Biotechnology', J. Wiley & Sons, 1998, with kind permission of the publisher)

3. SPECIFIC BIOPHARMACEUTICAL PRODUCTS

Thus far, in excess of 50 biopharmaceutical products have gained regulatory approval in the USA and/or the EU. All are proteins, although several nucleic acid-based therapeutic agents are currently in clinical trials. Most of the products approved may be categorized into specific families, depending upon their biological activity, or mode of action. These approved products are briefly reviewed below.

3.1 Blood products

Blood and blood products constitute a major group of traditional biologics (22). The major blood products which find therapeutic application include red blood cell and platelet concentrates, plasma and plasma protein fraction, albumin and blood clotting factors. While all these products are still sourced from healthy blood donations, associated with this practice is the potential for inadvertent transmission of blood-borne pathogens. Infectious agents which can be accidently transmitted via infected blood/blood products include: HIV, hepatitis B & C viruses, cytomegalovirus, human T cell lymphocytotrophic viruses (possible causative agents of lymphoma), as well as *Treponema pallidum* (causes syphilis), *Plasmodium protozoa* (causes malaria) and *Trypanosoma cruzi* (causes Chagas' disease).

A number of protein-based products, particularly blood clotting factors are now also produced by recombinant DNA technology. This essentially eliminates the risk of disease transmission and ensures a regular supply of product.

3.1.1 Blood clotting factors

The human body naturally produces 12 blood clotting factors, generally designated by Roman numerals (factors I-XIII; there is no factor VI). All but one are proteins and most are proteolytic precursors which become sequentially activated during the blood coagulation cascade. Any defect which impedes the biological activity of any blood factor can result in a severely retarded clotting ability. Genetic defects in all factors (except factor IV, i.e. calcium) have been characterized. However, up to 90% of such defects relate to factor VIII, while most of the remainder relate to factor IX. Poorly functional/dysfunctional factors VIII and IX result in haemophilia A and B, respectively, conditions treatable only by periodic administration of the appropriate clotting factor. Some recombinant blood clotting factors which have been approved for general medical use, or which are in clinical trials, are listed in Table 5.

Table 5. Recombinant blood factors which have gained marketing approval (in the US and/or the EU), as well as such products which are currently being assessed in clinical trials. Data sourced from PhRMA (http://www.phrma.org) and the EMEA (http://www.eudra.org/emea.html).

Product name	Company	Indication	Status
Benefix (r factor IX)	Genetics Institute, (MA, USA)	Haemophilia B	Approved 1997 (US)
	Genetics Institute, (Europe, France)	Haemophilia B	Approved 1997 (EU)
KoGENate (r factor VIII)	Bayer Corp (CT, USA)	Haemophilia A	Approved 1993 (USA)
Recombinate (r factor VIII)	Baxter Healthcare (CA, USA) and Genetics Institute (MA, USA)	Haemophilia A	Approved 1992 (USA)
NovoSeven (r VIIa)	Novo-Nordisk, Denmark	Prevents bleeding in patients with inhibitors to coagulation factor VIII or factor IX	Approved 1995 (EU)
	Novo-Nordisk, (NJ, USA)	Haemophilia A & B	Phase III trials (USA)

3.1.2 Anticoagulants

The inappropriate formation of a blood clot (thrombus) within a diseased blood vessel can have serious, if not fatal, medical consequences such as heart attacks and strokes. Anticoagulants are substances which can prevent blood clot formation and, hence, are applied therapeutically in cases where high risk of inappropriate blood clot formation is diagnosed (23). Traditional anticoagulants include heparin, dicoumarol and warfarin.

Heparin is a proteoglycan (highly glycosylated polypeptide) which is sourced commercially from beef lung or pig gastric mucosa. It functions by binding - and thus activating - a plasma protein: antithrombin III. The heparin-antithrombin III complex then binds a number of activated clotting factors. This binding inactivates the clotting factors, thus preventing clot formation. Although heparin is an effective and inexpensive anticoagulant, it can display a poorly predictable dose response and it can display a narrow benefit : risk ratio.

Dicoumarol and warfarin are low molecular weight, coumaran-based anticoagulants which can prevent the post-translational modification of various clotting factors, thus also rendering them inactive.

More recently, a protein-based anticoagulant has been developed. Refludan (lepirudin) is a hirudin-based anticoagulant which gained a marketing licence in the EU in 1997, and in the USA in 1998. It is approved for the treatment of adult patients with heparin-associated thrombocytopenia type II, and thromboembolic disease.

Hirudin was first noted in the 1880s as a major anticoagulant present in the saliva of leeches (24). It was purified in the late 1950s and found to be a 65 amino acid polypeptide containing a tyrosine residue at position 63 which is normally sulphated. Its anticoagulant activity is due to its ability to bind (and induce inactivation of) thrombin (factor IIa). The hirudin gene was cloned in the 1980s and expressed in various microbial systems. Refludan is produced commercially in yeast cells (*Saccharomyces cerevisiae*) transfected with an expression vector containing the hirudin gene. It is presented as a freeze dried powder which also contains the excipient mannitol as a bulking and tonicity agent. Unlike the native molecule, the recombinant form does not exhibit a sulphate group on tyrosine 63, but this has no major impact upon its anticoagulant activity.

3.1.3 Thrombolytic agents

In cases of inappropriate clot formation in a blood vessel, the level of tissue damage induced often depends upon how long the clot deprives the effected area of oxygen. Rapid clot removal can limit this damage, and a number of thrombolytic (clot degrading) agents are used medically for this purpose (25, 26). (In the USA alone, an estimated 1.5 million people suffer acute myocardial infarction each year, while an additional 0.5 million suffer strokes). Traditional thrombolytic agents include streptokinase (a protein produced by several strains of *Streptococcus haemolyticus* Group C), and urokinase (a serine protease produced in the kidney and which can be purified from urine).

The thrombolytic process, as it occurs naturally, is triggered by a 527 amino acid proteolytic enzyme, tissue plasminogen activator (tPA). tPA proteolytically converts the inactive protease plasminogen into active plasmin. Plasmin then proteolytically degrades fibrin, the major structural protein found in clots. The medical potential of tPA was obvious for many years, but its low levels of synthesis in the body precluded its medical use. The tPA gene was cloned in 1983, facilitating large-scale production of the protein. The gene has been expressed in both procaryotic systems (e.g. *E. coli*) and in an animal (CHO) cell line. Several recombinant tPA products have now gained marketing approval (Table 6) (27).

Table 6. Recombinant tissue plasminogen activator-based products which have gained marketing approval, or are in clinical trials. Data sourced from PhRMA (http://www.phrma.org) and the EMEA (http://www.eudra.org/emea.html).

Product name	Company	Indication	Status
Activase	Genentech (CA, USA)	Acute myocardial infarction	Approved 1987 (USA)
		Acute massive pulmonary embolism	Approved 1990 (USA)
		Acute myocardial infarction (accelerated infusion)	Approved 1995 (USA)
		Ischemic stroke	Approved 1996 (USA)
Retevase	Boehringer Manheim (MD, USA) and Centocor (PA, USA)	Acute myocardial infarction	Approved 1996 (USA)
Ecokinase	Galenus Mannheim (Germany)	Acute myocardial infarction	Approved 1996 (EU)
Rapilysin	Boerhinger Mannheim (Germany)	Acute myocardial infarction	Approved, 1996 (EU)
Lanoteplase	Bristol-Myers Squibb (NJ, USA)	Acute myocardial infarction	Phase III clinical trials
TNK	Genentech (CA, USA)	Acute myocardial infarction	Phase III clinical trials

3.2 Therapeutic enzymes

A variety of enzymes are used for therapeutic purposes (28, 29). Some (e.g. tPA and urokinase) have already been discussed. Traditional (non-recombinant) enzymes used for medical purposes includes asparaginase (used to treat some forms of leukaemia) as well as lactase, pepsin, papain and pancrelipase used as digestive aids. Proteolytic enzymes, such as trypsin, collagenase and pepsin, have also gained limited use as debriding and anti-inflammatory agents.

In the last few years, a number of recombinant enzymes have also gained marketing applications. These include DNase (Pulmozyme; dornase α) and glucocerebrosidase (cerezyme).

Pulmozyme, produced by Genentech, was first approved for treatment of Cystic Fibrosis in 1993. The most notable clinical symptom of Cystic Fibrosis (CF) is the production of an extremely viscous mucus in the lungs, which compromises respiratory function. The physiological changes induced in the lung of CF patients makes it susceptible to frequent, recurrent microbial infection. This, in turn, attracts phagocytes and other immune elements. The resultant destruction of the microbial (and some immune) cells results in a build-up of large quantities of free DNA which is extremely viscous. Until recently, the only way to successfully dislodge the mucus was by percussion therapy (physical pounding of the patient's chest to dislodge the mucus, allowing the patient to expel it). Delivery into the lung of recombinant DNase by aerosol technology promotes degradation of the free DNA, reducing its viscosity significantly. This allows the patient to expel it with greater ease (30, 31). The annual cost of treatment varies but often falls in the $10,000 - $15,000 range.

Gaucher's disease is a relatively rare genetic condition in which sufferers lack the enzyme, glucocerebrosidase. This compromises their ability to degrade glucocerebiosides (a specific class of lipid). Clinical consequences include enlargement and reduced function of the spleen and liver, bone damage and, on occasion, mental retardation.

The effects of this disease can be minimized by enzyme replacement therapy. Ceredase is a commercial glucocerebrosidase preparation extracted directly from plancentae obtained from maternity hospitals. Its low expression level in the placenta renders this product very expensive to produce. In 1994, a recombinant version (Cerezyme, produced by Genzyme) gained marketing approval. The current annual global market for this product is estimated at $200 million.

3.3 Recombinant therapeutic hormones

A number of recombinant therapeutic hormones have now gained marketing approval (Table 7). In fact, the first ever product of genetic engineering to gain regulatory approval as a medicine was Humulin (recombinant human insulin). Marketed by Eli Lilly, it was first granted regulatory approval in the USA in October 1982.

Insulin was first used medically in 1921 and, for the following 50 years or more, it was sourced from either porcine or bovine pancreatic tissue. In the 1970s, a method was developed which allowed the enzymatic conversion of porcine insulin into human insulin (insulin from these two species differ in sequence only by a single amino acid). Initially, recombinant insulin was produced by separate expression of insulin A and B chains in 2 different *E. coli* cells (both K12 strains) (32). After purification of the two chains, they were co-incubated under oxidizing conditions. This promotes interchain disulphide bond formation yielding mature insulin. Subsequently, an alternative method was developed which entails the expression in *E. coli* of a nucleotide sequence coding for human proinsulin. Purification of proinsulin is followed by *in vitro* proteolytic excision of the connecting (C) peptide, yielding mature insulin.

A more recently approved insulin product is Humalog (Eli Lilly). Humalog consists of insulin lispro, a human insulin analogue produced by recombinant DNA technology in *E. coli*. The amino acid sequence of insulin lispro is identical to that of human insulin except for an inversion of the natural proline-lysine sequence of the insulin B chain at positions 28 and 29. This modification produces an insulin product of quicker and shorter duration of therapeutic action. It is thus a short-acting insulin which can be administered to diabetics immediately before meals.

An additional recombinant hormone preparation which gained regulatory approval in the 1980s was Protropin (recombinant human growth hormone, hGH). It was approved by the FDA in 1985 for the treatment of growth deficiency in children. Since then, various additional recombinant hGH preparations have gained approval for this and additional supplementary indications (Table 7) (33, 34).

The approval of a recombinant form of hGH in 1985 coincided with the banning of the use of hGH preparations extracted directly from the pituitaries of deceased human donors. In that year, it was discovered that a young man who had died from Creutzfeldt-Jacob disease contracted this fatal condition from an infected batch of pituitary-derived hGH. (Unlike insulin, for example, growth hormone is relatively species specific, so animal-derived preparations exhibit little or no biological activity when administered to humans).

Recombinant follicle stimulating hormone (FSH) preparations have now also gained marketing approval (Table 7). FSH is a prominent member of the gonadotrophins, a family of hormones for which the gonads represent the primary target. The major activity of gonadotrophins is to regulate reproductive function and additional members of this family include luteinizing hormone (LH), (human) chorionic gonadotrophin (hCG), pregnant mare serum gonadotrophin (PMSG; horses only), inhibin and activin (35).

Table 7. Recombinant therapeutic hormones which have gained marketing approval or are in clinical trials. Data sourced from PhRMA (http://www.phrma.org) and the EMEA (http://www.eudra.org/emea.html)

Product name	Company	Indication	Status
Insulins			
Humulin	Eli Lilly (IN, USA)	Diabetes	Approved 1982 (USA)
Novolin (various presentations)	Novo-Nordisk (NJ, USA)	Diabetes	Approved 1991 (USA)

Product name	Company	Indication	Status
Humalog	Eli Lilly (IN, USA)	Diabetes	Approved 1996 (USA)
	Eli Lilly (The Netherlands)	Diabetes	Approved 1996 (EU)
Insuman	Hoechst AG (Germany)	Diabetes	Approved 1997 (EU)
Liprolog	Eli Lilly (UK)	Diabetes	Approved 1997 (EU)
Growth hormone			
Protropin	Genentech (CA, USA)	Human growth hormone deficiency in children	Approved 1985 (USA)
Humatrope	Eli Lilly (IN, USA)	Human growth hormone deficiency in children	Approved 1987 (USA)
Nutropin	Genentech (CA, USA)	Human growth hormone deficiency in children	Approved 1994 (USA)
		Turner's syndrome	Approved 1996 (USA)
		Growth hormone inadequacy in adults	Approved 1997 (USA)
BioTropin	Bio-Technology General (NJ, USA)	Human growth hormone deficiency in children	Approved 1995 (USA)
Genotropin	Pharmacia & Upjohn (ML, USA)	Human growth hormone deficiency in children	Approved 1995 (USA)
Saizen	Serono Laboratories (MA, USA)	Human growth hormone deficiency in children	Approved 1996 (USA)
Serostim	Serono Laboratories (MA, USA)	Treatment of AIDS-associated catabolism/wasting	Approved 1996 (USA)
		Paediatric HIV failure to thrive	Approved 1998 (USA)
Gonadotrophins			
Gonal F	Ares-Serono (UK)	Anovulation and superovulation	Approved 1995 (EU)
Gonal F	Serono Laboratories (MA, USA)	Female infertility	Approved 1997 (USA)
Puregon	N.V. Organon (The Netherlands)	Anovulation and superovulation	Approved 1996 (EU)
Hormones in clinical trials			
rDNA insulin	Inhale therapeutic systems (CA, USA)	Diabetes	Phase II clinical trials

Product name	Company	Indication	Status
Glucagen (rGlucagon)	Novo-Nordisk (NJ, USA)	Hypoglycemia	Phase III clinical trials
Glucagon for injection	Eli Lilly (IN, USA)	Hypoglycemia	Application submitted
Serostim (rhGH)	Serono Laboratories (MA, USA)	Cancer cachexia	Phase II clinical trials
Trovert (rhGH)	Sensus (TX, USA)	Diabetes related illness, acromegaly	Phase II clinical trials
Pralmorelin (rhGH)	Wyeth-Ayerst Labs (PA, USA)	Adult growth hormone deficiency	Phase I clinical trials
Prolease (rhGH)	Alkermes (MA, USA) & Genentech (CA, USA)	Growth hormone deficiency in children	Phase III clinical trials
Saizen	Serono Laboratories (MA, USA)	Adult growth hormone deficiency, intrauterine growth retardation in children, chronic renal failure in children	Phase III clinical trials
Gonal F (r-hFSH)	Serono Laboratories (MA, USA)	Male infertility	Phase III clinical trials
LhADI (r-hLH)	Ares-Serono & Serono Laboratories (MA, USA)	Female infertility	Phase III clinical trials
Ovidrel (r-hCG)	Ares-Serono & Serono Laboratories (MA, USA)	Female infertility Kaposi's sarcoma, AIDS-related hypogonadism	Phase III clinical trials Phase II clinical trials
Recombinant human parathyroid hormone	Allelix Biopharmaceuticals (Ontario) & Astra AB (Sweden)	Post-menopausal osteoporosis	Phase II clinical trials

FSH, along with hCG, is utilized medically to treat various reproductive disorders, such as anovulatory infertility. FSH preprations traditionally have been extracted from the urine of post-menopausal women, while hCG is purified from the urine of pregnant women. Urine is hardly an ideal source of any therapeutic agent, rendering attractive production of recombinant forms of gonadotrophins. Gonal F, for example, is a recombinant FSH preparation produced in Chinese Hamster Ovary cells which gained marketing approval in the EU in 1995 and the USA in 1997.

3.4 Haemopoietic growth factors

Haemopoietic growth factors are a group of polypeptide regulatory molecules which control the production of blood cells (and platelets) from haemopoietic stem cells (36-38). Several such factors, produced by recombinant means, have been approved for medical use (Table 8).

Erythropoietin (EPO) may be classified as a true endocrine hormone in that it is produced in the kidney and is primarily responsible for stimulating and regulating erythropoiesis (the production of red blood cells) in mammals (39, 40).

A human adult typically contains approx. 2.3×10^{13} erythrocytes, which are synthesized at a rate of about 2.3 million cells per second.

A variety of clinical conditions exist which are often characterized by a significantly depressed rate of erythropoiesis (and thus by anaemia). Examples include renal failure, various cancers, AIDS and some other infectious diseases, as well as bone marrow transplantations and rheumatoid arthritis. Many of those conditions appear responsive to administration of exogenous EPO, and recombinant EPO has been approved to treat various forms of anaemia (Table 8) (40, 41). Recombinant EPO preparations are usually produced in Chinese Hamster Ovary cell lines, which facilitates glycosylation of the polypeptide (native EPO is highly glycosylated). EPO was the first product of biotechnology whose annual global market value topped $1 billion. Its current annual sales value is now closer to $2 billion.

Colony stimulating factors (CSFs) are additional haemopoietic factors now approved for medical use (Table 8). In general, CSFs seem to stimulate the differentiation and maturation of specific white blood cell types from stem cell derived precursors. Two members of this family of regulatory proteins are granulocyte colony stimulating factor (G-CSF) and granulocyte-macrophage colony stimulating factor (GM-CSF). Both appear to function as growth and differentiation factors for neutrophils and their precursor cells (neutrophils are a sub-population of white blood cells capable of ingesting and killing bacteria). These factors, therefore, are valuable in the treatment of neutropenia (a condition characterized by the occurrence of frequent and serious infections due to a significantly decreased blood neutrophil count) (42, 43). These colony stimulating factors also likely target growth/maturation and/or activation of other cell types. GM-CSF, for example, is known to enhance the proliferation of macrophages, eosinophils, and erythrocytes and appears to activate phagocytes and augment the immune system's anti-tumour activity. As such, CSFs may also prove useful in the treatment of infectious diseases, some forms of cancer and the management of bone marrow transplants (44, 45).

Biopharmaceuticals, an overview

Table 8. Recombinant haemopoietic growth factors which have gained marketing approval, or are in clinical trials. Data sourced from PhRMA (http://www.phrma.org) and the EMEA (http://www.eudra.org/emea/html). (Note: rEPO = recombinant erythropoietin, rGM-CSF = recombinant granulocyte-macrophage colony stimulating factor, rG-CSF = recombinant granulocyte colony stimulating factor)

Product name	Company	Indication	Status
EPOGEN (rEPO)	Amgen (CA, USA)	Treatment of anemia associated with chronic renal failure or with Retrovir-treated AIDS patients	1989 (USA)
		Treatment of anemia caused by chemotherapy in patients with non-myeloid malignancies	1993 (USA)
		Prevention of anemia associated with surgical blood loss. Autologous blood donation adjuvant	1996 (USA)
Procrit (rEPO)	Ortho Biotech (NJ, USA)	Treatment of anemia associated with chronic renal failure or with Retrovir-treated AIDS patients.	1990 (USA)
		Treatment of anemia caused by chemotherapy in patients with non-myeloid malignancies.	1993 (USA)
		Prevention of anemia associated with surgical blood loss. Autologous blood donation adjuvant.	1996 (USA)
Neorecormon (rEPO)	Boehringer-Mannheim GmbH (Germany)	Treatment of anemia associated with chronic renal failure, prevention of anemia in premature infants, treatment of anemia in adults receiving platinum-based chemotherapy, increasing the yield of autologous blood from patients in a pre-blood donation programme.	1997 (EU)
Leukine (rGM-CSF)	Immunex (WA, USA)	Autologous bone marrow transplantation.	1991 (USA)
		Neutropenia resulting from chemotherapy in acute myelogenous leukaemia.	1995 (USA)
		Allogenic bone marrow transplantation, peripheral blood progenitor cell mobilization and transplantation.	1995 (USA)

Product name	Company	Indication	Status
NEUPOGEN (rG-CSF)	Amgen (CA, USA)	Chemotherapy induced neutropenia.	1991 (USA)
		Autologous or allogeneic bone marrow transplantation.	1994 (USA)
		Chronic severe neutropenia.	1994 (USA)
		Support peripheral blood progenitor cell transplantation.	1995 (USA)
Leukine (rGM-CSF)	Immunex (WA, USA)	Adjuvant to AIDS therapy, HIV infection, prevention of infection in HIV patients.	Phase II clinical trials
		Prophylaxis and treatment of chemotherapy-induced neutropenia and neutropenia in acute myelogenous leukaemia.	Applications submitted
NEUPOGEN (rG-CSF)	Amgen (CA, USA)	Treatment and prevention of neutropenia in HIV pateints.	Application submitted
		Acute myelogenous leukaemia.	Application submitted
		Multilobar pneumonia, pneumonia sepsis	Phase III clinical trials
EPREX (rEPO)	National Cancer Institute (MD, USA) & Ortho Biotech (NJ, USA)	Neuroblastoma	Phase II clinical trials
Leucotropin (rGM-CSF)	Cangene (Ontario)	Mobilization of peripheral blood stem cells in patients with adjuvant breast cancer.	Phase III clinical trials
Thrombopoietin	Genentech (CA, USA)	Thrombocytopenia related to cancer treatment.	Phase II clinical trials

3.5 Interferons and interleukins

Interferons (IFNs) and Interleukins (ILs) are two prominent sub-families of the cytokine group of regulatory proteins and several such products have gained regulatory approval (Table 9).

Humans produce at least 3 distinct IFN types: IFN-α, IFN-β and IFN-γ. At least 16 different (but closely related) IFN-α subtypes exist, while we produce a single type of IFN-β and a single IFN-γ. IFN-αs and IFN-β all display significant amino acid sequence homology, bind to the same receptor and induce very similar biological responses. As such, these are collectively termed type I IFN. IFN-γ, in contrast, is evolutionarily distinct from type I IFNs. It binds its own unique receptor and induces a range of biological activities which only partially overlap with type I IFNs. IFN-γ is thus classified as a type II IFN (46).

Table 9. Recombinant interferons and interleukins which have gained marketing approval or are in clinical trials. Data sourced from PhRMA (http://www.phrma.org) and the EMEA (http://www.eudra.org/emea/html). (Note: rIFN = recombinant interferon, rIL = recombinant interleukin)

Product name	Company	Indication	Status
Intron A (rIFN-α-2b)	Schering Plough (NJ, USA)	Hairy cell leukaemia	1986 (USA)
		Genital warts	
		AIDS-related Kaposi's sarcoma	1988 (USA)
		Hepatitis C	1988 (USA)
		Hepatitis B	1991 (USA)
		Malignant melanoma	1992 (USA)
		Follicular lymphoma in conjunction with chemotherapy	1995 (USA) 1997 (USA)
Roferon-A (rIFN-α-2a)	Hoffman-La Roche (NJ, USA)	Hairy cell leukaemia	1986 (USA)
		AIDS-related Kaposi's sarcoma	1988 (USA)
		Chronic myelogenous leukaemia	1995 (USA)
		Hepatitis C	1996 (USA)
Alferon N (rIFN-α-n3)	Interferon Sciences (NJ, USA)	Genital warts	1989 (USA)
Actimmune (rIFN-γ-1b)	Genentech (CA, USA)	Management of chronic granulomatous disease	1990 (USA)
Betaseron (rIFN-β-1b)	Berlex Laboratories (NJ, USA) & Chiron (CA, USA)	Relapsing, remitting multiple sclerosis	1993 (USA)
Avonex (rIFN-β-1a)	Biogen (MA, USA)	Relapsing multiple sclerosis	1996 (USA)
Infergen (rIFN-α)	Amgen (CA, USA)	Chronic hepatitis C	1997 (USA)
Avonex	Biogen (France)	Relapsing multiple sclerosis	1997 (EU)
Proleukin (rIL-2)	Chiron (CA, USA)	Renal cell carcinoma	1992 (USA)
		Metastatic melanoma	1998 (USA)
Alferon LDO (rIFNα-n3)	Interferon Sciences (NJ, USA)	AIDS-related complex, AIDS	Phase II clinical trials
Alferon N (rIFN-α-n3)	Interferon Sciences (NJ, USA)	HIV infection	Phase III clinical trials
		Papillomavirus infections	Phase II clinical trials
		Chronic hepatitis C infections	Phase III clinical trials
Ampligen	Hemispherix Biopharma	HIV infection, renal cancer	Phase II clinical trials

Product name	Company	Indication	Status
	(NY, USA)	Hepatitis, chronic fatigue syndrom	Phase II clinical trials
Actimmune (rIFN-γ-1b)	National Cancer Institute (MD, USA) & Genentech (CA, USA)	Cancer of the colon, lung, ovary, prostate and melanoma	Phase II clinical trials
Avonex (rIFN-β-1A)	Biogen (MA, USA)	Glioma	Phase II clinical trials
		Secondary, progressive multiple sclerosis	Phase III clinical trials
Betasteron (rIFN-β-1b)	National Cancer Institute (MD, USA) & Berlex Laboratories (NJ, USA)	Non-small-cell lung cancer; chronic, progressive multiple sclerosis	Phase III clinical trials
Intron A (rIFN-α-2b)	Schering Plough (NJ, USA)	Malignant melanoma, Hepatitis C, paediatric hepatitis B	Application submitted
Rebif (rIFN-β-1a)	Serono Laboratories (MA, USA)	Colorectal cancer, viral infections	Phase III clinical trials
		Multiple sclerosis	Application submitted
Roferon-A (rIFN-α-2a)	Hoffman La Roche (NJ, USA)	Malignant melanoma adjuvant	Phase III clinical trials
γ-interferon	Connectics (CA, USA)	Keloids	Phase II clinical trials
Interleukin-10	Schering Plough (NJ, USA)	HIV	Phase I clinical trials
		Rheumatoid arthritis, Crohn's disease, ulcerative colitis	Phase II clinical trials
		Ischemic reperfusion therapy, multiple sclerosis, acute lung injury, psoriasis	Phase I clinical trials
PEG-IL-2	Chiron (CA, USA)	HIV infection	Phase II clinical trials
Proleukin (rIL-2)	Chiron (CA, USA)	HIV infection, acute myelogenous leukaemia, non-Hodgkin's lymphoma	Phase III clinical trials
Quadrakine (rIL-4)	Schering Plough (NJ, USA)	Rheumatoid arthritis	Phase I clinical trials
Interleukin-4	National Cancer Institute (MD, USA) & Schering Plough (NJ, USA)	Malignant melanoma	Application submitted

Product name	Company	Indication	Status
Recombinant human Interleukin-12	Genetics Institute (MA, USA) & Wyeth-Ayerst (PA, USA)	Cancer, infectious diseases	Phase II clinical trials
Sigosix (rIL-6)	Ares-Serono (MA, USA)	Haematological conditions	Phase I clinical trials
Nuemega (rIL-11)	Genetics Institute (MA, USA)	Crohn's disease	Phase II clinical trials

In general, IFNs are produced by a range of cell types and their biological activities include induction of resistance to viral attack; moderating the immune response and regulating the growth and differentiation of various (immune and non-immune) cell types. Recently, a novel IFN (IFN-tau) has been discovered, which functions to sustain early pregnancy in some animal species.

Their range of biological activities suggests that IFNs could have multiple therapeutic applications, including priming the immune response against infectious agents (particularly viruses); treatment of some autoimmune conditions and treatment of some cancer types. As is evident from Table 9, several IFN preparations have now gained approval for such indications (46-49). As IFNs are produced naturally in exceedingly low concentrations, their direct extraction from native sources in large quantities is impractical. Some transformed cell lines are known to produce various IFNs in moderate quantities and culture of such cells provided most of the IFN initially used medically. A notable example was Wellcome's IFN-α producing 'Namalwa' lymphoblastoid cell line. Now all interferon preparations used medically are produced by recombinant means, mostly in *E. coli*.

Recombinant IFN-β has gained regulatory approval for the treatment of relapsing-remitting multiple sclerosis, an autoimmune condition characterized by the destruction of the myelin which surrounds the neurons of the central nervous sytem. While failing to cure the condition, administration of this IFN reduces the frequency of relapses in many patients. The molecular mechanisms by which it achieves this remains to be elucidated. However, it does appear to block synthesis/secretion of IFN-γ and tumor necrosis factor (TNF), both of which are believed to play a role in fueling progression of this disease.

IFN-γ is used medically to treat chronic granulomatous disease (CGD) (50). This is a rare genetic disease in which phagocytes of sufferers are poorly capable of ingesting and destroying foreign pathogens, particularly bacteria and protozoa. As a result, such persons suffer from repeated, usually serious infections. A prominent activity of IFN-γ is its ability to activate phagocytes, rendering its clinical application in CGD relatively obvious. Actimmune (IFN-γ from Genentech) was approved to treat CGD in 1990 (Table 9). It is produced in recombinant *E. coli* and, although devoid of the carbohydrate moiety present on native IFN-γ, it exhibits identical biological activity to the native molecule.

Interleukin-2 (IL-2) represents an additional cytokine that has gained regulatory approval for medical use (Table 9) (51, 52). At least 16 different interleukins have thus far been identified. Most are glycosylated (including IL-2, and display molecular weights ranging from 13-30 kDa. The range of biological activities of interleukins are extensive and very complex. They regulate virtually all aspects of immunity and inflammation and also modulate the growth of various cell types, including transformed cells.

IL-2 (also known as T cell growth factor) is the best characterized of the interleukins. This molecule acts as an autocrine growth factor for T-lymphocytes and also enhances antibody production in activated B lymphocytes. As such, it is a major regulator of both cell-mediated and humoral immunity. IL-2 also promotes differentiation and activation of natural killer (NK) cells, which play an important role in the destruction of transformed cells and virally infected cells and clinical trials continue to assess its potential for the treatment of various infectious diseases.

3.6   Recombinant vaccines

Genetic engineering allows large-scale production of any protein normally found on the surface of any pathogen. Such proteins, therefore, can be used as 'subunit' vaccines (53). This method of vaccine production displays a number of advantages, not least of which is that it eliminates the possibility of accidental transmission of disease. (Incomplete processing of, for example, inactivated or attenuated pathogens for use as vaccines can, in rare instances, result in accidental administration of active pathogen to the vaccine recipient).

Recombinant hepatitis B surface antigen (rHBsAg) was the first (and, thus far, only) subunit vaccine to gain regulatory approval. Merck's Recombivax HB was approved by the FDA in 1986. Since then, a number of additional recombinant HbsAg based products have gained approval. Some are combination vaccines, also containing non-recombinant constituents. Examples include SmithKline Beecham's Infanrix HepB, Twinrix paediatric and Tritanrix. Additional subunit vaccines remain in clinical trials.

3.7 Monoclonal antibody based products

Polyclonal antibodies have traditionally been used therapeutically to induce passive immunity. Monoclonal antibody production was made possible in the mid-1970s by the development of hybridoma technology. The unrivalled specificity of monoclonal antibodies renders them very attractive therapeutic tools and they are amongst the largest single category of biopharmaceuticals in clinical trials (54)

The first monoclonal antibody to be approved for medical use was Ortho Biotech's Orthoclone OKT3 (Table 10), used to promote a reversal of acute kidney transplant rejection. OKT3 recognizes the CD3 surface antigen found on T lymphocytes. Binding of the antibody to CD3 can induce destruction of these cells, which are the ones that mediate rejection of transplanted tissue.

More recently, greater emphasis has been placed upon development of monoclonal antibody preparations used to detect or treat various cancers (55, 56). Upon transformation, many cells express cell surface proteins which are either not normally present on the untransformed cell, or are expressed in ultra low quantities. Such proteins are often termed tumor associatedantigens (TAAs). A monoclonal antibody raised against such a TAA should bind only to the surface of transformed cell when injected into the body. Conjugation of a radioisotope to the antibody prior to its administration should, therefore, result in selective targeting of the radioactivity to the tumor surface. Conjugation of γ-emitting radioisotopes (which can penetrate outward through the body) to such antibodies could, therefore, be used for diagnostic purposes (immunoscintigraphy). Alternatively, conjugation of β emitters to the monoclonal allows targeted radiotherapy of the tumor (β particles will penetrate a thickness of several cells).

Although several radioactively labelled monoclonal antibodies/monoclonal antibody fragments are now approved for the detection/treatment of selected cancers (Table 10), many of the earlier attempts to develop such products failed. This was due to a number of reasons: (a) insufficient information was available regarding TAAs. Identification of additional TAAs represents an active area of research; (b) Murine monoclonals are themselves antigenic when administered to humans. This can be overcome (in part at least) by producing chimaeric or humanized antibodies using genetic engineering. These are hybrid antibodies containing sequences of human as well as murine origin, and are thus less immunogenic in man; (c) Monoclonal antibodies, due to their high molecular weight (i.e. large size) exhibit poor penetration of tumor mass. This difficulty can be overcome by using (antigen-binding) antibody fragments, rather than intact antibody. (The identification of tumor associated antigens also provides the possibility of developing specific cancer vaccines. Theoretically, administration of a TAA to an individual would immunize against cancer types which express that specific TAA).

Table 10. Monoclonal antibody based products approved for medical use and selected examples currently in clinical trials. Data sourced from PhRMA (http://www.phrma.org) and the EMEA (http://www.eudra.org/emea.html)

Product Name	Company	Indication	Status
Orthoclone	Ortho-biotech (NJ, USA)	Reversal of acute kidney transplant rejection.	1986 (USA)
		Reversal of heart and liver transplant rejection.	1993 (USA)
OncoScint	Cytogen (NJ, USA)	Detection, staging and follow-up of colorectal and ovarian cancers	1992 (USA)
CEA-scan	Immunomedics (NJ, USA)	Detection of recurrent and metastatic colorectal cancer	1996 (USA)
MyoScint	Centocor (PA, USA)	Myocardial infarction imaging agent	1996 (USA)
ProstaScint	Cytogen (NJ, USA)	Detection, staging and follow up of prostate adenocarcinoma	1996 (USA)
ReoPro	Centocor (PA, USA) and	Anti-platelet prevention of blood clots	1994 (USA)
	Eli Lilly (IN, USA)	Refractory unstable angina	1997 (USA)
Verluma	Boehringer-Ingelheim (CT, USA) & NeoRx (WA, USA)	Detection of small cell lung cancer	1996 (USA)
Zenapax	Hoffman-La Roche (NJ, USA)	Prevention of acute kidney transplant rejection	1997 (USA)
Rituxan	Genentech (CA, USA) and IDEC Pharmaceuticals (CA, USA)	Treatment of B-cell non-Hodgkin's lymphoma	1997 (USA)
Neumega	Genetics Institute (MA, USA)	Prevention of chemotherapy-induced thrombocytopenia	1997 (USA)
CEA-Scan	Immunomedics BV (The Netherlands)	Imaging for recurrance or metastases of carcinoma of the colon or rectum	1996 (EU)
Indimacis 125	CIS-bio (France)	Diagnosis of relapsing ovarian adenocarcinoma	1996 (EU)
Tecnemab-K-1	Sorin Biomedica Diagnostics SpA (Italy)	Detection of infection/ inflammation in bone in patients with suspected osteomyelitis	1997 (EU)
AD439/519	Tanox Biosystems (TX, USA)	HIV infection/AIDS	Phase II Clinical trials
Avakine	Centocor (PA, USA)	Rheumatoid arthritis	Phase III Clinical trials

Product Name	Company	Indication	Status
CD40 ligand antibody	Biogen (MA, USA)	Lupus, immune thrombocytopenic purpura	Phase II Clinical trials
Clenoliximab	IDEC Pharmaceuticals (CA, USA) and SmithKline Beecham (PA, USA)	Rheumatoid arthritis	Phase II Clinical trials
h5G 1.1	Alexion Pharmaceuticals (CT, USA)	Lupus, rheumatoid arthritis	Phase I Clinical trials
IDEC-131	IDEC Pharmaceuticals (CA, USA)	Systemic lupus erythematosus	Phase I Clinical trials
MDX-33	Medarex (NJ, USA)	Autoimmune disease, idiopathic thrombocytopenic purpura	Phase I Clinical trials
Orthoclone OKT4A	Orthobiotech (NJ, USA)	Treatment of CD4 mediated autoimmune diseases	Phase II Clinical trials
SMART Anti CD3	Protein design labs (CA, USA)	Autoimmune diseases	Phase I Clinical trials
YM-337 Mab	Yamanouchi (NY, USA) and Protein design labs (CA, USA)	Platelet aggregation	Phase I Clinical trials
AFP-Scan	Immunomedics (NJ, USA)	Extent of disease staging of liver and germ cell cancers	Phase II Clinical trials
Anti-idiotype monoclonal antibody	Novartis (NH, USA)	Cancer	Phase I Clinical trials
Anti-transferrin receptor Mab	National Cancer Institute (MD, USA)	Advanced, refractory solid tumors	Phase I Clinical trials
Anti-VEGF humanized Mab	Genentech (CA, USA)	Cancer	Phase I Clinical trials
C225, anti-EGFR chimeric Mab	Imclone Systems (NJ, USA)	Epidermal growth factor receptor positive cancers	Phase II Clinical trials
CEAcide humanized anti-CEA antibody	Immunomedics (NJ, USA)	Colorectal cancer	Phase II Clinical trials
CMB-401	Wyeth-Ayerst (PA, USA)	Ovarian cancer	Phase II Clinical trials
Herceptin; anti-HER-2 humanized Mab	Genentech (CA, USA)	Breast cancer	Phase III Clinical trials
Lymphocide anti-CD22 humanized Mab	Immunomedics (NJ, USA)	Non-Hodgkin's B-cell lymphoma	Phase II Clinical trials

Product Name	Company	Indication	Status
Avakine (chimeric anti-TNF antibody)	Centocor (PA, USA)	Crohn's disease	Application submitted
Anti-CD18 humanized MAb	Genentech (CA, USA)	Acute myocardial infarction	Phase II Clinical trials
Anti-TNF MAb	Chiron (CA, USA)	Sepsis	Phase III Clinical trials
ATM027 humanized MAb	T cell Sciences (MA, USA)	Multiple sclerosis	Phase I Clinical trials
Anti-IgE humanized MAb	Genentech (CA, USA)	Allergic asthma	Phase III Clinical trials
ICM-3	ICOS (WA, USA)	Psoriasis	Phase I Clinical trials
Simulect	Novartis (NJ, USA)	Transplantation	Application submitted
Zenapax	Hoffman-La Roche (NJ, USA)	Liver transplantation	Phase II Clinical trials
Enlimomab	Boehringer-Ingelheim (CT, USA)	Thermal injury	Phase II Clinical trials

In addition to imaging cancer, appropriate radiolabelled monoclonals can also be used to image additional conditions including cardiovascular disease, deep vein thrombosis and the site of bacterial infections. Monoclonals may also prove useful in the treatment of conditions such as septic shock and various autoimmune diseases.

4. NUCLEIC ACID THERAPEUTICS

Until now the term 'biopharmaceutical' has become virtually synonymous with 'proteins of therapeutic use produced by modern biotechnological techniques'. However, an additional class of biomolecule - nucleic acid - also exhibits great medical potential. Nucleic acid therapy centers around gene therapy and antisense technologies. While no nucleic acid-based biopharmaceutical product has yet been approved for general medical use, many are now in clinical trials. Some such products are likely to become a medical reality within the next few years.

4.1 Gene therapy

Gene therapy involves the introduction of a specific gene into the genetic complement of a cell, such that expression of this gene achieves a pre-defined therapeutic goal (57, 58). The gene inserted may, for example, replace a defective copy of a specific endogenous gene, or its expression may confer some novel ability/property upon the cell.

The most obvious application of gene therapy is in the treatment of genetic diseases (59). Well over 4,000 such diseases have been characterized to date. Many of these are caused by the lack of production of a single gene product (or the production of a defective/inactive gene product due to a mutation). Examples include haemophilia A & B (defective genes products; factors VIII and IX, respectively), as well as familial hypercholesterolaemia and Gaucher's disease (defective gene products; low density lipoprotein receptor and glucocerebrosidase, respectively). Gene therapy provides a theoretically straightforward and elegant method of correcting such diseases, simply by inserting a healthy copy of the gene in question into the appropriate cells of the sufferer.

Despite the simplicity of the concept, relatively few gene therapy products aimed at treating genetic diseases are currently being developed. This is due to a number of considerations, including: (a) the number of genetic diseases for which the actual gene responsible has been identified is still quite low; (b) Some genetic diseases are complex, involving more than one gene product and/or organ, or are caused by a lack of regulation of expression of the gene; (c) In some cases, the curative gene needs to be targeted to a specific organ/tissue type. Targeting specificity is technically challenging; (d) Many genetic diseases are quite rare and the small patient base makes the drug development process economically unattractive. The bulk of gene therapy protocols currently being assessed in clinical trials relate to the treatment not of genetic conditions, but of cancer and AIDS (60-63).

The annual incidence of cancer in the USA alone stands at close to 1.5 million cases. Treatment with conventional therapies yields a survival rate of the order of 50%. A number of gene-based anti-cancer therapeutic strategies have been developed and several such potential products are now being assessed in clinical trials.

Gene therapy is likely to prove useful in the treatment of infectious diseases. To date, most efforts in this area have focused upon the treatment of AIDS. One strategy entails introduction into viral sensitive cells of a gene coding for an altered (dysfunctional) HIV protein (e.g. gag or env). Intracellular synthesis of such products have been shown to inhibit viral replication (probably by interfering with correct assembly of HIV virions).

An additional application of gene therapy is the potential development of DNA-based vaccines (64). This would simply entail the introduction of a gene coding for a surface protein of the target pathogen into appropriate body cells. Use of a correct genetic construct would facilitate extracellular secretion of the pathogen-derived gene product, thereby exposing it to immune surveillance. Despite its great promise, a number of technical hurdles must be satisfactorily overcome before gene therapy is routinely applied in human medicine (65, 66).

4.2 Antisense technology

Antisense technology represents a (nucleic acid based) strategy which can down regulate or prevent the expression of particular genes (67, 68). A number of disease states are associated with the expression/over expression of specific genes. Examples include AIDS (expression of HIV genes) and some cancers (the expression of oncogenes). Additionally, increased expression of some genes can have negative medical consequences. For example, over expression of interferons, TNF or other cytokines (as sometimes occurs during an immunological response to an infectious agent), can actually worsen disease symptoms.

Antisense technology is based upon the synthesis of short, single-stranded stretches of nucleic acid (RNA or DNA-based), of specific nucleotide sequence. By choosing the appropriate sequence, such 'antisense oligonucleotides' or 'oligos' can bind DNA (at specific gene sites) or, more usually, form duplexes with a specific mRNA. The interaction in most cases is via standard nucleotide base pair complementarity. This interaction prevents either gene transcription or translation, either way preventing synthesis of the gene product.

The specificity of this technique makes it very attractive, although a number of technical hurdles must be satisfactorily resolved before it is likely to impact upon medical practice. Some problems include: how to get the oligos into individual cells and the sensitivity of such oligos to nucleases.

Although a few antisense-based products are now in early clinical trials, this technology is unlikely to impact upon medical practice for several years to come.

5. FUTURE TRENDS

Thus far, in the region of 54 products of biotechnology have gained regulatory approval for medical use. Biopharmaceuticals have thus become an established sector within the pharmaceutical industry. Currently, there are in the region of 350 additional such products undergoing clinical trials.

There are in the region of 85 monoclonal antibody preparations undergoing clinical evaluation, making them amongst the single largest category of biopharmaceutical in development. In addition, a range of vaccines, cytokines, gene therapy-based products, as well as some hormone and antisense-based products are being evaluated. More than one third of these products aim to treat cancer, particularly melanoma and calorectal cancer, as well as breast and prostate cancer. Of the almost 80 vaccines in development, most aim to prevent/treat AIDS or some form of cancer. It is thus likely that several new biopharmaceuticals will gain regulatory approval each year over the foreseeable future.

6. CONCLUSION

Modern biotechnological techniques continue to impact on every facet of our existance. Thus far, such technologies have impacted most notably upon healthcare practice and have made possible the development of a new class of drug: the biopharmaceutical.

The fact that the direct revenue from biopharmaceutical products has grown from zero in the early 1980s to an estimated $7 billion last year illustrates the medical importance of these drugs. Biomedical research continues to rapidly unveil the molecular principles underlining both health and disease and, as a consequence, more and more potential biopharmaceutical products are being identified. Such advances in basic and applied medical research will fuel continued growth of the biopharmaceutical industry for many years to come.

BIOGRAPHY

Gary Walsh is a lecturer in industrial biochemistry at the University of Limerick, Ireland. He was employed for several years in the pharmaceutical industry prior to his current appointment. In addition to co-editing this volume, Dr. Walsh is author of two related University text books. 'Protein Biotechnology' was published in 1994 by John Wiley & Sons, UK. 'Biopharmaceuticals: biochemistry and biotechnology' was published in 1998, also by Wiley.

REFERENCES

1. Drews, J. (1993). Into the 21st Century: biotechnology and the pharmaceutical industry in the next 10 years. Bio/Technology, 11, 516-520.
2. Klausner, A. (1993). Back to the future: biotech product sales, 1983-1993. Bio/Technology, 11, S35-S37.

3. Biotechnology in the US pharmaceutical industry: A special report, 4th edition (1995). Institute of Biotechnology Information, North Carolina.
4. Walsh, G. (1998). Biopharmaceuticals: Biochemistry and Biotechnology. J. Wiley & Sons Ltd. Chichester, U.K.
5. Carter, P. et al. (1992). High level *E. coli* expression and production of a bivalent humanized antibody fragment. Bio/Technology, 10, 163-167.
6. Chiswell, D. and McCafferty, J. (1992). Phage antibodies: Will new 'coliclonal antibodies' replace monoclonal antibodies? TIBTECH, 10, 80-84.
7. Hockney, R. (1994). Recent developments in heterologous protein production in *E. coli*. TIBS, 12, 456-463.
8. Walsh, G. and Headon, D. (1994). Protein Biotechnology. J. Wiley & Sons Ltd. Chichester, U.K.
9. Buckholz, R. and Gleeson, M. (1991). Yeast systems for the commercial production of heterologous proteins. Bio/Technology, 9, 1067-1071.
10. Kingsman, S. et al. (1985). Heterologous gene expression in *Saccharomyces cerevisiae*. In: Russell, G. (ed), Biotechnology and Genetic Engineering Reviews, Vol. 3, 377-416. Intercept.
11. Van Brund, J. (1986). Fungi: the perfect hosts? Bio/Technology, 4, 1057-1062.
12. Hu, W. and Peshwa, M. (1993). Mammalian cells for pharmaceutical manufacturing. Am. Soc. Microbiol. News, 59, 65-68.
13. HeLee, S. and De Boer, A. (1994). Production of biomedical proteins in the milk of transgenic dairy cows: the state of the art. J. controlled release, 29, 213-221.
14. Rosen, J. et al. (1996). The mammary gland as a bioreactor: factors resulting in the efficient expression of milk protein-based transgenes. Am. J. Clin. Nutr., 63, 627S-632S.
15. Ebert, K.M. et al. (1991). Transgenic expression of a variant of human tissue-type plasminogen activator in goats milk: generation of transgenic goats and analysis of expression. Bio/Technology, 9, 835-838.
16. Wright, G. et al. (1991). High level expression of active human alpha-1-antitrypsin in the milk of transgenic sheep. Bio/Technology, 9, 830-834.
17. Farrell, D. (1990). Purification of recombinant proteins for pharmaceutical use. Biochem. Soc. Trans., 18, 243-245.
18. Wheelwright, S. (1987). Designing downstream processes for large-scale protein purification. Bio/Technology, 5, 789-793.
19. Wang, Y.C. and Hanson, M. (1988). Parenteral formulations of proteins and peptides: stability and stabilizers. J. Parenter. Sci. Technol., 42, S4-S26.
20. Pilak, M. (1990). Freeze drying of proteins, Part I: process design. BioPharm., 3(8), 18-27.
21. Pilak, M. (1990). Freeze drying of proteins, Part II: formulation selection. BioPharm., 3(9), 26-30.
22. Prowse, C. (1992). Plasma and recombinant blood products in medical therapy. J. Wiley & Sons Ltd., Chichester, UK.
23. Tencate, H. et al. (1996). Developments in anti-thrombolic therapy: state of the art anno 1996. Pharmacy World Sci., 18(6), 195-203.
24. Dott, J. (1995). Anti-coagulatory substances of blood-sucking animals: from hirudin to hirudin mimetics. Angew. Chem. Int. Engl., 34, 867-880.
25. Emeis, J. et al. (1997). Progress in clinical fibrinolysis. Fibrinolysis Proteolysis, 11(2), 67-84.

26. Verstraete, M. et al. (1995). Thrombolytic agents in development. Drugs, 50(1), 29-42.
27. Gillis, J. et al. (1995). Alteplase: a reappraisal of its pharmacological properties and therapeutic use in acute myocardial infarction. Drugs, 50(1), 102-136.
28. Bickerstaff, G. (1987). Enzymes in Industry and Medicine. Edward Arnold, London.
29. Lauwers, A. and Scharpe, S. (Eds.)(1997). Pharmaceutical Enzymes. (Drugs in the Pharmaceutical Sciences, Volume 84). Marcel Dekker Inc., New York.
30. Edgington, S. (1993). Nuclease therapeutics in the clinic. Bio/Technology, 11, 580-582.
31. Conway, S. and Watson, A. (1997). Nebulized broncodialators, corticosteroids and rh DNase in adult patients with cystic fibrosis. Thorax, 52(2), 564-568.
32. Johnson (1983). Human insulin from recombinant DNA technology. Science, 219, 632-637.
33. Lippe, B. and Nakamoto, J. (1993). Conventional and non-conventional uses of growth hormone. Recent Prog. Horm. res., 48, 179-235.
34. Neely, E. (1994). Use and abuse of human growth hormone. Ann. Rev. Med., 45, 407-420.
35. De Doning, W. et al. (1994). Recombinant reproduction. Bio/Technology, 12, 988-992.
36. Dexter, M. (1991). Growth and differentiation in the haemopoietic system. Biochem. Soc. Trans., 19, 303-306.
37. Metcalf, D. (1992). Haemopoietic regulators. TIBS, 17, 286-289.
38. Spangrude, G. (1994). Biological and clinical aspects of haemotopoietic stem cells. Ann. Rev. Med., 45, 93-104.
39. Koury, M. and Bondurant, M. (1992). The molecular mechanism of erythropoietin action. Eur. J. Biochem., 210, 649-663.
40. Fried, W. (1995). Erythropoietin. Ann. Rev. Nutr., 15, 353-377.
41. Markham, A. and Bryson, H. (1995). Epoetin alfa: a review of its pharmacodynamic and pharmacokinetic properties and therapeutic use in non-renal applications. Drugs, 49(2), 232-254.
42. Frampton, J. et al. (1994). Filgrastim: a review of its pharmacological properties and therapeutic efficacy in neutropenia. Drugs, 48(5), 731-760.
43. Frampton, J. et al. (1995). Lenograstim, a review of its pharmacological properties and therapeutic efficacy in neutropenia and related clinical settings. Drugs, 49(5), 767-793.
44. Tabbara, I. et al. (1996). The clinical applications of granulocyte-colony stimulating factor in haematopoietic stem cell transplantation: a review. Anticancer Res., 16(6B), 3901-3905.
45. Harousseau, J. (1997). The role of colony stimulating factors in the treatment of acute leukaemia. Biodrugs, 7(6), 448-460.
46. Pestka, S. and Langer, J. (1987). Interferons and their actions. Ann. Rev. Biochem., 56, 727-777.
47. Haria, M. and Benfield, P. (1995). Interferon-α-2a. Drugs, 50(5), 873-896.
48. Simko, R. and Nagy, K. (1996). Interferon-alpha in childhood hematological malignancies. Postgrad. Med. J., 72(854), 709-713.
49. Woll, P. and Pettengell, R. (1997). Interferons in oncology. Br. J. Clin. Pract., 51(2), 111-115.
50. Todd, P. and Goa, K. (1992). Interferon gamma-1b. Drugs, 43(1), 111-222.

51. Jeal, W. and Goa, K. (1997). Aldesleukin (recombinant interleukin-2): a review of its pharmacological properties, clinical efficacy and tolerability in patients with renal cell carcinoma. Biodrugs, 7(4), 285-317.
52. Noble, S. and Goa, K. (1997). Aldesleukin (recombinant interleukin-2): a review of its pharmacological properties, clinical efficacy and tolerability in patients with metastatic melanoma. Biodrugs, 7(5), 394-422.
53. Martin, S. (1994). Vaccine design: future possibilities and potential. Biotech. Adv., 12, 619-624.
54. Carlsson, R. and Glad, C. (1989). Monoclonal antibodies into the 90s: the all-purpose tool. Bio/Technology, 7, 567-573.
55. Waldmann, T. (1991). Monoclonal antibodies in diagnosis and therapy. Science, 252, 1657-1662.
56. Heunnekens, F. (1994). Tumor targeting: activation of prodrugs by enzyme-monoclonal antibody conjugates. TIBTECH, 12, 234-239.
57. Morgan, R. and Anderson, W. (1993). Human gene therapy. Ann. Rev. Biochem., 62, 191-217.
58. Tolstoshev, P. (1993). Gene therapy, concepts, current trials and future directions. Ann. Rev. Pharmacol. Toxicol., 32, 573-596.
59. Kay, M. and Woo, S. (1994). Gene therapy for metabolic disorders. Trends Genet., 10(7), 253-257.
60. Blaese, R. (1997). Gene therapy for cancer. Sci. Am. (June), 91-95.
61. Vile, R. and Russell, S. (1994). Gene transfer technologies for the gene therapy of cancer. Gene Ther., 1, 88-98.
62. Gilboa, E. and Smith, C. (1994). Gene therapy for infectious diseases: the AIDS model. Trends Genet., 10(4), 139-143.
63. Yu, M. *et al.* (1994). Progress towards gene therapy for HIV infection. Gene ther., 1, 13-26.
64. Donnelly, J. (1997). DNA vaccines. Ann. Rev. immunol., 15, 617-648.
65. Friedmann, T. (1997). Overcoming the obstacles to gene therapy. Sci. Am. (June), 80-85.
66. Hodgson, C. (1995). The vector void in gene therapy. Bio/Technology, 13, 222-225.
67. Askari, F. (1996). Molecular medicine: antisense oligonucleotide therapy. N. Eng. J. Med., 334(5), 316-318.
68. Putnam, D. (1996). Antisense strategies and therapeutic applications. Am. J. Health Syst. Pharm., 53, 151-160.

Chapter 2

Abciximab: The First Platelet Glycoprotein IIb/IIIa Receptor Antagonist

Robert E. Jordan, Marian T. Nakada, Harlan F. Weisman
Centocor Inc., Malvern, Pennsylvania, USA

Key words: abciximab, GP IIb/IIIa, platelets, ReoPro, monoclonal antibody, antithrombotic therapy.

Abstract: Platelet aggregation plays a crucial role in the development of the life-threatening thrombosis responsible for such acute coronary syndromes as myocardial infarction and unstable angina pectoris. Although aspirin has traditionally been the mainstay of antiplatelet therapy, it is neither potent nor specific enough to provide adequate protection against thrombosis. The identification of glycoprotein (GP) IIb/IIIa as the key platelet receptor in the final common pathway of platelet aggregation and the development of therapeutic agents that block this receptor have opened up an entirely new dimension in cardiovascular medicine.

Abciximab is the first of an innovative new class of cardiovascular drugs, the GP IIb/IIIa receptor antagonists. As such, it is the first agent that specifically blocks the function of a cellular adhesion molecule, thereby interfering with the adhesion of platelets and inhibiting platelet aggregation.

This chapter will review the development of the first murine IgG_1 antibody directed against the GP IIb/IIIa receptor and describe how it was genetically engineered to become the less immunogenic but equally effective human/chimeric antibody fragment c7E3 Fab, or abciximab. The extensive studies that elucidated the mechanisms of action and binding characteristics of the new agent, as well as the pharmacology and toxicology studies that confirmed the safety and antithrombotic efficacy of antiplatelet therapy with an antibody fragment will be discussed. Finally, the chapter will explore in depth the findings of three large-scale clinical trials that established the role of abciximab in the prevention of thrombotic complications of percutaneous coronary intervention procedures. Ongoing trials investigating the efficacy of abciximab in the setting of acute myocardial infarction will also be discussed.

1. BRIEF REVIEW OF PLATELET PHYSIOLOGY

Platelet adhesion is a critical first step in the physiologic process of hemostasis following injury to the blood vessel wall. Damage to the vessel wall, whether resulting from trauma or from rupture of an atherosclerotic plaque, leads to the exposure of adhesive glycoproteins which are recognized by specific platelet-membrane receptors. A variety of physiologic stimuli may trigger the transformation of platelets from a quiescent state to an activated state. Platelet-activating stimuli exposed or released locally at the site of vascular injury include collagen, thrombin, adenosine diphosphate (ADP), epinephrine, and thromboxane A2. Platelets may also be activated by the adhesion process itself as well as by the high fluid shear forces produced at sites of vessel constriction (1-5).

Once the platelet is activated, the platelet GP IIb/IIIa receptor undergoes a conformational change that enables it to bind with high affinity to fibrinogen or von Willebrand factor. Fibrinogen that is bound to a GP IIb/IIIa receptor on one platelet may also bind to a GP IIb/IIIa receptor on another platelet, thus forming the bridges between adjacent platelets that are the structural basis of platelet aggregation. As more and more platelets are recruited and aggregate, a hemostatic platelet plug is formed. In addition, the aggregated platelets provide a surface for the assembly of coagulation factors leading to the conversion of prothrombin to thrombin.

The participation of platelets in the reparative process that prevents blood loss from damaged vessels also makes these discoid-shaped blood elements key culprits in the development of acute coronary syndromes that are manifested clinically as unstable angina or acute myocardial infarction (MI). Platelet aggregation at the site of either atherosclerotic plaque rupture or vessel injury during a percutaneous intervention procedure may culminate in thrombosis that partially or completely occludes a coronary vessel.

1.1 The GP IIb/IIIa receptor as a target for antithrombotic therapy

The fibrinogen-binding GP IIb/IIIa receptor belongs to the integrin family of receptors. Virtually all cells contain integrin receptors, which mediate contact between cells and structural protein elements (6-7). The GP IIb/IIIa receptor, however, is found exclusively on platelets and platelet progenitor cells. A single platelet has approximately 80,000 GP IIb/IIIa receptors on its surface (8), in addition to an internal pool of receptors that becomes externalized when the platelet is activated (9).

Independent of the nature of the precipitating injury to the vessel wall and regardless of the platelet-activating stimuli generated, the GP IIb/IIIa receptor is the final common pathway for platelet aggregation (10). Thus, it seemed reasonable to hypothesize that an agent that blocked this receptor would probably be a more potent inhibitor of platelet aggregation than, for example, aspirin, which inhibits only thromboxane-mediated aggregation.

The role of the GP IIb/IIIa receptor in platelet aggregation came to light during investigations of Glanzmann's thrombasthenia, a rare congenital disorder in which a deficiency of GP IIb/IIIa receptors gives rise to recurrent mucocutaneous bleeding (11, 12). However, the infrequency of visceral bleeding in patients with this condition suggested that pharmacologic antagonism of the GP IIb/IIIa receptor would probably not result in a dangerous excess of bleeding. Thus, the GP IIb/IIIa receptor emerged as a logical target in the development of therapeutic agents that could efficiently and safely counteract platelet aggregation (5).

1.2 Development of abciximab

The first step in the development of a therapeutic antagonist of the GP IIb/IIIa receptor was the isolation of antiplatelet antibodies from cell lines obtained from mice that had been immunized with human platelets (13). Coller and associates identified one such IgG1 antibody with a kappa light chain (designated 7E3), which was directed against the GP IIb/IIIa receptor, and would interfere with platelet-fibrinogen binding and inhibit platelet aggregation in response to a variety of stimuli (13, 14).

Since platelets coated with intact antibody could potentially be cleared or destroyed by the human immune system, murine 7E3 IgG was obviously inappropriate for *in vivo* application. To circumvent these problems, the antibody was subjected to enzymatic digestion by pepsin or papain (15). Proteolytic digestion produced two antigen-binding fragments, a bivalent 7E3 F(ab')$_2$ fragment derived from pepsin and a univalent 7E3 Fab fragment derived from papain (Figure 2). Both of these antibody fragments proved to have platelet-binding affinities comparable to those of the parent 7E3 IgG molecule. The univalent 7E3 Fab fragment was selected for further development for several reasons including the expectation that its smaller size (approximately Mr 50,000) would minimize immunogenicity.

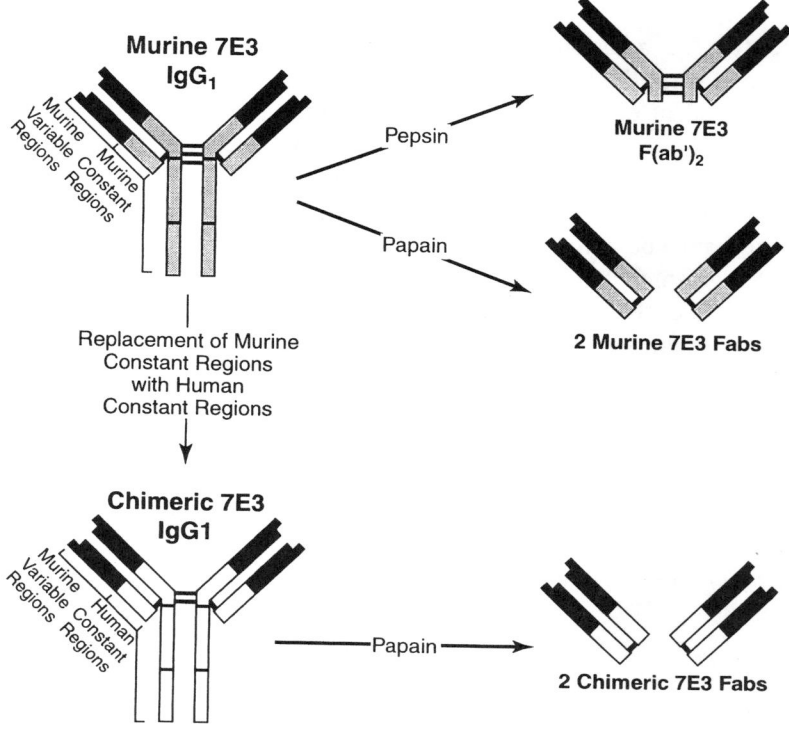

Figure 2. Schematic drawing of the development of the chimeric 7E3 antibody (c7E3 Fab, or abciximab). Murine 7E3 IgG$_1$ antibody directed against the GP IIb/IIIa receptor (top left) was proteolyzed with pepsin to yield a bivalent 7E3 F(ab′)$_2$ antigen-binding fragment and with papain to yield two univalent 7E3 Fab fragments (top right). To minimize immunogenicity, a genetic engineering approach was used to substitute human constant regions for the murine constant regions linked to the murine variable regions containing the antigen-binding sites (bottom left). Enzymatic digestion of the genetically engineered human/chimeric version of 7E3 IgG$_1$ yielded chimeric c7E3 Fab, known as abciximab (17).

To further reduce the possibility of a human antimurine antibody (HAMA) response, the molecular biologists at Centocor went on to produce a human/murine chimeric version of 7E3 Fab (16). This was achieved by re-engineering the genes that encode the murine 7E3 heavy and light antibody chains such that human constant domains would be substituted for the original mouse sequences. These human constant domains were linked to the murine variable regions that contained the antigen-combining sites (see Figure 2). The chimeric 7E3 (c7E3) Fab molecule that resulted from papain digestion of the chimeric IgG is a 47,600-dalton protein comprising roughly equal parts of murine variable region and human constant domain sequences and proved to have a KD for human platelet binding that was equivalent to that of the parent murine antibody. Importantly, the incidence of human immune responses to c7E3 Fab, known now as abciximab (marketed name ReoPro), was shown to be markedly reduced compared with murine 7E3 Fab. In large clinical trials, the rate of development of serologically-detectable immune responses was less than 6% of treated patients (17). No correspondence was found between the development of immune responses to abciximab and adverse clinical or safety outcomes. Thus, the dual objectives of equivalent functionality with reduced immunogenicity had been met.

2. PRECLINICAL PHARMACOLOGY

2.1 Binding Studies

Quantitative studies using radiolabeled antibodies revealed that approximately 80,000 abciximab molecules are bound to each human platelet at saturation (8). The equilibrium dissociation constant (Kd) for the binding of abciximab to human platelets is approximately 5 nM. Abciximab binds with similar affinity to primate platelets, binds less avidly to dog and rat platelets than to human platelets, and does not bind appreciably to platelets from mice, pigs, guinea pigs, or rabbits. A weak binding of 7E3 F(ab´)2 to rat platelets has recently been demonstrated (M. Nakada, unpublished observation).

The binding of abciximab to platelets is reversible. Under competitive conditions, about half of platelet-associated, radioactively labeled abciximab was shown to dissociate within about 4 to 5 hours (8). The dynamic binding of abciximab to platelets is also evident from flow cytometric studies. In a mixture of abciximab-coated platelets and noncoated platelets, the redistribution of abciximab from coated platelets to previously uncoated platelets is apparent at 30 minutes and is complete at 3 hours as evidenced by the single unimodal peak of intermediate abciximab binding (Figure 3) (17).

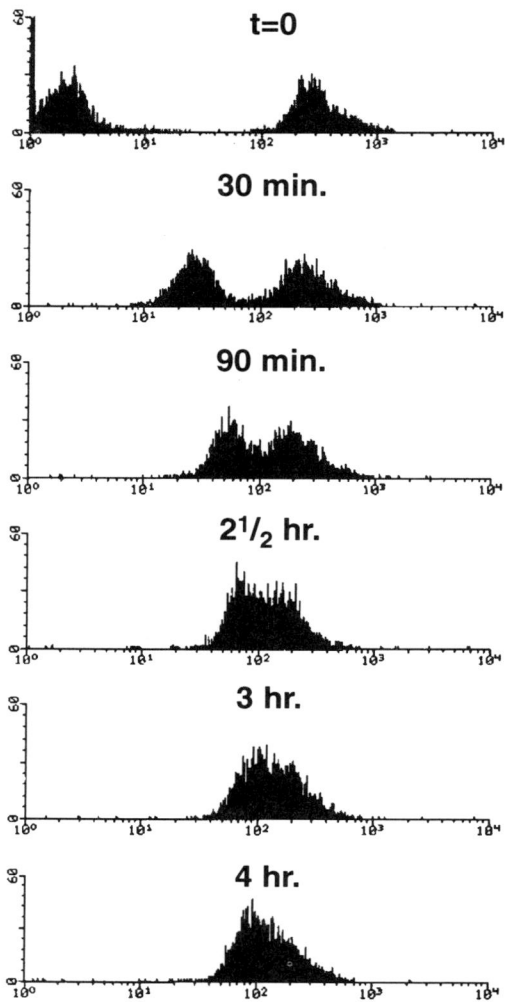

Figure 3. Platelet-bound abciximab redistributes to unlabeled platelets under *in vitro* mixing conditions. Equal volumes of washed abciximab-treated platelets and control (saline-treated) platelets were combined and continuously mixed at 37°C. Samples were periodically removed from the incubation and treated with a fluorescein-conjugated rabbit anti-abciximab antibody preparation and then fixed with 2% formalin. The smaples were evaluated for the presence of platelet-bound anti-abciximab on a Becton-Dickinson FACScan flow cytometer. From each sample, 5,000 events were analyzed in the forward-versus-side scatter gate that defined the platelet population. Individual platelet histograms are shown. The convergene of the two separte peaks into a unimodal pattern indicates that all platelets eventually acquired comparable amounts of abciximab. From Jakubowski J, Jordan RE, Weisman HF. Current antiplatelet therapy. In: Uprichard A, ed. Handbook of Experimental Pharmacology. In press.

In addition to binding to the GP IIb/IIIa receptor, abciximab also binds with comparable affinity to the related integrin receptor avb3, which is found on platelets albeit less abundantly than is GP IIb/IIIa (18). The avb3 receptor is also expressed by vascular endothelial cells (19), where it may play a role in wound healing and angiogenesis (20), and by vascular endothelial smooth muscle cells, where it may be involved in cellular migration and proliferation (21, 22). A recent study has demonstrated that abciximab freely redistributes between GP IIb/IIIa and avb3 receptors *in vitro* (23). This finding implies that abciximab may not only produce sustained blockade of platelet GP IIb/IIIa receptors but also may exert prolonged actions on vascular cell $\alpha v\beta 3$ receptors.

2.2 Dose-Response Studies

Both *in vitro* and *in vivo* studies have consistently demonstrated that platelet inhibition by abciximab is dose-dependent and correlates directly with GP IIb/IIIa receptor blockade. *In vitro* dose-response correlation studies have been performed by parallel determinations of receptor blockade, estimated by radiometric binding assay, and of platelet aggregation (24). Platelet aggregation was measured by the increase in visible light transmission through a stirred suspension of human platelets (25). At increasing concentrations of abciximab doses ranging from 0.75 to 2.0 mg/mL, increasing levels of receptor blockade ranging from 35% to 91% and corresponded to inhibition of ADP-induced platelet aggregation ranging from approximately 30% to 100% (Figure 4). Blockade of at least 80% of GP IIb/IIIa receptors was necessary to achieve complete or nearly complete inhibition of ADP-induced aggregation.

The results of these *in vitro* experiments were confirmed by *in vivo* studies in which sequential intravenous doses of abciximab, 0.05 mg/kg, were administered at intervals to a cynomolgus monkey (17). A cumulative total dose of 0.25 mg/kg produced 82% GP IIb/IIIa receptor blockade and nearly complete inhibition of ADP-induced platelet aggregation.

When abciximab was administered in a single bolus dose of 0.25 mg/kg to monkeys, it had a relatively short duration of platelet-inhibiting action. More than 80% blockade of GP IIb/IIIa receptors was achieved within 2 minutes of injection, but receptor occupancy by abciximab decreased to 75% at 1 hour, 69% at 2 hours, and 50% by 24 hours.

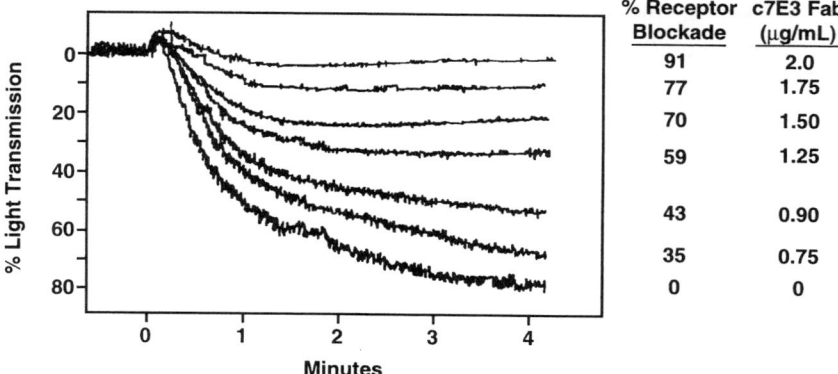

Figure 4 Concentration dependence of platelet inhibition by c7E3 Fab (abciximab) and correlation with GP IIb/IIIa receptor blockade. Platelet-rich plasma (250,000/mL) from a normal donor was incubated with different concentrations of c7E3 and then treated with ADP to induce platelet aggregation. Each tracing indicates the degree of platelet aggregation at a given c7E3 dose; greater inhibition of platelet aggregation corresponds with decreasing visible light transmission through the platelet suspension. Corresponding levels of GP IIb/IIIa receptor blockade for each c7E3 dose were determined by radiometric binding assay. From Jordan RE, Wagner CL, Mascelli MA, et al. Preclinical development of c7E3 Fab; a mouse/human chimeric monoclonal antibody fragment that inhibits platelet function by blockade of GPIIb/IIIa receptors with observations on the immunogenicity of c7E3 Fab in humans. In: Horton MA, ed. Adhesion Receptors as Therapeutic Targets. Boca Raton, Fla: CRC Press; 1996:281-305.

2.3 Antithrombotic Efficacy Studies In Animals

2.3.1 Abciximab Alone

The efficacy of abciximab has been explored in an array of animal models of coronary and carotid arterial thrombosis. Repeated experimental demonstrations that abciximab protected against thrombotic occlusion following vascular injury laid the foundation for subsequent clinical trials of GP IIb/IIIa receptor blockade in patients undergoing percutaneous intervention. It should be noted that because of the low affinity binding of abciximab to dog platelets, the divalent murine 7E3 F(ab')2 fragment was used in canine studies instead of the chimeric antibody.

In dogs whose coronary arteries had been mechanically injured and constricted by an adjustable cylinder surrounding the artery, 7E3 F(ab')2 prevented platelet deposition and vascular occlusion (26). 7E3 F(ab')2 produced similarly encouraging results in a variation of this experiment performed in the carotid arteries of monkeys (27). Importantly, the bleeding time was only modestly prolonged at doses that blocked 80% of the GP IIb/IIIa receptors, almost completely inhibited platelet aggregation, and prevented thrombosis.

Other experiments in dogs showed that pretreatment with 7E3 F(ab')2, but not with aspirin or heparin or both, was able to prevent coronary artery thrombosis and occlusion after electrical injury to a mechanically stenosed artery (28). Protection against platelet deposition and acute thrombosis at the site of deep arterial injury lasted for as long as 5 hours. Similar protective effects were documented with abciximab after electrolytic injury to the carotid artery in cynomolgus monkeys (29). Another study, with obvious clinical ramifications, showed that 7E3 F(ab')2 was more effective than aspirin in preventing thrombosis and reducing mortality for up to 6 days after balloon angioplasty in dogs (30).

2.3.2 Abciximab Combined with Fibrinolytic Agents

Since thrombosis represents the culmination of processes involving both platelet aggregation and coagulation, ample theoretical rationale supports the combination of abciximab and fibrinolytic agents. In experiments in dogs, the combination of 7E3 F(ab')2 with tissue plasminogen activator (tPA) accelerated the dissolution of thrombin-induced thrombus and prevented reocclusion in stenotic coronary arteries (31). In addition, since platelet-rich clots are implicated as factors underlying the resistance of coronary occlusive thrombus to fibrinolytic therapy, overcoming this resistance might be possible by the adjunctive use of a GP IIb/IIIa receptor blocker. Indeed, pretreatment with an intravenous bolus dose of 7E3 F(ab')2 prior to tPA administration was more effective than either aspirin or dipyridamole in preventing reocclusion in dogs (32). In this study, seven of eight control animals who did not receive the GP IIb/IIIa antagonist experienced reocclusion.

In another highly thrombogenic model in which a canine coronary artery segment was surgically removed, everted, and re-anastomosed, the administration of tPA alone proved to be of limited efficacy in restoring coronary blood flow after thrombotic occlusion (33). Reperfusion was successfully achieved, however, when 7E3 F(ab')2 was administered prior to fibrinolytic therapy. Another study which employed the same model showed that 7E3 inhibited the development of thrombosis in the everted grafts for a period extending beyond the duration of blockade of platelet aggregation (34).

Similarly, in studies using the electrolytic model of vascular injury, administration of 7E3 F(ab')2 prior to fibrinolytic therapy accelerated thrombolysis (35), prevented reocclusion and reduced infarct size (36), and improved 5-day survival as compared with animals pretreated with aspirin or hirudin (29, 37).

2.4 Animal Toxicology Studies

Toxicology studies in monkeys indicated that abciximab, administered in bolus doses as high as 8 mg/kg, has a favorable safety profile. Monitoring of the animals for 2 weeks following abciximab treatment revealed only mild, transient mucocutaneous bleeding, such as gingival bleeding, epistaxis, and bruising. However, these signs were believed to be largely the result of laboratory restraint procedures and frequent blood collections. Administration of a bolus dose of up to 0.6 mg/kg, followed by a continuous 96-hour infusion of 0.8 mg/kg/min, was likewise well tolerated, with no toxicity noted either during the infusion or during 3 to 6 weeks of observation thereafter.

In an effort to more closely simulate the clinical situation in which abciximab likely would be applied, monkeys were treated with a bolus dose of 0.3 mg/kg, followed by a continuous 48-hour infusion at 0.45 or 0.5 mg/kg/min, in combination with aspirin, heparin and either tPA or streptokinase. This study confirmed that pairing an abciximab bolus and infusion with standard antiplatelet, antithrombotic, and fibrinolytic therapy was well tolerated both acutely and for at least 3 weeks following treatment. The combination regimen was not associated with any signs of acute adverse reactions, such as hypersensitivity, hemorrhage, or thrombocytopenia.

2.5 Other Actions of Abciximab

2.5.1 Inhibition of Granule Release

Abciximab exerts a dose-dependent inhibitory effect on the release of platelet granule constituents (38). At a dose that produced 100% inhibition of platelet aggregation, abciximab resulted in 81% inhibition of the release of adenosine triphosphate (ATP) from dense platelet granules. Abciximab also acted on the release of constituents from platelet α-granules, inhibiting the release of β-thromboglobulin (β-TG) by 87% and the release of plasminogen activator inhibitor 1 (PAI-1) by 81%. These findings suggested an additional mechanism through which abciximab might counter the formation of mural thrombin and reduce mitogen release at the site of vascular injury.

2.5.2 Inhibition of inflammatory processes

The IgG form of 7E3 was shown to bind to the leukocyte integrin receptor Mac-1 (39), although with apparent lower affinity than has been shown for

GPIIb/IIIa or αvβ3. Nevertheless, this binding pointed to a potential inhibitory effect of abciximab on the recruitment of inflammatory cells to injured blood vessel walls. Incursion of excessive numbers of inflammatory cells such as monocytes and macrophages likely contributes to intimal hyperplasia and restenosis at sites of vessel injury. Recent *in vitro* observations confirmed that relatively high concentrations of abciximab effectively blocked monocyte and neutrophil adhesion to vessel wall ligands fibrinogen and ICAM-1 (40). In related observations in treated patients, abciximab reduced the activated-platelet mediated activation of leukocytes and the upregulation of Mac-1 on those cells in circulation (41). Thus, both a direct and indirect inhibition of Mac-1-mediated inflammatory pathways and the potent antithrombotic action of abciximab may converge to aid in the regulation of vascular repair and to sustaining clinical benefit.

2.5.3 Inhibition of platelet-mediated thrombin generation

Activated platelets are catalytic centers for the generation of thrombin, the enzyme responsible for causing fibrin deposition within the thrombus. The anticoagulant heparin inhibits thrombin and slows the coagulation process. In the EPIC trial (42) noted that patients receiving abciximab had longer activated clotting times (ACT) than placebo patients receiving similar doses of heparin but no abciximab. This suggested an anticoagulant action of abciximab that was confirmed in a direct *in vitro* study in which abciximab was shown to block the generation of thrombin on activated platelet surfaces (43). This effect was a direct consequence of the blockade of both platelet GPIIb/IIIa and αvβ3. Thus, the anti-thrombotic benefits of abciximab may contain an anticoagulant component in addition to the inhibition of platelet aggregation.

3. CLINICAL PHARMACOLOGY

3.1 Dose-Response Studies in Humans

Studies in healthy volunteers and in patients with stable coronary artery disease confirmed that an abciximab bolus dose of 0.25 mg/kg was sufficient to block at least 80% of GP IIb/IIIa receptors and virtually abolish ADP-induced platelet aggregation (44, 45). Although the onset of action of abciximab is immediate, the above effects were reported at 2 hours after administration. Concomitant with a decrease in receptor blockade to below 80% at 4 to 6 hours, the degree of platelet inhibition began to wane gradually.

At 24 hours following the bolus dose, the level of GP IIb/IIIa receptor blockade had fallen to 50% to 60% and platelet aggregation was inhibited by only 60%.

Since it is likely to take more than 8 hours for an injured atherosclerotic blood vessel to undergo passivation (24), that is, to become nonthrombogenic and nonplatelet-reactive, it was suspected that the duration of platelet inhibition achieved with a single bolus dose might not be therapeutically adequate. A bolus dose followed by a continuous infusion proved to be more effective in producing sustained platelet inhibition. Studies conducted in patients with stable coronary artery disease and in patients undergoing high-risk percutaneous transluminal coronary angioplasty (PTCA) showed that a bolus dose of 0.25 mg/kg, followed by a 24-hour infusion of 10 mg/min, yielded greater than 80% blockade of GP IIb/IIIa receptors and virtually complete abolition of platelet aggregation for the entire 24-hour infusion period (Figure 5) (46, 45). (As discussed below, an abciximab regimen consisting of a 0.25 mg/kg bolus and a 12-hour infusion of 10 mg/min was later to be deployed for large-scale clinical trials in patients undergoing percutaneous intervention). Partial recovery of platelet aggregation, to 50% of baseline levels, was seen within 6 hours of cessation of the infusion. However, full recovery of platelet function, in parallel with the fall-off in GP IIb/IIIa receptor blockade, is much more gradual. Abciximab remains bound to and homogeneously distributed among platelets for prolonged periods of up to 2 weeks (47, 48). Since the duration of GPIIb/IIIa binding exceeds the average platelet life span, it appears that abciximab is continuously redistributed and re-equilibrated among all circulating platelets, including those newly entering the circulation from the bone marrow. The duration of platelet inhibition following abciximab treatment is also prolonged and varies depending on the particular stimulus of platelet aggregation. Recovery of platelet function after abciximab in response to a strong agonist such as 20 mM ADP takes 2 days, but when aggregation is induced by weaker stimuli, such as lower doses of ADP or certain types of shear force, full recovery of platelet function takes at least 7 days (49).

Abciximab: The First Platelet Glycoprotein IIb/IIIa Receptor Antagonist

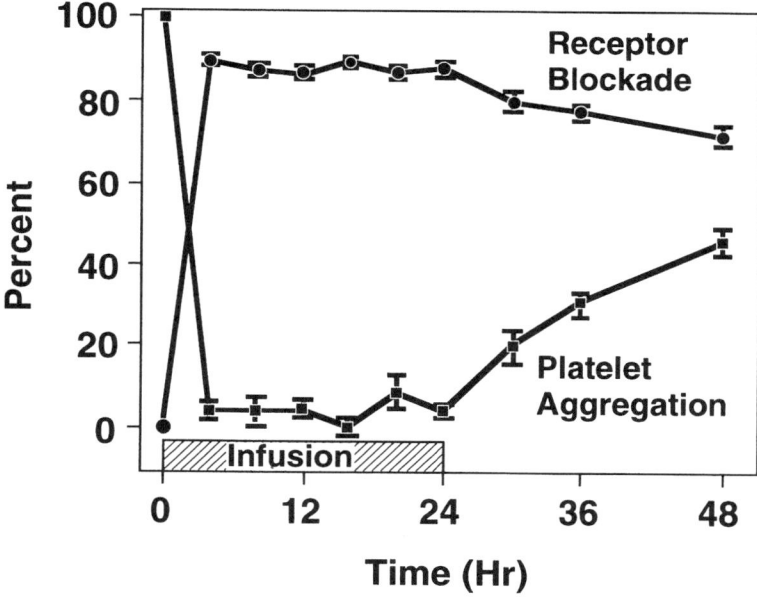

Figure 5. Blockade of GP IIb/IIIa receptors and inhibition of ex vivo platelet aggregation in patients with stable coronary artery disease who received abciximab, in a bolus dose of 0.25 mg/kg followed by a 24-hour infusion of 10 mg/min. From Jordan RE, Wagner CL, Mascelli MA, et al. Preclinical development of c7E3 Fab; a mouse/human chimeric monoclonal antibody fragment that inhibits platelet function by blockade of GPIIb/IIIa receptors with observations on the immunogenicity of c7E3 Fab in humans. In: Horton MA, ed. Adhesion Receptors as Therapeutic Targets. Boca Raton, Fla: CRC Press; 1996:281-305.

In another recent study using blood that was collected from patients undergoing angioplasty and then subjected *ex vivo* to laminar shear stress in a cone-and-plate viscometer, a bolus injection of abciximab, 0.25 mg/kg, rapidly and almost completely blocked GP IIb/IIIa receptors and inhibited this type of shear-induced platelet aggregation by 50% (50). In addition, abciximab completely eliminated the formation of large platelet aggregates. Platelet aggregation recovered partially within 2 days but remained inhibited to some degree for 1 week, at which time abciximab still blocked 35% of GPIIb/IIIa on circulating platelets.

In contrast to the prolonged circulation of platelet-bound abciximab, plasma levels of unbound abciximab disappear rapidly from the circulation following a bolus dose. The initial half-life of free abciximab is about 30 minutes and at 2 hours, less than 4% of the administered dose is present in plasma in the unbound form (24). When abciximab is administered as a bolus followed by an infusion, a median plasma concentration of more than 100 ng/mL is maintained for the duration of the infusion. The pharmacokinetic profile of free abciximab probably reflects several factors including the rapidity and extent to which the antibody binds to platelets and clearance by the kidneys (51).

The rapid clearance of free abciximab means that there is no reservoir of unbound antibody to interact with new platelets entering the circulation. This pharmacokinetic property, taken together with the observation that bound abciximab continually redistributes among the entire pool of circulating platelets, provided the rationale for the use of packed platelet transfusions to rapidly reverse the effects of abciximab when necessary. The possibility of reversing the effects of GP IIb/IIIa receptor blockade was investigated in abciximab-treated monkeys, who exhibited partial normalization of the bleeding time and partial restoration of platelet aggregation within 15 minutes of platelet transfusion (52). The efficacy of this approach in restoring hemostatic function has also been borne out in limited numbers of patients treated with abciximab who have been able to safely undergo coronary bypass after platelet transfusion (53).

4. CLINICAL TRIALS

4.1 Prevention of PTCA Complications

The dreaded complications of abrupt vessel closure or acute MI following PTCA are the consequences of platelet adherence and aggregation at the site of balloon-induced injury to the vascular wall. Thus, the interventional cardiology setting represents a logical context for the clinical evaluation of the antithrombotic actions of abciximab.

4.2 Early Studies

An early pharmacodynamic study established the safety of abciximab treatment during PTCA in combination with heparin and aspirin (45). Other early studies investigated the ability of abciximab to treat incipient platelet-mediated complications of PTCA. These early studies were open-label and did not usually contain randomized control groups. One such study revealed that abciximab was able to resolve transient formation of platelet thrombus, as reflected by cyclic variations in coronary flow in patients undergoing PTCA

(54, 55). In this study, abciximab stabilized coronary flow in patients who developed post-PTCA flow variations despite treatment with aspirin, high-dose intravenous heparin, and nitroglycerin.

One randomized, placebo-controlled Phase II PTCA trial was conducted in 60 patients with severe unstable angina that was unresponsive to treatment with heparin, aspirin and nitrates (56). Patients who received a bolus plus 24-hour infusion of abciximab had 50% less recurrent myocardial ischemia from the time the treatment was started until PTCA was performed 18-24 hours later. Moreover, 7 of 30 placebo patients (23.3%) exhibited a major PTCA-related clinical event (death, myocardial infarction or urgent coronary intervention) whereas 1 of 30 abciximab-treated patients (3.3%) had an event.

4.3 The EPIC Trial: High-Risk Patients

The Evaluation of 7E3 Fab (abciximab) for the Prevention of Ischemic Complications (EPIC) trial was the first large-scale investigation of whether treatment targeted at the GP IIb/IIIa receptor could avert platelet-mediated thrombotic complications in high-risk patients undergoing coronary angioplasty or directional coronary atherectomy (57). The 2,099 participants in this 56-center, randomized, double-blind trial were deemed high-risk because they either had unstable angina, were undergoing primary or rescue intervention within 12 hours of acute MI, or had unfavorable coronary lesion morphology.

4.3.1 Methods.

Patients were randomly assigned to receive either a bolus dose of abciximab (0.25 mg/kg) followed by a 12-hour infusion of abciximab (10 µg/min), or the abciximab bolus followed by a placebo infusion, or a placebo bolus followed by a placebo infusion. Thus, the study was designed to test not only the efficacy of abciximab versus placebo but also to weigh the relative efficacy and safety of an active bolus alone as compared with an active bolus plus infusion. In addition to the study regimen, all patients received standard therapy with oral aspirin and high-dose intravenous heparin.

The primary study endpoint was the composite incidence of death, acute MI, or urgent reintervention occurring within 30 days of randomization. Urgent reintervention encompassed repeat PTCA, coronary artery bypass grafting (CABG), intracoronary stent placement, or intra-aortic balloon pump insertion. Safety analysis was based on a comparison of the frequencies of major bleeding, intracerebral hemorrhage, and minor bleeding among the three study arms. To ascertain whether any benefits of abciximab treatment would be sustained over the long-term, the study also employed a secondary efficacy endpoint. The secondary composite endpoint was defined as the incidence of death, acute MI, and both urgent and nonurgent reintervention within 6 months of randomization. An independent, blinded, clinical endpoint committee reviewed and confirmed all efficacy and safety endpoints.

4.3.2   30-Day Efficacy Results.

Intention-to-treat analysis was performed, based on data from all EPIC enrollees regardless of whether they actually underwent revascularization or received treatment. The landmark outcome of EPIC was the finding that the abciximab bolus coupled with the abciximab infusion reduced the 30-day incidence of death, MI, or urgent reintervention by a striking 35% relative to placebo (8.3% versus 12.8%, $P = 0.009$). The difference between active treatment and placebo was evident as early as the first day after randomization and were sustained over the course of 30 days. The individual components of the overall 30-day benefit of abciximab bolus plus infusion were a 39.4% reduction in the incidence of MI (5.2% versus 8.6% with placebo, $P = 0.014$) and a 49.1% reduction in the incidence of urgent reintervention (mostly urgent PTCA and CABG; 4.0% versus 7.8% with placebo, $P = 0.003$). In all three treatment arms, 30-day mortality was low with no significant differences among the treatment groups. The event-free survival curves expressed as the proportion of patients in each treatment group who did not experience a primary endpoint at 30 days is shown in Figure 6.

The effect of the abciximab bolus alone was intermediate between that of bolus plus infusion and that of placebo (11.5%), but the difference relative to placebo did not reach statistical significance.

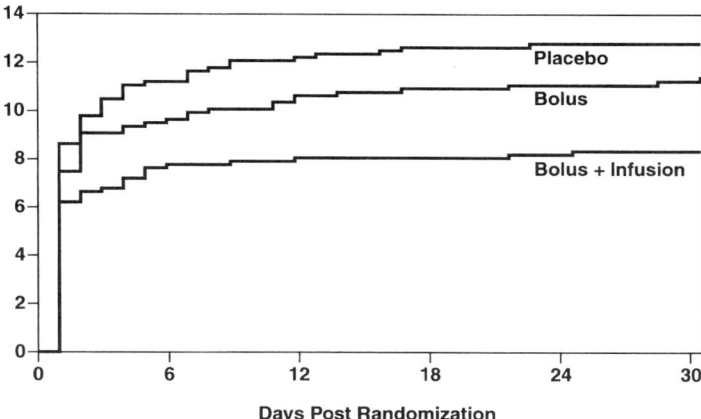

Figure 6. The EPIC trial: Kaplan-Meier plot of the 30-day composite endpoint of death, MI, or urgent reintervention. Event rates are shown as the proportion of patients in each treatment group who did not experience one of these events during the 30-day period following randomization. Events that occurred during randomization were classified as day 0 events.

4.3.3   30-Day Safety Results.

The principal drawback of GP IIb/IIIa receptor blockade in the EPIC trial was a 14% incidence of major bleeding events in the bolus plus infusion group, which was approximately twice the incidence observed in the placebo arm. Such bleeding events, two thirds of which occurred within 36 hours of treatment initiation, were also associated with a higher rate of blood transfusion among abciximab-treated patients. However, importantly most of these episodes occurred at a vascular access site and that abciximab was not linked with any increased risk of intracerebral hemorrhage or bleeding necessitating surgical intervention. In addition, the risk of bleeding was most pronounced among patients with the lowest body weights who, because of the use of standard fixed-dose heparin boluses, received the proportionately greatest heparin doses. Taken together, these findings suggested that the use of lower dose heparin regimens, coupled with meticulous attention to the vascular access site, might minimize the risk of bleeding. This supposition was eventually borne out in subsequent clinical trials of abciximab (see below).

Thrombocytopenia (platelet counts < 50,000/mm^3) developed in 1.6% of abciximab-treated patients and in 0.7% of subjects assigned to placebo.

The incidence of immune responses to abciximab, as measured by human antichimeric antibody (HACA) assays, was 5.8%. The majority of these HACA responses were of low or moderate titer.

4.3.4   6-Month Efficacy Results.

Follow-up of the EPIC participants confirmed that the short-term benefits of abciximab were durable for at least 6 months, as evidenced by a 22.9% reduction in the incidence of death, MI, or repeat intervention in the bolus plus infusion group (27.0% versus 35.1% with placebo, P = 0.001) (Figure 7). This decrease in the composite event rate reflected principally a 34.2% reduction in the incidence of MI (P = 0.018) and a 31% reduction in the need for repeat PTCA (P = 0.01) (Figure 8). Separate analyses performed for the interval between 2 days and 6 months indicated a statistically significant 24.5% reduction in the composite endpoint, even when the 2 days during which most acute events were concentrated and during which therapeutic platelet inhibition was most intense are excluded from consideration. An additional benefit accruing between 30 days and 6 months was also suggested by the results. Overall, these findings spotlighted abciximab as the first pharmacologic therapy to improve the clinical outcome following percutaneous coronary intervention beyond that of conventional therapy with aspirin and heparin.

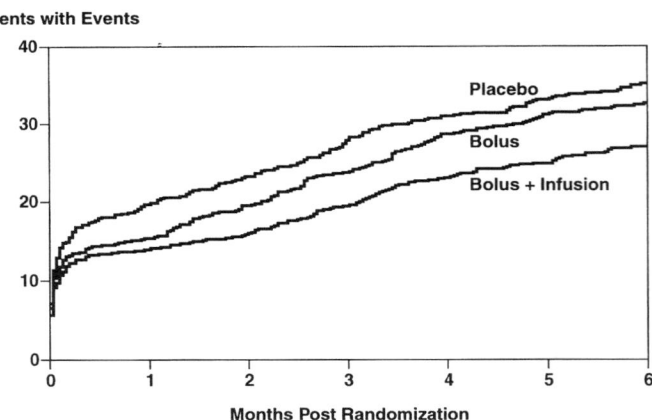

Figure 7. The EPIC trial: Kaplan-Meier plot of the 6-month composite endpoint of death, MI, or coronary revascularization. The reduction in events achieved with the abciximab bolus plus infusion was sustained during 6 months of follow-up. From Topol, E.J. *et al.* Randomised trial of coronary intervention with antibody against platelet IIb/IIIa integrin for reduction of clinical restenosis: results at six months. Lancet, 1994; 343: 881-886.

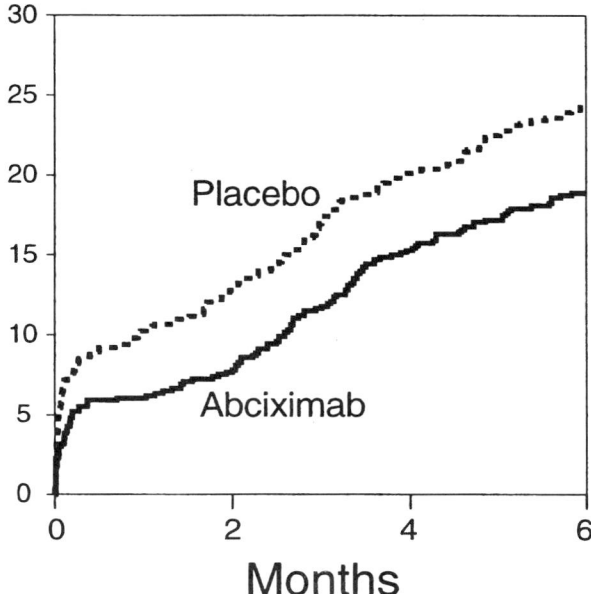

Figure 8. The EPIC trial: effect of abciximab on the need for repeat target vessel revascularization during 6 months of follow-up. Treatment with an abciximab bolus plus infusion significantly reduced the incidence of coronary angioplasty or coronary artery bypass grafting.

4.3.5 3-Year Efficacy Results.

EPIC participants were followed with the double blind maintained for a median duration of over 3 years. Follow-up analysis revealed that the improvement in outcome observed at 30 days and at 6 months among patients who received the abciximab bolus and infusion was sustained for this extended 3-year period (58). The bolus plus infusion regimen reduced the composite endpoint by 13%, from 47.2% to 41.1% with placebo (Figure 9) (P = 0.009). The incidence of acute MI was reduced by 21% (from 13.6% to 10.7%; P = 0.08), and the need for repeat revascularization was reduced by 13% (from 40.1% to 34.8%; P = 0.02). The latter difference translated into an avoidance of 9 procedures per 100 patients treated with abciximab bolus and infusion. The predefined subgroup of highest risk patients, which included individuals with evolving acute MI or unstable angina, exhibited a 3-year mortality reduction of 60%. The improvement in long-term survival was found to be related to the abciximab-related reduction in periprocedural MI; the EPIC investigators reported that the greater the magnitude of the periprocedural rise in creatine kinase (CK) levels, the higher the risk of mortality.

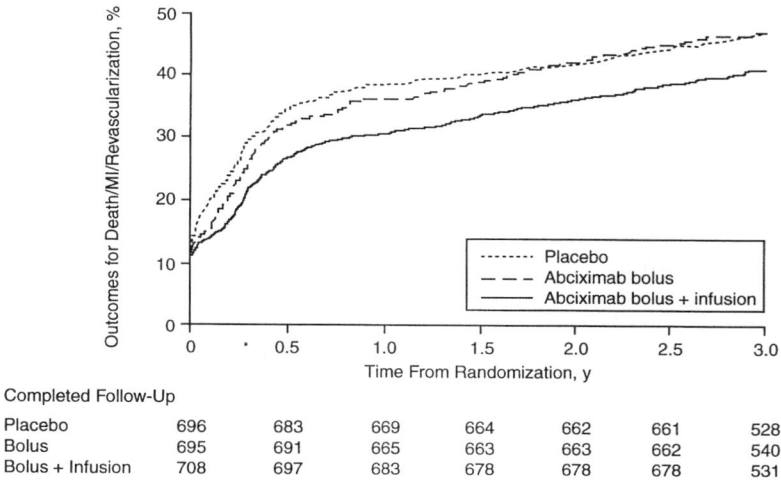

Figure 9. The EPIC trial: influence of abciximab on the composite outcome of death, myocardial infarction, or revascularization at 3 years. From Topol E.J. et al. Long-term protection from myocardial ischemic events in a randomized trial of brief integrin β_3 blockade with percutaneous coronary intervention. JAMA, 1997; 278: 479-484.

These long-term results made EPIC the first clinical trial to demonstrate that an adhesion molecule antagonist could exert a long-term salutory influence on the natural history of coronary atherosclerotic disease. The EPIC investigators proposed that the apparent ability of abciximab to prevent restenosis might have stemmed, at least in part, from this agent's blockade of the $\alpha v \beta_3$ receptor, with consequent inhibition of smooth muscle cell migration.

4.3.6 Subgroup Analyses.

Analyses of various EPIC subpopulations confirmed that the benefits of abciximab extended virtually across the board. In the 470-patient subgroup with unstable angina, for example, abciximab bolus plus infusion reduced the 30-day composite clinical event rate by 70% (59). Six-month follow-up demonstrated that among these patients, the bolus plus infusion regimen reduced mortality from 6.7% to 0.7% and decreased the incidence of acute MI from 11.3% to 1.3%.

The 42 EPIC patients who had primary PTCA for acute MI and the 22 patients who underwent rescue PTCA after thrombolytic therapy proved ineffective were at particularly high risk of ischemic events. Here, the abciximab bolus and infusion reduced the 6-month composite endpoint of death, MI, or repeat coronary intervention from 47.8% to 4.5% (60).

In the placebo arm, the incidence of non-Q-wave MI was twice as high among patients who underwent atherectomy as in their angioplasty-treated counterparts. Treatment with the abciximab bolus plus placebo eliminated this doubling of risk in the atherectomy subgroup (61).

Two hundred thirty-seven patients in the EPIC study were undergoing PTCA for a lesion that had already been subjected to this procedure within the preceding 12 months and subsequently restenosed. In this subgroup, the bolus plus infusion reduced the 30-day composite event rate from 14% to 3% and decreased the 6-month event rate from 39% to 26% (62). Similar findings were observed for the subpopulation of 114 patients who underwent revascularization of saphenous vein grafts. Here, the 30-day clinical event rate fell from 13% to 7.5%, although this effect was not maintained at 6 months (63).

The only EPIC participants who did not appear to benefit from active treatment were those 506 high-risk individuals with diabetes, in whom the risk of bleeding posed by GP IIb/IIIa receptor blockade outweighed the clinical advantage of platelet inhibition (64). According to 3-year follow-up data, abciximab treatment did not produce a long-term reduction in clinical events in diabetic patients (65).

4.3.7 Anticoagulant Effect of Abciximab.

The EPIC investigators also reported longer activated clotting times (ACTs) in patients treated with abciximab plus heparin relative to those who received similar doses of heparin alone (42). The suggestion that abciximab might have anticoagulant as well as antiplatelet activity has, in fact, been substantiated by recent work showing that abciximab prolongs the ACT *in vitro* (66).

The most obvious mechanisms for an anticoagulant effect of abciximab would be a reduction in the activated platelet surfaces available for the assembly of coagulation factors and the prevention of the surface expression of the coagulation cofactor, platelet factor V. However, it has also recently been shown that abciximab leads to a 48% reduction in the generation of thrombin induced by tissue factor in the presence of platelets (43). On the other hand, thrombin generation was inhibited only half as much by an antibody that acts with equal potency on the GP IIb/IIIa receptor but does not block the $\alpha v \beta 3$ receptor. This raised the possibility that dual blockade of the GP IIb/IIIa and $\alpha v \beta 3$ receptors by abciximab might have led to enhanced inhibition of thrombin generation.

4.4   The PROLOG Trial: Maximizing Safety

The next logical step in the clinical evaluation of abciximab was to ascertain whether the benefits documented in high-risk EPIC participants undergoing percutaneous intervention could also be achieved in a broader population of patients. First, however, it was necessary to address the safety issues raised in the EPIC trial.

To this end, a multicenter, double-blind pilot study was conducted to compare the efficacy and safety of a standard weight-adjusted heparin regimen with that of a low-dose, weight-adjusted heparin regimen in percutaneous intervention patients treated with the abciximab bolus plus infusion regimen (67, 68). The 103 participants in the PROLOG (Precursor to EPILOG) trial were randomized to receive heparin either as a 100 U/kg bolus with additional doses administered to achieve an activated clotting time (ACT) > 300 seconds or as a 70 U/kg bolus.

The PROLOG results indicated that the incidence of bleeding complications in patients assigned to low-dose heparin was comparable to that seen in the placebo arm in the EPIC trial. This probably reflected the fact that the median ACT in the low-dose heparin group was only 256 seconds, as compared with 330 seconds in the standard weight-adjusted arm and 398 seconds in the EPIC non-weight-adjusted heparin abciximab bolus plus infusion arm. Early removal of the catheter sheath from the femoral artery may have also played a role in minimizing access-site bleeding in PROLOG.

Most notably, this reduced incidence of bleeding complications was accomplished without any compromise in the efficacy of abciximab. The incidences of the composite clinical endpoint of death, acute MI, or urgent reintervention at 7 days and at 30 days in PROLOG were comparable to those documented in EPIC with the abciximab bolus plus infusion regimen.

4.5   The EPILOG Trial: Low-Risk and High-Risk Patients

The Evaluation in PTCA to Improve Long-Term Outcome with Abciximab GP IIb/IIIa Blockade (EPILOG) trial was designed to determine (1) whether abciximab would reduce ischemic complications in all patients undergoing percutaneous intervention, regardless of their risk status; and (2) to confirm the PROLOG finding that the safety of abciximab treatment could be enhanced without sacrifice of efficacy by the use of low-dose, weight-adjusted heparin (The EPILOG Investigators, 1997).

4.5.1 Methods.

This 69-center randomized, double-blind North American trial assigned patients to one of three treatment regimens: (a) placebo plus aspirin and standard weight-adjusted heparin (100 U/kg with boluses added as needed to maintain the ACT ($\geq$ 300 seconds); (b) abciximab bolus plus 12-hour infusion, along with aspirin and standard weight-adjusted heparin; or (c) abciximab bolus plus infusion, together with aspirin and low-dose, weight-adjusted heparin (70 U/kg with bolus doses added as needed to achieve an ACT ($\geq$ 200 seconds). The protocol also mandated that heparin treatment be withdrawn immediately after the interventional procedure and that vascular sheaths be removed as soon as the ACT reached 175 seconds. Because of the dramatic benefits of GP IIb/IIIa receptor blockade observed in EPIC enrollees with acute ischemic syndromes, EPILOG excluded patients with electrocardiograph (ECG)-documented unstable angina and patients who had experienced an acute MI within the past 24 hours. Other noteworthy exclusion criteria were planned stent implantation or rotational atherectomy.

4.5.2 30-Day Efficacy Results.

EPILOG was terminated prematurely when only 2,792 of the planned 4,800 patients had been enrolled because interim analysis revealed evidence of unequivocal benefit with both abciximab-containing regimens. At 30 days, the primary composite endpoint of death, acute MI, or urgent revascularization was reduced by approximately 56% in both the abciximab plus standard weight-adjusted heparin group (5.4% versus 11.7% with placebo; $P < 0.001$) and the abciximab plus low-dose weight-adjusted heparin group (5.2% versus 11.7% with placebo; $P < 0.001$) (Figure 10). Similar reductions in risk were likewise noted for the components of the composite endpoint. The 30-day incidence of MI fell 58%, from 8.7% with placebo to 3.8% with abciximab ($P < 0.001$); the 30-day incidence of death or MI was reduced from 9.1% to 4.0% ($P < 0.001$); and the 30-day frequency of urgent revascularization fell 68%, from 5.2% to 2.0% ($P < 0.001$). Overall, for every 1,000 patients treated with abciximab, 65 fewer experienced acute ischemic complications of percutaneous intervention. This degree of risk reduction is, to date, unmatched by any other agent described in the coronary revascularization literature.

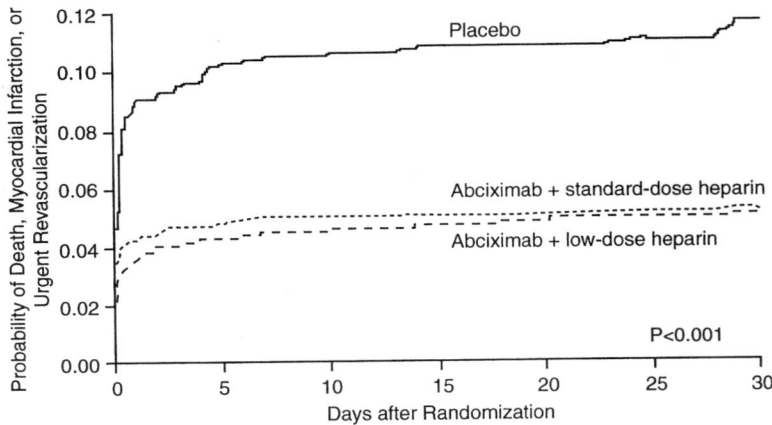

Figure 10. The EPILOG trial: Kaplan-Meier plot of the probability of the composite endpoint of death, myocardial infarction, or urgent revascularization within 30 days of randomization. From The EPILOG Investigators. Platelet glycoprotein IIb/IIIa receptor blockade and low-dose heparin during percutaneous coronary revascularization. N. Engl. J. Med., 1997; 336: 1689-1696.

The improvements in clinical outcome associated with abciximab treatment were comparable in all patient groups, including low-risk patients, high-risk patients who had unfavorable lesion morphology or had experienced an acute MI within the 7 days prior to randomization, and patients requiring bailout stenting or atherectomy. Although EPILOG excluded patients with unstable angina and associated ECG changes, about half of the study population had clinical signs of unstable angina; abciximab with low-dose heparin was beneficial in these patients as well.

4.5.3 30-Day Safety Results in Contrast to EPIC Results.

The risk of major bleeding in EPILOG did not differ significantly among the three treatment arms. This improvement in safety may be because even those patients treated with the standard weight-adjusted heparin regimen used in EPILOG received less heparin than did their EPIC counterparts, who were treated with a non-weight-adjusted heparin regimen.

The incidence of minor bleeding was significantly higher, however, among patients receiving abciximab plus standard heparin (7.4%) than among those receiving abciximab plus low-dose heparin (4.0%) or placebo plus standard heparin (3.7%, $P < 0.001$). The need for red cell transfusion was also significantly greater in the placebo plus standard heparin group than in the abciximab plus low-dose heparin arm (3.9% versus 1.9%, P < 0.01). Thrombocytopenia was an infrequent complication, occurring at a comparable incidence of less than 1% in all three treatment groups.

Thus, the EPILOG results provided proof that with the use of low-dose heparin, early discontinuation of heparin, and meticulous care of the vascular access site, it is indeed possible to uncouple the cardioprotective benefits of GP IIb/IIIa receptor blockade from the risk of bleeding complications.

4.5.4   6-Month Efficacy Results.

Six months of follow-up revealed that the reduction in ischemic complications achieved with abciximab was sustained. At 6 months, the composite endpoint of death, MI, or any revascularization was reduced by 11.6%, from 25.8% with placebo to 22.8% with abciximab and low-dose heparin (P = 0.07), and by 13.6%, to 22.3%, with abciximab and standard heparin (P = 0.04); the composite endpoint of death, MI, or urgent revascularization was reduced by 43%, from 14.7% with placebo to 8.4% with abciximab (P < 0.001). Abciximab significantly decreased both the 6-month incidence of myocardial infarction, from 9.9% to 5.2%, and the 6-month rate of urgent revascularization, from 6.7% to 3.3%. In contrast to the EPIC trial, however, abciximab did not reduce the need for non-urgent target vessel revascularization at 6 months.

4.5.5   1-Year Efficacy Results.

Follow-up analysis revealed that the 58% relative risk reduction documented at 30 days was sustained without attenuation at 1 year (70). Treatment with abciximab reduced the 1-year incidence of the composite endpoint from 16.1% to 9.6% (P < 0.001), the 1-year frequency of death or MI from 12.3% to 6.6% (P < 0.001), and the 1-year rate of urgent intervention from 7.2% to 4.0% (P < 0.001). Not surprisingly, in light of the 6-month results, the 1-year incidence of elective target vessel revascularization procedures was not reduced by abciximab.

Thus, EPILOG demonstrated that the durable long-term benefits proven for high-risk patients in the EPIC trial extend across the full spectrum of coronary intervention patients.

4.6   Unstable Angina: The CAPTURE Trial

The c7E3 Fab Antiplatelet Therapy in Unstable Refractory Angina (CAPTURE) study was the first large-scale, randomized, placebo-controlled trial expressly designed to ascertain whether pretreatment with abciximab would reduce the risk of ischemic complications before, during and after angioplasty in patients with refractory unstable angina and concomitant ECG abnormalities (71), as had been suggested in a previous pilot study (56).

4.6.1 Methods.

Patients recruited in 69 centers in 12 European countries were assigned to treatment with abciximab, in a bolus dose of 0.25 mg/kg followed by a continuous infusion of 10 μg/min, or with placebo. The infusion was administered for 18 to 24 hours preceding and 1 hour following PTCA. In addition, all participants received routine medical treatment with aspirin, standard heparin, intravenous nitroglycerin, and, where appropriate, other antianginal agents.

4.6.2 Efficacy Results.

An interim analysis pointing to the clear superiority of active treatment with abciximab led to the discontinuation of the CAPTURE study when only 1,266 of the planned 1,400 patients had been enrolled. The primary endpoint of death, MI, or urgent intervention at 30 days was reduced by 28.9% with abciximab pretreatment, from 15.9% to 11.3% ($P = 0.012$) (Figure 11). The benefit of active treatment was attributable primarily to a 50% decrease in the incidence of both Q-wave and non-Q-wave acute MIs (4.1% with abciximab versus 8.2% with placebo; $P = 0.002$). The advantages of abciximab were consistent across all patient subgroups, regardless of age, gender, entry ECG findings, or the presence of diabetes, peripheral vascular disease, or renal dysfunction.

Patients in the abciximab group had a 70% lower incidence of acute MI during the 24 hours pre-PTCA treatment period and a 50% lower incidence during or within 24 hours after PTCA than did their placebo-treated counterparts. This finding suggested that abciximab produced stabilization or passivation of unstable plaque during this period and raised the possibility that treatment might even have averted the need for PTCA altogether. In addition, abciximab significantly lessened the incidence of recurrent ischemia and the total ischemic burden during the 24 hours preceding and 6 hours following intervention (72).

Follow-up analysis indicated that the advantages of GP IIb/IIIa receptor blockade were maintained at 6 months, although the differences no longer reached statistical significance. The investigators suggested that the benefits of treatment might have been more durable had the abciximab infusion been continued for 12 hours after percutaneous intervention.

Abciximab: The First Platelet Glycoprotein IIb/IIIa Receptor Antagonist

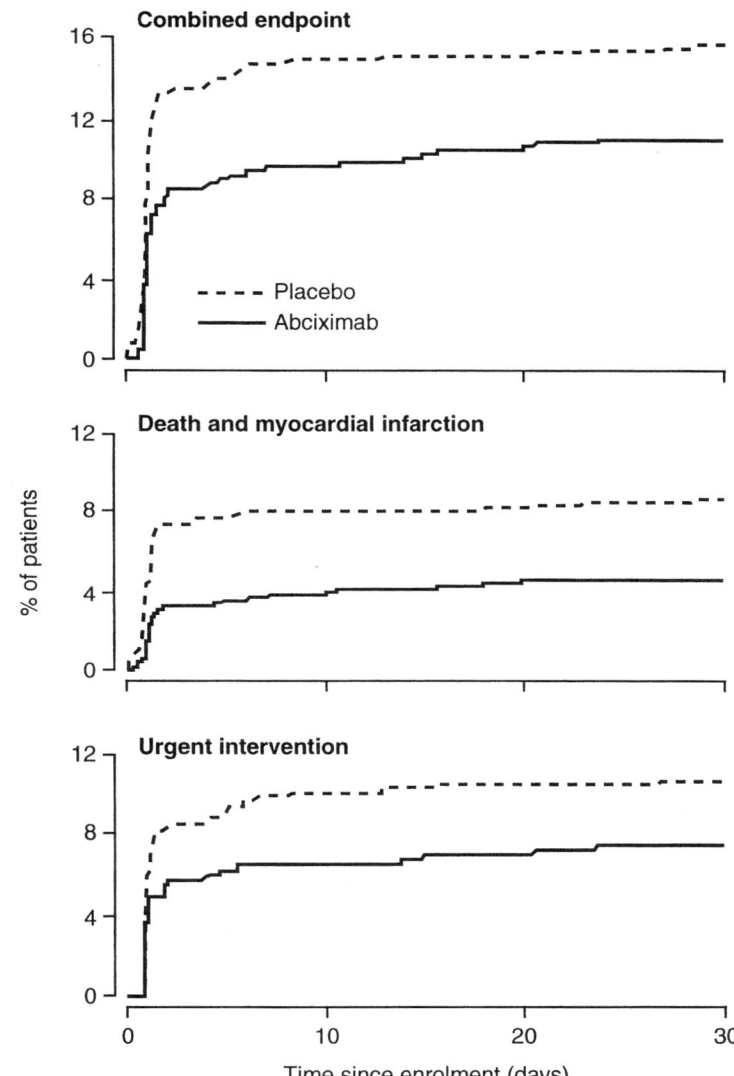

Figure 11. The CAPTURE trial: time course of the combined primary endpoint (death, myocardial infarction, or urgent intervention) and its components. From The CAPTURE Investigators. Randomised placebo-controlled trial of abciximab before and during coronary intervention in refractory unstable angina: the CAPTURE study. Lancet. 1997; 349: 1429-1435.

4.6.3 Safety Results.

The frequency of major and minor bleeding events was somewhat more common in patients treated with abciximab than in those who received placebo, and was most common in those who received high doses of heparin during PTCA and in those with low body weight. However, the incidence of major bleeding was considerably lower in CAPTURE than in EPIC (3.8% versus 10.6%, respectively), probably because of the lower heparin doses and greater attention to the vascular access site in CAPTURE. Greater reductions in heparin doses and early sheath removal, as in the EPILOG study, might have further enhanced safety.

4.7 Acute MI Abciximab in Combination With Primary Angioplasty: RAPPORT

The striking benefits of abciximab in the small EPIC subgroup of acute MI patients undergoing primary or rescue angioplasty were apparently commensurate with the extremely high risk of complications in this population. The potential use of abciximab as an adjunct to primary angioplasty in patients with acute MI was further explored in the ReoPro in Acute Myocardial Infarction and Primary PTCA Organization and Randomized Trial (RAPPORT).

4.7.1 Methods.

The RAPPORT investigators randomized 483 patients with acute MI documented by ST-segment elevation to receive, prior to primary angioplasty, either an abciximab bolus (0.25 mg/kg) plus a 12-hour infusion (0.125 mg/kg/min up to a maximum of 10 µg/min) or a placebo bolus and placebo infusion in addition to standard medical therapy (Brener, 1997).

4.7.2 30-Day Results.

Intention-to-treat analysis revealed a 48% reduction in the primary composite endpoint of death, MI, or urgent revascularization at 30 days (5.8% with abciximab versus 11.2% with placebo) (73). The benefit of abciximab was evident as early as 7 days after randomization, at which point the incidence of death or MI was half that in the placebo group and the need for emergency revascularization was 25% less. When only the 89% of enrollees who actually received the study treatment were considered, the reduction in the 30-day combined endpoint with abciximab was 52% (4.9% versus 10.3% with placebo; $P = 0.03$) and the reduction in the need for urgent target vessel revascularization was a striking 68% (1.8% versus 5.6%, P = 0.03) (74). The difference in the need for emergency revascularization was independent of acute angiographic results, suggesting that passivation of the vessel wall by abciximab contributes to the prevention of recurrent ischemic events.

In patients who received the study treatment, the need for bailout stenting during primary angioplasty was reduced by 38% (11.5% with abciximab versus 18.3% with placebo; P = 0.03) (75). Among those who received the infusion for the entire 12-hour period, the effect of abciximab was even more pronounced; the need for bailout stenting was reduced by 41%, from 20.6% to 12.3% (P = 0.02). This effect probably stemmed from the prevention of platelet aggregation by abciximab at the site of balloon inflation.

4.7.3 6-Month Results.

At 6 months, the 35% reduction in the rate of death, MI, or urgent revascularization with abciximab was of borderline statistical significance (73). When the analysis was limited only to those patients who actually received the study medication and angioplasty, abciximab was demonstrated to produce a significant 49% reduction in this 6-month combined endpoint. However, the 6-month follow-up data revealed no differences between the two groups with respect to the composite endpoint of death, MI, or any revascularization.

4.8 Abciximab in Combination With Rescue Angioplasty

4.8.1 GUSTO-III.

Of the subgroup of 387 patients from the large-scale Global Use of Strategies to Open Occluded Coronary Arteries (GUSTO)-III trial who underwent rescue angioplasty after receiving fibrinolytic therapy, 81 (21%) were treated with abciximab (76). Thirty-day mortality was significantly lower in patients who received abciximab than in those who did not (3.7% versus 9.8%, P = 0.04). This improvement in survival with abciximab was associated with only a slightly increased risk of bleeding. The GUSTO-III investigators recommended further studies to elucidate the appropriate doses of fibrinolytic agent, heparin, and abciximab in acute MI patients.

4.9 Abciximab as an Adjunct to Fibrinolysis

A growing body of both experimental and clinical data suggests that the combination of abciximab with fibrinolytic therapy may be a promising approach to the pharmacologic management of acute MI. In fact, recent work has demonstrated that abciximab, in the presence of heparin and aspirin, can initiate coronary reflow in some patients within 10 minutes without fibrinolytic therapy (77).

4.9.1 TAMI-8.

A pilot study from the Thrombolysis and Angioplasty in Myocardial Infarction (TAMI)-8 investigators was the first clinical study to demonstrate the efficacy and safety of combining a GP IIb/IIIa receptor blocker with a fibrinolytic agent (78). In this study, 60 patients with acute MI received escalating bolus doses of murine-derived 7E3 antibody 3, 6, and 15 hours after the start of a 100-mg infusion of tPA. More than 90% of patients with available coronary angiograms exhibited patency of the infarct artery, as compared with only five of nine control patients treated with tPA alone. In addition, 87% of patients who received the combination of abciximab and tPA were free of recurrent ischemia. Perhaps most importantly, the pairing of GP IIb/IIIa receptor blockade and fibrinolysis did not lead to any excess of bleeding complications relative to fibrinolysis alone.

4.9.2 Abciximab Alone Versus Abciximab Plus Low-Dose Plasminogen Activator.

In an ongoing multicenter trial, patients with acute MI and ST-segment elevation are being randomized to receive either an intravenous bolus dose of abciximab, 0.25 mg/kg, or placebo. After undergoing angiography 60 to 90 minutes after initial therapy, patients then are crossed over to a bolus injection of whichever agent was not administered initially. If TIMI (thrombolysis in myocardial infarction) 3 flow is not evident on a second angiogram obtained 10 minutes after crossover therapy, patients are then randomized to adjunctive low-dose tPA or placebo. Preliminary data from 26 patients have revealed that 31% of patients achieved TIMI 3 flow and 50% of patients achieved TIMI 2 or 3 flow after treatment with abciximab alone (79).

4.9.3 TIMI-14.

In an attempt to determine whether abciximab enhances fibrinolysis, the TIMI-14 investigators are comparing TIMI 3 flow at 90 minutes in patients with acute MI and ST-segment elevation who were randomly assigned to one of the following regimens: (1) accelerated full-dose tPA; (2) abciximab, in a bolus dose of 0.25 mg/kg followed by a 12-hour infusion of 0.125 µg/kg/min; (3) abciximab plus reduced doses of streptokinase; or (4) abciximab plus reduced doses of tPA. Results from the dose-finding phase, which included 35 patients in each group, indicated that abciximab alone achieves TIMI 3 flow in about one third of patients (80). However, abciximab in combination with low-dose fibrinolytic therapy produces TIMI 3 flow rates as high as or higher than those observed with full doses of tPA alone. In fact, in one treatment arm that combined full-dose abciximab with a 60-minute infusion of 50 mg tPA, TIMI 3 patency at 90 minutes was 79%—notably greater than the 58% patency rate recorded for accelerated tPA in the trial. This synergistic effect was achieved without an increase in bleeding risk. These promising observations will continue to be evaluated as the trial proceeds.

5. CONCLUSIONS: CURRENT INDICATIONS AND FUTURE PROSPECTS

The validity of GP IIb/IIIa receptor blockade as a therapeutic strategy has now been confirmed in large-scale randomized, double-blind, placebo-controlled clinical trials involving thousands of patients. These trials have consistently demonstrated that abciximab can prevent the development of thrombotic complications following percutaneous coronary intervention procedures, in high- and low-risk patients alike, in those with unstable angina, and in those undergoing primary angioplasty for acute MI. Moreover, they have confirmed that a few simple, straightforward measures can minimize the risk of bleeding associated with potent antiplatelet therapy.

Abciximab first received regulatory approval for use in patients undergoing high-risk percutaneous coronary intervention procedures. More recently, abciximab has been approved for use in all patients undergoing percutaneous intervention and in patients with unstable angina not responding to conventional medical therapy when percutaneous intervention is planned within 24 hours. Perhaps the most exciting potential application of the GP IIb/IIIa receptor antagonists is as pharmacologic therapy for unstable angina, whether or not intervention is planned. The GUSTO-IV trial should provide further insights. Another avenue of investigation concerns the use of abciximab as an adjunct to fibrinolytic therapy in patients with acute MI, and studies also are under way to evaluate the efficacy and safety of abciximab in this setting. This approach is particularly promising since it offers the hope that adjunctive abciximab treatment will not only help overcome resistance to fibrinolytic therapy but also allow the use of lower and safer fibrinolytic doses.

The development of abciximab and its translation into the clinic represent the sum total of enormous leaps in our knowledge of molecular biology, receptor biochemistry, immunology, and cardiovascular pathophysiology. We can anticipate that future studies integrating laboratory and clinical science will engender new insights into the mechanisms underlying atherosclerosis and occlusive thrombotic disease and their therapeutic modulation by antagonists of the platelet GP IIb/IIIa receptor.

BIOGRAPHY

Robert E. Jordan, Ph.D. is Director of Cardiovascular Pharmacology in the Department of Clinical Research at Centocor. Harlan F. Weisman, M.D. is Vice President of Clinical Research at Centocor. Marian T. Nakada, Ph.D. is Assistant Director of Cardiovascular Pharmacology in the Department of Clinical Research at Centocor.

ReoPro was developed by Centocor Inc., and is manufactured by Centocor B.V. in the Netherlands. Centocor's mission is to develop and commercialize novel therapeutic products and services that solve critical needs in human health care. The company concentrates on research and development, manufacturing and market development, with a primary technology focus on monoclonal antibodies and DNA-based products.

REFERENCES

1. Coller, B.S. (1992). Platelets in cardiovascular thrombosis and thrombosis. In: Fozzard HA, ed. The Heart and Cardiovascular System. New York, Raven Press. pp. 219-273.
2. Colman, R.W. and Walsh, P.N. (1987). Mechanisms of platelet aggregation. In: Colman, R.W., Hirsh, J., Marder, V.J. and Salzman, E.W., (eds.) Hemostasis and Thrombosis: Basic Principles and Clinical Practice. Philadelphia, J.B. Lippincott Co. pp. 594-605.
3. White, J.G. (1987). Anatomy and structural organization of the platelet. In: Colman, R.W., Hirsh, J., Marder, V.J. and Salzman, E.W. (eds.). Hemostasis and Thrombosis: Basic Principles and Clinical Practice. Philadelphia, J.B. Lippincott Co. pp. 537-554.
4. Lefkovits, J. et al. (1995). Platelet glycoprotein IIb/IIIa receptors in cardiovascular medicine. N. Engl. J. Med., 332, 1553-1559.
5. Coller, B.S. (1995). The role of platelets in arterial thrombosis and the rationale for blockade of platelet GP IIb/IIIa receptors as antithrombotic therapy. Eur. Heart J., 16(suppl L), 11-15.
6. Hynes, R.O. (1992). Integrins: versatility, modulation and signaling in cell adhesion. Cell, 69, 11-25.

7. Ruoslahti, E. (1991). Integrins. J. Clin. Invest., 87, 1-5.
8. Wagner, C.L. et al. (1996). Analysis of GPIIb/IIIa receptor number by quantification of 7E3 binding to human platelets. Blood, 88, 907-914.
9. Niiya, K. et al. (1987). Increased surface expression of the membrane glycoprotein IIb/IIIa complex induced by platelet activation. Relationship to the binding of fibrinogen and platelet aggregation. Blood, 70, 475-483.
10. Plow, E.F. and Ginsberg, M.H. (1989). Cellular adhesion: GP IIb/IIIa as a prototypic adhesion receptor. Prog. Hemost. Thromb., 9, 117-156.
11. Nurden, A.T. and Caen, J.P. (1974). Abnormal platelet glycoprotein patterns in three cases of Glanzmann's thrombasthenia. Br. J. Haematol., 28, 253-260.
12. Phillips, D.R. and Agin, P.P. (1979). Platelet membrane defects in Glanzmann's thrombasthenia: evidence for decreased amounts of two major glycoproteins. J. Clin. Invest., 60, 535-545.
13. Coller, B.S. et al. (1983). A murine monoclonal antibody that completely blocks the binding of fibrinogen to platelets produces a thrombasthenic-like state in normal platelets and binds to glycoproteins IIb and/or IIIa. J. Clin. Invest., 73, 325-338.
14. Coller, B.S. (1985). A new murine monoclonal antibody reports an activation-dependent change in the conformation and/or microenvironment of the platelet glycoprotein IIb/IIIa complex. J. Clin. Invest., 76, 101-108.
15. Parham. P. (1986). Preparation and purification of active fragments from mouse monoclonal antibodies. In: Weir, D.M. (ed.). Handbook of Experimental Immunology. 4th Edition. Blackwell Scientific Publications, Oxford. pp. 14.1 – 14.23.
16. Knight, D.M. et al. (1995). The immunogenicity of the 7E3 murine monoclonal Fab antibody fragment variable region is dramatically reduced in humans by substitution of human for murine constant regions. Mol. Immunol., 32, 1271-1281.
17. Jordan, R.E. et al. (1996). Preclinical development of c7E3 Fab; a mouse/human chimeric monoclonal antibody fragment that inhibits platelet function by blockade of GPIIb/IIIa receptors with observations on the immunogenicity of c7E3 Fab in humans. In: Horton, M.A., (ed.) Adhesion Receptors as Therapeutic Targets. Boca Raton, Fla., CRC Press. pp. 281-305.
18. Coller, B.S. et al. (1991). Platelet vitronectin receptor expression differentiates Iraqi-Jewish from Arab patients with Glanzmann thrombasthenia in Israel. Blood, 77, 75-83.
19. Charo, I.F. et al. (1986). Platelet glycoproteins IIb and IIIa: evidence for a family of immunologically and structurally related glycoproteins in mammalian cells. Proc. Natl. Acad. Sci. U.S.A., 83, 8351-8355.
20. Brooks, P.C. et al. (1994). Requirement for vascular integrin $\alpha_v\beta_3$ for angiogenesis. Science, 264, 569-571.
21. Choi, E.T. et al. (1994). Inhibition of neointimal hyperplasia by blocking $\alpha_v\beta_3$ integrin with a small peptide antagonist GpenGRGDSPCA. J. Vasc. Surg., 19, 125-134.
22. Srivatsa, S.S. et al. (1997). Selective $\alpha v \beta 3$ integrin blockade potently limits neointimal hyperplasia and lumen stenosis following deep coronary arterial stent injury: Evidence for the functional importance of integrin $\alpha v \beta 3$ and osteopontin expression during neointima formation. Cardiovascular Res., 36, 408-428.
23. Nakada, M.Y. et al. (1998). Abciximab can freely redistribute between GP IIb/IIIa and $\alpha_v\beta_3$. J. Am. Coll. Cardiol., 31(suppl A), 236A-237A. Abstract 1088-98.
24. Coller, B.S. (1997). GP IIb/IIIa antagonists: pathophysiologic and therapeutic insights from studies of c7E3 Fab. Thromb. Haemost., 78, 730-735.

25. Chanarin, I. (1989). Laboratory Haematology. London, Churchill Livingstone. pp. 371-399.
26. Coller, B.S. et al. (1986). Antithrombotic effect of a monoclonal antibody to the platelet glycoprotein GP IIb/IIIa receptor in an experimental animal model. Blood, 68, 783-786.
27. Coller, B.S. et al. (1989). Abolition of *in vivo* platelet thrombus formation in primates with monoclonal antibodies to the platelet GP IIb/IIIa receptor. Circulation, 80, 1766-1774.
28. Mickelson, J.K. et al. (1989). Antiplatelet monoclonal F(ab')$_2$ antibody directed against the platelet GP IIb/IIIa receptor complex prevents coronary artery thrombosis in the canine heart. J. Mol. Cell Cardiol., 231, 393-405.
29. Rote, W.E. et al. (1994). Prevention of rethrombosis after coronary thrombolysis in a chronic canine model. I. Adjunctive therapy with monoclonal antibody 7E3 F(ab')$_2$ fragment. J. Cardiovasc. Pharmacol., 23, 194-202.
30. Bates, E.R. et al. (1991). A monoclonal antibody against the platelet glycoprotein IIb/IIIa receptor complex prevents platelet aggregation and thrombosis in a canine model of coronary angioplasty. Circulation, 84, 2463-2469.
31. Gold, H.K. et al. (1988). Rapid and sustained coronary artery recanalization with combined bolus injection of recombinant tissue-type plasminogen activator and monoclonal antiplatelet GP IIb/IIIa antibody in a canine preparation. Circulation, 77, 670-677.
32. Yasuda, T. et al. (1988). Monoclonal antibody against the platelet glycoprotein (GP) IIb/IIIa receptor prevents coronary artery reocclusion after reperfusion with recombinant tissue-type plasminogen activator in dogs. J. Clin. Invest., 81, 1284-1291.
33. Yasuda, T. et al. (1990). Lysis of plasminogen activator-resistant platelet-rich coronary artery thrombus with combined bolus injection of recombinant tissue-type plasminogen activator and antiplatelet GP IIb/IIIa antibody. J. Am. Coll. Cardiol., 16, 1728-1735.
34. Kiss, R.G. et al. (1994). Time course of the effects of a single bolus infection of F(ab')$_2$ fragments of the antiplatelet GP IIb/IIIa antibody 7E3 on arterial eversion graft occlusion, platelet aggregation, and bleeding time in dogs. Arteriosclerosis Thromb., 14, 367-374.
35. Fitzgerald, D.J. et al. (1991). Systemic lysis protects against the effects of platelet activation during coronary thrombolysis. J. Clin. Invest., 88, 1589-1595.
36. Mickelson, J.K. et al. (1990). Antiplatelet antibody [F(ab')$_2$] prevents rethrombosis after recombinant tissue-type plasminogen activator-induced coronary artery thrombolysis in a canine model. Circulation, 81, 617-627.
37. Rote, W.E. et al. (1994). Prevention of rethrombosis after coronary thrombolysis in a chronic canine model. II. Adjunctive therapy with r-hirudin. J. Cardiovasc. Pharmacol., 23, 203-211.
38. Mascelli, M.A. et al. (1998). Abciximab inhibits release of platelet granule constituents. J. Am. Coll. Cardiol., 31(suppl A), 24A-25A. Abstract 1009-17.
39. Altieri, D. and Edgington, T.S. (1988). A monoclonal antibody reacting with distinct adhesion molecules defines a transition in the functional state of the receptor CD11b/CD18 (Mac-1). J. Immunol., 141, 2656-2660.
40. Simon, D.I. et al. (1997). 7E3 monoclonal antibody directed against the platelet glycoprotein IIb/IIIa Cross-reacts with the leukocyte integrin Mac-1 and blocks adhesion to fibrinogen and ICAM-1. Arterioscler. Thromb. Vasc. Biol., 17, 528-535.

41. Mickelson, J.K. et al. (1996). Chimeric 7E3 Fab (ReoPro) decreases detectable CD11b on neutrophils from patients undergoing coronary angioplasty. Circulation, 94(suppl 1), 1-42. Abstract 0233.
42. Moliterno, D.J. et al. (1995). Effect of platelet glycoprotein IIb/IIIa integrin blockade on activated clotting time during percutaneous transluminal coronary angioplasty or directional atherectomy (the EPIC Trial). Am. J. Cardiol., 75, 559-562.
43. Reverter, J.C. et al. (1996). Inhibition of platelet-mediated, tissue factor-induced thrombin generation by the mouse/human chimeric 7E3 antibody. J. Clin. Invest., 98, 863-874.
44. Bhattacharya, S. et al. (1995). Blockade of the human platelet GP IIb/IIIa receptor by a murine monoclonal antibody Fab fragment (7E3): potent dose-dependent inhibition of platelet function. Cardiovasc. Drugs Ther., 9, 665-675.
45. Tcheng, J.E. et al. (1994). Pharmacodynamics of chimeric glycoprotein IIb/IIIa integrin antiplatelet antibody Fab 7E3 in high-risk coronary angioplasty. Circulation, 90, 1757-1764.
47. Christopoulos, C. et al. (1993). Flow cytometric observations on the in vivo use of Fab fragments of a chimaeric monoclonal antibody to platelet glycoprotein IIb-IIIa. Blood Coag. Fibrinol., 4, 729-737.
48. Mascelli, M.A. et al. (1998). Pharmacodynamic profile of short-term abciximab treatment demonstrates prolonged platelet inhibition with gradual recovery from GP IIb/IIIa receptor blockade. Circulation, 97, 1680-1688.
49. Jordan, R.E. et al. (1997). Abciximab causes profound, immediate inhibition of platelet function that recovers gradually after PTCA. Circulation, 96(suppl I), I-721. Abstract 4041.
50. Konstantopoulos, K. et al. (1995). Shear-induced platelet aggregation is inhibited by in vivo infusion of an anti-glycoprotein IIb/IIIa antibody fragment, c7E3, in patients undergoing coronary angioplasty. Circulation, 91, 1427-1431.
52. Wagner, C.L. et al. (1995). Reversal of the anti-platelet effects of chimeric 7E3 Fab by platelet transfusion in cynomolgus monkeys. Thromb. Haemost., 73, 1313. Abstract #1586.
54. Anderson, H.V. et al. (1994). Cyclic flow variations after coronary angioplasty in humans: clinical and angiographic characteristics and elimination with 7E3 monoclonal antiplatelet antibody. J. Am. Coll. Cardiol., 23, 1031-1037.
55. Anderson, H.V. et al. (1992). Intravenous administration of monoclonal antibody to the platelet GP IIb/IIIa receptor to treat abrupt closure during coronary angioplasty. Am. J. Cardiol., 69, 1373-1376.
56. Simoons, M.L. et al. (1994). Randomized trial of a GP IIb-IIIa platelet receptor blocker in refractory unstable angina. Circulation, 89, 596-603.
57. The EPIC Investigators. (1994). Use of a monoclonal antibody directed against the platelet glycoprotein IIb/IIIa receptor in high-risk coronary angioplasty. N. Engl J. Med., 330, 956-961.
58. Topol, E.J. et al. (1997). Long-term protection from myocardial ischemic events in a randomized trial of brief integrin β_3 blockade with percutaneous coronary intervention. JAMA, 278, 479-484.
59. Lincoff, A.M. et al. (1994). Striking clinical benefit with platelet GP IIb/IIIa inhibition by c7E3 among patients with unstable angina: outcome in the EPIC trial. Circulation, 90(suppl I), I-21. Abstract 0104.

60. Lefkovits, J. et al. (1996). Effects of platelet glycoprotein IIb/IIIa receptor blockade by a chimeric monoclonal antibody (abciximab) on acute and six-month outcomes after percutaneous transluminal coronary angioplasty for acute myocardial infarction. Am. J. Cardiol., 77, 1045-1051.
61. Lefkovits, J. et al. (1996). Increased risk of non-Q wave MI after directional atherectomy is platelet dependent: evidence from the EPIC trial. J. Am. Coll. Cardiol., 28, 849-855.
62. Lefkovits, J. et al. (1995). Can conjunctive platelet glycoprotein IIb/IIIa receptor blockade improve outcomes of coronary interventions for restenotic lesions. Circulation, 92(suppl I), I-607. Abstract 2907.
63. Challapalli, R.M. et al. (1995). Platelet glycoprotein IIb/IIIa monoclonal antibody (c7E3) reduces distal embolization during percutaneous intervention of saphenous vein grafts. Circulation, 92(suppl I), I-607. Abstract 2908.
64. Moliterno, D.J. et al. (1995). Special considerations for diabetic patients receiving platelet IIb/IIIa antagonists during coronary interventions: results from the EPIC trial. J. Am. Coll. Cardiol., 25(suppl A), 155A-156A. Abstract 935-34.
65. Narins, C.R. et al. (1997). Does abciximab improve outcome following angioplasty in diabetics? Long-term follow-up results from the EPIC study. Circulation, 96(suppl 1), 1-162. Abstract 904.
66. Ammar, T. et al. (1997). *In vitro* effects of the platelet glycoprotein IIb/IIIa receptor antagonist c7E3 Fab on the activated clotting time. Circulation, 95, 614-617.
67. Lincoff, A.M. et al. (1995). A multicenter, randomized, double-blind pilot trial of standard versus low dose weight-adjusted heparin in patients treated with the platelet glycoprotein IIb/IIIa receptor antibody fragment abciximab (c7E3 Fab) during percutaneous coronary revascularization. J. Am. Coll. Cardiol., 80A, 711-713.
68. Lincoff, A.M. et al. (1997). Standard versus low dose weight-adjusted heparin in patients treated with the platelet glycoprotein IIb/IIIa receptor antibody fragment abciximab (c7E3 Fab) during percutaneous coronary revascularization. Am. J. Cardiol., 79, 286-291.
69. The EPILOG Investigators. (1997). Platelet glycoprotein IIb/IIIa receptor blockade and low-dose heparin during percutaneous coronary revascularization. N. Engl. J. Med., 336, 1689-1696.
70. Lincoff, A.M. et al. (1997). Durable inhibition of ischemic complications by abciximab during percutaneous coronary revascularization: one-year results of the EPILOG trial. Circulation, 96(suppl I), I-162. Abstract 902.
71. The CAPTURE Investigators. (1997). Randomised placebo-controlled trial of abciximab before and during coronary intervention in refractory unstable angina: the CAPTURE study. Lancet, 349, 1429-1435.
72. Klootwijk, P. et al. (1997). Reduction of recurrent ischemia with abciximab during continuous ECG-ischemia monitoring in patients with unstable angina refractory to standard treatment (CAPTURE). J. Am. Coll. Cardiol., 29(suppl A), 367A. Abstract 781-3.
73. Brener, S.J. et al. (1997). A randomized, placebo-controlled trial of abciximab with coronary angioplasty for acute MI: the RAPPORT trial. Presented at the 70th Scientific Sessions of the American Heart Association, Orlando, Fla.
74. Brener, S.J. et al. (1998). Abciximab reduces urgent target vessel revascularization at 30 days after primary angioplasty, independently of acute angiographic results: the RAPPORT trial. J. Am. Coll. Cardiol., 31(suppl A), 54A. Abstract 802-1.

75. Barr, L.A. et al. (1998). Abciximab reduces the need for bail-out stenting during primary angioplasty: the RAPPORT trial. J. Am. Coll. Cardiol., 31(suppl A), 237A. Abstract 1088-101.
76. Miller, J.M. et al. (1998). Survival benefit of abciximab administration during early rescue angioplasty: analysis of 387 patients from the GUSTO-III trial. J. Am. Coll. Cardiol., 31(suppl A), 191A. Abstract 842-2.
77. Gold, H.K. et al. (1997). Restoration of coronary flow in myocardial infarction by intravenous chimeric antibody without exogenous plasminogen activators: observations in animals and humans. Circulation, 95, 1755-1759.
78. Kleiman, N.S. et al. (1993). Profound inhibition of platelet aggregation with monoclonal antibody 7E3 Fab after thrombolytic therapy: results of the Thrombolysis and Angioplasty in Myocardial Infarction (TAMI) 8 pilot study. J. Am. Coll. Cardiol., 22, 381-389.
79. Gold, H.K. et al. (1997). A randomized, placebo-controlled crossover trial of ReoPro alone or combined with low-dose plasminogen activator for coronary reperfusion in patients with acute myocardial infarction: preliminary results. Circulation, 96(suppl 1), 1-474. Abstract 2648.
80. Antman, E.M. et al. (1998). Abciximab (ReoPro) potentiates thrombolysis in ST elevation myocardial infarction: results of TIMI 14 trial. J. Am. Coll. Cardiol., 31(suppl A), 191A. Abstract 842-1.

Chapter 3

Recombinant Coagulation Factor IX (BeneFix®)

John Edwards, Neil Kirby
Genetics Institute Inc., 87 Cambridge Park Drive, Cambridge, Mass.

Key words: Hemophilia, regulatory, manufacturing, clinical, development

Abstract: Coagulation factor IX is a zymogen that is an essential component of the clotting process; deficient factor IX activity results in hemophilia B. To overcome problems associated with plasma-derived blood products, Genetics Institute developed a recombinant coagulation factor IX (rFIX, BeneFix®) produced using a Chinese hamster ovary cell line. The production cell line includes an rFIX expression vector as well as another plasmid expressing PACE-SOL for necessary posttranslational processing. Extensive testing of master cell bank, working cell bank, and end-of-production cells has demonstrated that these plasmids are accurately integrated into the genome of the CHO cell line and that rFIX is correctly and stably expressed across multiple generations. No blood or plasma products are used in the manufacturing process or formulation of rFIX. Additionally, to ensure product purity, the manufacturing process includes four chromatographic separation procedures and two filtration procedures to remove impurities and potential contaminants. The identity, purity, potency, safety, and quality of rFIX drug substance are ensured by a series of test procedures, including gel electrophoresis, size-exclusion chromatography, peptide mapping, carbohydrate fingerprinting, and biological activity in a one-stage clotting assay. Preclinical and clinical testing of rFIX has shown it to be safe and effective for treating hemorrhagic episodes and preventing bleeding during surgery in patients with hemophilia B.

1. INTRODUCTION

Hemophilia B is a hemorrhagic disorder that results from a deficiency of factor IX (FIX) coagulant activity. This genetic disease was first described more than 40 years ago (1, 2) and has traditionally been treated by replacement of the deficient clotting factor with a product pooled from thousands of blood donors. Although currently available factor IX concentrates are effective in controlling bleeding in patients with hemophilia, it has proven difficult to completely remove or inactivate blood-borne infective agents. To overcome the problems associated with plasma-derived factor IX (pdFIX) products, Genetics Institute developed a recombinant factor IX (rFIX). This product is manufactured without the use of blood or plasma products and provides high purity, lot-to-lot consistency, and high specific activity.

Hemophilia B results from mutations of the factor IX gene that reduce expression of functional factor IX protein (3, 4). The disease occurs almost exclusively in males since the gene coding for factor IX is located on the X chromosome. Males have only a single X chromosome, and a defect in their factor IX gene results in symptoms of hemophilia B. Since females have two X chromosomes, a defective copy of one of their factor IX genes still allows production of sufficient factor IX to permit normal blood clotting. However, such women are carriers of the disease and may pass symptomatic hemophilia B on to their sons and carrier status on to their daughters. Hemophilia B occurs in 1 of 25,000 to 30,000 male births, and it accounts for approximately 20% of all patients with hemophilia (5-7). The disease varies in severity according to the amount of factor IX activity that remains in the patient's blood. Patients with <1% of normal factor IX activity are considered to have severe disease, patients with 1% to 5% of normal factor IX activity have moderate disease, and those with factor IX activity >5% are considered to have mild disease (8).

The consequences of hemophilia B, especially in patients with severe disease, can be quite serious. Individuals with severe hemophilia B often have spontaneous bleeding, particularly in their muscles and joints (hemarthrosis), that may lead to joint destruction (4, 9). Clearly, safe and effective control over bleeding and prevention of bleeding that might occur during surgery are critically important for patients with hemophilia B.

The original factor concentrates used to treat hemophilia B were prothrombin complex concentrates (PCCs). These products made treatment easier and more effective than treatment with whole plasma, but they contained other clotting factors (factors II, VII, and X) in addition to factor IX. These clotting factors could cause too much clotting when PCCs were given in large amounts to treat serious bleeding episodes or given repeatedly to control a bleed. Recently, scientists discovered a way to separate the factor IX from the rest of the factors. This made treating factor IX deficiency much safer. However, since these factor concentrates are made from pooled human plasma, the risk of contamination from viruses or other pathogens was never completely eliminated.

In order to overcome the risks of plasma-derived products, a fundamental change had to occur in the way factor IX concentrate was made. This change took place when recombinant DNA technology was applied to the manufacturing of factor concentrates. It is this technology that led to the production of BeneFIX®, the first licensed recombinant (genetically engineered) factor concentrate for the treatment of hemophilia B.

Recombinant factor IX is indicated for the control and prevention of hemorrhagic episodes in patients with hemophilia B (congenital factor IX deficiency or Christmas disease). This indication includes control and prevention of bleeding in surgical settings.

2. MECHANISM OF ACTION

As noted above, hemophilia B results from a deficiency of factor IX, a key element in the clotting cascade. Factor IX, a vitamin-K–dependent plasma glycoprotein, is an essential component of the early phase of the coagulation process and can be activated by extrinsic clotting pathways (3, 10). Activated factor IX (factor IXa) interacts with a complex composed of factor VIII, factor X, calcium, and phospholipids that are usually located on the surface of platelet membranes to activate factor X. Activated factor X (factor Xa) then interacts with factors V and II (prothrombin) to convert factor II into factor IIa (thrombin). In the final stage of the coagulation cascade, factor IIa proteolyses fibrinogen into fibrin monomers that undergo polymerization. These fibrin polymers are stabilized by the cross-linking activity of factor XIII in association with thrombin and calcium (3, 10).

Factor IX is a single-chain glycoprotein of approximately 55 kDa and is composed of five structural domains (Figure 12). The structural domains are a γ-carboxyglutamic acid (Gla) domain, two epidermal growth factor-like (EGF-like) domains, an activation peptide (AP) sequence, and a serine protease domain (5). In addition, there is a 46-residue propeptide before the NH_2-terminus of the mature protein. The Gla domain includes 12 glutamic acid residues that are normally γ-carboxylated and function in calcium binding, which is important in activation of the zymogen (11). The mature, secreted FIX protein consists of 415 amino acids with multiple posttranslational modifications. Mature FIX is a zymogen that is converted to the enzymatically active form *in vivo* by proteolytic cleavage at Arg-145 and

Arg-180. The cleavage yields a disulfide-linked heterodimer, designated activated FIX (FIXa), and a 35-residue activation peptide.

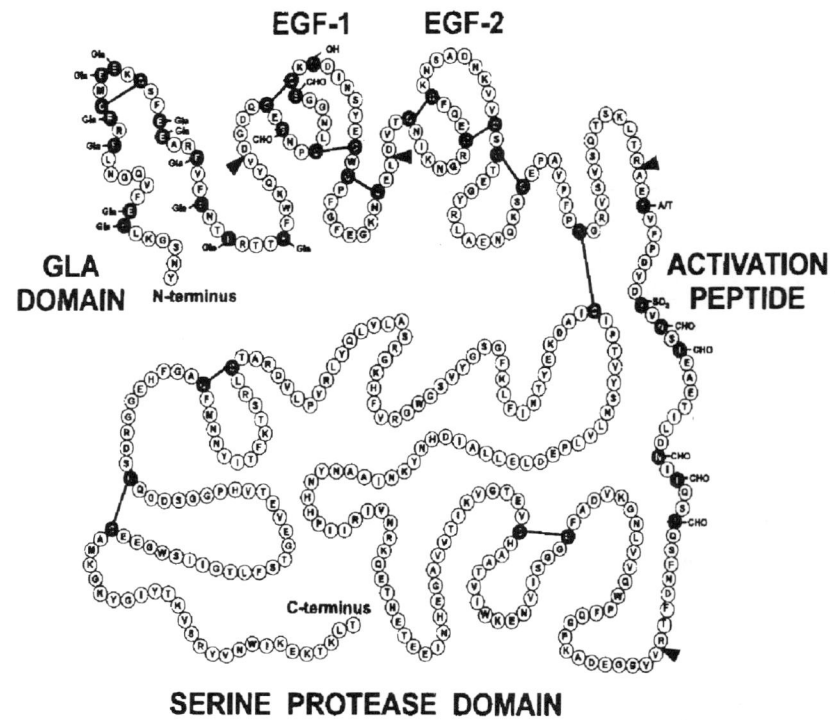

Figure 12. Primary structure of human factor IX

3. DEVELOPMENT OF THE PRODUCTION CELL LINE

Recombinant factor IX is produced using a Chinese hamster ovary (CHO) cell line, FIX.1F. This cell line has multiple copies of a DNA expression vector containing sequences encoding rFIX and a DNA expression vector containing sequences encoding a truncated, secreted form of human recombinant paired basic amino acid-cleaving enzyme (PACE-SOL).

The cell system chosen for the production of rFIX is the Chinese hamster ovary cell. A number of considerations were taken into account in selecting a

CHO cell expression system to produce rFIX: 1) the ability of CHO cells to perform the complex posttranslational processing necessary for the secretion of biologically active rFIX, 2) the successful use of CHO cells by the biopharmaceutical industry to produce therapeutics that have been used to treat a large number of patients without evidence of viral transmission, and 3) the significant experience of Genetics Institute in large-scale cell culture of CHO cells in the manufacture of recombinant proteins.

The strategy for developing the rFIX production cell line was to create an expression construct containing a DNA segment encoding human rFIX and to express rFIX in CHO cells under conditions where suitable amounts of functional protein are secreted. Additionally, this strategy included development of the rFIX/CHO cell line such that the phenotype was suitable for rFIX production in a suspension-culture, stirred-tank bioreactor format using a cell culture medium lacking any human- or animal-derived proteins (12, 13).

The vitamin-K–dependent coagulation proteins, including FIX, undergo extensive posttranslational modification during synthesis and secretion into the blood. These modifications include signal peptide cleavage, N- and O-linked glycosylation, β-hydroxylation of aspartic acid, tyrosine sulfation, serine phosphorylation, γ-carboxylation, and cleavage of an 18-amino-acid NH_2-terminal propeptide (14, 15). The propeptide directs vitamin-K–dependent γ-carboxylation of 12 glutamic acid residues, located in a region adjacent to the propeptide. γ-Carboxylation of 10 to 12 of these residues is essential for the calcium- and phospholipid-binding properties of the protein (15). Subsequent removal of the propeptide, mediated by an endoprotease, is required for functional activity (16). When rFIX is expressed at very high levels in CHO cells, much of the protein produced is not biologically active, due to limitations in the capacity of the cells to posttranslationally modify high levels of rFIX (17). CHO cell limitations in rFIX processing capacity result in incomplete γ-carboxylation and incomplete propeptide cleavage, as was observed in early attempts to produce rFIX in highly amplified CHO cell lines (17, 18). These early efforts produced CHO cell lines expressing very high levels of total rFIX protein with very low percentages of active rFIX (0.2% to 4.4%). Even after monoclonal antibody purification, the rFIX produced by these cells had specific activities ranging from only 35 to 75 IU/mg (as compared with approximately 200 IU/mg for plasma-derived FIX).

The subsequent strategy for rFIX production cell line development was to create CHO cell lines in which the rFIX gene was amplified (using the DHFR-selectable, amplifiable marker, described below) only to a limited extent, and to screen for rFIX-producing cell clones in which the rFIX protein production level and cellular capacity to posttranslationally modify rFIX are balanced. Such a strategy was intended to produce cell lines that secrete a high percentage of biologically active rFIX. Using this strategy, cell lines were identified that appeared to secrete rFIX of a much higher specific activity than previously reported (17, 18). However, a substantial fraction of the secreted rFIX still contained the 18-amino-acid propeptide. To circumvent this limitation, the rFIX/CHO cells were further engineered to coexpress a protease capable of augmenting the host cell capacity for propeptide cleavage.

Paired basic amino-acid cleaving enzyme (PACE) is a calcium-dependent serine protease capable of processing rFIX (18). PACE belongs to a family of subtilisin-like proteases that share considerable structural homology with the yeast protease Kex2, which is involved in the proteolytic processing of propeptides (19). Laboratory experiments demonstrated that coexpression of rFIX with PACE or with PACE-SOL, an engineered secreted form of PACE, in stably transfected CHO cells resulted in the secretion of mature rFIX with the correct NH_2-terminus (18).

Accordingly, the PACE-SOL gene was introduced into candidate rFIX cell lines to augment their capacity for propeptide cleavage. The gene was introduced using a second selectable, amplifiable marker, the human adenosine deaminase (ADA) gene (20). Candidate cell lines coexpressing rFIX and PACE-SOL were selected and the ability of PACE-SOL to effectively supplement the cellular capacity for propeptide removal was confirmed.

Candidate cell lines were adapted over a period of months to grow in a culture medium lacking any human- or animal-derived proteins, and the lead candidate (the FIX.1F cell line) was further adapted to bioreactor conditions similar to those expected to be used in the rFIX manufacturing process. Additionally, vitamin K analog and dose optimization studies revealed that relatively low concentrations of vitamin K_1 could maximize the capacity of the FIX/PACE-SOL coexpressing FIX.1F cells to produce biologically active rFIX.

3.1 Preparation of the rFIX Coding Sequence

The coding sequence for rFIX was assembled from clones derived from a human liver cDNA library and a human XXXXY chromosome genomic library (17). A human liver cDNA library in the bacteriophage vector λGT10, prepared using oligo-d(T) primed human liver poly(A)$^+$ RNA, was screened with oligonucleotide probes that were based on the published sequence of human FIX (21, 22). Three overlapping but incomplete cDNA recombinants were isolated and characterized extensively. Two of these recombinants were combined to create a partial cDNA that contained the coding region of human preproFIX from codon 11 (within the signal peptide sequence) to the COOH-terminus, as well as the 3′ untranslated sequence (23, 24).

The sequence encoding the first 10 codons of human preproFIX was obtained by cloning a human genomic fragment that contained the FIX promoter and all of exon I of the FIX gene, encoding the first 29 amino acids of preproFIX (23, 24). A human XXXXY chromosome cell line (human lymphoblastoid cell line GM1202A from the NIGMS mutant cell repository) genomic library was prepared in the bacteriophage vector Charon 4A (25) and screened with a nick-translated FIX cDNA probe. Five recombinant clones were obtained and characterized by restriction mapping, Southern blotting, and DNA sequencing. These five clones were found to have overlapping DNA sequences encompassing the entire 35-kb FIX gene (24). One recombinant clone containing a 5' segment of the gene was used to complete the construction of the full-length gene for preproFIX.

The assembly of the complete coding sequence for rFIX (17) from the partial cDNA clone and the 5' genomic clone was accomplished by a method exploiting homologous reciprocal recombination in *E. coli* (26, 27). The principle of this method involves homologous recombination between a recombinant bacteriophage, harboring amber mutations, and a recombinant plasmid containing a tyrosine tRNA amber-suppressor gene, followed by selection for bacteriophage that acquire an integrated copy of the recombinant plasmid by growth on a suppressor-free host strain. Using this procedure, a recombinant bacteriophage was isolated that contained the human FIX promoter, a short 29-bp 5' untranslated segment, 1245 base pairs of FIX coding sequence, and a 1390-bp 3' untranslated segment. No introns were present in this recombinant bacteriophage.

To prepare the rFIX coding segment for insertion into eukaryotic expression plasmids, the 5' flanking segment containing the human FIX promoter was subsequently removed. This was accomplished by restriction enzyme and exonuclease digestion, followed by cloning into the bacteriophage plasmid M13mp11. A clone was selected in which the rFIX gene was present on a 2.5-kb *Pst*I fragment that contained 18 nucleotides of 5' untranslated sequence, the entire FIX coding segment, and 1028 nucleotides of 3' untranslated sequence.

3.2   Cell Line Development

The expression system is based on the stable integration and amplification of expression plasmids within the genome of a CHO cell line, known as DUKX-B11, that is deficient in dihydrofolate reductase (DHFR). The DHFR-deficient CHO cell line, DUKX-B11, is a derivative of the CHO-K1 cell line (28, 29).

The FIX.1F cell line was derived by transfection of two expression plasmids into the genome of DUKX-B11 cells: the rFIX expression plasmid (pMT2-IX) and the PACE-SOL expression plasmid (pEA-PACE-SOL). These expression plasmids were amplified by standard methods to increase their copy number and to permit production and secretion of appropriately high levels of rFIX and levels of PACE-SOL sufficient for processing the expressed rFIX.

The production cell line was selected from among several dozen candidate cell lines based on superior rFIX volumetric productivity (active rFIX titer/liter of culture/day) in a suspension culture free of serum and other animal and human proteins (subsequently described as serum-free), consistent secretion of higher percentage of active rFIX product, and consistent secretion of highly γ-carboxylated rFIX (>11 mol Gla/mol rFIX).

3.3 Construction of the rFIX Expression Vector

For expression in CHO cells, the 2.5-kb *Pst*I fragment of M13mp11-IX containing the rFIX coding region was initially subcloned into the mammalian expression plasmid p91023(B) (30). The resultant plasmid, designated p91023-IX, was used for studies demonstrating biological activity of rFIX produced in CHO cells and allowed analysis of the requirements for expression, post-translational modification and secretion of active rFIX (17).

For development of the production cell line, the 2.5-kb *Pst*I fragment from plasmid p91023-IX, containing the rFIX DNA, was isolated and subcloned into the *Pst*I site of the expression vector pMT2 (31). The resultant plasmid was designated pMT2-IX. The expression vector pMT2-IX was sequenced in entirety. The nucleotide sequence data confirmed the orientation and integrity of the rFIX DNA segment in the expression construct pMT2-IX.

3.4 Preparation of the PACE-SOL Coding Sequence

The coding sequence for PACE-SOL was prepared by modification of a composite, full-length human PACE cDNA (32). The cloning of the human PACE cDNA from HepG2 (human hepatoma cell line, American Tissue Type Culture, Rockland, Maryland) mRNA was carried out by Chiron Corporation (19). A 3.3-kb PACE cDNA was isolated from a HepG2 cDNA library using oligonucleotide probes based on the sequence of the human *fur* gene, a transcription unit found upstream of the *fes/fps* proto-oncogene (33, 34) and reported to be homologous to the yeast Kex2 protease (35, 36). This 3.3-kb human PACE cDNA lacked the 5' end of the PACE coding sequence. A second library was constructed from HepG2 poly(A)$^+$ mRNA in the bacteriophage Lambda Zap II Vector® (Stratagene Cloning Systems, La Jolla, California) using PACE-specific internally primed mRNA and screened with a PACE probe. The longest clone obtained from this library encoded the 5' end of PACE and was used, in conjunction with the 3.3-kb PACE cDNA, to construct a 4.4-kb PACE composite cDNA. This 4.4-kb composite PACE cDNA contained 388 base pairs of 5' untranslated sequence, a 2382-bp segment encoding the 794-amino-acid protein, and 1597 base pairs of 3' untranslated sequence (19).

4. CELL BANKING AND STABILITY OF THE PRODUCTION CELL LINE

A cell bank is maintained which contains hundreds of frozen vials of CHO cells from which the factor IX is expressed. This is the source for initiating each manufacturing campaign, which can produce up to 20 batches of drug substance. Genetics Institute's currently established cell banks are designed to have a sufficient number of vials to meet worldwide commercial demand for factor IX for over 100 years.

The banking process for the cells used to manufacture rFIX involved establishing a master cell bank (MCB), developing a working cell bank (WCB), inoculating cells and operating the production process, retaining samples of cells from every harvest, and establishing an extended cell bank for postprocess analysis. Extensive analysis of the MCB, WCB, and end-of-production (EOP) cells has demonstrated the purity of the CHO cell line used for the production of rFIX. Extensive testing has demonstrated that the plasmids with the coding sequences for rFIX and PACE-SOL are accurately integrated into the DNA of these cells and that rFIX expression is stable across multiple generations of these cells in a given production campaign. This section presents the characterization of the identity of the cell line in terms of its genotype and phenotype to confirm its origin and to provide a basis for comparing preproduction cells with end-of-production cells.

4.1 Characterization of the Cell Banks and End-of-Production Cells

The identity of the MCB cells used in the production of rFIX has been studied in terms of morphology, growth rate, cellular productivity, and species-specific immunofluorescence and isoenzyme analysis. Additionally, the rFIX and PACE-SOL expression vectors and transcripts have been analyzed. The cell line exhibits the appropriate phenotypic and genotypic characteristics for a DUKX-B11 derivative that has incorporated the rFIX and PACE-SOL DNA expression vectors.

Following extensive testing of the master cell bank, the working cell bank, and samples of cells taken from a bioreactor at the completion of a production run (i.e., end-of-production cells) were analyzed. Since the test results for the MCB were similar to the results obtained with the WCB and EOP cells, only the results of the WCB and EOP analysis are presented below.

4.2 Analysis of the Working Cell Bank and End-of-Production Cells

Testing is performed on rFIX WCB and rFIX EOP cells to examine the intrinsic performance characteristics of the extended rFIX MCB culture. The single passage, 72-hour orbiting dish secretion assay used for this purpose is better suited to measure intrinsic performance than is a batch-refeed bioreactor culture. The cells used in the dish assay are appropriately staged by a 24-hour prepassage step in which the cells are recovered from suspension by centrifugation and refed with 100% fresh medium. Bioreactor culture measurements reflect cumulative performance characteristics acquired over successive passages in which approximately 75% of the culture volume is replaced with fresh medium.

Results from this testing are compared with results from the rFIX MCB cell tests and historical (post-MCB) FIX.1F cell tests to provide assurance that rFIX WCB and rFIX EOP cells are FIX.1F CHO cells and to assess their stability with respect to growth rate and cellular productivity.

4.2.1 Morphology

This analysis provides assurance that rFIX WCB and rFIX EOP cells have physical characteristics expected for CHO-derived cells adapted to suspension culture. Cells were examined by light microscopy for cell shape.

Examination of rFIX WCB and rFIX EOP cells by light microscopy revealed that the cells are essentially round and highly refractile. These observations are consistent with those made when rFIX MCB cells were similarly examined and support the conclusion that FIX.1F cell morphology is grossly similar from MCB to EOP. rFIX WCB and rFIX EOP cell morphology is as expected for CHO-derived, suspension-culture-adapted cells and compares well with rFIX MCB cells.

4.2.2 Growth Rate

This analysis provides assurance that rFIX WCB and rFIX EOP cell growth rates are within the expected range for FIX.1F cells, based on the performance of the rFIX MCB and on the historical performance of FIX.1F cells assayed in a 72-hour, serum-free orbiting dish secretion assay. As for the MCB cells, the assay protocol utilizes thawed rFIX WCB or rFIX EOP cells that have been grown in Vitamin K_1-free medium for two to four passages and then prepassaged for 24 hours in 100% fresh medium prior to use in the assay. Replicate 6-mL Petri dish cultures are seeded at 1.5×10^6 cells per mL with prepassaged rFIX WCB or rFIX EOP cells and vitamin K_1 is added. The cultures are then grown on an orbiting shaker for approximately 72 hours in the dark in a 7% CO_2, temperature- and humidity-controlled incubator. The dishes are sampled for cell counting by hemocytometer or automated cell counter. Additional samples are retained for determinations of active rFIX and total rFIX protein levels and to confirm PACE-SOL protein expression.

growth rate (μ) = (ln X_2 – ln X_1)/hours in culture

Growth rate is reported in reciprocal hours, where X_2 = final cell density (10^6 cells/mL), and X_1 = initial cell density (10^6 cells/mL). Note that doubling time (t_d) is calculated from μ: t_d = (ln 2)/μ.

Table 11 presents the growth rates of rFIX WCB and EOP cells. rFIX WCB and EOP cell growth rates are within expected range based on historical analysis of FIX.1F cells and compare well with rFIX MCB cells.

Table 11. rFIX WCB and EOP Growth Performance

FIX.1F Cells	Growth Rate (hours^{-1})	N
rFIX WCB	0.027 ± 0.002	4
rFIX EOP	0.028 ± 0.002	8
rFIX MCB	0.028 ± 0.001	8
FIX.1F cells (historical data)*	0.025 ± 0.002	104

*Individual 6-mL dishes were assayed using FIX.1F cells grown out to ≤72 CPD from the rFIX MCB.

4.2.3 Cellular Productivity

This analysis provides assurance that rFIX WCB and rFIX EOP cells express both rFIX and PACE-SOL and that the rFIX secretion rate is within the expected range for FIX.1F cells, based on the established performance of rFIX MCB cells and on the historical performance of FIX.1F cells assayed in a 72-hour secretion assay. The assay protocol utilizes rFIX WCB or rFIX EOP cells that have been grown and assayed as described in this section for growth rate determination. The assay is designed to measure the intrinsic performance characteristics of cells in a 72-hour serum-free suspension culture. Active rFIX levels are determined using the one-stage clotting assay performed with FIX-deficient plasma. Total rFIX protein levels are determined using the rFIX ELISA. PACE-SOL expression is confirmed by

Western blot analysis. Cellular productivity (q_p) is calculated by the following method:

$$q_p = (P_2 - P_1)/(X_2 - X_1) \times \mu \times 24 \text{ hours/day}$$

Cellular productivity is reported as mU active rFIX/10^6 cells/day, µg total rFIX/10^6 cells/day, or ng PACE-SOL/10^6 cells/day, where P_2 = final concentration (of active rFIX or total rFIX), P_1 = initial concentration (of active rFIX or total rFIX), X_2 = final cell density (10^6 cells/mL), X_1 = initial cell density (10^6 cells/mL), and μ = growth rate = ($\ln X_2 - \ln X_1$)/hours in culture.

In summary, the rFIX WCB and the 95-2 rFIX EOP cells express rFIX, the rFIX secretion rates are within the expected range based on historical analysis of FIX.1F cells, and the WCB cells are comparable to rFIX MCB cells.

4.2.4   Species Identity Test

rFIX EOP cells were found to express an isoenzyme pattern consistent with that expressed by CHO cells and were found to express hamster antigens. There was no evidence that cells other than those of Chinese hamster origin were present in the culture.

4.2.5   Characterization of Expression Vectors and Transcripts

The complex pMT2-IX plasmid integrant structure is stably maintained in the FIX.1F host cell genome following expansion and full-scale culture (representing 72 CPD from the MCB). The pMT2-IX plasmid copy number in rFIX EOP cells is essentially indistinguishable from that in rFIX MCB cells. Consistent with this, the abundance of rFIX mRNA in rFIX EOP cells is comparable to that of rFIX MCB cells. There is no evidence for the generation of gross rearrangements within rFIX mRNAs during full-scale production, and consistent with this, no evidence for gross rearrangement in any of the integrated pMT2-IX plasmids within the predicted rFIX-DHFR transcript. DNA sequence analysis of the rFIX coding region of cDNAs derived from the expressed rFIX genes, reveal only the sequence predicted. Together these data confirm that pMT2-IX plasmids with the expected rFIX coding sequence are accurately integrated and stably maintained in the genome of rFIX EOP cells during inoculum build up and full-scale production (to 72 CPD from the rFIX MCB).

Additionally, the data presented indicate that the pEA-PACE-SOL plasmid copy number in rFIX EOP cells is indistinguishable from that in rFIX

MCB cells. This confirms that pEA-PACE-SOL plasmids are stably integrated and maintained in the genome of rFIX EOP cells during inoculum build-up and full-scale production (to 72 CPD from the rFIX MCB).

4.3 Evaluation of Genotypic and Phenotypic Stability

The primary criteria for acceptable cell culture process stability relate directly to product quality. Therefore, the principal criteria for assessing stability involve direct indicators and predictors of product quality, so that the highest priority is given to the drug substance release data, DNA sequence data, and data indicating RNA integrity. Secondary criteria include the indirect predictors of product quality such as cellular productivity, growth rate, and DNA copy number.

In-process cellular performance data, Northern and Southern blot and DNA sequence data, purification yields, and rFIX product characteristics demonstrate the stability of the cell culture process. The in-process data are evaluated as a function of cumulative population doublings (CPD) over the full duration of the batch-refeed production culture to facilitate analysis of trends in the data set. Results of the analyses provide assurance of the stability of the cell culture process, and therefore of the rFIX protein produced by that process, for cells "at the limit of *in vitro* age used in production," as defined in the ICH Viral Safety Evaluation of Biotechnology Products Derived from Cell Lines of Human or Animal Origin (Step 2 draft, December 1, 1995).

Several indicators of process performance have been evaluated to establish that the cell culture process is stable to at least 58 CPD from rFIX WCB. These indicators include measures of cell culture performance (specific growth rate, viability, rFIX cellular productivity by clotting assay, rFIX cellular productivity by ELISA, proportion of ELISA-reactive rFIX that is active, and PACE-SOL expression) and assays of drug substance (peptide maps, carbohydrate fingerprints, γ-carboxyglutamic acid stoichiometry, proFIX content, and specific activity).

Five measures of cell culture performance show no statistically significant correlation with CPD and thus indicate stability. These measures are specific growth rate, viability, rFIX cellular productivity by clotting activity, rFIX cellular productivity by ELISA, and proportion of ELISA-reactive rFIX that is active.

A sixth measure of cell culture performance, PACE-SOL expression, has tended to increase (approximately twofold to threefold over 50 CPD). However, this trend has been found to have no significant impact on quality of the drug substance with respect to either PACE-SOL content or proFIX content.

Overall purification recovery has shown a modest decrease with CPD (9% over 50 CPD) which is not statistically significant at the 90% confidence level. Therefore, we conclude that the process is stable as judged by purification recovery.

Two assays of drug substance provide further evidence of stability. The first is peptide mapping. Peptide maps have consistently conformed to specifications, and a comparison of peptide maps of batches spanning a production run shows no significant differences (Figure 13A). Peptide mapping is one of the most sensitive analytical techniques for detecting small covalent modifications to protein structure, such as amino acid substitution or change in posttranslational modification. The second assay of drug substance that provides evidence of stability is carbohydrate fingerprinting. As with peptide maps, carbohydrate fingerprints have consistently conformed to specifications, and a comparison of carbohydrate fingerprints of batches spanning a production run shows no significant differences (Figure 13B).

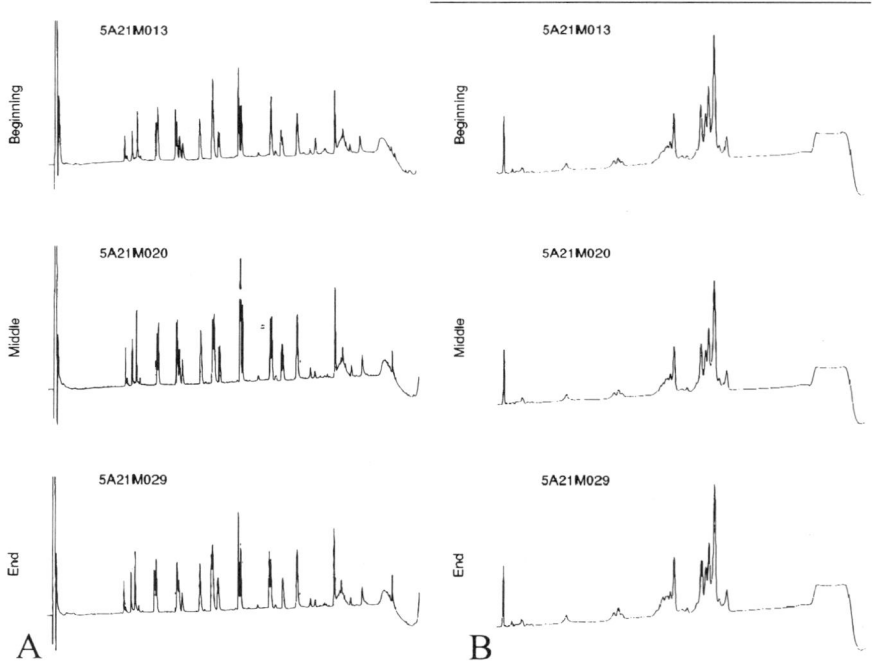

Figure 13. Peptide (A) and carbohydrate (B) maps demonstrating consistency from the beginning, middle, and end of an rFIX production campaign

5. MANUFACTURING PROCESS

The manufacturing process for rFIX consists of production of recombinant factor IX from cultured cells, followed by purification and concentration of the product secreted by these cells. Finally, the purified material is formulated, filled into vials, lyophilized, and then packaged.

5.1 Cell-Culture Manufacturing Steps

The cell-culture manufacturing process for rFIX consists of the following steps: 1) thawing vials of cells from the WCB and expanding the culture in spinner flasks, 2) inoculating a 250-L bioreactor from spinner flasks, 3) expanding the volume of the cell culture in the 250-L bioreactor, 4) inoculating 2500-L bioreactors from a 250-L bioreactor, 5) growing the cells in 2500-L bioreactors, 6) harvesting the cell culture, 7) separating the cells by microfiltration, and finally, 8) concentrating and buffer exchanging the cell-free conditioned medium by a combined ultrafiltration/diafiltration step.

Once the cell-culture inoculum has been expanded into a 2500-L production bioreactor, a batch of rFIX is produced by growing the cells for approximately 3 days, until they have reached high cell densities. At this point, approximately 80% of the cell suspension is removed from the bioreactor, the cells are separated by filtration, and the resulting cell-free conditioned medium is taken through purification. The remaining 20% of the cell suspension left in the bioreactor is resuspended in fresh medium and acts as the source of cells for the next batch production cycle. This process continues for up to 20 batch-refeed cycles, at which time the production run is terminated.

5.2 Concentration and Purification of rFIX from Conditioned Medium

The conditioned medium containing rFIX secreted by the CHO cells is purified by a process that consists of four chromatographic separation procedures, as well as membrane-based filtration (37).

5.2.1 Ultrafiltration/Diafiltration

Cell-free conditioned medium is concentrated by ultrafiltration and then diafiltered against a Tris/NaCl buffer to remove low-molecular-weight components and to provide a consistent buffer matrix for loading onto the first chromatography column (38).

5.2.2   Q-Sepharose FF Chromatography

Q-Sepharose FF is used as the capture step for binding rFIX in the ultrafiltration-diafiltration retentate pool and serves to remove impurities in the conditioned media. This process step is operated in pseudoaffinity mode. As such, rFIX binds to the resin via charge interactions (traditional anion exchange) at pH 8.0. The column is washed with a buffer of increased conductivity (0.2 M NaCl), higher than that used for loading but insufficient to elute bound rFIX. The effluent conductivity is then decreased to a level (0.1 M NaCl) below that used during column loading, followed by pseudoaffinity elution of bound rFIX by the addition of calcium chloride (final concentration 10 mM) to the buffer at pH 8.0. The addition of calcium creates a conformational change unique to the factor IX molecule that causes it to detach from the Q-Sepharose resin. Operation of the Q-Sepharose FF column in pseudoaffinity mode provides significant purification with high yield (>90%) and obviates the need for immunoaffinity purification and its associated potential for introduction of contaminants (37, 38).

5.2.3   Matrex Cellufine Sulfate Chromatography

The Q-Sepharose FF product peak pool is loaded directly onto the second chromatography column used for purification, Matrex Cellufine Sulfate, a heparin analogue. Matrex Cellufine Sulfate is used for affinity purification of proteins with heparin-binding domains and also behaves as a cation exchange resin due to the negatively charged sulfate groups. Upon completion of loading, the column is washed with buffer (50 mM Tris, 0.2 M NaCl, pH 8.0) to remove loosely bound contaminants, and the bound rFIX is eluted by increasing the sodium chloride concentration of the buffer (0.5 M NaCl). Matrex Cellufine Sulfate chromatography is used to remove low levels of residual impurities present in the Q-Sepharose FF product peak pool (37).

5.2.4   Ceramic Hydroxyapatite Purification

Ceramic hydroxyapatite (HA) is a synthetic form of calcium phosphate consisting of spheroidal macroporous particles with high mechanical strength. Ceramic-HA is used to separate proteins of varying charges on the basis of specific interaction with the resin. The pH of the Matrex Cellufine Sulfate product peak pool is adjusted to pH 7.2 prior to loading onto the Ceramic-HA column. The column is washed with a buffer containing a low concentration of potassium phosphate (0.05 M) prior to elution by increasing the phosphate concentration (0.5 M). The Ceramic-HA column is used for rFIX purification to provide additional capacity for removal of remaining trace impurities. In addition, the Ceramic-HA process step provides an elution pool in a buffer appropriate for loading onto the final chromatography resin.

5.2.5 Chelate-EMD-Cu(II) Purification

Chelate-EMD resin is composed of a methacrylate polymer derivatized with iminodiacetic acid functional groups to which transition-state metal ions can be bound. Proteins that interact with the immobilized metal ions are retained by the resin. The rFIX purification process uses copper (II) as the immobilized metal ion. The Ceramic-HA product peak pool is loaded directly onto the Chelate-EMD-Cu(II) column. The column is washed to complete the loading and bound rFIX is eluted with imidazole as the displacer. This final chromatography step is effective in removing trace contaminants, including low levels of residual host cell proteins.

When used together, these independent and complementary affinity chromatographic separation methods used to purify and concentrate rFIX comprise a robust process that consistently produces a product with high structural integrity, purity, and specific activity.

5.2.6 Nanofiltration

The process used for purification of rFIX employs a nanofiltration step, included as an added layer of viral safety. The membrane used for this purpose, Viresolve-70®, has the ability to retain molecules with apparent molecular weights greater than 70,000 Da, such as large proteins and viral particles, while smaller molecules (such as rFIX: molecular weight, 55,000 Da) pass through the membrane.

5.2.7 Final ultrafiltration/diafiltration

The final ultrafiltration/diafiltration step for rFIX is designed to exchange the Chelate-EMD-Cu(II) product peak pool buffer to the formulation buffer and to concentrate rFIX further.

5.3 Product Formulation

The goal for the formulation of rFIX was to develop a lyophilized dosage form having a shelf life of at least 24 months when stored at 2° to 8°C and at least 6 months when stored at ≤25°C. As with all other phases of the manufacturing process, the formulation was developed without the use of blood or plasma products, including albumin. Only nonproteinaceous excipients were used in the development of the rFIX product, and they were required to provide optimal protection from the stresses of lyophilization as well as good lyophilized cake morphology at all dosage strengths. In addition, the final product was required to reconstitute quickly in sterile water for injection, resulting in a solution at physiological pH.

Extensive testing of numerous excipients resulted in the development of a final formulation that is stable and contains no preservatives or materials derived from blood or plasma (e.g., albumin is not used). The formulation contains 10 mM histidine, 260 mM glycine, 1% sucrose, and 0.005% polysorbate-80 (pH 6.8). Polysorbate-80 provides protection for the protein from freezing-induced damage (e.g., aggregation). Sucrose provides protection to the protein in the freeze-dried state. Glycine provides for a high-quality cake morphology. Histidine provides optimal buffering stability at the desired pH and minimizes aggregate formation upon storage in the lyophilized state. This combination of excipients has resulted in a product that is easy to reconstitute and demonstrates excellent stability (39).

6. BIOCHEMICAL CHARACTERIZATION

Characterization of the rFIX molecule employed traditional biochemical techniques, including sodium dodecyl sulfate–polyacrylamide gel electrophoresis (SDS-PAGE), peptide mapping, and carbohydrate fingerprinting, as well as more recent procedures such as matrix-assisted laser desorption time-of-flight (MALDI-TOF), electrospray ionization (ESI), and liquid chromatography/electrospray ionization (LC/ESI) mass spectrometry (MS). The same procedures were also applied to a monoclonal-purified pdFIX preparation (designated as high-purity pdFIX-1 here) to establish a basis of comparison with a previously approved product.

6.1 Primary Structure and Posttranslational Modifications

The primary structure of rFIX (see Figure 12) was examined using peptide mapping, mass spectrometry, NH_2-terminal sequencing, and COOH-terminal analysis (40). Using these procedures, the entire amino acid sequence of rFIX was confirmed to be identical to the Ala-148 allotype of pdFIX. The observed mass of rFIX as determined by MALDI-TOF MS was 55,290 Da, which is higher than the predicted value calculated for the amino acid sequence alone (47,054 Da) and which reflects the presence of posttranslational modifications that add additional mass.

As described below, the posttranslational modifications of rFIX were examined and found to be similar, but not identical, to those observed for pdFIX. Posttranslational modifications were detected in the Gla, EGF-1, EGF-2, and activation peptide regions and were characterized by peptide mapping and carbohydrate fingerprinting followed by mass spectrometry, NH_2-terminal sequencing, and enzymatic subdigestion. As was the case for pdFIX, no posttranslational modifications were detected in the serine protease domain of rFIX.

The Gla domain of factor IX contains 12 potential sites for γ-carboxylation. In pdFIX, all 12 sites are occupied (i.e., γ-carboxylated). In rFIX, 10 of the 12 sites are fully occupied, and 2 sites are partially occupied, resulting in an average of 11.5 Gla per molecule. The 11- and 10-Gla isoforms of rFIX were shown to be undercarboxylated at residue 40 or at residues 36 and 40, respectively (41). Thus, the recombinant protein exists as a mixture of predominantly three isoforms containing, in order of prevalence, 12, 11, and 10 Gla per molecule. The three Gla-related isoforms show similar clotting activities and are indistinguishable in terms of phospholipid binding, endothelial cell binding, and activation of factor X via the tenase complex (42). Based on these data as well as sequence homology with other vitamin-K–dependent factors and three-dimensional structure information (11, 42, 43), the Gla-36 and Gla-40 residues appear to be unimportant in terms of protein structure and function.

The EGF-1 domain of both rFIX and pdFIX contains three posttranslational modification sites. The modifications identified by peptide mapping and mass spectrometry in rFIX were Xyl-Xyl-Glc- at Ser-53, NeuAc-Gal-GlcNAc-Fuc- at Ser-61, and β-hydroxylation at Asp-64 (40). The same modifications were also detected in high-purity pdFIX-1, but Ser-53 contained Xyl-Glc- in addition to Xyl-Xyl-Glc-, and the relative proportion of Asp-64 β-hydroxylation was slightly lower than that in rFIX (approximately 46% and 37%, for rFIX and pdFIX, respectively) (40, 44, 45). Both rFIX and pdFIX showed evidence of a low level of deamidation in the EGF-2 domain at Asn-92, and the relative amounts were similar.

The activation peptide region of rFIX contains numerous posttranslational modifications, all of which are present in high-purity pdFIX-1. These include sulfation at Tyr-155; phosphorylation at Ser-168; O-glycosylation at Thr-159, Thr-169, and Thr-172; and N-glycosylation at Asn-157 and Asn-167 (46, 47). The degree of occupancy of the O-glycosylation (partial) and N-glycosylation (essentially complete) sites is similar for the recombinant and plasma-derived proteins, the sulfation site is largely unfilled in rFIX and largely filled in pdFIX, and the phosphorylation site is unfilled in rFIX and largely filled in pdFIX. The structures of the O-glycans appear to be the same in the two proteins, and the N-glycans are similar. In both cases, the N-glycans are of the complex type, exist predominantly in tri- and tetra-antennal forms, and can contain additional fucose and poly-*N*-acetyllactosamine repeat units (48-50). There are differences, however, and the *N*-glycans from pdFIX are generally more complex and contain a broader range of linkages, fucosylation, and poly-*N*-acetyllactosamine repeat structures. The plasma-derived protein contains one modification that is not present in rFIX, a largely filled phosphorylation site at Ser-158 (47).

6.2 Higher-Order Structure

The higher-order structure of rFIX was evaluated by examining the disulfide bond structure and by analysis with three biophysical techniques. Proteolytic digestion of rFIX followed by MALDI-TOF MS analysis definitively identified 4 of the 11 disulfide bonds and tentatively identified the remaining 7; all assignments were consistent with the proposed disulfide bond structure for factor IX based on sequence homologies for the Gla, EGF-like, and serine protease domains, as well as porcine FIXa crystal structure (51). Further structural analysis employed fluorescence spectroscopy, circular dichroism spectroscopy, and analytical ultracentrifugation. For all experiments, pdFIX was purified before use in order to remove aggregated protein, which interfered with the analyses. The results showed that the secondary and tertiary structure of rFIX was similar to that of pdFIX and that the addition of calcium promoted similar changes in the spectroscopic profiles and sedimentation velocities of both molecules (52, 53). Therefore, the higher order structure of rFIX is comparable with that of pdFIX.

6.3 Specific Activity

The *in vitro* biological activity of rFIX was examined using a one-stage clotting assay with factor-IX-deficient plasma and activated partial thromboplastin reagent. The average specific activity of rFIX (260 IU/mg) in this study was determined to be higher than that observed for high-purity pdFIX-1 (230 IU/mg). The activity difference was consistent with the presence of an inactive, HMW material in the plasma-derived preparation, which was identified as noncovalently aggregated factor IX. Other commercially available pdFIX preparations were examined in this study for clotting activity as well and showed specific activities lower than those for rFIX. The observed lower specific activities of plasma-derived products reflected the amount of HMW material present, as determined by size-exclusion chromatography and SDS-PAGE. These results have not been shown to have clinical significance.

6.4 Purity

Using SDS-PAGE, size-exclusion high-performance liquid chromatography (SEC-HPLC), reverse-phase HPLC, and N-terminal sequencing, the purity of rFIX in this study was determined to be higher than that of four different high-purity pdFIX concentrates. SDS-PAGE analysis under nonreducing conditions revealed a single major band for both rFIX and high-purity pdFIX-1, whereas the other pdFIX preparations contained numerous additional bands in significant amounts. NH_2-terminal sequence analysis of the additional pdFIX product bands after electroblotting revealed the presence of non-factor IX protein species in the less pure pdFIX preparations, including prothrombin and protein C. Under reducing conditions, both rFIX and high-purity pdFIX-1 displayed faint secondary bands at 45, 32, 28, and 14 kDa. These were identified by NH_2-terminal sequencing as FIX-related species corresponding to (predominantly): heavy-chain activation peptide (residues 146-415) and light-chain activation peptide (residues 1-180), heavy chain (residues 181-415), light chain (residues 1-145), and COOH-terminal peptide (residues 319-415), respectively.

Size-exclusion high-performance liquid chromatography of rFIX revealed a single, major peak essentially free of high- and low-molecular-weight impurities (54). Similar analysis of high-purity pdFIX detected a major peak for factor IX but also showed several early eluting peaks that correspond to HMW material; the average amount observed for three separate high-purity pdFIX-1 lots corresponded to 10% of the total protein (Table 12). This material, which was identified in separate experiments as noncovalently aggregated factor IX, was found to be inactive in clotting assays and antigenically distinct from pdFIX (i.e., it was reactive with some anti-factor IX monoclonal antibodies but not others). All of the pdFIX preparations analyzed contained readily detected amounts of this and other HMW material by SEC-HPLC, and preparations with the higher amounts of HMW material showed the lower specific activities. The average proportion of HMW material in rFIX was only 0.5%. These results have not been shown to have clinical significance.

Activated FIX (FIXa) is present in rFIX and pdFIX preparations at low levels as evidenced by immunological and enzymatic analysis (55, 56). Analysis of rFIX production batches using a nonactivated partial thromboplastin time assay (as described in the European Pharmacopoeia) shows that FIXa levels are consistently below the European Pharmacopoeia test limits. For example, analysis of 10 lots of rFIX drug product by a different NAPTT clotting assay detected levels of FIXa ranging from 0.03% to 0.07% (units FIXa/IU FIX), which is similar to values (0.01% to 0.04%) obtained in another laboratory for four lots of high-purity pdFIX (57).

Table 12. Observed specific activity and %HMW of rFIX and pdFIX preparations

FIX	Specific Activity (IU/mg ± SD)	Average %HMW ± SD	Number of Lots Tested
rFIX drug product	260 ± 12	0.5 ± 0.2	14
High-purity pdFIX-1	230 ± 10	10 ± 0.9	3
High-purity pdFIX-2	160 ± 4	30 ± 0.7	3
High-purity pdFIX-3	170 ± 9	26 ± 3.9	3
High-purity pdFIX-4	90 ± 19	50 ± 3.9	2

Note: Protein concentration was determined by SEC-HPLC in comparison to a FIX reference sample of known concentration.

7. PRODUCT QUALITY EVALUATIONS

7.1 Drug Substance Test Procedures

The specifications and test methods that have been established for release of rFIX drug substance were selected to ensure the identity, purity, potency, safety, and quality of the protein obtained from the cell culture and purification processes. The utility of each test method with respect to product identity, purity, safety, and quality, in addition to the rationale for the specification for each test method, is briefly described below.

rFIX consists of both zymogen and cleavage products of the mature zymogen. The cleavage products of both rFIX and pdFIX are similar; both preparations contain FIX and FIX-related species, and although the relative proportions may vary, essentially the same species are observed for both preparations and the overall amounts are approximately the same. Therefore, all of the FIX-related species described above, even though some are not active, are not considered impurities. Taken together, the mature FIX (zymogen) and the FIX-related species that are expected to be present define rFIX. Impurities are defined as all non-FIX species that are derived from the host cell or the production process, including proteins, carbohydrates, and small molecules.

7.1.1 Identity

Specifications for identity of rFIX drug substance have been established on the basis of the biological activity of the molecule, electrophoretic mobility, the Gla content, peptide maps, and carbohydrate fingerprinting.

Identity of the rFIX drug substance is ensured by performing the clotting assay for the determination of biological activity and the SEC assay for protein concentration, and subsequent calculation of the specific activity for the protein. The determination of specific activity is required because other clotting factors demonstrate activity in a one-stage clotting assay using FIX-deficient plasma. Only FIX would demonstrate a specific activity in the range of the specification established.

Gel electrophoretic analysis (SDS-PAGE, reduced) of rFIX drug substance is performed to determine identity. The SDS-PAGE analysis of individual batches when compared with a reference material provides confirmation of drug substance identity.

The rFIX anion exchange chromatography (anion exchange HPLC) method is used routinely to determine the γ-carboxyglutamic acid (Gla) content of rFIX drug substance samples. The Gla residues are present in the amino-terminal region of the FIX molecule. These residues are important in the binding of calcium ions to FIX and thus to the activity of the FIX molecule. A major fraction of the rFIX molecules are γ-carboxylated at all 12 sites, but some fraction of the molecules are also present as an 11-Gla isoform and a smaller fraction are present as a 10-Gla isoform. Gla isoforms with <10 Gla residues are present at very low levels. Anion exchange HPLC partially separates the different isoforms on the basis of Gla content, and integration of the different peaks allows an accurate assessment of the total Gla content of the isoform mixture. The specification for the total Gla content is 11.0 to 12.0 mols Gla/mole rFIX. Full rFIX activity by one-stage clotting assay is maintained across this range.

Peptide map analysis is performed routinely to assess the batch-to-batch consistency of manufacture of rFIX drug substance. In doing so, product identity is also confirmed. Peptide map analysis may be sensitive to modifications of amino acid residues (posttranslational modifications, degradation) or changes in amino acid sequence (for example, those resulting from point mutations) through changes in peak area or retention times of certain peptides. In the peptide map analysis of rFIX, relative retention times (the difference between the absolute retention time of a peak and that of an internal reference peptide) and peak area ratios (the ratio of the area of a peak to that of the internal reference peptide) are reported for each peptide peak of the test article and compared with those of a reference material to assess for any significant differences.

N-linked oligosaccharide fingerprint analysis is performed routinely for monitoring oligosaccharides for changes, for loss of sialic acid, and for modifications to the expected glycan structure. The primary use of *N*-linked carbohydrate fingerprint analysis is to monitor the consistency of the *N*-linked oligosaccharide distribution and the presence and absence of expected glycan structure, reflecting the consistency of rFIX manufacture of the glycoprotein, including the degree of sialylation. Thus, for rFIX *N*-linked carbohydrate fingerprint analysis, peak areas for groups of peaks representing each sialylated *N*-linked glycan region are recorded and the tetra-sialyl region peak area ratio (total peak area of the tetra-sialyl region divided by the total peak area of neutral, mono-, di-, and tri-sialylated regions) is assessed. In addition, relative retention times of specific peaks (absolute retention time of a peak minus the retention time of an added standard, stachyose) are reported for the sample and compared with those with rFIX reference material.

7.1.2   Purity

Purity of the rFIX drug substance is ensured by performing the following analyses: the SEC-HPLC analysis for determination of high molecular weight material, the SDS-PAGE analysis under reducing conditions, and the RP-HPLC analysis.

Analysis of rFIX drug substance by size-exclusion chromatography (SEC-HPLC) is performed to determine the amount of high-molecular-weight (HMW) material present in the test article. HMW material is defined as the sum of all peaks eluting prior to the rFIX peak.

Gel electrophoretic analysis (SDS-PAGE, reduced) of rFIX drug substance is performed to detect impurities. Species that are neither zymogen nor zymogen-related are classified as impurities and are quantified. The method is orthogonal to chromatographic techniques. Specifications have been established for total impurity and are expressed as a percent of total protein observed on SDS-PAGE by integrated optical density using a gel scanner. The method is selective and quantitative for PACE (limit of quantitation is 0.6%) but is highly variable; precision is limited by the inherent variability of electrophoretic methods and gel scanning.

Analysis of rFIX drug substance by reversed-phase high-pressure liquid chromatography (RP-HPLC) is performed as a measure of impurity. Species that are neither zymogen nor zymogen-related are classified as impurities and are quantified. The method is orthogonal to electrophoretic techniques. Specifications have been established for total impurity and are expressed as a percent of total protein observed in percent relative peak area. All peaks not eluting at the expected positions of FIX and FIX-related species are defined as impurities and can be quantified. The method has been demonstrated to be capable of quantitating 0.5% host cell protein (HCP). The method is sufficiently sensitive to detect process failure in terms of host cell protein removal in the event of a process deviation.

7.1.3   Potency

The activity of rFIX drug substance (IU/mL) is ensured by performing a traditional one-stage clotting assay using FIX-deficient plasma. rFIX drug substance is manufactured to an activity that will ensure successful further processing into drug product.

Total protein concentration of rFIX drug substance is determined by SEC-HPLC and is reported in mg/mL. This assay is also used to determine HMW species. Total protein concentration is determined for the calculation of specific activity, and both the FIX and HMW species are included in the calculation. As both activity in IU/mL and specific activity are routinely monitored and controlled by specifications, no specification is established for total protein concentration.

7.1.4 Safety

Control of the endotoxin and bioburden levels in rFIX drug substance is dependent upon the processing and handling of the material; therefore, the specifications for endotoxin and bioburden are process-driven with due consideration for purity and microbiological quality.

7.2 Drug Product Test Procedures

In general, the specifications for the rFIX drug product have been established on the basis of manufacturing process performance that has been shown to produce a consistent, safe, and efficacious product. The assays performed to establish product and process performance are validated test methods. The method variability has been considered in establishing specifications. Additionally, the stability profile has been considered in the specifications development.

7.2.1 Identity

Specifications for identity of rFIX drug product have been established on the basis of the biological activity of the molecule and SDS-PAGE reducing analysis.

7.2.2 Purity

Purity of the rFIX drug product is assured by performing SEC-HPLC analysis for determination of high molecular weight material, SDS-PAGE analysis under reducing conditions. In addition to these tests, analysis of level of activated factor IX (FIXa) is performed.

Activated factor IX (FIXa) is a potential by-product of the process described in this application. Analysis of FIXa is performed in accordance with the method described in European Pharmacopoeia monograph 554 (58).

The specification for the activated FIX (FIXa) is also in accordance with European Pharmacopoeia 554.

7.2.3 Potency

The specifications for potency of rFIX drug product have been established to assure adequate delivery of the dose of rFIX consistent with the product labeling.

Potency of the rFIX drug product (IU/vial) is assured by performing a traditional one-stage clotting assay using FIX-deficient plasma with an international reference material for FIX clotting activity. The specification for potency is in accordance with European Pharmacopoeia monograph 554 (58). The potency specification for the three strengths of rFIX is as follows: the estimated potency is not less than 80% and not more than 125% of the stated potency. The fiducial limits of error of the estimated potency ($P=0.95$), also known as the 95% confidence interval, are not less than 64% and not more than 156% of the stated potency.

Protein concentration is also determined by size-exclusion chromatography as an additional measure of product strength and to enable calculation of specific activity for the product.

7.2.4 Safety

The safety of rFIX drug product, meaning the relative freedom from harmful effect to the patient, is ensured by adherence to the following requirements: freedom from viable contaminating microorganisms as determined by sterility tests, relative freedom from pyrogenic substances as determined by bacterial endotoxin testing, and freedom from visible particles and relative freedom from subvisible particles as determined by particulate testing.

7.2.5 Quality

The quality of rFIX drug product is assessed by the appearance of the product before and after reconstitution, pH, residual moisture, and activated FIX. rFIX drug product is a lyophilized product; control of residual moisture is essential to the suitability and stability of the product. The residual moisture is monitored using a validated coulometric titration procedure.

7.3   Summary

The test methods described above for bulk drug substance and finished drug product are performed on every lot of rFIX produced. Over 150 tests are performed on each lot of rFIX before it is released for sale. These quality control procedures have been established to ensure that the rFIX manufacturing process consistently produces a safe and efficacious product.

8.   PRECLINICAL TESTING

Recombinant factor IX (rFIX) has been extensively evaluated in preclinical studies. Studies in a dog model of hemophilia B indicated that rFIX was as effective as a highly purified plasma-derived replacement factor in normalizing indices of hemostasis. Pharmacokinetic studies indicated a dose-proportional profile for rFIX. Pharmacokinetic/pharmacodynamic analysis showed that increases in the plasma concentration of rFIX following administration were closely correlated with measured factor IX activity in the plasma. Appropriate *in vitro* and *in vivo* toxicology studies have been performed to support the clinical use of rFIX for the treatment of hemophilia B.

Because rFIX is an essentially pure recombinant protein product with a high degree of structural similarity to the endogenous protein, a limited but appropriate preclinical safety assessment has been performed to support its clinical use. *In vitro* studies conducted have included Ames bacterial mutagenicity and human lymphocyte chromosomal aberrations assays. Acute to subchronic *in vivo* toxicology studies in Swiss ICR and CD-1 mice, Sprague-Dawley rats, and beagle dogs have also been completed. The only toxicologic finding in these studies was the occurrence of thrombosis and consumptive coagulopathy in mice administered high doses of rFIX (>500 IU/kg/day) intraperitoneally for 1 to 7 days. Similar findings have been previously reported with plasma-derived FIX products. Based upon pharmacokinetic data demonstrating higher exposures in dogs and humans than achieved in mice, the absence of similar effects in other species, and the excellent clinical safety profile of rFIX, the mouse appears to be uniquely susceptible to this phenomenon. Therefore, thrombogenic effects in the mouse are not considered predictive of human risk. In addition, thrombogenicity studies of rFIX in rabbits found a low thrombogenic potential with thrombosis occurring in only 1 of 28 animals tested with rFIX (at 50–1000 IU/kg), compared with 6 of 12 animals tested with lower-purity factor IX PCC (at 15 and 50 IU/kg) and 5 of 18 animals tested with high-purity factor IX (at 1000 IU/kg). On the basis of these preclinical safety and efficacy studies, rFIX was considered safe for proceeding with clinical trials in the treatment of hemophilia B.

9. CLINICAL STUDIES

After laboratory and animal testing was completed, clinical studies with humans using rFIX were started. The first studies began in early 1995 in the United States, Europe, and Canada. To date, four clinical studies have evaluated the effectiveness of rFIX. These include a completed study comparing recovery and half-life of rFIX to those of a plasma-derived factor IX concentrate, a study of rFIX efficacy and safety for the prevention and control of bleeding in patients already exposed to factor concentrates, a completed study of patients using rFIX during surgery, and a trial of rFIX in patients who have never received any type of factor concentrate or blood product.

9.1 Efficacy in Spontaneous Bleeding Episodes

Results of the previously treated patient clinical trial that included 56 participants (47 with severe hemophilia B and 9 with moderate hemophilia B) showed that treatment with rFIX resolved spontaneous bleeding episodes rapidly and completely. During the course of the study, a total of 1456 bleeds were treated; 80% of these hemorrhages resolved with one infusion. Of the 2180 infusions given for the treatment of bleeding episodes, 88% were rated as providing excellent or good clinical response. These results are consistent with similar information from clinical studies using other hemophilia products.

9.2 Efficacy in Surgery

Safe and effective prevention of bleeding during and after surgery is a major concern for people with hemophilia. Recombinant factor IX has been used successfully in a number of surgical procedures, including liver transplantation, orthopedic procedures, dental procedures, inguinal hernia repairs, and skin biopsy. The estimated blood loss during and after surgery did not differ from that seen in persons without hemophilia undergoing similar procedures. Overall, clinical responses during surgery were rated as good or excellent in 97% of the surgical procedures. No thrombotic complications occurred in the surgical setting.

9.3 Safety

In the ongoing trial involving 56 previously treated patients, 2180 infusions have been given. Fifty-one moderate or mild and transient adverse events were reported as being definitely related to the study drug. One patient who continues to use recombinant factor IX developed a renal infarct which was judged by the investigator as unlikely to be related to treatment with rFIX. In the higher-dose surgical studies where the risk of thrombotic complications is of most concern, there were no thrombotic events, despite the fact that several patients had a prior history of thrombosis associated with surgery using prothrombin complex concentrates. rFIX has a safety profile comparable with that of other currently available factor IX concentrates. As with the IV administration of any protein product, common side effects include nausea, headache, fever, chills, altered taste, discomfort at the IV site, and dizziness.

Patients who have previously received factor concentrates provide the best evaluation of the safety of a new product with regard to the development of antibodies (inhibitors). Most patients who develop an inhibitor do so during childhood, usually after about 10 to 15 infusions. Therefore, when a young child develops an inhibitor, it is hard to determine if the inhibitor formation is related to the recombinant factor or if it would also have occurred with the use of a plasma-derived factor product. In an adult who has been exposed numerous times to blood products without developing an inhibitor, inhibitor development after receiving recombinant factor IX would indicate that the body is able to identify rFIX as foreign and produce antibodies (inhibitor) against it. Only 1 of 56 previously treated patients developed a transient, low-titer inhibitor.

Studies of replacement factor concentrates in people who have not been exposed previously to blood or plasma products are an effective means to evaluate the incidence of inhibitor formation and viral transmission. A study of rFIX in untreated hemophilia B patients is currently ongoing. As of September 1997, 42 patients were enrolled in this study, 31 of whom had received rFIX. Ninety-two percent of the responses to the infusion for bleeding episodes have been rated as excellent or good. One of the 31 previously untreated patients developed a high-titer inhibitor (maximum titer 30 Bethesda units). This patient tested positive for an inhibitor after 6 exposure days and is now being treated with recombinant factor VIIa.

9.4 Summary

The clinical experience with recombinant factor IX has clearly documented its safety and efficacy in patients with hemophilia B in a variety of clinical settings worldwide.

10. REGISTRATION STRATEGY

The registration strategy developed for recombinant factor IX was based on the concept of performing a comprehensive global clinical program. Although this offered considerable efficiency, it was not clear whether this would be acceptable to the various regulatory agencies throughout the world. In fact, one of the most debated subjects over the past several years has been the issue of acceptance of "foreign" clinical data by regulatory agencies. This relates not only to the issue of comparable medical practices but to a more fundamental issue of how different human populations may respond to a biopharmaceutical product. Although pharmaceutical products may be metabolized differently by different populations, it is not clear that such effects are evident with biological products. One of the drivers for developing a global clinical strategy was to minimize the need to repeat basically the same studies in each region—a practice that is both expensive and wasteful of both resources and time for companies and regulatory agencies. Despite the efficiency that is possible with global clinical trials, there are very few examples of simultaneous clinical programs using essentially identical protocols in the United States, Europe, and Japan, especially for biopharmaceutical products.

An important step in developing a valid global clinical development plan, especially for novel products and therapies such as recombinant factor IX, was to obtain early, science-based advice from the regulatory agencies that would ultimately review the regulatory submission. This was successfully accomplished with the FDA, HPB, and the UK MCA within 6 months of filing the IND.

Once the decision was made to pursue a global development strategy, the next step was to define an appropriate clinical program that would be acceptable in all regions. During development of the clinical plan, Genetics Institute utilized a draft CPMP guidance document describing the clinical trials required for authorization of factor VIII and factor IX products. Although this guideline specifically excludes recombinant DNA products, the patient numbers and end points specified in this document were used as the basis for defining the clinical plan for rFIX.

The clinical plan utilized pharmacokinetics (half-life and recovery) parameters as surrogate end points for efficacy, with the initial pharmacokinetics study being designed in line with the International Society for Thrombosis and Haemostasis (ISTH) guidelines. Following completion of this study, three additional protocols covering on-demand therapy, surgical prophylaxis, and previously untreated patients were then initiated (see Section 8.0).

During the finalization of the clinical plan, discussions were held between Genetics Institute and the Medicines Control Agency in the United Kingdom (at this time the CPMP Scientific Advice route was not available), FDA, Koseisho, and HPB in Canada. During these meetings we obtained scientific input and a degree of buy-in to the proposed clinical designs and registration strategy. The pivotal pharmacokinetic study was performed exclusively in the United States, with clinical sites in Canada and Europe (seven countries) to be involved in the subsequent three clinical studies.

The strategy in Japan was to use the North American and European clinical data, together with a pharmacokinetic study in a small number of Japanese patients using essentially the same protocol as that used in the United States. This strategy constitutes a "bridging study" as defined in the recent draft ICH guidelines on ethnic factors in acceptability of clinical data. In addition, Genetics Institute was granted orphan drug status for rFIX in Japan, which allowed the company much greater access to the regulators during the development of the clinical strategy for Japan.

The ability to interact with the regulatory agencies during the development of rFIX was of great value in defining an appropriate clinical plan. Genetics Institute had ongoing discussions with the FDA, which allowed the agency to comment upon the data in a timely manner. In Europe, there were discussions with the MCA, but it was an accepted risk that their opinions may not be representative of the total European view. In order to facilitate the review process in Canada, Genetics Institute allowed the FDA and HPB to discuss the licensure submission with each other. This did not constitute a joint review per se, but allowed the two agencies to share opinions about the data.

Another novel element to this product's development was the decision to pursue simultaneous filings in Europe, the United States, and Canada using a common format. Unfortunately, due to the timing for the bridging study in Japan it was not possible to submit in Japan within the same time frame. The common format chosen for the application was the European form as defined in the Notice to Applicants. The only differences in the application submitted were in the administrative information (application forms, etc.) and specification changes to account for European Pharmacopoeia test methods replacing USP methods as required. The data presented in these submissions were identical, with the clinical study reports being formatted according to the

ICH guideline. This strategy has allowed the harmonization of the labeling in these regions, which is an important issue for many companies from a product liability perspective.

The time from submission of the clinical trial application to initiation of the clinical trials ranged from 0 to 17 weeks in the three regions. The time from submission of the product license application to approval (or CPMP positive opinion) ranged from 22 to 37 weeks, indicating that the strategy of using a common format did not cause any noticeable delay to the approval process. Overall, the time from submission of the IND in the United States to approval of the BLA was only 2 years. Compared to all previously approved biopharmaceuticals, rFIX has had the most efficient time from initiation of clinical trials to licensure submission and approval.

The rFIX global development plan has allowed a very rapid, high-quality clinical development program, leading to an efficient registration in North America and Europe—and an acceptable pathway to approval in Japan.

ACKNOWLEDGEMENTS

The authors would like to acknowledge the contributions of the Recombinant Factor IX collaborators—K. Brinkhouse (University of North Carolina, Chapel Hill, NC), J. Lusher (Children's Hospital of Michigan), M. Ragni, (University of Pittsburgh, Pittsburgh, PA), A. Shapiro (University of Indiana, Indianapolis, IN), A. Thompson (Puget Sound Blood Center, Seattle, WA), and the rFIX Study Group—as well as the Recombinant Factor IX Project team leaders and task force members: S. Adamson, T. Ahern, G. Amphlett, L. Bartlett, D. Bonam, M. Bond, P. Bouchard, S. Brodeur, B. Burnett, L. Bush, T. Charlebois, B. Clancy, R. Costigan, S. Courter, D. Drapeau, W. Foster, E. Fritsch, P. Garzone, D. Gates, N. Gencarella, J. Goodfellow, M. Hamilton, K. Hanley, S. Harrison, M. Janowski, S. Karnik, J. Kaye, J. Keith, B. Kelley, T. Keutzer, A. Knight, S. Koza, M. Krane, G. Larsen, M. Leonard, B. Letwin, M. Magill, K. McCarthy, M. McCarthy, P. Oakes, B. O'Connell, H. Patel, J. Rouse, B. Rup, J. Ryan, R. Schaub, H. Scoble, J. Steckert, K. Sterl, A. Strang, M. Switzer, G. Timony, K. Tubridy, R. Walsh, C. Webb, B. Xu, and R. Zollner.

BIOGRAPHY

Genetics Institute Inc., a wholly-owned subsidiary of American Home Products Corporation, is a research-based company dedicated to the discovery, development and commercialization of breakthrough biopharmaceuticals. Genetics Institute has been a leader in the development of safe therapies through recombinant technology. Founded in 1980 by a group of visionary scientists from Harvard University, Genetics Institute began with a promising initial goal: to assemble a first-class team of scientists and business professionals to develop and commercialize protein-based therapeutic products through genetic engineering.

REFERENCES

1. Aggeler, P. M. et al. (1952). Plasma thromboplastin component (PTC) deficiency: a new disease resembling hemophilia. Proceedings of the Society for Experimental Biology and Medicine, 79, 692-694.
2. Biggs, R. et al. (1952). Christmas disease, a condition previously mistaken for haemophilia. British Medical Journal, 2, 1378-1382.
3. Bloom, A. L. (1982). Introduction. In: Bloom, A. L. (Ed.) The hemophilias. Churchill Livingstone, Edinburgh. pp. 1-17.
4. Forbes, C. D. (1984). Clinical aspects of the hemophilias and their treatment. In: Ratnoff, O. D. and Forbes, C. D. (Eds.) Disorders of hemostasis. Grune & Stratton, Orlando, Florida. pp. 177-239.
5. Thompson, A. R. (1986). Structure, function, and molecular defects of factor IX. Blood, 67, 565-572.
6. Aledort, L. M. (1991). Introduction and overview of treatment. Seminars in Hematology, 28(suppl 6), 1-2.
7. Yao, S.-N. et al. (1991). Expression of human factor IX in rat capillary endothelial cells: toward somatic gene therapy for hemophilia B. Proceedings of the National Academy of Sciences of the United States of America, 88, 8101-8105.
8. Hirsh, J. and Brain, E. A. (1983). Inherited coagulation disorders. In: Hemostasis and thrombosis, a conceptual approach, 2nd ed. Churchill Livingstone, New York. pp. 66-78.
9. Buzzard, B. M. and Heim, M. (1995). A study to evaluate the effectiveness of (Air-Stirrup) splints as a means of reducing the frequency of ankle haemarthroses in children with haemophilia A and B. Haemophilia, 1, 131-136.
10. Harmening, D. M. (1992). Introduction to hemostasis: an overview of hemostatic mechanism, platelet structure and function, and extrinsic and intrinsic systems. In: Harmening, D. M. (Ed.) Clinical hematology and fundamentals of hemostasis. F. A. Davis, Philadelphia. pp. 427-429.
11. Freedman, S. J. et al. (1995). Structure of the calcium ion-bound γ-carboxyglutamic acid-rich domain of factor IX. Biochemistry, 34, 12126-12137.
12. Harrison, S. et al. (1995). Development of a serum-free process for recombinant factor IX expression in Chinese hamster ovary cells. Abstract, XVth Congress of the International Society of Thrombosis & Haemostasis, June 11-16, 1995, Jerusalem, Israel.
13. Adamson, S. R. (1994). Recombinant factor IX concentrate development. Symposium Proceedings, Advances in the Treatment of Hemophilia and von Willebrand Disease, June 25-26, 1994, Berkeley, California.

14. Davie, E. W. (1987). The blood coagulation factors: their cDNAs, genes, and expression. In: Colman, R. W., Hirsch, J., Marder, V. J., and Salzman, E. W. (Eds.) Hemostasis and thrombosis, 2nd ed. J. B. Lippincott, Philadelphia. pp. 242-267.
15. Furie, B. and Furie, B. C. (1990). Molecular basis of vitamin K-dependent γ-carboxylation. Blood, 75, 1753-1762.
16. Bristol, J. A. et al. (1994). Profactor IX: the propeptide inhibits binding to membrane surfaces and activation by factor XIa. Biochemistry, 33, 14136-14143.
17. Kaufman, R. J. et al. (1986). Expression, purification, and characterization of recombinant γ-carboxylated factor IX synthesized in Chinese hamster ovary cells. Journal of Biological Chemistry, 261, 9622-9628.
18. Wasley, L. C. et al. (1993). PACE/furin can process the vitamin K-dependent pro-factor IX precursor within the secretory pathway. Journal of Biological Chemistry, 268, 8458-8465.
19. Wise, R. J. et al. (1990). Expression of a human proprotein processing enzyme: correct cleavage of the von Willebrand factor precursor at a paired basic amino acid site. Proceedings of the National Academy of Sciences of the United States of America, 87, 9378-9382.
20. Kaufman, R. J. et al. (1986). Selection and amplification of heterologous genes encoding adenosine deaminase in mammalian cells. Proceedings of the National Academy of Sciences of the United States of America, 83, 3136-3140.
21. Choo, K. H. et al. (1982). Molecular cloning of the gene for human anti-haemophilic factor IX. Nature, 299, 178-180.
22. Kurachi, K. and Davie, E. W. (1982). Isolation and characterization of a cDNA coding for human factor IX. Proceedings of the National Academy of Sciences of the United States of America, 79, 6461-6464.
23. Anson, D. S. et al. (1984). The gene structure of human anti-haemophilic factor IX. EMBO Journal, 3, 1053-1060.
24. Yoshitake, S. et al. (1985). Nucleotide sequence of the gene for human factor IX (antihemophilic factor B). Biochemistry, 24, 3736-3750.
25. Toole, J. J. et al. (1984). Molecular cloning of a cDNA encoding human antihaemophilic factor. Nature, 312, 342-347.
26. Seed, B. (1983). Purification of genomic sequences from bacteriophage libraries by recombination and selection in vivo. Nucleic Acids Research, 11, 2427-2445.
27. Maniatis, T. et al. (1982). Molecular cloning: a laboratory manual, 1st ed. Cold Spring Harbor Laboratory Press, Cold Spring Harbor, New York.
28. Graf, L. H. Jr. and Chasin, L. (1982). Direct demonstration of genetic alterations at the dihydrofolate reductase locus after gamma irradiation. Molecular and Cellular Biology, 2, 93-96.
29. Urlaub, G. and Chasin, L. A. (1980). Isolation of Chinese hamster cell mutants deficient in dihydrofolate reductase activity. Proceedings of the National Academy of Sciences of the United States of America, 77, 4210-4220.
30. Wong, G. C. et al. (1985). Human GM-CSF: molecular cloning of the complimentary DNA and purification of the natural and recombinant proteins. Science, 228, 810-815.
31. Kaufman, R. J. et al. (1989). The phosphorylation state of eukaryotic initiation factor 2 alters translational efficiency of specific mRNAs. Molecular and Cellular Biology, 9, 946-958.
32. Rehemtulla, A. and Kaufman, R. J. (1992). Preferred sequence requirements for cleavage of pro-von Willebrand factor by propeptide-processing enzymes. Blood, 79, 2349-2355.

33. Roebroek, A. J. M. et al. (1986). Characterization of human c-fes/fps reveals a new transcription unit (fur) in the immediately upstream region of the proto-oncogene. Molecular Biology Reports, 11, 117-125.
34. Roebroek, A. J. M. et al. (1986). Evolutionary conserved close linkage of the c-fes/fps proto-oncogene and genetic sequences encoding a receptor-like protein. EMBO Journal, 5, 2197-2202.
35. Fuller, R. S. et al. (1989). Intracellular targeting and structural conservation of a prohormone-processing endoprotease. Science, 246, 482-486.
36. van den Ouweland, A. M. et al. (1990). Structural homology between the human fur gene product and the subtilisin-like protease encoded by yeast Kex2. Nucleic Acids Research, 18, 664.
37. Foster, W. B. et al. (1995). Development of a process for purification of recombinant human factor IX. Abstract, 37th Annual Meeting, American Society of Hematology, December 1-5, 1995, Seattle, WA.
38. Bonam, D. et al. (1995). Purification of recombinant human factor IX by pseudoaffinity anion exchange chromatography. Abstract, 37th Annual Meeting, American Society of Hematology, December 1-5, 1995, Seattle, WA.
39. Bush, L. et al. (1996). Rational development of albumin-free rFIX lyophilized dosage form. Abstract, XXII International Congress of the World Federation of Haemophilia, June 23-28, 1996, Dublin, Ireland.
40. Bond, M. D. et al. (1994). Structural analysis of recombinant human factor IX. Blood, 84(suppl 1), 194a. [Abstract]
41. Bond, M. D. et al. (1995). Structural and biological characterization of Gla-related isoforms of recombinant human factor IX shows that Gla36 and Gla40 are not essential for in vitro clotting activity. Thrombosis and Haemostasis, 73, 1167-1168.
42. Gillis, S. et al. (1997). γ-Carboxyglutamic acids 36 and 40 do not contribute to human factor IX function. Protein Science, 6, 185-196.
43. Soriano-Garcia, M., Padmanabhan, K. et al. (1992). The Ca^{2+} ion and membrane binding structure of the Gla domain of Ca-prothrombin fragment 1. Biochemistry, 31, 2554-2566.
44. Nishimura, H. et al. (1989). Identification of a disaccharide (Xyl-Glc) and a trisaccharide (Xyl_2-Glc) O-glycosidically linked to a serine residue in the first epidermal growth factor-like domain of human factors VII and IX and protein Z and bovine protein Z. Journal of Biological Chemistry, 264, 20320-20325.
45. Harris, R. J. et al. (1993). Identification and structural analysis of the tetrasaccharide NeuAcα(2→6)Galβ(1→4)GlcNAcβ(1→3)Fucα1→O-linked to serine 61 of human factor IX. Biochemistry, 32, 6439-6547.
46. Agarwala, K. L. et al. (1994). Activation peptide of human factor IX has oligosaccharides O-glycosidically linked to threonine residues at 159 and 169. Biochemistry, 33, 5167-5171.
47. Bond, M. D. et al. (1994). Identification of O-glycosylation, sulfation and phosphorylation sites in the activation peptide of human plasma factor IX. Blood, 84(Suppl 1), 531a. [Abstract]
48. Huberty, M. C. et al. (1994). Structural characterization of glycopeptides from the activation region of human factor IX. Glycobiology, 4, 85. [Abstract]
49. Rouse, J. C. et al. (1995). The characterization of N-glycans from recombinant and naturally derived glycoproteins: combining the use of HPAE, mass spectrometry and glycosidase digestion. Glycobiology, 5, 711. [Abstract]

50. Strang, A.-M. et al. (1995). Structural characterization of the N-glycans of recombinant human factor IX. Glycoconjugate Journal, 12, 412. [Abstract]
51. Brandstetter, H; et al. (1995). X-ray structure of clotting factor IXa: active site and module structure related to Xase activity and hemophilia B. Proceedings of the National Academy of Sciences of the United States of America, 92, 9796-9800.
52. Steckert, J. and Amphlett, G. (1996). Comparative biophysical characterization of recombinant human factor IX and human plasma-derived factor IX. Haemophilia Journal, 2(suppl 1), 421. [Abstract]
53. Philo, J. S. (1994). Measuring sedimentation, diffusion, and molecular weights of small molecules by direct fitting of sedimentation velocity concentration profiles. In: Schuster, T. M. and Laue, T. M. (Eds.) Modern analytical ultracentrifugation: acquisition and interpretation of data for biological and synthetic polymer systems. Birkhäuser, Boston. pp. 156-170.
54. Rodrigues, H. et al. (1995). Analytical characterization of recombinant human factor IX (rhFIX). Thrombosis and Haemostasis, 73, 1206. [Abstract]
55. Limentani, S. A. et al. (1995). In vitro characterization of high purity factor IX concentrates for the treatment of hemophilia B. Thrombosis and Haemostasis, 73, 584-591.
56. Gray, E. et al. (1995). Measurement of activated factor IX in factor IX concentrates: correlation with in vivo thrombogenicity. Thrombosis and Haemostasis, 73:675-679.
57. Hrinda, M. E. et al. (1991). Preclinical studies of a monoclonal antibody-purified factor IX, Mononine™. Seminars in Hematology, 28(suppl 6), 6-14.
58. European Department for the Quality of Medicines. (1987). Freeze-dried human coagulation factor IX. European Pharmacopoeia 554, Strasbourg, France.

Chapter 4

Biopharmaceutical Drug Development: A Case History
Filgrastim (NEUPOGEN, GRAN)

MaryAnn Foote, and Thomas Boone
Amgen Inc., USA

Key words: Filgrastim, r-metHuG-CSF, colony-stimulating factors, haematopoiesis, neutrophils, Amgen Inc.

Abstract: In the 1980s, human and murine forms of many hematopoietic colony-stimulating factors were cloned. One factor that was purified, cloned, and produced in commercial quantities was granulocyte colony-stimulating factor (G-CSF, Filgrastim), a protein that acts on the neutrophil lineage. Neutrophils are the body's major defence against infections. The initial licensing indication for Filgrastim was the amelioration of chemotherapy-induced neutropenia. It has been approved in more than 75 countries for a variety of uses, including aplastic anaemia, severe chronic neutropenia, and the mobilisation of peripheral blood progenitor cells for transplantation. Filgrastim has an excellent safety profile with the only common side effect of administration reported being mild to moderate bone pain. Current issues being investigated include evaluation of the cost benefit of Filgrastim in various clinical settings and impact on survival in dose-intensive chemotherapy with Filgrastim given as an adjunct treatment.

1. INTRODUCTION

Haematopoietic growth factors are glycoproteins that act on cells at various stages to produce mature haematopoietic (i.e. blood) cells. There are many hematopoietic growth factors (Figure 14); one factor that has been isolated, purified, cloned, and produced in commercial quantities is granulocyte colony-stimulating factor (G-CSF), a protein that acts on the neutrophil lineage to stimulate the proliferation, differentiation, and function of committed progenitor cells and functionally active mature neutrophils.

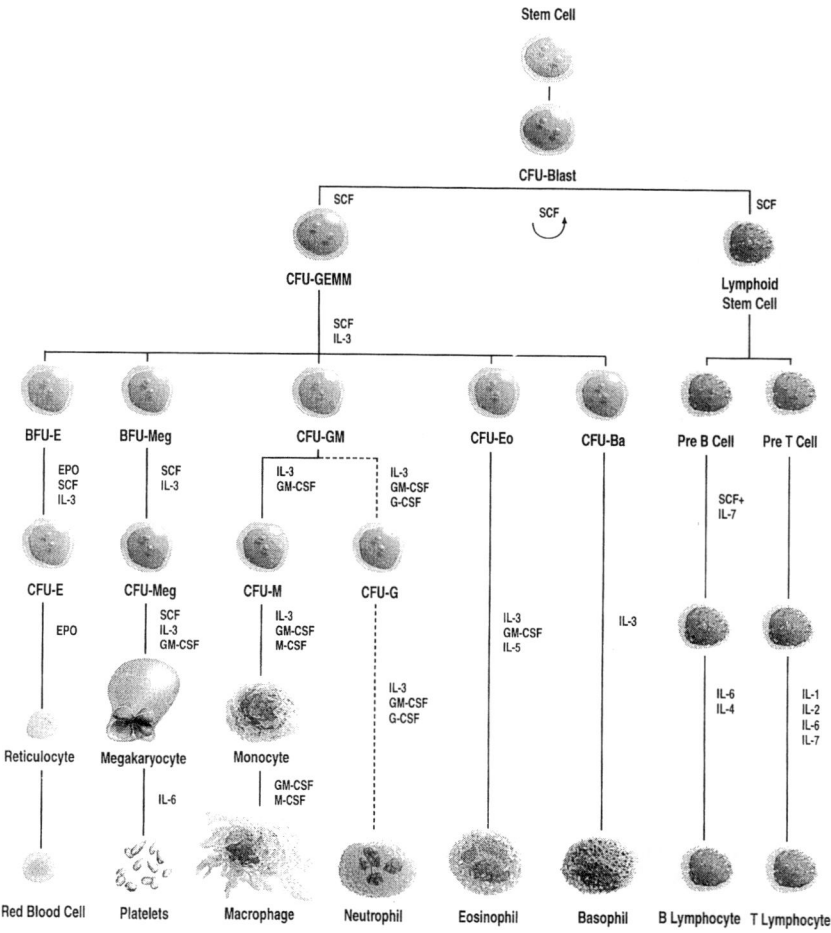

Figure 14. Haematopoietic Growth Factors. More than 20 factors have been identified. They work at various stages of the haematopoietic cascade to produce mature functional blood cells from precursor and stem cells. (Figure courtesy of Amgen Inc.)

The amount of circulating endogenous G-CSF in the blood increases in response to an infection or to neutropenia (1), and this has been shown to occur in a variety of pathological conditions including exposure to endotoxin (2-6). The existing data suggests that, within the human body, G-CSF is the primary factor mediating neutrophilic response to infection and neutropenia (1). In cases of gram-negative and fungal infections, the amount of G-CSF are elevated in the blood (4). The highest G-CSF levels are found in neutropenic patients and are correlated with fever (1).

Mice lacking endogenous G-CSF have chronic neutropenia and impaired neutrophil function (7). This suggests that G-CSF is indispensable for maintaining the normal quantitative balance of neutrophil production during "steady state" granulopoiesis *in vivo*, and indicates that G-CSF has a role in "emergency" granulopoiesis during infection.

Filgrastim, the non-glycosylated recombinant human methionyl form of G-CSF, is produced in *Escherichia coli*. It is marketed by three companies: Amgen Inc. (Thousand Oaks, CA), Kirin Brewery (Tokyo, Japan), and F Hoffmann-La Roche (Basel, Switzerland) under the tradenames NEUPOGEN® and GRAN®. Because of local marketing partners in Italy (Dompé) and Spain (Esteve-Pensa), there is a second tradename (NEUPOGEN® and GRANULOKINE®) in these countries. The name "Filgrastim" is the generic name given to the compound by the United States Adopted Name Council, and refers to the function of the product. The "-grastim" suffix means that the product acts on granulocytes; the "fil-" prefix refers phonetically to neutrophils.

Filgrastim was initially approved for marketing in the indication of chemotherapy-induced neutropenia. Since 1991, Filgrastim has been approved in more than 75 countries for treatment of myelosuppression after bone marrow transplantation, severe chronic neutropenia, acute leukaemia, aplastic anaemia, AIDS, myelodysplastic syndromes, and the mobilisation of peripheral blood progenitor cells for transplantation. Not all countries have received approval for all indications.

Filgrastim produces rapid, specific, dose-dependent increases in neutrophils and macrophages at high concentrations, and augments their functions (8-12). It reduces neutrophil maturation time from 5 days to 1 day, causing a rapid release of mature neutrophils from the bone marrow into the peripheral circulation (13). Neutrophils treated with Filgrastim have normal survival (13,14), enhanced chemotaxis through increased binding of f-Met-Leu-Phe (15), and enhanced phagocytic activity of neutrophils in patients with HIV infection (16,17). Filgrastim appears to protect rodents against lipopolysaccharide (LPS)-induced toxicity by suppression of tumour necrosis factor (TNF)-α release (18).

1.1   Producer Cell Line

In 1985, Amgen investigators were successful in cloning the gene for G-CSF (19). To do so, they first purified the native human G-CSF from a cell line and identified a portion of the amino acid sequence of the molecule. They used this information to synthesise a series of oligonucleotides that contained the genetic code for these amino acids of the G-CSF molecule. Using advanced hybridisation techniques, they were able to clone the human G-CSF gene. This gene was then inserted into *Escherichia coli* for the production of Filgrastim.

The producer cell line was a common laboratory strain of *E coli* K12, chosen because of its robust growth properties and its approval by the Recombinant Advisory Committee in the United States for use at the greater than 10-L fermentation scale. The production strain harbouring the recombinant G-CSF-expression plasmid displayed excellent stability properties, including high plasmid retention during both the growth phase and protein-expression phase in the fermentor, and genetic stability within the G-CSF-coding region on the plasmid.

E coli from a master seed lot containing the gene for Filgrastim are grown in a fermentor, in medium optimised for product synthesis, to a specified cell density. The cells are harvested and lysed and the product extracted. The product is allowed to oxidise to its native state, and is purified by several chromatographic and filtration steps.

1.2   Initial Product Characterisation

Filgrastim is a sterile, clear, colourless, preservative-free liquid. The highly purified bulk product consists of a 175-amino acid recombinant-derived protein with a molecular weight of 18,800 daltons. The purified product contains an N-terminal methionine and lacks O-linked glycosylation.

The product was originally formulated in a 10 mM sodium acetate buffer at pH 4.0 containing 5% mannitol and 0.004% Tween®80. Filgrastim is now formulated with sorbitol rather than mannitol, a modification that has enhanced the cryoprotective properties of the formulation. Filgrastim can now withstand accidental freezing and subsequent thawing at temperatures as low as -20°C.

1.3   Patent Issues

The Patent and Trademark Office awarded Amgen the first U.S. patent on Filgrastim in March 1989.

U.S. Patent No. 4,810,643 (issued 7 March 1989) and U.S. Patent No. 4,999,291 (issued 12 March 1991) relate to different aspects of DNAs, vectors, and processes relating to recombinant G-CSF. U.S. Patent No. 5,580,755 (issued 3 December 1996) relates to G-CSF polypeptides, and U.S. Patent No. 5,582,823 (issued 10 December 1996) relates to methods of treatment using G-CSF polypeptides.

U.S. Patent Nos. 4,810643 and 4,999,291 expire 7 March 2006; U.S. Patent No. 5,580,755 expires 3 December 2013, and U.S. Patent No. 5,582,823 expires 10 December 2013.

2. PRECLINICAL STUDIES

Filgrastim was administered to monkeys, dogs, hamsters, rats, and mice as part of a comprehensive preclinical toxicology program. Single-dose acute, repeated-dose subacute, and chronic studies were carried out. Single-dose administration of Filgrastim by the oral, intravenous, subcutaneous, or intraperitoneal routes did not produce significant toxicity in mice, rats, hamsters, or monkeys.

When injected *in vivo*, Filgrastim increases the number of mature neutrophils in the circulation. This has been shown in rats (20), mice (21), hamsters (22), dogs (23), non-human primates (24,25), pigs (26), guinea pigs (27), and rabbits (28,29). The increase in neutrophil counts occurs after a transitory neutropenia, which occurs within 1 hour of injecting Filgrastim; thereafter numbers increase rapidly (22,30).

Cynomolgus monkeys showed dramatic (up to 10-fold) increases in blood neutrophils when injected with between 1 and 100 μg Filgrastim/kg body weight/day for up to 27 days (31). The neutrophils of these animals also showed normal or elevated functional activities, and there was evidence of enhanced myelopoiesis in bone marrow, lymph nodes, and spleen, but not in other organs.

Enhanced recovery from cyclophosphamide treatment also has been noted in monkeys treated with Filgrastim (31), where the period of neutropenia was significantly reduced, and marrow cellularity recovered more quickly.

In a canine model of cyclic neutropenia, administration of Filgrastim did not eliminate cycling in neutrophil counts, but the neutrophil counts were maintained at a higher value, i.e., without the neutropenia (23). In grey collie dogs with cyclic neutropenia (32) the number and activity of G-CSF receptors are normal. It may be concluded from these data that an event downstream of G-CSF receptor/ligand binding is responsible for the unusual response to Filgrastim and that production of the regulator is not the fundamental problem.

Pharmacodynamic studies in rabbits have been reported by Stevens *et al.* (11). The pharmacodynamics of a single subcutaneous dose of Filgrastim 10 μg/kg was monitored and correlated with serum concentrations, specific oxygenation capacity of circulating polymorphonuclear leukocytes (PMN) as measured by luminol-dependent chemiluminescence using C3b/C3bi-opsonised zymosan as a stimulus, absolute neutrophil count, and estimation of

PMN-myeloperoxidase content by peroxidase staining. After administration of Filgrastim, the concentration peaked at 2 to 4 hours with an accompanying 5-fold decrease in absolute neutrophil count and a 2.5-fold increase in chemiluminescence, but no change in PMN-myeloperoxidase content. The neutrophil count returned to normal by 6 hours, at which time the chemiluminescence was within normal range. Between 24 and 48 hours with Filgrastim concentrations at baseline values, the absolute neutrophil count increased 6-fold, chemiluminescence increased a second time to 3-fold greater than normal, and the PMN-myeloperoxidase content concomitantly increased 6-fold. The results of this preclinical study showed a biphasic increase in PMN oxygenation capacity that consisted of an early phase which likely occurs because of increased CR1 and CR3 opsonin-receptor expression due to the direct exposure of the peripheral blood PMN to Filgrastim and a second phase which correlates with increased PMN-myeloperoxidase. In addition to the increase in the absolute neutrophil count, the effects of Filgrastim to increase the PMN oxygenation capacity may provide a proportional therapeutic advantage with regard to PMN microbicidal capacity.

3. CLINICAL TRIAL FINDINGS

Reviews of clinical experience with Filgrastim can be found in several papers (33-38).

In phase 1 and phase 2 studies involving 96 patients with a variety of non-myeloid malignancies (i.e., urothelial cancer, breast cancer), administration of Filgrastim resulted in a dose-dependent increase in circulating neutrophil counts over the dose range of 1 to 70 µg/kg/day (39-42). This increase in neutrophil counts was observed when Filgrastim was administered intravenously, subcutaneously, or by continuous subcutaneous infusion. Discontinuation of Filgrastim caused the neutrophil counts to return to baseline values in most patients within 4 days. These studies demonstrated the utility of using Filgrastim to decrease the incidence of infection, as manifested by febrile neutropenia, in patients with non-myeloid malignancies receiving myelosuppressive chemotherapy.

Phase 3 trials have demonstrated the beneficial effect of Filgrastim on neutropenia after standard-dose chemotherapy. Two randomised, placebo-controlled, double-blind studies involving more than 300 patients with small-cell lung cancer receiving cyclophosphamide, adriamycin, etoposide (CAE) chemotherapy showed that Filgrastim significantly decreased the incidence, severity, and duration of severe neutropenia (43,44). In other randomised, placebo-controlled, double-blind trials, Filgrastim allowed increases in dose intensity of doxorubicin (45) and cyclophosphamide, adriamycin, 5-fluorouracil (CAF) chemotherapy (46).

A randomised, phase 3 trial in patients with non-Hodgkin's lymphoma showed that Filgrastim significantly improved delivery of full-dose chemotherapy compared with control patients (47).

Placebo-controlled studies showed that Filgrastim accelerates neutrophil recovery after allogeneic bone marrow transplantation. An absolute neutrophil count recovery to >0.5 x 10^9/L was achieved in 14 days or less in studies with Filgrastim (48-50).

Studies have been carried out in patients with myelodysplastic syndromes, and treatment with Filgrastim has been associated with a sustained improvement in neutrophil function, but without increased adherence or impaired chemotaxis (51). In a phase 3 randomised study involving 102 patients with RAEB or RAEB-t subtypes of myelodysplastic syndromes, Filgrastim was shown to be efficacious in increasing neutrophil counts (52).

A phase 3 trial of Filgrastim in patients with severe chronic neutropenia has shown long-term efficacy (>200 patient-years experience) and tolerance, and the hematologic and clinical benefits were sustained during maintenance treatment (53,54). In one study, when children with severe chronic neutropenia did not respond to treatment with another hematopoietic growth factor, they were switched to treatment with Filgrastim with a resulting increase in neutrophil counts (55).

Filgrastim has been shown to improve tolerance to ganciclovir allowing delivery of full doses (56). Filgrastim given for 2 weeks at doses of 0.3 to 3.6 µg/kg/day, increased neutrophil numbers 9-fold, and maintained this increase during concomitant therapy with erythropoietin and zidovudine (57). Use of Filgrastim permitted some patients to receive full doses of antiviral therapy and these patients had preserved or improved neutrophil function. Use of Filgrastim in HIV-infected patients rapidly reversed their neutropenia and maintained normal neutrophil counts (58). This allowed greater use of myelosuppressive medications without the life-threatening complication of neutropenia. In a randomised, multicenter, controlled trial of 201 evaluable HIV-infected patients, severe neutropenia or death was less frequent in patients who received daily (12.8%) or intermittent (8.2%) doses of Filgrastim compared with control patients (34.1%) (59). Filgrastim-treated patients developed 31% fewer bacterial infections and 54% fewer severe bacterial infections than did control patients. This translated to 26% fewer hospital days including 45% fewer hospital days for bacterial infections and 28% fewer days of intravenous antibacterial therapy. Neutrophils from HIV-infected patients treated with Filgrastim have been shown to have improved function (12,16). Seventy-six HIV-infected patients and 28 normal uninfected volunteers were used in this study, but only the patients were treated with

Filgrastim. Filgrastim treatment caused a significant increase in chemiluminescence and bacterial killing and decrease in *in vivo* activation of circulating neutrophils.

Filgrastim alone or in combination with chemotherapy is an effective agent for recruiting peripheral blood progenitor cells with long-term reconstituting ability (60-66). In a historically controlled study, Filgrastim-generated PBPC in conjunction with autologous BMT and Filgrastim accelerated recovery of neutrophil and platelet count (62). Use of Filgrastim for mobilisation resulted in a significantly accelerated time to recovery of granulocytes when compared with non-mobilised PBPC recipients in a study of 85 patients with relapsed HD (60). The use of mobilised PBPC resulted in a significantly accelerated time to platelet engraftment when compared with non-mobilised PBPC recipients. There was a statistically significant reduction of costs in patients who received Filgrastim-mobilised PBPC.

Filgrastim given after chemotherapy significantly reduces neutropenia and infection-related complications and maintains survival benefits for patients receiving induction and/or consolidation therapy for acute myeloid leukaemia (67). Patients receiving Filgrastim have a statistically significant shorter time to neutrophil recovery compared with patients receiving placebo, and significant reductions in infection-related complications after induction and consolidation therapy. These reductions in clinical signs and symptoms translated into significantly shorter periods of hospitalisation, with a median of 20 days for Filgrastim-treated patients and 25 days for placebo-treated patients.

Filgrastim has an excellent safety profile. The most consistently observed adverse event with Filgrastim is mild-to-moderate bone pain which is easily controlled with non-prescription analgesics. Filgrastim did not produce dose-limiting toxicities even when administered at a dose of 115 µg/kg, a dose that can cause marked leukocytosis (50×10^9/L) (68). In some clinical trials, there have been reports of elevated alkaline phosphatase, lactate dehydrogenase, and uric acid; and transient decreases in platelets at high doses; as well as a few cases of Sweet's syndrome and general allergic-type reactions (69). Although there have been reports of possible pulmonary toxicity associated with bleomycin, a review of two controlled studies did not show an increased frequency of serious pulmonary toxicity in the setting of lymphoma (70). There are no published reports of formation of antibodies against Filgrastim (69)

Filgrastim is generally administered as a subcutaneous injection, the dose depending upon the indication (i.e., 5 µg/kg/day for chemotherapy-induced neutropenia and 10 µg/kg/day after bone marrow transplantation). Filgrastim may be diluted for intravenous administration with 5% dextrose, but the concentration must be ≥15 µg/mL. If the concentration is >2 but <15 µg/mL, human serum albumin must be added to prevent adsorption to plastic materials of infusion systems.

Single-use prefilled syringes of Filgrastim are available in some parts of the world. The syringes contain either 300 µg in 1.0 mL or 480 µg in 1.6 mL.

4. VALIDATION STUDIES

Manufacturing controls and in-process testing at multiple stages are performed to assure product purity, identity, strength, and quality. The purified bulk Filgrastim is tested for identity and purity by SDS-PAGE, amino acid analysis, N-terminal sequence analysis, peptide mapping, HPLC, isoelectric focusing, and Western blot analysis. The product is also tested for the presence of *E coli* proteins and DNA. The final product is also tested to ensure the identity, quality, safety, purity, strength, potency, and excipient chemical content of the final dosage form.

5. SUMMARY OF DOSSIER PREPARATION AND EXPERIENCE OF GETTING IT THROUGH REGULATORY AGENCIES

Filgrastim is a biological response modifier, a term used to distinguish a class of agents composed of native and altered endogenous proteins that cause a specific desired cellular response. The Biological Response Modifier Program of the United States National Institute of Cancer stipulates that biological response modifiers must show therapeutic efficacy in preclinical models as a requirement to continue to clinical testing (71).

The regulation of Filgrastim, as a biological response modifier, is governed by different laws than those for traditional pharmaceutical drugs. In the United States, drugs are covered under the Food, Drug and Cosmetic Act (FD&C), and biological response modifiers have additional laws under the Public Health Services Act (PHS). The approval requirements for a drug include more than one adequate and well-controlled study demonstrating both safety and efficacy. Approval requirements for Filgrastim also included an adequate and well-controlled study as well as supporting studies.

As with traditional pharmaceutical products, there are a number of regulatory processes to be completed, beginning with an application filing for an Investigational New Drug (IND). The IND contains preclinical and other studies, and must also include the clinical protocol for the initial phase 1 study. Rather than a New Drug Application (NDA), Filgrastim and other biological response modifiers use a Biological License Application (BLA). A BLA is an application submitted to the United States Food and Drug Administration (FDA) to obtain approval for marketing. A BLA addresses the processes by which the product is manufactured and tested. It also contains the results of the preclinical testing and clinical studies conducted under the IND. The actual buildings where the biological product is manufactured, tested, and/or stored must also be approved. In the case of Filgrastim, a BLA was not filed; rather the separate pathways of Product License Application (PLA) and Establishment License Application (ELA), now both combined in the BLA, were filed.

Amgen received FDA approval on the manufacturing and marketing license applications for Filgrastim in February 1991. The CPMP recommended approval of Filgrastim for the European Community in February 1991; and, in the following month, the United Kingdom became the first country in the European Community to approve Filgrastim for marketing. The Health Protection Branch of Health and Welfare Canada in February 1992 approved Filgrastim for use in Canada, and, in May 1992, the Therapeutic Goods Administration approved Filgrastim for use in Australia.

In 1994, Amgen received FDA approval for two additional indications for Filgrastim. It is now used for the reduction in the duration of neutropenia in patients with non-myeloid malignancies undergoing myeloablative chemotherapy followed by allogeneic or autologous bone marrow transplantation. Filgrastim is also approved for chronic administration to reduce the incidence and duration problems related to neutropenia in patients with severe chronic neutropenia. This was followed by European Member States approval of Filgrastim for the harvesting of early acting peripheral blood progenitor cells that can speed the recovery of infection-fighting white blood cells after cancer chemotherapy. In 1995, the FDA approved it for use in harvesting peripheral blood progenitor cells and local approval by European Members States was obtained for this indication. In 1998, there was approval of Filgrastim in the setting of acute myeloid leukaemia in the United States and Europe.

6. ADDITIONAL TECHNICAL/BUSINESS MATTERS

From its founding, corporate and academic collaborations were seen as the way to accelerate Amgen into an independent commercial enterprise by accessing research, patients, and markets in areas where Amgen did not have entry or alone could not invest sufficient resources.

In September 1988, Amgen entered into a 5-year agreement with the Swiss pharmaceutical company F. Hoffmann-La Roche & Co. Ltd., under which both companies developed and marketed the product within the European Community. Amgen retains sole marketing rights in the United States, Canada, and Australia. Roche has marketing rights in Latin America, Africa, the Middle East, New Zealand, and the remainder of Europe. Kirin Brewery of Japan has marketing rights in Japan and the Pacific countries. In 1997, under an extended agreement, Filgrastim is marketed by affiliates of Amgen Inc., although F Hoffmann-La Roche still co-markets in certain European Union countries.

The company site in California was chosen for its convenience to University of California at Los Angeles, University of California at Santa Barbara, and California Institute of Technology. Strong corporate and academic collaborations continue.

BIOGRAPHY

Mary ann Foote and Thomas Boone are employees of Amgen, the world's largest independent biotechnology company. It was founded in Thousand Oaks, California as AMGen (Applied Molecular Genetics) in 1980. Amgen uses recombinant DNA technology and molecular biology to develop highly specialised human therapeutics in the area of hematopoietic growth factors, neuroendocrinology, inflammation, and tissue growth factors.

ACKNOWLEDGEMENTS

The authors wish to thank Michael Mann and Liz Jenkins for their helpful assistance in preparing this chapter.

REFERENCES

1. Cebon J. et al. (1994). Endogenous haemopoetic growth factors in neutropenia and infection. Br J Haematol. 86, 265-274.
2. Morstyn G. et al. (1991). Cytokines in infections and as anticancer agents. Ann Haematol. 29A, 96a.

3. Watari K. et al. (1989). Serum granulocyte colony-stimulating factor levels in healthy volunteers and patients with various disorders as estimated by enzyme immunoassay. Blood. 73, 117-122.
4. Kawakami M. et al. (1990). Levels of serum granulocyte colony-stimulating factor in patients with infections. Blood. 76, 1962-1964.
5. Metcalf D. (1988). The Molecular Control of Blood Cells. Harvard University Press. Cambridge, MS.
6. Hartung T. et al. (1995). Effect of granulocyte colony-stimulating factor treatment on ex vivo blood cytokine response in human volunteers. Blood. 85, 2482-2489.
7. Lieschke GJ. et al. (1994). Mice lacking granulocyte colony-stimulating factor have chronic neutropenia, granulocyte and macrophage progenitor cell deficiency, and impaired neutrophil mobilization. Blood. 84, 1737-1746.
8. Moore MA. et al. (1987). Biological activities of recombinant granulocyte colony stimulating factor (rhG-CSF) and tumor necrosis factor: in vivo and in vitro analysis. Hamatol Bluttransfus. 31, 210-220.
9. Tamura M. et al. (1987). Induction of neutrophilic granulocytosis in mice by administration of purified human native granulocyte colony-stimulating factor (G-CSF). Biochem Biophys Res Commun. 142, 454-460.
10. Welte K. et al. (1985). Purification and biochemical characterization of human pluripotent hematopoietic colony-stimulating factor. Proc Natl Acad Sci USA. 82, 1526-1530.
11. Stevens P. et al. (1991). Pharmacodynamics of recombinant human G-CSF with respect to an increase of neutrophil oxidative metabolism. J Leucocyte Biol. 2, 40.
12. Hartung T. et al. (1998). Filgrastim (r-metHuG-CSF) restores IL-2 production of blood from advanced HIV patients.. J Infect Dis. (in press).
13. Lord BI. et al. (1989). The kinetics of human granulopoiesis following treatment with granulocyte colony-stimulating factor in vivo. Proc Natl Acad Sci USA. 86, 9499-9503.
14. Bronchud MH. et al. (1988). In vitro and in vivo analysis of the effects of recombinant human granulocyte colony-stimulating factor in patients. Br J Cancer. 58, 64-69.
15. Colgan SP. et al. (1992). Neutrophil function in normal and Chediak-Higashi syndrome cats following administration of recombinant canine granulocyte colony-stimulating factor. Exp Hem. 20, 1229-1234.
16. Pitrak DL. et al. (1996). Filgrastim (r-methuG-CSF) treatment of HIV-infected patients improves neutrophil function. Int Conf AIDS. 282, Abstract ThB4181.
17. Roilides E. et al. (1991). Granulocyte colony-stimulating factor enhances the phagocytic and bactericidal activity of normal and defective human neutrophils. J Infect Dis. 163, 579-583.
18. Gorgen I. et al. (1992). Granulocyte colony-stimulating factor treatment protects rodents against lipopolysaccharide-induced toxicity via suppression of systemic tumor necrosis factor-alpha. J Immunol. 149, 918-924.
19. Souza LM. et al. (1986). Recombinant human granulocyte colony-stimulating factor: effects on normal and leukemic myeloid cells. Science. 232, 61-65.
20. Ulich TR. et al. (1988). Kinetics and mechanisms of recombinant human granulocyte-colony stimulating factor-induced neutrophilia. Am J Pathol. 133, 630-638.
21. Broxmeyer HE. et al. (1988). Recombinant human granulocyte-colony stimulating factor and recombinant human macrophage-colony stimulating factor synergize in vivo to enhance proliferation of granulocyte-macrophage, erythroid, and multipotential progenitor cells in mice. J Cell Biochem. 38, 127-136.

22. Cohen AM. et al. (1987). In vivo stimulation of granulopoiesis by recombinant human granulocyte colony-stimulating factor. Proc Natl Acad Sci USA. 84, 2484-2488.
23. Lothrop CJ. et al. (1988). Correction of canine cyclic hematopoiesis with recombinant human granulocyte colony-stimulating factor. Blood. 72, 1324-1328.
24. Welte K. et al. (1988). Recombinant human granulocyte-colony stimulating factor: in vivo effects on myelopoiesis in primates. Behring Inst Mitt. 83, 102-106.
25. Gillio AP. et al. (1987). Effects of recombinant human granulocyte-colony stimulating factor on hematopoietic reconstitution after autologous bone marrow transplantation in primates. Transpl Proc. 19, 153-156.
26. Fink MP. et al. (1993). Effect of granulocyte colony-stimulating factor on systemic and pulmonary responses to endotoxin in pigs. J Trauma. 34, 571-577.
27. Kanazawa M. et al. (1992). Granulocyte colony-stimulating factor does not enhance endotoxin-induced acute lung injury in guinea pigs. Am Rev Respir Dis. 145, 1030-1035.
28. Smith WS. et al. (1995). Granulocyte colony-stimulating factor versus placebo in addition to penicillin G in a randomized blinded study of gram-negative pneumonia sepsis: analysis of survival and multisystem organ failure. Blood. 86, 1301-1309.
29. Gratwohl A. et al. (1995). Transplantation of G-CSF mobilized allogeneic peripheral blood stem cells in rabbits. Bone Marrow Transplant. 16, 63-68.
30. Tanaka H. et al. (1991). Pharmacokinetics of recombinant human granulocyte colony-stimulating factor conjugated to polyethylene glycol in rats. Cancer Res. 51, 3710-3714.
31. Welte K. et al. (1987). Recombinant human granulocyte colony-stimulating factor: effects on hematopoiesis in normal and cyclophosphamide-treated primates. J Exp Med. 165, 941-948.
32. Avalos BR. et al. (1994). Abnormal response to granulocyte colony-stimulating factor (G-CSF) in canine cyclic hematopoiesis is not caused by altered G-CSF receptor expression. Blood. 84, 789-794.
33. Hollingshead LM, Goa KL. (1991). Recombinant granulocyte colony-stimulating factor (rG-CSF). A review of its pharmacological properties and prospective role in neutropenic conditions. Drug Evaluation. 42, 300-330.
34. Lieschke GJ, Burgess AW. (1992). Granulocyte colony-simulating factor and granulocyte-macrophage colony-stimulating factor (2). N Eng J Med. 327:28-35.
35. Steward WP. (1993). Granulocyte and granulocyte-macrophage colony-stimulating factor. Lancet. 342, 153-157.
36. Frampton JE. et al. (1994). Filgrastim. A review of its pharmacological properties and therapeutic efficacy in neutropenia. Drug Evaluation. 48, 731-760.
37. Welte K. et al. (1996). Filgrastim (r-metHuG-CSF): the first 10 years. Blood. 88, 1907-1929.
38. Foote MA. et al. Granulocyte colony-stimulating factor. IN: Cytokines, edited by AR Mire-Sluis, R Thorpe; Academic Press, London; pages 231-244, 536.
39. Gabrilove JL. et al. (1988). Effect of granulocyte colony-stimulating factor on neutropenia and associated morbidity due to chemotherapy for transitional cell carcinoma of the urothelium. N Engl J Med. 111, 887-892.
40. Gabrilove JL. et al. (1988). Phase I study of granulocyte colony-stimulating factor in patients with transitional cell carcinoma of the urothelium. J Clin Invest. 82, 1454-1461.
41. Morstyn G. et al. (1988). Effect of granulocyte colony-stimulating factor on neutropenia induced by cytotoxic chemotherapy. Lancet. 1, 667-672.

42. Bronchud MH. et al. (1987) Phase I/II study of recombinant human granulocyte colony-stimulating factor in patients receiving intensive chemotherapy for small cell lung cancer. Br J Cancer. 56, 809-813.
43. Crawford J. et al. (1991). Reduction by granulocyte colony-stimulating factor of fever and neutropenia induced by chemotherapy in patients with small cell lung cancer. N Engl J Med. 325, 164-170.
44. Trillet-Lenoir V. et al. (1993). Recombinant granulocyte colony stimulating factor reduces the infectious complications of cytotoxic chemotherapy. Eur J Cancer. 29A, 319-324.
45. Bronchud MH. et al. (1989). Phase I/II study of recombinant human granulocyte colony-stimulating factor to increase the intensity of treatment with doxorubicin in patients with advanced breast and ovarian cancer. Br J Cancer. 60, 121-128.
46. Demetri GD. et al. (1991). Recombinant methionyl granulocyte-CSF (r-metHuG-CSF) allows an increase in the dose intensity of cyclophosphamide/doxorubicin/5-fluorouracil (CAF) in patients with advanced breast cancer. Proc ASCO. 10, 70a.
47. Pettengell R. et al. (1992). Granulocyte colony-stimulating factor to prevent dose-limiting neutropenia in non-Hodgkin's lymphoma: a randomized controlled trial. Blood. 80, 1430-1436.
48. Sheridan WP. et al. (1989). Granulocyte colony-stimulating factor and neutrophil recovery after high-dose chemotherapy and autologous bone marrow transplantation. Lancet. 2, 891-895.
49. Peters WP. et al. (1989). Comparative effects of rHuG-CSF and rHuGM-CSF on hematopoietic reconstitution and granulocyte function following high dose chemotherapy and autologous bone marrow transplantation (ABMT). Proc ASCO. 18, 18A.
50. Taylor KM. et al. (1989). Recombinant human granulocyte colony-stimulating factor hastens granulocyte recovery after high-dose chemotherapy and autologous bone marrow transplantation in Hodgkin's disease. J Clin Oncol. 7, 1791-1799.
51. Negrin RS. et al. (1990). Maintenance treatment of patients with myelodysplastic syndromes using recombinant human granulocyte colony-stimulating factor. Blood. 7, 36-43.
52. Greenberg P. et al. (1993). Phase III randomized multicenter trial of G-CSF vs observation for myelodysplastic syndromes (MDS). Blood. 82, 196a.
53. Dale DC. et al. (1993). A randomized controlled phase III trial of recombinant human granulocyte colony-stimulating factor (Filgrastim) for treatment of severe chronic neutropenia. Blood. 81, 2496-2502.
54. Dale DC. et al. (1990). Long term treatment of severe chronic neutropenia with recombinant human granulocyte colony-stimulating factor (r-metHuG-CSF). Blood. 76, 545a.
55. Welte K. et al. (1990). Differential effects of granulocyte colony-stimulating factor and granulocyte-macrophage colony-stimulating factor in children with severe congenital neutropenia. Blood. 75, 1056-1063.
56. Jacobsen MA. et al. (1992). Ganciclovir with recombinant methionyl human granulocyte colony-stimulating factor for treatment of cytomegalovirus disease in AIDS patients. AIDS. 6, 515-517.
57. Miles SA. et al. (1991). Combined therapy with recombinant granulocyte colony-stimulating factor and erythropoietin decreases hematologic toxicity from zidovudine. Blood. 77, 2109-2117.
58. Hermans P. et al. (1996). Filgrastim to treat neutropenia and support myelosuppressive medication dosing in HIV infection. G-CSF 92105 Study Group. AIDS. 10, 1627-1633.

59. Kuritzkes DR. et al. (1998). Filgrastim prevents severe neutropenia and reduces infective morbidity in patients with advanced HIV infection: results of a randomized, multicenter, controlled trial. AIDS. 12, 65-74.
60. Chao NJ. et al. (1993). Granulocyte colony-stimulating factor "mobilized" peripheral blood progenitor cells accelerate granulocyte and platelet recovery after high-dose chemotherapy. Blood. 81, 2031-2035.
61. Hohaus S. et al. (1993). Successful autografting following myeloablative conditioning therapy with blood stem cells mobilized by chemotherapy plus rhG-CSF. Exp Hematol. 21, 508-514.
62. Sheridan WP. et al. (1992). Effect of peripheral-blood progenitor cells mobilised by filgrastim (G-CSF) on platelet recovery after high-dose chemotherapy. Lancet. 339, 640-644.
63. Sheridan WP. et al. (1990). Granulocyte colony-stimulating factor (G-CSF) in peripheral blood stem cell (PBSC) and bone marrow transplantation. Blood. 76, S1.
64. Dürhsen U. et al. (1988). Effects of recombinant human granulocyte colony-stimulating factor on hematopoietic progenitor cells in cancer patients. Blood. 72, 2074-2081.
65. Faucher C. et al. (1996). Autologous transplantation of blood stem cells mobilized with filgrastim alone in 93 patients with malignancies: the number of $CD34^+$ cells reinfused is the only factor predicting both granulocyte and platelet recovery. J Hematother. 5, 663-670.
66. Schmitz N. et al. (1996). Randomised trial of filgrastim-mobilised peripheral blood progenitor cell transplantation versus autologous bone-marrow transplantation in lymphoma patients. Lancet. 347, 353-357.
67. Heil G. et al. (1997). A randomized, double-blind, placebo-controlled, phase III study of Filgrastim in remission induction and consolidation therapy for adults with de novo acute myeloid leukemia. Blood. 90, 4710-4718.
68. Lieschke GJ, Morstyn G. (1990). Role of G-CSF and GM-CSF in the prevention of chemotherapy-induced neutropenia. In: Hematopoietic Growth Factors in Clinical Applications, R Mertelsmann and F Herrmann, eds, pages 191-223.
69. Patterson KL. et al. (1998). Safety profile of r-metHuG-CSF (Filgrastim). In: Morstyn G, Dexter TM, Foote MA (eds). Filgrastim (r-metHuG-CSF) in Clinical Practice. Marcel Dekker, Inc. New York
70. Bastion Y. et al. (1994). Possible toxicity with the association of G-CSF and bleomycin. Lancet. 343, 1221-1222.
71. Oldham RK. (1982) Biological Response Modifiers Programme and cancer chemotherapy. Int J Tissue React. 4, 173-188.

Chapter 5

Follitropin beta (Puregon)
Recombinant human follicle-stimulating hormone

Henk J. Out
N.V. Organon, P.O. Box 20, 5340 BH Oss, The Netherlands

Key words: FSH, infertility, IVF, follitropin, Puregon

Abstract: Follitropin beta (Puregon) is a human follicle stimulating hormone (FSH) produced by means of recombinant DNA technology. Chinese hamster ovary cells were transfected with genes encoding for the α- and β- subunit of FSH. The production of the amino-acid backbone and the very essential post-translational glycosylation process results in a human FSH virtually identical to natural human pituitary FSH. In clinical trials it was shown that this recombinant FSH is more effective and efficient in ovarian stimulation, eventually leading to a higher chance for a pregnancy, than the traditional FSH derived from postmenopausal urine. In this chapter, various aspects on the molecular biology, pharmacokinetics, preclinal tests, and clinical trials are addressed.

1. INTRODUCTION

Follicle stimulating hormone (FSH) is produced by the gonadotropic cells of the anterior pituitary and released into the circulation. FSH together with the luteinizing hormone (LH), which is produced by the same cells of the pituitary, controls oocyte maturation in females and spermatogenesis in males. Human FSH is used for treatment of women with normogonadotropic anovulatory disorders and in assisted reproduction technologies, such as *in-vitro* fertilization (IVF) and intracytoplasmic sperm injection (ICSI).

FSH and LH belong to a family of glycoproteins that are heterodimers, containing two non-covalently linked α- and β subunits. The subunits are encoded by separate genes. The other members of the glycoprotein hormone family are the thyroid stimulating hormone (TSH) and human chorionic gonadotrophin (hCG). Within an animal species the amino acid sequence of the α-subunits are identical, whereas the β-subunits differ and confer biological specificity on the individual gonadotropins. Both the α- and β-subunits are glycosylated. The α- and β-subunit of FSH each have two potential asparagine-linked glycosylation sites, characterized by the consensus sequence Asn-X-Ser, on positions α52, α78 and β7, β24, respectively.

After initial glycosylation in the Endoplasmic Reticulum (ER) further processing involves trimming by glucosidases and mannosidases, and remodeling of the carbohydrates in a complex series of biochemical reactions. The final structure of the Asn-linked carbohydrates on glycoproteins is dependent on the protein itself and the tissue in which it is produced. In particular the terminal residues on the carbohydrate antenna's may differ: LH carbohydrates terminate with sulfate-4-N-Acetyl-galactosamine, whereas FSH bears more highly branched sialylated structures. As a result of the extensive biochemical processing each of the glycan chains demonstrates considerable microheterogeneity resulting in numerous glycoforms, which can be resolved by isoelectric focussing. The carbohydrates on the gonadotropins serve many important functions. They are required for proper folding, assembly and secretion of the gonadotropins. Furthermore, carbohydrates are also highly relevant for the biological activity. It is well established that glycosylation determines the half-life of the gonadotropins. Furthermore, alterations in the carbohydrate structures may result in molecules with decreased ability to stimulate adenylate cyclase and steroidogenesis, but with unaffected receptor affinity. Thus, post-translational modifications, such as glycosylation is an absolute requirement for proper expression and full biological activity of the glycoprotein hormones.

FSH preparations for clinical use have traditionally been isolated from human post-menopausal urine and therefore contain some LH and other contaminating proteins from human origin. The use of such a natural source implies limited product availability and consistency. Most of the available urinary preparations contain more than 95% impurities. In the early eighties NV Organon in the Netherlands started to look for possibilites to produce FSH by means of recombinant DNA technology. Production of a recombinant FSH would have the advantage that it can be extensively purified and is not contaminated with other fertility hormones or proteins.

In this chapter, some of the most important features of follitropin beta (Org 32489, Puregon®, or Follistim®, NV Organon, Oss, The Netherlands) will be described.

2. CLONING STRATEGY AND RESULTS

Production of glycoproteins presents a particular challenge because of the need for proper post-translational modification of the protein backbone. Moreover, the gonadotropins are composed of two subunits, which are transcribed from separate genes. Although it has been attempted to recombine subunits into the intact hormone *in vitro*, this process is not very efficient. Another potential problem in this respect is that individually produced subunits may exhibit altered glycosylation. Therefore, it was an obvious choice to co-express the α- and β-subunit genes in a single mammalian cell, analogous to the co-expression of the heavy-and light immunoglobulin chains.

2.1 Host cells

In the early eighties a number of cell lines (such as mouse fibroblasts, Vero cells, Baby Hamster Kidney cells and Chinese Hamster Ovary (CHO) cells) were available which were considered as suitable hosts for secreted glycoprotein production. Main selection criteria were the possibility to transfect cells with heterologous DNA, the growth characteristics in cell culture, the glycosylation patterns observed with other recombinant glycoproteins and the potential safety hazards connected with the use of mammalian cells. In this respect the CHO cell line proved to be most promising. The CHO cell line K1 used in our experiments originates from a parental CHO cell line, biopsied from an ovary of an adult Chinese hamster, and is probably from epithelial origin. The CHO K1 cell line can be easily transfected, and transformants can be grown on a large scale. Moreover, many strong promoters to direct transcription of foreign genes are active in CHO cells. Assurance of proper glycoprotein expression and safety of the cell line was given by promising preliminary reports on the cloning and expression in CHO-cells of other glycoproteins such as human tissue Plasminogen Activator (tPA) and Erythropoietin (EPO). Thus, the CHO-K1 cell line was the obvious choice as the host cell line for the co-expression of the human α- and β-FSH subunit genes.

2.2 Gene cloning

The α-subunit gene is common to all four glycoprotein hormones and is encoded by a single gene. Cloned α-subunit cDNAs and genes of several species, including man, had been isolated previously. The ß subunits of the glycoprotein family are encoded by different genes and determine the biological activity of each hormone. At the time of the construction no reports were available on the nucleotide sequence of the human FSHß gene. On the other hand, the amino acid sequences of the FSHß protein from human, equine, porcine and ovine sources had been published. With these data as a guide, synthetic oligonucleotide probes were used for screening of cDNA and genomic libraries.

By preparing complementary DNA libraries and screening with DNA probes that covered portions of the reported amino acid sequence of βFSH, βFSH-specific clones were obtained . During the course of our work the nucleotide sequence of a human FSHβ gene was confirmed.

2.3 Construction of the expression vector and selector plasmid

Cloning vectors should enable the efficient transfer of heterologous genes into recipient cells and ensure stable inheritance of the genes, preferably in high copy number. Furthermore, expression vectors should contain all the regulatory elements, such as promoters and enhancers needed for transcription. Efficient expression vectors had been designed for this purpose but were not considered appropriate for large scale production of recombinant FSH. Alternatively, the subunit genes were transfected by physico-chemical methods into the recipient CHO cells. This integrative process was known to result in the integration of multiple copies of the expression vector. The presence of a high copy number of the cloned genes is advantageous for high expression levels.

The α- and βFSH genes were cloned into a single vector (pKMS.FSH$\alpha_g\beta_g$) in order to assure that the copy number of the α- and βFSH genes in the transfected cell is the same.

Furthermore, the vector contains a number of DNA elements that are needed for further vector assembly and efficient expression of the FSH subunit genes. The correct assembly of all the DNA elements in pKMS.FSH$\alpha_g\beta_g$ was confirmed by digestion of the vector with restriction endonucleases and determination of the size of the fragments obtained by agarose gel electrophoresis. The experimentally obtained values matched perfectly well with the theoretically deduced fragment lengths.

We directly selected for high copy number clones by co-transfection of the FSH-expression plasmid with the selector plasmid pAG60/MTIIa. This plasmid contains a neomycin resistance gene (neor) and the human metallothionein gene (MT-II$_A$). Co-transfection of the FSH expression vector together with a neomycin resistance gene allows the selection of transfectants in media containing the antibiotic geneticin (G418). The human MT-II$_A$ gene confers resistance to the toxic effects of heavy metal ions such as Cd^{2+}. Since co-transfected plasmids often integrate at the same position into the chromosome, and thus are located in the same expression environment, a high expression of MT-II$_A$ is likely to be associated with a high expression of FSH.

2.4 Transfection and selection of FSH-producing CHO cells

Transfection of CHO-K1 cells with the FSH expression plasmid pKMS.FSHα$_g$β$_g$ and the selector plasmid pAG60/MT2 (ratio of 10 :1) was carried out essentially following a calcium phosphate-DNA co-precipitation protocol. Geneticin resistant cells were subsequently grown in the presence of different concentrations of cadmium chloride (up to 10 µM). A number of pools of Cd^{2+} resistant cells indeed produced significant amounts of recombinant FSH, as determined by a FSH-specific ELISA. These pools were subjected to single cell cloning procedures. Finally, a CHO clone coded CHO.FSH.30 was selected for further characterisation of genetic stability, cell growth and FSH productivity.

2.5 Genetic stability

Genetic stability is an important requirement for cell clones to be used in a production process for a therapeutic protein. Thus, the CHO.FSH.30 line was subjected to detailed genetic analysis. Since the vectors used do not contain an origin of replication, they must integrate into the chromosome of the CHO cell in order to be stably passed to the daughter cells during growth. Proof of integration was obtained by Fluorescence *In Situ* Hybridisation (FISH). Metaphasic chromosomes and interphase nuclei of CHO.FSH.30 were hybridised with a biotin-labeled probe containing the bacterial DNA sequences present in the original expression plasmid. Hybridisation of probe and metaphasic chromosome was visualized by incubation with fluorescein-coupled avidine. This strongly indicates the integration of the vector DNA at a single position of the CHO cell genome.

Since this CHO cell proved to be stable, we decided to develop a production process for recombinant FSH.

3. PRODUCTION AND PURIFICATION OF RECOMBINANT FSH

The culturing of mammalian cells on production scale is technically much more difficult compared to microbial fermentation. This is caused by the low growth rate (usually with population doubling times in the order of 16-24 h) and the fragility of mammalian cells. Furthermore, they usually require growth factors, such as those present in fetal calf serum. Because of the complexity of sera and the potential presence of adventitious agents the use of such a component in production media should be avoided. Therefore, cells must be adapted to FSH production in serum-free media.

The CHO cell line is an anchorage-dependent cell line, which implicates that a proper surface must be provided for growth of the cells. In order to obtain a favorable surface/volume ratio cells are grown on small beads with a diameter of approx. 0.2 mm. The use of microcarriers in cell culture provides also an opportunity for easy physical separation of the cells from the culture supernatant. This enabled the development of a perfusion-type continuous culture of CHO cells. Compared to batch cultures, perfusion cultures have the advantages of a high cell concentration, due to retention of the cells, easy separation of supernatant and a short residence time preventing product degradation by proteases.

Clone CHO.FSH.30 was grown to high cell density in a continuously stirred bioreactor. The bioreactor was designed for aseptic operation and maintenance of optimal growth conditions. The perfusion of the culture was started at the end of the exponential growth phase with serum containing medium at a dilution rate of 0.5 volume/day. After increasing the perfusion rate to 1.0 volume/day a maximum cell concentration of 10^7 cells/ml was obtained. At this stage the supply medium was changed to a serum-free formulation. After a number of perfusions for removal of serum components the culture supernatant was collected and used as source for recombinant FSH. Under conditions of serum-free productions the cultures could be maintained for at least three months, without significant loss of productivity. Recombinant FSH was isolated from pooled culture supernatant by a series of chromatographic steps including anion and cation exchange chromatography, hydrophobic interaction chromatography and size exclusion chromatography. The overall recovery was approximately 50 %. The final product was stored as a lyophilized powder.

4. CHARACTERISATION: PURITY AND IDENTITY

Complex biotechnological products such as recombinant FSH can only be identified and characterised properly by the combined results of several physico-chemical methods such as gel electrophoresis, HPLC, amino acid analysis, Edman degradation analysis, peptide mapping, mass spectrometry and carbohydrate analysis. Residual DNA and CHO cell-derived proteins were analysed by hybridization and enzyme immunoassay, respectively.

4.1 Purity

The purity of recombinant FSH was determined by several complementary methods including sodium dodecyl sulphate-polyacrylamide gel electrophoresis (SDS-PAGE), western blotting, high performance size exclusion chromatography (HP-SEC), enzyme immuno assays (ELISA) and DNA hybridisation. These assays showed the absence of FSH aggregates (oligomers), no detectable amounts of free subunits, and no occurrence of CHO cell-derived proteins and culture medium components in the product. The level of contaminating DNA was less than 10 pg DNA per 500 IU.

4.2 Identity

The molecular mass of the recombinant FSH-dimer as determined by SDS-PAGE was found to be 40 to 45 KDa and the α- and ß-subunits migrated at a molecular mass of approximately 25 to 30 KDa. These masses correspond well with the calculated masses of the protein plus oligosaccharides. The molecular size was determined by size exclusion chromatography and was of approximately 45 KDa.

Peptide mapping is routinely used to compare the protein structure of a product to that of a reference material or to those of previous lots to confirm the primary structure and to ensure lot-to-lot consistency of protein structure. It also supports the genetic stability of FSH genes in CHO cells during culture. Peptide mapping of recombinant FSH is accomplished by reducing and modifying all disulfide bonds in each of the two subunits. This is followed by cleaving the subunits into a number of small fragments by digestion with the endoproteinase LysC, which cleaves the protein at specific lysine residues. Peptide mapping in combination with amino acid analysis, Edman degradation and ES MS has been used to verify the entire primary structure of recombinant FSH, which was shown to be in full agreement with the cDNA-derived sequence. In addition, it has provided additional information on the positions of the glycosylation sites *viz*. in the α-subunit at positions 52 and 78 and in the ß-subunit at positions 7 and 24 "gaps" were sequenced where asparagine was expected. This is indicative of the attachment of carbohydrate side chains. Consequently, this technique is particularly important in the quality control of recombinant FSH.

Oligosaccharide sequencing of recombinant FSH has been performed and showed only minor differences in the structure of carbohydrate antenna's between recombinant and pituitary FSH, most notably the absence of intersecting N-acetyl glucosamine moieties in recombinant FSH. Moreover, N-acetyl neuraminic acid is linked in an α2-3 conformation while in natural FSH both α2-3 and α1-6 occur.

5. SUMMARY OVERVIEW OF THE DEVELOPMENTAL HISTORY OF THE DRUG

5.1 Toxicology

Toxicity was tested in rats by means of single-dose administration of high doses (up to 2500 IU/kg) and by multiple-dose IM administration of doses up to 500 IU/kg/day during 14 days. No drug-related mortalities or any other drug-related effects with respect to clinical signs, body weight, food and water consumption were observed. It was concluded that no toxic effects were observed with doses up to 100-fold in excess of the anticipated maximum daily human dose (5 IU/kg). In addition, follitropin beta displayed no mutagenic potential.

5.2 Pharmacology

5.2.1 *In vitro* studies

Follitropin beta displayed good receptor-binding affinity and compared well with classical urinary preparations in *in vitro* studies. Aromatase induction in Sertoli and granulosa cells was comparable to urinary preparations, both in terms of dose-dependency and maximum responses obtained. Structural and functional similarity between the active substance in Follitropin beta and urinary preparations was confirmed by a similar pattern of inhibition of induced aromatase activity by specific monoclonal antibodies. No relevant intrinsic LH bioactivity was measured in a Leydig cell assay. The isohormone distribution of recombinant FSH in Puregon ranged between pI values of 5.7 and 3.2, and was more basic than that of urinary FSH.

5.2.2 *In vivo* studies

In the standard ovarian weight augmentation assay, Follitropin beta displayed a specific bioactivity of about 10,000 IU/mg protein. In hypophysectomized rats, Follitropin beta was effective in increasing ovarian weight and aromatase activity, without increasing plasma estradiol levels. Follitropin beta caused a gradual shift from small antral to large preovulatory follicles and reduced the number of atretic follicles in a dose-dependent way. Co-administration of hCG was required to make Follitropin beta effective in the elevation of plasma estradiol levels. The capability of Follitropin beta to increase ovarian weight and plasma estradiol levels compared well to those of urinary preparations, when it was supplemented with the proper quantities of hCG.

5.3   Pharmacokinetics

5.3.1   Introduction

With respect to pharmacokinetics, FSH differs from classical chemical substances in two aspects:
1. A given dose is expressed in terms of *in vivo* bioactivity as determined in the rat Steelman-Pohley assay.
2. On the basis of their heterogeneous isohormone character, gonadotropins can not be considered as single-component drugs.

Natural gonadotropins display so-called microheterogeneity because they occur in various isoforms. This is due to differences in carbohydrate chain structure especially in the degree of sialylation.

Isohormones can be separated by chromatofocusing or electrofocusing techniques on the basis of differences in isoelectric points (pI). Acidic isohormones combine relatively low receptor- binding affinity and intrinsic bioactivity with a long plasma residence time, whereas the basic isoforms display relatively high receptor binding and intrinsic bioactivity together with a short plasma residence time. It was shown that human pituitary FSH could be separated into at least 20 isohormone fractions, which displayed 7 discrete levels of FSH receptor-binding activities.

In a comparative study of 13 batches of recombinant FSH and 10 batches of urinary FSH, it was shown that recombinant FSH contains an approximately twofold higher proportion of basic (pI>4.7) isoforms (32.0% vs 17.0%) and a twofold lower proportion of acidic (pI<4.1) isoforms (14.7% vs 31.4%). In *in vitro* assays it was shown that recombinant FSH had a higher potency than urinary FSH.

Differences in isohormone composition especially in sialylation, have a direct effect on the kinetics of gonadotropin preparations and may therefore influence their bioactivity.

Gonadotropins can be quantified with four essentially different types of assays, all having their own specific merits, ie immunoassays, receptor-

binding assays, *in vitro* bioassays and *in vivo* bioassays. These assays measure four different basic characteristics of gonadotropin molecules.
1. Immunoassays measure a structural feature of a gonadotropin molecule. It is general belief that immunoassays provide a 'relative' measure for the mass of gonadotropins. In other words, immunoassays measure the number of molecules present.
2. Receptor-binding assays provide information on the proper conformation for receptor binding.
3. *In vitro* bioassays measure, in contrast to the two previous assays, a functional aspect of gonadotropins, namely their intrinsic biological activity in terms of second messenger activation and subsequent steroid biosynthesis.
4. *In vivo* assays measure the overall bioactivity of gonadotropins. This *in vivo* bioactivity is determined by the number of molecules injected, the pharmacokinetic behaviour of these molecules, their receptor-binding affinity and intrinsic bioactivity.

5.3.2 Single-dose studies

A single-dose of 300 IU recombinant FSH was administered intramuscularly (buttock) to female gonadotropin-deficient, but otherwise healthy subjects. After a wash-out period of at least two weeks, patients from this group received a single intramuscular injection of 300 IU urinary FSH.

The extent of absorption of immunoreactive FSH was significantly higher for urinary FSH than for recombinant FSH (C_{max} and AUC both about 65% of those after urinary FSH). However, serum bioactive FSH (as determined in the *in vitro* bioassay at 6, 24, and 72 hours after drug administration) indicated that the circulating intrinsic FSH bioactivity was higher after recombinant FSH than after urinary FSH injection. Apparently, administration of recombinant FSH leads to low FSH immunolevels combined with high FSH bioactive levels.

In an open randomized three-way cross-over study, the absolute bioavailability was assessed after single-dose intramuscular (buttock) and subcutaneous (abdominal wall) administration of 300 IU recombinant FSH to healthy female subjects, pituitary suppressed with a high-dose oral contraceptive. Based on serum immunoreactive FSH, C_{max}, t_{max}, area under the serum-level-vs-time curve up to 312 hours after dosing (AUC_{0-312}) and the absolute bioavailability were calculated.

With respect to the extent of absorption (AUC) of immunoreactive FSH, intramuscular and subcutaneous administration proved to be equivalent. For both routes the absolute bioavailability was found to be about 77%. Although the means were almost identical, no bioequivalence could be proven with respect to C_{max} and t_{max}. For the latter, this may have been caused by the high intrasubject variability.

These data indicate that single-dose intramuscular and subcutaneous administration lead to highly similar immunoreactive FSH serum concentration curves.

5.3.3 Multiple-dose studies

The pharmacokinetics of recombinant FSH after repeated administration was investigated in three open group-comparative studies:
- in weekly rising doses of 75, 150, and 225 IU, in gonadotropin-deficient but otherwise healthy volunteers (seven females) and
- in four groups of healthy female volunteers, whose endogenous gonadotropin production was suppressed by a high-dose contraceptive pill (2.5 mg lynestrenol + 50 µg ethinylestradiol) who received 75, 150, or 225 IU recombinant FSH, or 150 IU urinary FSH intramuscularly. Nine patients per group were recruited.
- and, similarly, 75, 150, or 225 IU recombinant FSH given subcutaneously, compared to recombinant FSH 150 IU given intramuscularly. In all groups, 12 volunteers were recruited.

Relatively high concentrations of FSH were reached within 12 hours of administration. Steady state was reached after four daily doses. Due to the relatively long elimination half-life, serum concentrations of FSH at steady state are higher than after single administration. Based on C_{min} (FSH concentration just prior to each dosing), a cumulation factor of approximately 1.5-2.5 can be estimated.

Bioequivalence in bioavailability of immunoreactive FSH (reflected by AUD and C_{max}) after repeated administration of 150 IU recombinant FSH and urinary FSH could not be proven. With respect to $t_{1/2}$ and Cl/kg, recombinant FSH and urinary FSH proved to be bioequivalent.

In the multiple rising dose study, growth of one or more ovarian follicles was observed in six out of seven women, and in accordance with number and size of follicles, median serum inhibin concentrations rose from 30 U/l (at baseline) to 581 U/l. LH and androgen concentrations remained very low throughout treatment. Estradiol concentrations only showed minor rises, due to insufficient LH-induced androgen production.

In the pituitary-suppressed volunteers (by means of the high oral contraceptive), treatment with 75 IU recombinant FSH for seven days was insufficient to induce significant follicular growth; daily doses of 150 and 225 IU recombinant FSH induced a clear and roughly equal response (as expressed by the total number of follicles per size). In the pharmacodynamic comparison of seven days of treatment with 150 IU of recombinant FSH and urinary FSH, recombinant FSH treatment seemed to be more effective, in that it induced more follicles with a diameter > 8mm.

5.4 Clinical trials

5.4.1 Controlled ovarian hyperstimulation

In order to reliably assess the safety and efficacy of Organon's recombinant FSH, a large-scale clinical trial program in IVF was organised. This included studies which were all prospective and randomised. Most of them were multicentre. Comparative drugs involved hMG and urinary FSH, and various GnRH-agonists were applied.

For technical reasons these studies were not double-blind, since recombinant FSH was provided in vials and urinary FSH in ampoules. Instead, the studies were conducted *assessor*-blind, implying that the person making decisions upon treatment regimens was not aware of the medication used.

Statistical analysis took into account centre effects in multicentre studies, by weighing the standard error of the treatment difference of the two comparative drugs. The higher the standard error (e.g. because of a small number of subjects recruited) the smaller the weight of that centre in the analysis. Ninety-five percent confidence intervals of the estimated treatment difference were calculated enabling investigators to assess whether clinically relevant differences were detected.

All analyses were carried out on an intent-to-treat basis, including all subjects who received at least one dose of FSH. The main advantages of this rule are that more patients are available for final analysis of efficacy and that it more closely reflects how physicians evaluate a therapeutic agent in the clinical setting, outside an experimental control.

5.4.1.1 Pilot efficacy study in IVF

In an open, pilot efficacy study the effects of recombinant FSH in assisted reproduction were investigated. The approach used various down-regulation protocols to assess whether the administration of an FSH-only preparation in these circumstances would be sufficient for follicular development and adequate steroidogenesis.

Follitropin beta (Puregon)

In total, 51 infertile women were treated with recombinant FSH alone (group I), or with recombinant FSH in conjunction with buserelin intranasal spray 4 x 150 µg daily in a short protocol (group II) or in a long protocol (group III), or using triptorelin in a long protocol, giving a single dose of 3.75 mg intramuscularly (group IV) or daily subcutaneous injection of 200 µg (group V).

In all women, treatment with recombinant FSH resulted in multiple follicular growth and rises of serum inhibin and estradiol. The latter indicated that the amount of remaining LH was sufficient to support FSH-induced estrogen biosynthesis (see Table 13). Eight ongoing pregnancies were achieved resulting in the birth of nine healthy children.

Table 13. Median serum LH and estradiol levels in each group at baseline starting Puregon treatment and on the day of hCG administration

		LH (IU/l)	E_2 (pmol/l)
Group I: recFSH alone	baseline	6.7	143
	day hCG	5.1	4042
Group II: buserelin short	baseline	4.9	121
	day hCG	2.3	6971
Group III: buserelin long	baseline	3.2	121
	day hCG	1.3	6509
Group IV: triptorelin im	baseline	2.8	99
	day hCG	1.2	6490
Group V: triptorelin sc	baseline	2.4	88
	day hCG	1.6	5620

5.4.1.2 Pivotal trial

The pivotal trial in the program encompassed 1000 cycles. This study was set up in order to detect even small differences between both groups. The primary endpoints, as defined prior to starting the study, were the number of oocytes retrieved and the ongoing pregnancy rate, defined as the maintenance of pregnancy at least 12 weeks after embryo transfer. With a randomisation in a 3:2 ratio between recombinant and urinary FSH, 80% power and a two-sided significance level of 5%, a difference of 1.2 oocytes retrieved (assuming SD=6) and 6% in pregnancy rates could be detected.

The trial was designed as a prospective, randomised, assessor-blind, multicenter study. Eighteen centres from eleven European countries participated.

Selection criteria were age between 18 and 39 years, good physical health, normal weight, at least one year of infertility, no male factor, no endocrine abnormalities, and normal regular ovulatory cycles. The protocol included intranasal buserelin down-regulation in a long protocol; 150 or 225 IU for the first four days after which the dose was adapted according to ovarian response; hCG administration when at least 3 follicles ≥ 17 mm were seen; and a maximum replacement per transfer of three embryos. The results are given in Table 14.

Table 14. Mean results of recombinant vs urinary FSH

Parameter	recFSH	urinary FSH	95% CI of treatment difference	p-value
Number of subjects treated	585	396	not applicable	not applicable
Total number of oocytes retrieved	10.8	9.0	1.2 to 2.6	p<0.0001
Number of mature oocytes retrieved	8.6	6.8	1.1 to 2.4	p<0.0001
Total FSH dose (IU)	2138	2385	-338 to -158	p<0.0001
Duration of treatment (days)	10.7	11.3	-0.9 to -0.3	p<0.0001
Number of follicles ≥17 mm	4.6	4.4	-0.0 to 0.5	p=0.09
Number of follicles ≥15 mm	7.5	6.7	0.4 to 1.2	p=0.0002
Maximum serum estradiol (pmol/l)	6084	5179	494 to 1317	p<0.0001
Number of high quality embryos	3.1	2.6	0.2 to 0.8	p=0.003
Ongoing pregnancy rate per attempt	22.2%	18.2%	-1.1 to 9.0	p=0.13
Ongoing pregnancy rate per transfer	26.0%	22.0%	-1.9 to 9.8	p=0.19
Ongoing pregnancy rate per attempt including frozen embryo cycles	25.6%	20.4%	0.0 to 10.6	p=0.05

CI = confidence interval

The main efficacy parameter, i.e. the number of oocytes retrieved was consistently higher after recombinant FSH treatment in all 18 participating centres (see Figure 15).

Follitropin beta (Puregon)

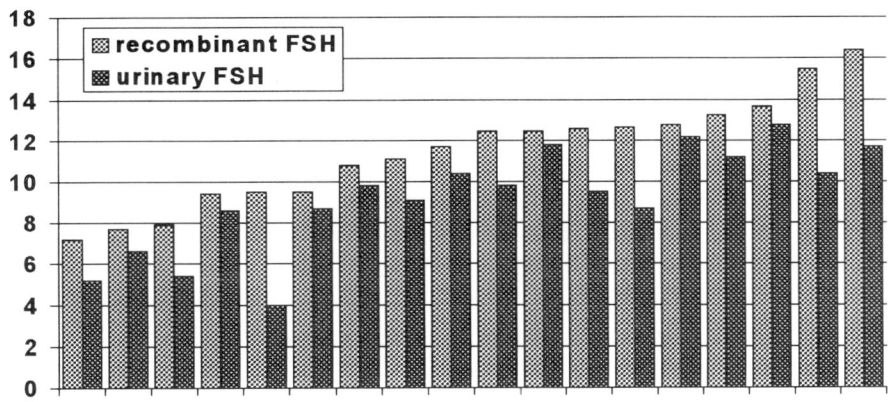

Figure 15. Number of oocytes retrieved in 18 centres participating in a randomized trial comparing recombinant and urinary FSH in IVF

The incidence of OHSS leading to hospitalisation was seen in 19 out of 585 recombinant FSH-treated subjects (3.2%) *vs* 8 out of 396 urinary FSH-treated subjects (2.0%), which was not significantly different.

In conclusion, this study which was the largest prospective randomised clinical trial ever performed in IVF demonstrated a significantly higher number of oocytes, embryos and ongoing pregnancies (efficacy endpoints), using a lower total dose during a shorter treatment period (efficiency endpoints) with a similar incidence of OHSS (safety endpoint) for recombinant FSH when compared with urinary FSH.

5.4.1.3 Supportive trials

In a multicentre (n=6) French study comparing recombinant and urinary FSH with 100 µg daily subcutaneous administration of triptorelin in a long protocol, 90 subjects were randomised and treated in a 3:2 ratio between recombinant and urinary FSH. No significant differences were found.

In a randomised single-centre study recombinant FSH and hMG were compared in non-pituitary-suppressed women. Eighty-nine women were treated. Recombinant FSH compared favourably to hMG with respect to the main efficacy parameters (i.e. number of oocytes retrieved, 11.2 vs 8.3 and ongoing pregnancy rates per attempt and per transfer, 22.2% vs 17.1% and 30.8% vs 22.2%, respectively). However, none of the differences in this study was significant.

In a randomised study, efficacy and local side-effects of recombinant FSH administered either SC or IM were investigated in 118 and 77 subjects, respectively. Local side-effects were assessed daily by the patients on the presence (mild, moderate, severe) of swelling, itching, redness, pain, and bruising at the injection site. There were no significant differences in efficacy, efficiency and overall safety endpoints. The only significant difference detected was the occurrence of bruising, which was seen more often ($p=0.019$) in the SC group. This was attributed to the more superficial injection of the SC route, enabling better inspection of the occurrence of bruising as compared with the IM route.

When the three recombinant FSH versus urinary gonadotropins studies are combined, an combined analysis is possible. In this way, the whole clinical trial programme of recombinant FSH is considered to be one large multicentre trial (25 centres).

In this analysis, ongoing pregnancy rates directly following the fresh embryo transfer are significantly higher after recombinant FSH treatment as compared to urinary gonadotropins (22.9% vs 17.9%, $p=0.04$). When the results of the cryoprogramme were included, this difference was even more significant (26.3 vs 19.9%, $p=0.01$).

5.4.2 Ovulation in clomiphene resistant normogonadotropic anovulatory women

The safety and efficacy of recombinant FSH in comparison with urinary FSH was assessed in a prospective, randomised, assessor-blind, multicentre study in patients with clomiphene-resistant chronic anovulation (WHO group II). Twelve centers in 9 different European countries participated in this study.

Selection criteria included age between 18 and 39 years, positive progestagen induced withdrawal bleeding or spontaneous menstrual bleeding, clomiphene resistance, normal levels of FSH, prolactin, TSH and DHEAS, no excessively high levels of androgens, no male factor, at least one open Fallopian tube, and a body mass index between 19 and 32 kg/m^2. Clomiphene-resistance was defined as failure to ovulate during three cycles, or to conceive during six cycles with clomiphene treatment. At cycle day 3, a low-dose step-up dose regimen was started (initial dose 75 IU/day intramuscularly) including a fixed-dose for the first two weeks in the first treatment cycle and, when needed, weekly upward adjustments by half an ampoule. In the second and third cycles, upward adjustments were already allowed in the second treatment week. Randomisation was in a 3:2 ratio between recombinant FSH and urinary FSH. Ovulation was triggered with intramuscular administration of 10 000 IU hCG. hCG was given when one follicle $\geq$ 18 mm, or two or three follicles $\geq$ 15 mm were seen on ultrasound. The cycle was cancelled when more than three follicles $\geq$ 15 mm had developed, or when no ovarian response after 42 days of treatment was noticed. Ovulation was confirmed by a midluteal serum progesterone concentration of 25 nmol/l or more on at least one occasion. Measurements of serum endocrinological parameters were carried out at the local hospital according to local standards. Predefined primary efficacy endpoint was the cumulative ovulation rate, i.e. the chance to ovulate within three cycles using life-table analysis (Kaplan-Meier).

A total of 178 subjects (recombinant FSH: n=109, urinary FSH: n=69) were randomised. One hundred seventy-two subjects (recombinant FSH: n=105, urinary FSH: n=67) were treated in the first treatment cycle, 111 (recombinant FSH: n=69, urinary FSH: n=42) subjects in the second, and 78 subjects (recombinant FSH: n=49, urinary FSH: n=29) in the third treatment cycle.

The cumulative ovulation rates did not differ significantly between both treatment groups and were 95% and 96% for recombinant FSH and urinary FSH, respectively. Taking all cycles together, 155 out of 223 recombinant FSH cycles were ovulatory (69.5%) compared to 92 out of 138 urinary FSH cycles (66.7%). Cumulative pregnancy rates were 27% in the recombinant FSH group, compared to 24% in the urinary FSH group. The miscarriage rates were 31% and 32%, respectively, in the recombinant FSH and urinary FSH groups. The higher efficiency of recombinant FSH was demonstrated by a significantly shorter treatment duration to reach ovulation: a median of 10 days (range 4-27) compared to 13 days (range 4-49) after urinary FSH treatment ($p<0.001$). Correspondingly, the total dose used in the recombinant FSH group was 750 IU (range 300-2738) and 1035 IU (range 300-7350) in the urinary FSH group ($p<0.001$).

The higher activity of recombinant FSH was also demonstrated by a significantly increased number of follicles > 12 mm and serum estradiol levels. Serum FSH levels were significantly lower after recombinant FSH treatment as compared with urinary FSH. The higher potency of recombinant FSH was not correlated with an increased incidence of cycle cancellation due to too many follicles and (or) too high serum estradiol levels. Also, multiple gestation rates were low and similar when compared to urinary FSH. In both groups one twin (4.2% and 7.7% in the recombinant FSH and urinary FSH group, respectively) was seen and in the recombinant FSH-group only one triplet (4.2%). All cases of OHSS were mild or moderate in severity. In only one case (recombinant FSH), the occurrence of OHSS led to hospitalisation.

5.5 Discussion of findings in the development

The clinical trial programme with recombinant FSH quite clearly showed a higher bioactivity of this compound as compared to the traditional urinary preparations.

Although the content of ampoules of gonadotropins is standardised based on *in vivo* bioactivity, apparently this *in vivo* assay is not able to reliably predict the FSH activity in the human. *In vivo* bioactivity units of recombinant FSH as determined in the rat Steelman-Pohley assay have a higher activity in the human than corresponding units of urinary FSH. This is not surprising since *in vivo* bioactivity is measured as the weight augmentation of ovaries from rats treated with FSH and a surplus of hCG in a comparative way with an international reference preparation. Microscopical examination of these ovaries reveals many luteinized unruptured follicles which contain large volumes of follicular fluid which contribute disproportionally to the weight augmentation.

It is unlikely that the known relatively large inter- and intra-assay variation of the Steelman-Pohley assay leading to co-incidental differences in the number of international units per ampoule, account for the differences found. Pharmaceutical companies can easily overcome this problem by multiple analyses and careful standardisation of their procedures.

The reason for the higher bioactivity of recombinant FSH is unknown. It might be speculated that the different isohormone profile of recombinant FSH, i.e. more basic isoforms as compared to urinary FSH accounts for a higher receptor binding and accordingly enhanced signal transduction. However, this has not been proven and it might also be that small differences at the carbohydrate level or the pharmaceutical formulation explain the higher efficacy and efficiency of recombinant FSH. In addition, FSH inhibiting substances might be present in the fraction of contaminating proteins in urinary FSH preparation used, which accounts for more than 95% of the preparation. Further research is needed to elucidate the precise role of these factors in the comparison of gonadotropin activities.

A number of reasons can be hypothesised as cause for the increased pregnancy rate in IVF after recombinant FSH treatment. First, the higher pregnancy rates directly following the fresh transfer suggests a higher embryo quality after recombinant FSH treatment. This may be related to an increased number of available embryos enabling the embryologist to replace the morphologically best embryos. It is believed that embryo morphology correlates well with the chance for implantation and therefore pregnancy. However, one can not exclude the possibility that oocyte quality and therefore embryo quality are influenced by the type of gonadotropin preparation used. It has been shown in *in vitro* experiments with mammalian oocytes that meiotic progression, polar body emission, cumulus-oocyte interactions, and oocyte cytoskeletal organisation are influenced by the presence or absence of gonadotropins in the culture medium. One can speculate that the different nature of follitropin beta compared to traditional urinary preparations as manifested in the absence of impurities and a relatively more basic isohormone profile may influence the ability of the embryo to implant. Second, a quantitative advantage in the number of oocytes retrieved and embryos obtained after follitropin beta treatment leading to a surplus of embryos stored in the freezer will ultimately lead to more pregnancies because of the higher availability of embryos that can be replaced in natural cycles. This will increase the chance for a pregnancy per stimulation cycle and decrease the gonadotropin consumption per cumulative IVF treatment. Therefore, the treatment difference in ongoing pregnancy rates was even more in favour of recombinant FSH when the results of the cryoprogramme were included ($p=0.01$).

It seems unlikely that differences in endometrial development at the time of the fresh embryo transfer account for the higher pregnancy rates obtained with follitropin beta, because these rates were also higher after frozen-embryo replacements in natural cycles.

No antibodies against FSH or Chinese hamster ovary cell-derived proteins were detected. Also no clinically relevant changes in biochemical or haematological parameters were found or significant trends in vital signs, as assessed before and after treatment. The subcutaneous route enabling self-administration was well-tolerated as prospectively assessed in comparison with the intramuscular route.

6. QUALITY CONTROL

In contrast to the impure urinary gonadotropins, it now is possible to control the whole production process and the contents of an ampoule. Numerous quality control tests are performed on the master cell bank, the master working cell bank and postproduction cells (See Table 15).

Table 15. Quality control test of follitropin beta

	MCB	MWCB	PPC
Tumourgenicity			+
Virology cell culture	+		+
Virology animals/eggs	+		+
Map test	+		
XC plaque assay	+		
S + L-focuts assay	+		
Reversed transcriptase	+		+
Mycoplasma	+	+	+
Bovine viruses	+		+
Transmission EM level I			+
Transmission EM level II	+		
Karyology	+		
Microbiological contamination	+		
DNA fingerprinting	+		+

MCB = master cell bank, MWCB = master working cell bank, PPC = postproduction cells (extended)

7. FINAL PRODUCT FORMULATION AND FORMAT

Recombinant FSH distinghuishes itself from urinary FSH by its high purity. Next to its apparent advantages it also poses the pharmaceutical scientist with a problem: How can it be effectively administered? The problems arise because pure proteins are relatively unstable and difficult to process. The physical and chemical characteristics of a protein hormone are determined by its complex three-dimensional structure. Many chemical and physical processes, including aggregration, adsorption, hydrolysis, deamidation, or oxidation, can alter this structure. Since these deactivation processes occur more rapidly in an aqueous environments than in dry ones, most protein products are freeze-dried. However, the freeze-drying process itself (lyophilization) is often found to cause serious damage to the product. For this reason the proteins must be protected during the lyophilization stage. Urinary FSH consist of 98% foreign proteins. These proteins play a major role in stabilizing the active material. Recombinant FSH, however, does not have these protective impurities and the activity yield after conventional freeze drying is significantly less than 100 per cent. To complicate matters, many pure proteins adsorb to glass and polymers. As a result, losses through adsorption are often dramatic.

The goal therefore was to develop a formulation which (a) guarantees full activity recovery after manufacturing, (b) exhibits optimum stability, and (c) prevents the loss of protein through adsorption. Obviously, the use of an additional protein such as HSA or gelatine was not possible. Although this could lead to a stable product, the goal was to develop a formulation which allowed monitoring of the product's quality. This is not possible for urinary FSH products: the exact composition of the foreign proteins is not known and the proteins interfere with the monitoring of the quality of the FSH.

The answer to this problem was the development of the lyosphere. The lyosphere is a freeze-dried spherical presentation of FSH. Lyosphere technology makes use of specially designed equipment. The procedure together with the formulation have yielded a product which can be manufactured without activity loss and which exhibits superior stability characteristics. Moreover, no foreign proteins are present, enabling extensive monitoring of the product's quality.

8. CONCLUSION

The availability of follitropin beta (recombinant FSH) has a number of distict advantages:

- Due to the recombinant process where culture condition are kept very constant, a high batch-to-batch consistency can be expected.

- In principle, there is an unlimited source of human FSH available. Shortages of urinary FSH were experienced in the last 10 years due to unsufficient availability of urinary sources.

- The human recombinant FSH produced by the Chinese hamster ovary cells is very similar to pituitary FSH as produced by the woman during her reproductive years, in contrast to urinary FSH which is derived from (the urine of) postmenopausal women.

- Recombinant FSH is characterized by a more relatively basic isohormone profile compared to urinary FSH. Since various isohormones have different biological activities, this is pharmacodynamically important.

- The different isohormone profile is reflected in different pharmacokinetics for recombinant FSH.

- In clinical trials, it was consistently shown that follitropin beta had a higher bioactivity.

- This did not only result in a higher pharmacodynamic response (more follicles, more oocytes), but also in a quantitative advantage (more pregnancies).

- The high purity of the compound enables subcutaneous administration.

BIOGRAPHY

Henk J Out, MD PhD, is a clinical scientist who works as an International Medical Adviser on infertility for NV Organon.

NV Organon develops and produces pharmaceutical products in fields such as gynaecology, psychiatry, athero-thrombosis, and auto-immune diseases. Major product groups are oral contraceptives, infertility treatments, and preparations for menopausal complaints, depression, and psychosis. The company employs approximately 9,000 people worldwide. Annually, between 15 and 20 per cent of Organon's sales income is invested in its drug discovery and development programmes. NV Organon is one of the pharmaceutical business units of Akzo Nobel. Akzo Nobel, headquartered in Arnhem, the Netherlands, is a market-driven and technology-based company, serving customers throughout the world with healthcare products, coatings, chemicals, and fibers. The company employs 69,000 people and has activities in more than 60 countries. In 1997, consolidated sales aggregrated NLG 24.1 billion.

REFERENCES

All data described in this Chapter have been published previously in numerous reports. An extensive list of references can be obtained from the author.

The following literature on follitropin beta is especially recommended:

Preclinical
1. Hård, K. *et al.* (1990). Isolation and structure determination of the intact sialylated N-linked carbohydrate chains of recombinant human follitropin expressed in Chinese hamster ovary cells. Eur. J. Biochem., 193, 263-271.
2. Mannaerts, B. *et al.* (1991). Comparative *in vitro* and *in vivo* studies on the biological characteristics of recombinant human follicle stimulating hormone. Endocrinology 1991, 129, 2623-2629.
3. Matikainen, T. *et al.* (1994). Circulating bioactive and immunoreactive recombinant follicle stimulating hormone (Org 32489) after administration to gonadotropin-deficient subjects. Fertil. Steril., 61, 62-69.
4. Mannaerts, B. *et al.* (1994). Folliculogenesis in hypophysectomized rats after treatment with recombinant human follicle-stimulating hormone. Biol. Reprod.,51, 72-81.
5. Lambert,A. *et al.* (1995). *In-vitro* potency and glycoform distribution of recombinant human follicle stimulating hormone (Org 32489), Metrodin and Metrodin-HP. Molec. Hum. Reprod., 10, 1928-1935.
6. De Leeuw, R. *et al.* (1996). Structure-function relationship of recombinant FSH (Org 32489). Molec. Hum. Reprod., 2, 361-369.
7. Olijve, W. *et al.* (1996). Molecular biology and biochemistry of human recombinant follicle stimulating hormone (Puregon). Molec. Hum. Reprod., 2, 371-382.

8. Harris, S.D. et al. (1996). Internal carbohydrate complexity of the oligosaccharide chains of recombinant human follicle-stimulating hormone (Puregon, Org 32489): a comparison with Metrodin and Metrodin-HP. Molec. Hum. Reprod., 2, 807-811.

Clinical

9. Schoot, D.C. et al. (1992). Human recombinant follicle-stimulating hormone induces growth of preovulatory follicles without concomitant increase in androgen and estrogen biosynthesis in a woman with isolated gonadotropin deficiency. J. Clin. Endocrinol. Metab., 74, 1471-1473.
10. Devroey, P. et al. (1992). Successful *in-vitro* fertilisation and embryo transfer after treatment with recombinant human FSH. Lancet, 339, 1170-1171.
11. Mannaerts, B. et al. (1993). Single-dose pharmacokinetics and pharmacodynamics of recombinant human follicle-stimulating hormone (Org 32489) in gonadotropin-deficient volunteers. Fertil. Steril., 59, 108-114.
12. Devroey, P. et al. (1994). Clinical outcome of a pilot efficacy study on recombinant human follicle-stimulating hormone (Org 32489) combined with various gonadotrophin-releasing hormone agonist regimens. Hum. Reprod., 9, 1064-1069.
13. Schoot, D.C. et al. (1994). Recombinant human follicle-stimulating hormone and ovarian response in gonadotrophin-deficient women. Hum. Reprod., 9, 1237-1242.
14. Dessel, van H.J.H.M. et al. (1994). First established pregnancy and birth after induction of ovulation with recombinant human follicle stimulating hormone in polycystic ovary syndrome. Hum. Reprod., 9, 55-56.
15. Kliesch, S. et al. (1995). Recombinant human follicle-stimulating hormone and human chorionic gonadotropin for induction of spermatogenesis in a hypogonadotropic male. Fertil. Steril., 63, 1326-1328.
16. Out, H.J. et al. (1995). A prospective, randomized, assessor-blind, multicentre study comparing recombinant and urinary follicle-stimulating hormone (Puregon *vs* Metrodin) in *in-vitro* fertilization. Hum. Reprod., 10, 2534-2540.
17. Hedon, B. et al. (1995). Efficacy and safety of recombinant FSH (Puregon) in infertile women pituitary-suppressed with triptorelin undergoing *in-vitro* fertilisation: A prospective, randomised, assessor-blind, multicentre trial. Hum. Reprod., 10, 3102-3106.
18. Mannaerts, B. et al. (1996). Serum hormone concentrations during treatment with multiple rising doses of recombinant follicle stimulating hormone (Puregon) in men with hypogonadotropic hypogonadism. Fertil. Steril., 65, 406-410.
19. Geurts, T.B.P. et al. (1996). Puregon - (Org 32489) - recombinant human follicle-stimulating hormone. Drugs of Today, 32, 239-258.
20. Mannaerts, B.M.J.L. et al. (1996). Clinical profiling of recombinant follicle stimulating hormone (rFSH; Puregon): relationship between serum FSH and efficacy. Hum. Reprod. Update, 2, 153-161.
21. Out, H.J. et al. (1996). Recombinant follicle stimulating hormone (rFSH; Puregon) in assisted reproduction: More oocytes, more pregnancies. Results from five comparative studies. Hum. Reprod. Update, 2, 162-171.
22. Mitchell, R. et al. (1996). Oestradiol and immunoreactive inhibin-like secretory patterns following controlled ovarian hyperstimulation with urinary (Metrodin) or recombinant follicle stimulating hormone (Puregon). Hum. Reprod., 11, 962-967.
23. Ubaldi, F. et al. (1996). Premature luteinization in *in vitro* fertilization cycles using gonadotropin-releasing hormone agonist (GnRH-a) and recombinant follicle-stimulating hormone (FSH) and GnRH-a and urinary FSH. Fertil. Steril., 66, 275-280.

24. Albano, C. et al. (1996). Pregnancy and birth in an *in-vitro* fertilization cycle after controlled ovarian hyperstimulation in a woman with a history of allergic reaction to human menopausal gonadotrophin. Hum. Reprod., 11, 1632-1634.
25. Out, H.J. et al. (1997). A prospective, randomized, study to assess the tolerance and efficacy of intramuscular and subcutaneous administration of recombinant follicle-stimulating hormone (Puregon). Fertil. Steril., 67, 278-283.
26. Out, H.J. et al. (1997). Recombinant follicle-stimulating hormone (follitropin beta, Puregon) yields higher pregnancy rates in *in vitro* fertilization than urinary gonadotropins. Fertil. Steril., 68 138-142.
27. Jones, H.W. Jr. et al. (1997). Cryopreservation: The practicalities of evaluation. Hum. Reprod., 12, 1522-1524.
28. Coelingh Bennink, H.J.T. et al. for the European Puregon Collaborative Anovulation Study Group (1998). Recombinant FSH (Puregon) is more efficient than urinary FSH (Metrodin) in clomiphene-resistant normogonadotropic chronic anovulatory women: A prospective, multicenter, assessor-blind, randomised, clinical trial. Fertil. Steril., 69, 19-25.

Chapter 6

Insulin Lispro (Humalog)

Ronald E. Chance, N. Bradly Glazer and Kathleen L. Wishner
Eli Lilly and Company, Indianapolis, USA

Key words: Insulin, insulin analogues, insulin lispro, lispro, Humalog, diabetes, diabetes mellitus, rapid acting insulin.

Abstract: Initially discovered in 1921, insulin was first made commercially available in 1923. Up until the early 1980s, all insulin preparations used medically were obtained by direct extraction from the pancreatic tissue of animals. In 1982, Humulin ® (recombinant human insulin) became the first recombinant therapeutic product to gain marketing approval. By the mid-1980s, efforts to develop insulin analogues displaying improved therapeutic properties were well underway. Insulin LISPRO (Humalog(R)) is such an analogue which has gained regulatory approval for general medical use. It is identical to human insulin except that the Pro-Lys amino acid sequence at positions B28 and B29 of the native molecule are reversed.

Insulin lispro has a more rapid onset of activity and a shorter duration of action when compared to regular human insulin while maintaining equal glucose lowering ability. Insulin lispro provides better postprandial glucose control at a more convenient time relative to consumption of a meal.

1. INTRODUCTION

Diabetes mellitus was identified more than 2000 years ago. Until relatively recent times, the medical community was faced with diagnosing a fatal disease without any effective treatment options. Banting, Best and Macleod changed this grim prognosis in 1921 with the discovery of insulin (1, 2). Lauded by many as one of the most miraculous medical achievements of the twentieth century (3), the discovery of insulin earned Banting and Macleod the Nobel Prize in Physiology or Medicine in 1923, one of the fastest recognitions of a medical discovery in its history (4). Frederick G. Banting shared his prize money with his collaborator and co-worker, Charles H. Best; Professor J. J. R. Macleod divided his prize money with J. B. Collip, the biochemist who developed the acid-alcohol method for successfully extracting insulin from pancreatic tissues. But, to literally millions of patients with diabetes, the discovery of insulin meant much more than a Nobel prize; through the discovery of insulin, it meant life itself.

Eli Lilly and Company, a pharmaceutical manufacturer based in Indianapolis, IN was quick to act on this dramatic discovery. By 1923 Lilly, in collaboration with the University of Toronto insulin team, introduced the first commercially available insulin of animal origin. Over the next 57 years, this was followed by modifications to insulin formulations of animal origin to improve purity and clinical effectiveness. However, despite these improvements, animal insulin differs structurally from human insulin (beef by three amino acids and pork by one amino acid) and the impetus to produce native human insulin formulations was strong of a concern for future limitations in animal insulin supplies. In addition, the use of beef and pork insulins by some individuals may result in insulin allergy, insulin resistance, or insulin lipodystrophy (5).

In 1982, using breakthrough recombinant DNA (rDNA) technology, Lilly commercially synthesised and produced an insulin identical in structure to human insulin (recently reviewed by Chance and Frank) (6). Recombinant human insulin marked a significant improvement over animal insulin. Although structurally identical, exogenously-injected human insulin still failed to perfectly mimic the pharmacokinetic profile of endogenously-secreted insulin.

Compared to endogenous insulin, subcutaneously-administered insulin has a slower onset and longer duration of action (7). Even unmodified human insulin (regular) has a high propensity to self-associate which slows absorption as pointed out by Brange and coworkers (8). Exogenous insulin must be given as a fixed dose with an attempt to match the timing and dose of the injection to the expected timing and dose of calories consumed. Consequently, although injected insulin allows patients with diabetes to thrive, it does not control the blood glucose profile as tightly as endogenously secreted insulin.

Until the last decade, the greatest concern with uncontrolled blood glucose levels centred on severe acute problems including diabetic ketoacidosis, hyperosmolar nonketotic syndrome, or hypoglycaemic coma. However, in 1993, the landmark study by the Diabetes Control and Complications Trial Research Group (DCCT) definitively linked the long-term complications of retinopathy, neuropathy, and nephropathy to poor glycaemic control (9).

Importantly, the DCCT study, as well as a smaller Swedish study (10) both involving patients with type 1 diabetes and a similar trial in patients with type 2 diabetes (11) showed that intensive insulin therapy could dramatically reduce the incidence or severity of complications. Nevertheless, intensive therapy was not without risk because it increased the likelihood of severe hypoglycaemia.

The inability of available insulin to perfectly mimic endogenous insulin was addressed at an international conference convened in Monaco in 1985 by the Juvenile Diabetes Foundation International in collaboration with the World Health Organisation (12). One of the recommendations from the Insulin Therapy Study Sub-Group stated, "Improved absorption of insulins, or new insulins are required to mimic more accurately physiological insulin profiles. Monomeric insulins and insulin derivatives should be tested."

In a continuing effort to meet the needs of patients with diabetes, Lilly tested numerous modifications of the insulin molecule. This process led to the development of insulin lispro, a rapid-acting insulin analogue.

1.1 Biotechnology Leading to Insulin Lispro

The 75 years between the discovery of insulin and the commercialisation of insulin lispro were marked by great scientific and technologic strides (7). The earliest days of harvesting animal pancreas were followed by procedures to enhance the purity of animal pancreatic insulin (13, 14). This was, in turn, supplanted by rDNA technology which ultimately led to the first DNA human health care product, Humulin®, (human insulin of recombinant DNA origin, Lilly).

The first evidence for preparation of insulin using rDNA technology was obtained by Goeddel and coworkers at City of Hope and Genetech on 24 August 1978. A small, but detectable amount of insulin (~20 ng) was found by radioimmunoassay following the combination of A and B chains that were individually expressed in *Escherichia coli* using chemically synthesised genes (6, 15-17). Key to this synthesis was a newly developed method for rapid chemical synthesis of DNA coupled with the emerging technology of reversed phase-high performance liquid chromatography (HPLC) which aided in both the purification and preparation of appropriate DNA fragments (18-21), as well as the detection and characterisation of the expressed proteins (17, 22). This accomplishment initiated a major shift in the direction of insulin production and marked the beginning of independence from pancreatic-derived insulins.

Shortly after this first biosynthesis of human insulin chains in *E. coli*, Lilly and Genentech entered into a contractual agreement allowing Lilly to develop and commercialise the preparation of the recombinant human insulin using the Genentech plasmids containing the synthetic A- and B- chain genes (23) and, subsequently, the gene for natural human proinsulin (6).

Even though the production of recombinant human insulin signified the epitome of success in insulin therapy during the 1980's, it also represented new research technologies which dramatically expanded the field of protein structure-activity studies, particularly among hormones. Lilly scientists were quick to view recombinant technology as a means to make related, always scarce, insulin-like molecules ie, human proinsulin (24, 25), insulin-like growth factor-I (IGF-I) (26), and insulin-like growth factor-II (IGF-II) (27) to further improve diabetes therapy. This, coupled with the expertise gained in preparing regular human insulin by either chain combination, (28), or transformation of proinsulin to insulin, (24) formed a basis for subsequent insulin analogue studies.

To understand the direction of insulin analogue research, it is necessary to understand insulin's structure and physicochemical properties. The human insulin molecule is a heterodimer consisting of an A-chain with 21 amino acids and a B-chain with 30 amino acids. Two interchain disulphide bonds covalently link the two peptide chains with an additional intrachain disulphide bond in the A-chain (see Figure 16).

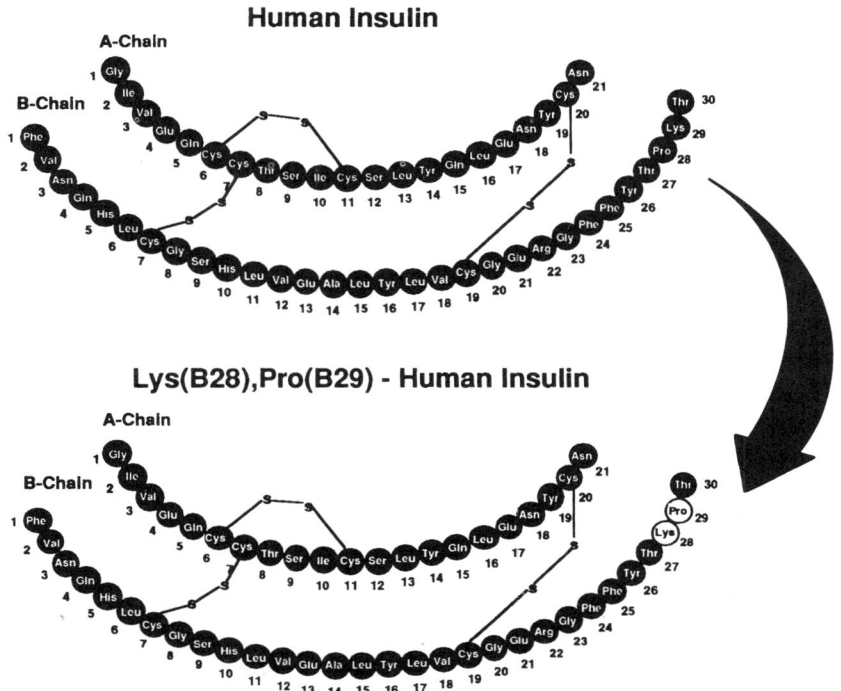

Figure 16. Primary structure of human insulin and LysB28ProB29-human insulin (insulin lispro) with emphasis on the sequence inversion of positions B28 and B29. (Adapted from Anderson *et al.* (116) with permission from the authors and the publisher of Diabetes News).

This structure of the insulin molecule predisposes it to closely associate with other insulin molecules. In pharmaceutical formulations of insulin, the high affinity of one insulin molecule for another causes self-association into dimers and hexamers (29, 30). The amino acid residues in the C-terminus of the B-chain of each individual insulin molecule align in an antiparallel fashion and form hydrogen bonds between molecules. Two molecules form a dimer and three dimers aggregate into a hexamer stabilised by coordination with two zinc ions. Unfortunately, in the hexameric state, insulin is not readily absorbed. Insulin must first dissociate into dimers and monomers, and this dissociation is viewed as the rate-limiting step for absorption (31). The time required for dissociation and subsequent absorption necessitates that regular human insulin be administered 30 to 45 minutes before a meal so that it can be most effective. Because this is inconvenient and not realistic for the person with diabetes, this timing is often ignored.

To identify analogues that could overcome the self-association phenomenon, Lilly chemists chose to focus on the structural changes within the C-terminal pentapeptide portion of the B-chain (32, 33). Not only does this portion of the molecule play an important role in self-association, it is also a region that can generally be modified without significantly affecting insulin's recognition by its *in vivo* receptor. The process of systematically removing amino acid residues elicited many interesting analogues, but none met the experimental goals.

Unexpectedly, an answer was found to self-association in Lilly's research with another recombinant DNA-derived homologue of insulin, insulin-like growth factor-I (IGF-I) (26). Structurally, IGF-I is very similar to human insulin. Approximately 50% of the residues within the A-and B-domains of IGF-I are identical to those in comparable positions in the A-and B-chains of insulin, (including the critical C-terminal region of the B-chain) and IGF-I exhibits the same disulphide configuration. However, IGF-I self-associates to a much lesser degree than insulin. Furthermore, the Pro-Lys sequence in insulin at positions B28 and B29 is reversed in IGF-I (Lys-Pro). Hypothesising that a similar reversal in insulin amino acid positions could generate an analogue that would not self-associate (34, 35), we (The Lilly Insulin Team) explored structure-activity studies in which proline was moved to position B29 and various amino acids substituted into position B28 (36-39).

Although more than 50 analogues were synthesised in this series, the original one, $Lys^{B28}Pro^{B29}$-human insulin, best fulfilled the research criteria for several reasons. First, it demonstrated identical hypoglycaemic potency to regular insulin *in vivo* (40). Second, it had a rapid onset of activity with a short duration of action in both dogs (40-42) and pigs (43, 44). Third, although it was not strictly monomeric, analytical ultracentrifugation showed it had significantly reduced self-association (33, 37). Finally, and perhaps, most importantly, the transposition of the Pro-Lys sequence to Lys-Pro represents a natural modification. This insulin analogue has become known as insulin lispro (45), nomenclature that accommodates consistent global registration of the drug since some languages do not have the letter "y" in their alphabets.

As we experimented with insulin lispro, we discovered that it had a faster onset of action and a shorter duration than regular insulin in both the dog model (40-42) and multiply-catheterised pigs (43, 44). Further studies revealed its pharmacokinetic profile was indistinguishable from that of a truly monomeric insulin that has served as a model insulin analogue for various physical and solution structure studies [ie, $Asp^{B10}Lys^{B28}Pro^{B29}$-human insulin (46)]. Insulin lispro's onset of action was virtually identical to $Asp^{B10}Lys^{B28}Pro^{B29}$-human insulin (44).

The rapid onset of insulin lispro was surprising considering it is formulated as a well-ordered, stable hexamer in the presence of zinc and meta-cresol, which gives it a 2-year shelf life at 4°C comparable to regular human insulin. To determine the mechanism of insulin lispro's rapid absorption we evaluated regular human insulin and insulin lispro with and without Zinc *in vitro* using X-ray crystal structure analysis (47), stopped-flow spectroscopy (48), static light scattering (43) and other physical measurements. From these studies, it was theorised that the absorption profiles of U-100 formulated regular human insulin and insulin lispro differ due to varying stabilities between the zinc hexamers which remain after the lipophilic preservative (phenol and/or meta-cresol) dissipates following subcutaneous administration. More specifically, when the preservative in the insulin lispro formulation dissipates, the remaining zinc hexamer is relatively unstable and dissociates directly to monomer subunits as depicted schematically in Figure 17. By comparison, when the preservative in the regular human insulin formulation dissipates, the remaining zinc hexamer is relatively more stable. Consequently, the absorption of regular human insulin is delayed until its more stable zinc hexamer dissociates into dimers and monomers. Experimental studies indicate that the dimerization constant for insulin lispro is about 300 times less than that of human regular insulin (49). This is consistent with the theory that movement of proline from position B28 in insulin to B29 in insulin lispro eliminates critical hydrophobic interactions between the monomers that normally self-associate in an antiparallel alignment (47).

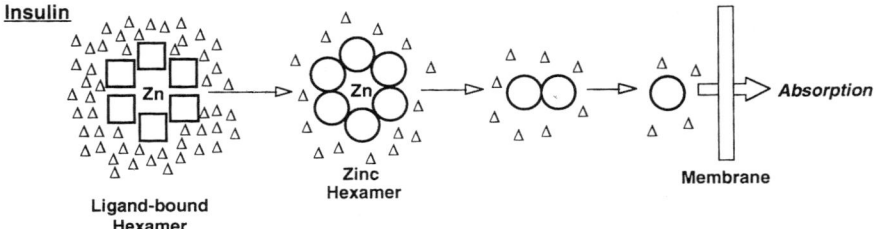

Δ = phenolic ligand

Figure 17. Schematic diagram illustrating the physicochemical basis for the faster absorption of insulin lispro after s.c. administration compared to regular human insulin. Although both insulins exist in their respective formulations as zinc hexamers intercalated with phenolic ligands, the insulin lispro hexamer dissociates virtually instantaneously upon injection as a result of the phenolic ligand (meta-cresol) rapidly dissipating into the tissues, leaving the remaining unstable zinc complex to dissociate directly to monomer subunits. This contrasts to the insulin hexamer which binds to zinc more strongly than lispro, thus necessitating further dilution before insulin hexamers will dissociate into absorbable subunits. The ligand-bound zinc hexamers of both insulin and insulin lispro are represented with squares to symbolize the monomer subunits in a general R-state conformation in contrast to a T-state conformation (depicted by circles) once the phenolic ligand is no longer present (48). From Chance *et al.* (32) as adapted from Bakaysa *et al.* (43).

1.2 Manufacturing Process for Insulin Lispro

Lilly's experience in manufacturing human insulin in *E. coli* using rDNA technology is well documented (6, 23, 50-54). Commercial recombinant human insulin was initially made by a chain combination procedure (28), then four years later in 1986 via the proinsulin route (24). In this latter process a 277-residue chimeric fusion protein (Trp LE'- Methionine-Human Proinsulin) is expressed in *E. coli* followed by chemical cleavage at methionine with cyanogen bromide to release human proinsulin mixed disulphides. Human insulin is obtained subsequent to appropriate folding, enzymatic transformation, large-scale purification and crystallisation (55, 56).

Insulin lispro is manufactured in essentially the same manner as human insulin with a few exceptions. The expression product from fermentation is a smaller precursor molecule with a shorter amino terminal extension that is removed enzymatically to yield $Lys^{B28}Pro^{B29}$-human proinsulin. Insulin lispro is liberated from proinsulin by hydrolysis with trypsin and carboxypeptidase B (24), chromatographically purified, and finally crystallised in the presence of Zinc and phenol (47). As with recombinant human insulin (6), a complex battery of analytical tests were used to evaluate insulin lispro during the research, development, and analytical control of the drug. These are listed below. Those tests with an asterisk are used routinely as quality control checks for each batch of bulk crystals to assure high product quality.

1.2.1 Evaluative Tests for Insulin Lispro (* denotes routine tests)

- *Rabbit hypoglycaemia assay for bioidentity
- Insulin receptor binding (39)
- IGF-I receptor binding (39)
- Cell growth studies (57)
- Amino acid composition and sequence (58)
- Peptide mapping by RP-HPLC (28, 59)
- *Purity and identity by RP-HPLC and SE-HPLC (59)
- *Potency by RP-HPLC (60)
- Crystallisation and X-ray crystal structure (47)
- Absorption and circular dichroic spectra (33)
- NMR spectroscopy (61)
- Aggregation behaviour by hydrodynamic methods (33, 43, 48)
- *Endotoxin and pyrogen tests
- *E. coli host proteins by immunoassay
- Residual DNA (62)
- Residual enzyme activities by immunoassays (54)
- *Proinsulin and C-peptide immunoassays (54)

1.3 Toxicopharmacologic Effects

Prior to clinical trials, insulin lispro was compared to regular human insulin in a variety of receptor-binding and cell-binding functional assays. These studies demonstrated that insulin lispro was equipotent or slightly less potent than regular human insulin in binding to the human insulin receptor. Conversely, studies also found insulin lispro was equipotent or only slightly more potent than regular human insulin in binding to the IGF-I receptor (39, 57, 63, 64). These binding considerations were important considering an earlier insulin analogue, Asp^{B10}-human insulin (8) was shown to bind to a significantly greater extent to both insulin receptors (65) and IGF-I receptors (66) when compared with human insulin. Asp^{B10}-human insulin was also found to induce mammary tumours in female rats in a 12-month toxicity study (66-69). Subsequent literature suggests that this analogue's disproportionately enhanced mitogenic to metabolic potency may be due to its slower dissociation kinetics from the insulin receptor compared to human insulin (70-76).

Experiments were also conducted to compare the dissociation kinetics of ^{125}I-human insulin, ^{125}I-insulin lispro, and ^{125}I-Asp^{B10}-human insulin from HepG2 cells (a minimal deviation human hepatoma cell line that expresses high levels of insulin receptors) (39, 77). These studies showed that the dissociation rate of insulin lispro equalled that of regular human insulin, but the dissociation rate of Asp^{B10}-human insulin was only 50% of regular human insulin which was consistent with Asp^{B10}-human insulin's observed increased affinity for the insulin receptor (78). Cellular metabolism studies were also consistent with these dissociation kinetic studies (79). A comparison between insulin lispro and regular human insulin in the stimulation of glucose and amino acid transport and the activation of insulin signalling pathways in L6 skeletal muscles demonstrated insulin lispro was equipotent to regular human insulin (80).

Also investigated was insulin lispro's relative potency in cell growth assays in human aortic smooth muscle cells (HSMC) and human mammary epithelial cells (HMEC). Because IGF-I receptors are present in far greater numbers than insulin receptors in these tissues, these assays largely represent an IGF-I mediated event. Using [^{3}H] thymidine incorporation in HSMCs to index cell growth, an initial set of experiments found insulin lispro to be more potent than regular human insulin. A second set, however, found insulin lispro was less potent (81). A follow-up study (82) showed insulin lispro and insulin to have equal potency with regard to [^{3}H] thymidine incorporation into the DNA of cultured rat liver cells from a H4 hepatoma cell line. Subsequent HMEC testing (believed to be more quantifiable and reproducible than [^{3}H] thymidine incorporation), which measured the actual increase in cell number, determined insulin lispro to be essentially equipotent with regular human insulin (57). These studies provided confirmation that, biologically, insulin lispro could be expected to behave like regular human insulin.

The next step was to broaden experiments to include testing insulin lispro in animals. To evaluate cardiovascular effects, insulin lispro was administered to dogs by intravenous bolus injection. No toxicologically important changes in either cardiovascular or respiratory parameters were found. The slight prolongations of the QRS duration and the Q-Tc interval mimicked those of an equivalent dose of regular human insulin and were probably attributable to insulin-induced hypoglycaemia (83).

Rat studies also showed no significant changes in mean arterial pressure, heart rate, systolic and diastolic pressure, and pulse pressure following subcutaneous injection of insulin lispro (83, 84). When we evaluated the pharmacologic activity of insulin lispro on smooth and cardiac muscle function at concentrations of $\leq 1 \times 10^{-5}$ M, insulin lispro was found to have no effect on the ileum, atrium, estrogen-primed uterus, or vas deferens. At 1×10^{-5} M, insulin lispro may exhibit a slight antagonism of cholinergic, adrenergic, and angiotensin receptors, but this was not found to be clinically relevant (84).

Data from acute, subchronic, and chronic toxicity studies in both rats and dogs showed nothing that would preclude the chronic use of insulin lispro in humans (77, 85-87). Importantly, results from the 1-year rat study using 200U/kg (a high dose regimen) daily subcutaneous doses showed no association with mammary tumours. This result contrasted sharply with the earlier results with the Asp^{B10}-human insulin. No changes were found in any lesions including neoplasms, hematologic or urinalysis values due to insulin lispro.

1.4 Reproduction and Teratology Studies

To evaluate the effect of insulin lispro on reproduction, Fischer 344 rats were injected daily with subcutaneous doses of the analog at 0, 1, 5 or 20 U/kg during a chronic 6-month study. Mating indices, fertility indices, preimplantation and postimplantation losses were not adversely affected (88). Insulin lispro was further evaluated in rats in both the parent (F0) and the offspring (F1) generation (88). Treatment with insulin lispro did not affect mating or fertility of the F0 generation and no treatment-related effects were observed on live birth index, litter size, or F1 growth and survival to Day 21 after birth. Additionally, there was no indication of teratogenicity.

Insulin lispro was studied in New Zealand White rabbits at doses ranging from 0.1-0.75U/kg/day on gestation Days 7-19. Based on fetal viability, weight, and morphology on Day 28, no toxicity from insulin lispro in the developing rabbit conceptus was found (83).

Mutagenic potential was tested by induction of reverse mutations in *Salmonella typhimurium* and *Escherichia coli*, induction of unscheduled DNA synthesis in primary cultures of adult rat hepatocytes, induction of mammalian cell mutation in the L5178YTK±mouse lymphoma assay, *in vivo* bone marrow micronuclei induction in bone marrow of male and female ICR mice, and chromosomal aberration induction in Chinese hamster ovary (CHO) cells. In all tests, insulin lispro showed no mutagenic potential (77, 83, 86).

Rhesus monkey tests using a protocol involving weekly immunisation in Freund's adjuvant (89) were used to evaluate insulin lispro's immunological potential. It was found that insulin lispro had extremely weak immunologic potential which was consistent with earlier findings in the 1-year rat study (83).

1.5 Clinical Studies on Insulin Lispro

1.5.1 Pharmacokinetics and Glucodynamics

Howey and coworkers at the Lilly Clinic in Indianapolis administered the first dose of insulin lispro to humans on July 11, 1990 (90). Immediately, there was great excitement about the potential for this new innovative insulin to make a difference in the lives of people with diabetes. Its more rapid absorption and shorter duration of action compared to regular human insulin was impressive and highly suggestive that insulin lispro could provide better postprandial glucose control at a more convenient injection time (91). These seminal studies were soon followed by numerous other clinical pharmacology investigations comparing insulin lispro with regular human insulin. In their review on the pharmacokinetics and glucodynamics of insulin lispro, Heinemann and Woodworth (92) note that after subcutaneous administration of a clinically relevant range of doses, insulin lispro consistently has a more rapid absorption and elimination profile than regular human insulin. Both insulins are essentially equivalent in biological potency when administered intravenously. One clinical pharmacology study with practical significance was performed in a randomised, double-blinded fashion to evaluate the blood glucose response to a carbohydrate-rich meal in 10 patients with type 1 diabetes (93). Participants were maintained on low-dose intravenous insulin overnight then administered subcutaneous insulin lispro or regular human insulin immediately before a 140 g carbohydrate meal (see Figure 18). The results documented the effectiveness of insulin lispro on the blood glucose profile. After the insulin lispro injection, the area under the blood glucose curve was 22% smaller ($p<.01$), and the peak rise in blood glucose was 2 mmol/L less ($p<.05$) than following the injection of regular human insulin (92, 93).

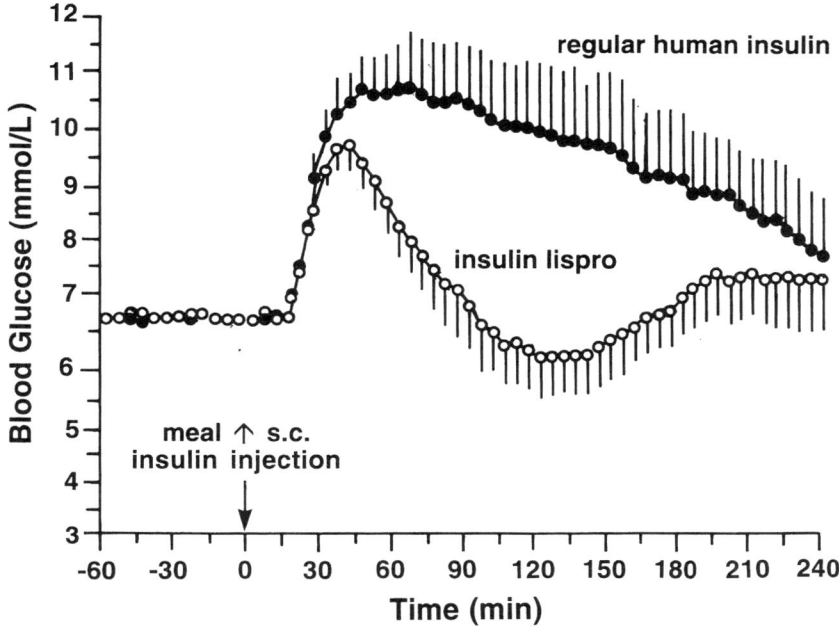

Figure 18. Blood glucose profile of 10 patients with type 1 diabetes in good metabolic control after s.c. injection of insulin lispro (self-selected dose) or regular human insulin immediately prior to a meal rich in rapidly absorbable carbohydrates. Doses and meal given at time = 0. [Reproduced with permission (93).]

1.5.2   Global Registration Studies

Insulin lispro became available for clinical use worldwide during 1996-1997. Prior to drug registration approvals and launches, several multicenter clinical trials were conducted involving more than 3000 patients with diabetes (94). Several clinical studies of insulin lispro in people with diabetes between the ages of 12 and 70 years have been published (95-97). These studies have been comprehensively reviewed recently by Anderson and Kovisto (94). Study designs included a parallel, 12-month study of 336 patients with type 1 diabetes and 295 patients with type 2 diabetes, and two 6-month crossover trials of 1008 patients with type 1 diabetes and 722 patients with type 2 diabetes.   Studies were randomised, open-label comparisons of mealtime treatment with insulin lispro and regular human insulin (98-100). To allow optimal timing of insulin injections, studies were not blinded—regular human insulin was administered 30 to 45 minutes before meals while insulin lispro was injected within 15 minutes of meals. Study participants used human NPH or ultralente insulin as a basal regimen. Following a 2-to 4-week lead-in period, participants were randomised to one of two treatment sequences. Those in the parallel trial were randomised to pre-meal multiple dose therapy with either insulin lispro or regular human insulin. Those in the crossover studies were treated with pre-meal therapy with either insulin lispro or regular human insulin for 3 months, then crossed over to the other short-acting insulin. To compare the effects of insulin lispro and regular insulin on postprandial glycaemic control, one- and two-hour blood glucose values were determined after test meals at baseline and at regular intervals. Both trials showed better postprandial glucose control with insulin lispro than with regular human insulin in patients with type 1 and type 2 diabetes. These results were consistent with a 2-month crossover study of 379 patients with type 1 diabetes and 328 patients with type 2 diabetes who administered their insulin in two daily injections (101).

1.5.3 Insulin Lispro In Continuous Subcutaneous Infusion Therapy

Insulin treatment regimens designed to mimic natural physiologic responses to meals (bolus insulin) coupled with the need to suppress hepatic glucose production between meals (basal insulin) are very demanding and imprecise. Continuous subcutaneous insulin infusion, or CSII, via programmable external infusion pumps, is an alternate method for insulin administration and theoretically provides the opportunity for more ideal basal/bolus therapy. The advent of insulin lispro has stimulated investigations comparing this new insulin analogue with regular human insulin in CSII regimens (102-106). Zinman et al. (102) studied 30 patients with type 1 diabetes in a double-blind crossover study (3 months each insulin) in which the before-meal boluses of both insulins were given immediately before breakfast, lunch and supper. The 1-hour postprandial blood glucose levels were lower during insulin lispro therapy and HbA_{1c} was significantly reduced when compared to regular human insulin ($8.00 \pm 0.16\%$ versus $7.66 \pm 0.13\%$, p=0.0041).

In a study by Melki *et al.* (103), the efficacy of insulin lispro was compared with regular human insulin in 39 patients with type 1 diabetes who had been treated with external pump therapy for about 5 years. The study was an open-label, randomised, crossover, multicenter design comparing 3 months of treatment with each insulin. This study differed from the one above in that bolus insulin administration was given 0 to 5 minutes prior to meals for insulin lispro but 20 to 30 minutes prior to meals for regular human insulin. Even when the timing of regular human insulin was optimized, insulin lispro provided better glycaemic control than regular insulin without increasing the frequency of hypoglycaemia. At the end of the study, 95% of the patients chose insulin lispro for the extension phase. Similarly, Campbell et al. (106) surveyed insulin pump users and reported that insulin lispro is highly accepted and preferred over regular human insulin.

1.5.4 Postprandial Administration

As previously discussed, optimal injection time of exogenous insulin is not always practical, and in some cases, impossible; insulin lispro was further evaluated when injected after a meal (107). Though this dose schedule is not common, it does occur particularly when meal composition is extremely variable or when parents are trying to match a child's erratic eating habits to an appropriate insulin dose. The results showed that insulin lispro injected 20 minutes or immediately before a meal was significantly more effective in controlling postprandial blood glucose excursion than any of the regular human insulin treatments injected from 40 to 0 minutes before the meal. In addition, the overall glucodynamic effect when insulin lispro was injected 15 minutes after a meal was numerically greater than with all of the regular human insulin treatments. This study proved that satisfactory postprandial control can be achieved with insulin lispro even when injected shortly after a meal.

1.5.5 Paediatric Use

Insulin lispro has been tested in different paediatric populations with type 1 diabetes (83, 108-113). Holcombe *et al.* (108) studied 61 prepubertal children between the ages of 2.9 and 11.4 years in an open-label, randomised, crossover study divided into three groups (regular human insulin injected 30 to 45 minutes before meals; insulin lispro injected within 15 minutes before meals; and insulin lispro injected after meals). Each regimen lasted 3 months and the basal insulin was constant throughout the 9-month study. Results showed that insulin lispro administered within 15 minutes before meals was associated with lower postprandial glucose levels and glucose excursions compared with the other two treatments. In addition, children receiving insulin lispro after meals had glucose levels and glucose excursions comparable to those who received regular human insulin 30 to 45 minutes before meals. There was no significant difference in the incidence of hypoglycaemia among the three treatments. Rutledge *et al.* (109) studied 10 children ranging in age from 22 to 58 months. Children received either regular human insulin before breakfast on Day 1 and insulin lispro on Day 2 after breakfast, or the converse. Breakfasts were identical both days, and the children continued their basal insulin during the study. Glucose excursions on postprandial insulin lispro administration were as good or better than those obtained with an equivalent dose of regular human insulin administered before the meal.

Holcombe *et al.* (110-112) have also compared insulin lispro versus regular human insulin in 481 pubertal children with type 1 diabetes ranging in ages from 9.1 to 18.9 years. Treatment periods of 4 months in this randomised, crossover, open-label study consisted of either insulin lispro or regular human insulin injected just before meals. Both insulin regimens were in conjunction with NPH insulin given up to 3 times per day. The insulin lispro treatment resulted in fewer hypoglycaemic episodes, particularly between midnight and 6 a.m. More than 82% of the patients and their parents indicated lispro made the patients' activities easier, and more than 85% of the patients indicated their preference to use insulin lispro in the future. Similarly, Rami and Schober (113) studied 12 children with type 1 diabetes during a diabetes summer camp of 2-weeks duration (age range 12 to 16.5 years). Insulin lispro injected immediately before the meal led to lower postprandial blood glucose levels.

1.5.6 Insulin Lispro and Improved Quality of Life

The more rapid onset and shorter duration of action of insulin lispro provides for more convenient insulin therapy. This characteristic is vital for lifestyle benefits. Being able to inject insulin closer to meals not only gives patients better postprandial glucose control, it also allows flexibility to a person's lifestyle. A major quality-of-life study was conducted by Kotsanos *et al.* (114) comparing health-related quality-of-life parameters in patients receiving either insulin lispro or regular human insulin. Primary analyses showed that treatment satisfaction scores and treatment flexibility scores were higher for insulin lispro in patients with type 1 diabetes. Similar results were obtained in a smaller study by Desmet *et al.* (115).

1.6 Summary

Insulin lispro, a human insulin analogue created by reversing the native amino acid sequence from $Pro^{B28}Lys^{B29}$ to $Lys^{B28}Pro^{B29}$, is a rapid-acting insulin due to its weakened propensity to self-associate into dimers. This new insulin has recently been developed particularly for mealtime therapy through injection immediately before, or, in some cases, just after the meal. Its rapid absorption rate and short duration of action provide several advantages for patients on insulin therapy. Insulin lispro consistently reduces the postprandial rise in blood glucose as compared to regular human insulin in patients with type 1 or type 2 diabetes. Insulin lispro offers patients increased flexibility and convenience while decreasing postprandial hyperglycaemia and decreasing the risk for hypoglycaemia.

ACKNOWLEDGEMENTS

We would like to thank Ms. Peggy Campbell for her significant editorial and collaborative contributions to this manuscript.

BIOGRAPHY

Eli Lilly and Company is a global research-based pharmaceutical corporation dedicated to creating and delivering innovative pharmaceutical-based health care solutions that enable people to live longer, healthier and more active lives. Ron Chance, Ph.D. is a research fellow at Lilly Research Laboratory. Brad Glazer, Pharm. D. is the communication associate for the Insulins Product Team. Kathleen Wishner, Ph.D., M.D. is the medical director for the Insulins Product Team. The authors are located at Eli Lilly and Company.

REFERENCES

1. Banting, F.G. *et al.* (1922). The internal secretion of the pancreas. American Journal of Physiology, 59, 479.
2. Banting, F.G. and Best, C.H. (1922). The internal secretion of the pancreas. Journal of Laboratory and Clinical Medicine, 7, 251-266.
3. Bliss, M. (1996). The discovery of insulin. 75th Anniversary Edition. McClelland and Stewart, Inc., Toronto 11.
4. Bliss, M. (1993). The history of insulin. Diabetes Care, 16(Suppl. 3), 4-7.
5. Sherwin, R.S. (1996). Diabetes Mellitus. In: Bennett, J.C. and Plum, F. (eds.) Cecil Textbook of Medicine, 20th ed. W.B. Saunders Company, Philadelphia. p. 1265.
6. Chance, R.E. and Frank, B.H. (1993). Research, development, production, and safety of biosynthetic human insulin. Diabetes Care, 16(Suppl. 3), 133-142.
7. Galloway, J.A. and Chance, R.E. (1994). Improving insulin therapy: achievements and challenges. Hormone And Metabolic Research, 26(12), 591-598.
8. Brange, J. *et al.* (1990). Monomeric insulins and their experimental and clinical implications. Diabetes Care, 13(9), 923-954.
9. The Diabetes Control and Complications Trial Research Group (DCCT). (1993). The effect of intensive treatment of diabetes on the development and progression of long-term complications in insulin-dependent diabetes mellitus. New England Journal of Medicine, 329, 977-986.

10. Reichard, P. et al. (1993). The effect of long-term intensified insulin treatment on the development of microvascular complications of diabetes mellitus. The New England Journal of Medicine, 329, 304-309.
11. Ohkubo, Y. et al. (1995). Intensive insulin therapy prevents the progression of diabetic microvascular complications in Japanese patients with non-insulin-dependent diabetes mellitus: a randomized prospective 6-year study. Diabetes Research and Clinical Practice, 28(2), 103-117.
12. JDFI World Conference on Diabetes Research. (1985). Current Status, Future Directions. Report of the Juvenile Diabetes Foundation International World Conference on Diabetes Research (Nov2-6, Monaco). p. 42.
13. Schlichtkrull, J. et al. (1974). Monocomponent insulin and its clinical implications. Hormone and Metabolic Research, (Suppl. Ser)5, 134-143.
14. Chance, R.E. et al. (1976). The immunogenicity of insulin preparations. Acta Endocrinologica, 83(Suppl. 205), 185-196.
15. Hall, S.S. (1988). Invisible Frontiers. The race to synthesise a human gene. Sidgwick and Jackson Limited, London.
16. Beckman, A.O. and Roberts, E. (1988). Genetically Engineered Insulin Videotape. National Academy of Sciences and Beckman Research Institute of the City of Hope December 1, 1988. Copy of the videotaped proceedings available through the History of Medicine Division of the National Library of Medicine, Bethesda MD and can be obtained via interlibrary loan per personal communication with John Parascandola, Chief, History of Medicine Division.
17. Goeddel, D.V. et al. (1979). Expression in *Escherichia coli* of chemically synthesized genes for human insulin. Proceedings of the National Academy of Sciences of the United States of America, 76(1), 106-110.
18. Itakura, K. et al. (1977). Expression in *Escherichia coli* of a chemically synthesized gene for the hormone somatostatin. Science, 198, 1056-1063.
19. Itakura, K. and Riggs, A.D. (1980). Chemical DNA synthesis and recombinant DNA studies. Science, 209, 1401-1405.
20. Itakura, K. et al. (1984). Synthesis and use of synthetic oligonucleotides. Annual Review of Biochemistry, 53, 323-356.
21. Riggs, A.D. et al. (1980). Synthesis, cloning, and expression of hormone genes in *Escherichia coli*. Recent Progress in Hormone Research, 36, 261-276.
22. Crea, R. et al. (1978). Chemical synthesis of genes for human insulin. Proceedings of the National Academy of Sciences of the United States of America, 75(12), 5765-5769.
23. Johnson, I.S. (1983). Human insulin from recombinant DNA technology. Science, 219, 632-637.
24. Frank, B.H. et al. (1981). The production of human proinsulin and its transformation to human insulin and C-peptide. In: Rich, D.H. and Gross, E. (eds.) PEPTIDES: Synthesis-Structure-Function. Proceedings of the Seventh American Peptide Symposium. Pierce Chemical Company, Rockford. pp. 729-738.
25. Galloway, J.A. et al. (1992). Biosynthetic human proinsulin. Review of chemistry, in vitro and in vivo receptor binding, animal, and human pharmacology studies, and clinical trial experience. Diabetes Care, 15(5), 666-692.
26. DiMarchi, R. et al. (1989). Synthesis of insulin-like growth factor I through recombinant DNA techniques and selective chemical cleavage at tryptophan. In: Tam, J.P. and Kaiser, E.T. (eds.) Synthetic Peptides: Approaches to Biological Problems. Alan R Liss, Inc., New York. pp. 283-294.

27. Furman, T.C. et al. (1987). Recombinant human insulin-like growth factor II expressed in *Escherichia coli*. Biotechnology, 5, 1047-1051.
28. Chance, R.E. et al. (1981). The production of human insulin using recombinant DNA technology and a new chain combination procedure. In: Rich, D.H. and Gross, E. (eds.) PEPTIDES: Synthesis-Structure-Function. Proceedings of the Seventh American Peptide Symposium. Pierce Chemical Company, Rockford. pp. 721-728.
29. Blundell, T. et al. (1972). Insulin: the structure in the crystal and its reflection in chemistry and biology. Advances in Protein Chemistry, 26, 279-402.
30. Baker, E.N. et al. (1988). The structure of 2Zn pig insulin crystals at 1.5 Å resolution. Philosophical Transactions Of The Royal Society Of London. B: Biological Sciences, 319(1195), 369-456.
31. Mosekilde, E. et al. (1989). Modeling absorption kinetics of subcutaneous injected soluble insulin. Journal of Pharmacokinetics and Biopharmaceutics, 17(1), 67-87.
32. Chance, R.E. et al. (1998). Discovery and development of insulin lispro. Drugs of Today, 34(Suppl. C), 1-9.
33. Brems, D.N. et al. (1992). Altering the association properties of insulin by amino acid replacement. Protein Engineering, 5(6), 527-533.
34. DiMarchi, R.D. et al. (1992). Synthesis of a fast-acting insulin based on structural homology with insulin-like growth factor I. In: Smith, J.A. and Rivier, J.E. (eds.) Peptides. Chemistry and Biology. Proceedings of the Twelfth American Peptide Symposium, ESCOM, Leiden. pp. 26-28.
35. DiMarchi, R.D. et al. (1994). Preparation of an insulin with improved pharmacokinetics relative to human insulin through consideration of structural homology with insulin-like growth factor I. Hormone Research, 41(Suppl. 2), 93-96.
36. Long, H.B. et al. (1992). Human insulin analogs with rapid onset and short duration of action. In: Smith, J.A. and Rivier, J.E. (eds.). Peptides. Chemistry and Biology. Proceedings of the Twelfth American Peptide Symposium. ESCOM, Leiden. pp. 88-90.
37. Frank, B.H. et al. (1991). Manipulation of the position of proline in the B-chain produces monomeric insulins. Diabetes, 40(Suppl. 1), 423A.
38. Chance, R.E. et al. (1996). Insulin analogs modified at position 29 of the B chain. United States Patent Number 5,514,646. May 7, 1996.
39. Slieker, L.J. et al. (1997). Modifications in the B10 and B26-30 regions of the B chain of human insulin alter affinity for the human IGF-I receptor more than for the insulin receptor. Diabetologia, 40(Suppl. 2), S54-S61.
40. Shaw, W.N. and Su, K.S.E. (1991). Biological aspects of a new human insulin analog: [Lys(B28), Pro(B29)]-human insulin. Diabetes, 40(Suppl. 1), 464A.
41. Galloway, J.A. et al. (1991). Human insulin and its modifications. In: Reidenberg, M.M. (ed). The clinical pharmacology of biotechnology products. Elsevier, Amsterdam. pp. 23-34.
42. Su, K.S. et al. (1994). Using dog model for comparing time action of insulins after subcutaneous (s.c.) injection: prediction of rapid onset of a new insulin analog [Lys(B28), Pro(B29)]-human insulin (KP). Pharmaceutical Research, 11(10)(Suppl.), S357.
43. Bakaysa, D.L. et al. (1996). Physicochemical basis for the rapid time-action of $Lys^{B28}Pro^{B29}$-insulin: dissociation of a protein-ligand complex. Protein Science, 5(12), 2521-2531.
44. Radziuk, J. et al. (1997). Bioavailability and bioeffectiveness of subcutaneous human insulin and two of its analogs—$Lys^{B28}Pro^{B29}$-human insulin and $Asp^{B10}Lys^{B28}Pro^{B29}$-human insulin—assessed in a conscious pig model. Diabetes, 46, 548-556.

45. USAN Council. (1995). New Names. Clinical Pharmacology and Therapeutics, 57, 98.
46. Hua, Q.-X. et al. (1996). Mapping the functional surface of insulin by design: structure and function of a novel A-chain analogue. Journal of Molecular Biology, 264, 390-403.
47. Ciszak, E. et al. (1995). Role of C-terminal B-chain residues in insulin assembly: the structure of hexameric $Lys^{B28}Pro^{B29}$-human insulin. Structure, 3, 615-622.
48. Birnbaum, D.T. et al. (1997). Assembly and dissociation of human insulin and $Lys^{B28}Pro^{B29}$-insulin hexamers: a comparison study. Pharmaceutical Research, 14(1), 25-36.
49. Frank, B.H. et al. (1995). $Lys^{B28}Pro^{B29}$-human insulin (insulin lispro): solution properties of a rapid-acting insulin. Diabetologia, 38(Suppl. 1), A189.
50. Chance, R.E. et al. (1981). Chemical, physical, and biological properties of recombinant human insulin. In: Gueriguian, J L. (ed.). Insulins, Growth Hormone, and Recombinant DNA Technology. Raven Press, New York. pp. 71-86.
51. Chance, R.E. et al. (1981). Chemical, physical, and biologic properties of biosynthetic human insulin. Diabetes Care, 4(2), 147-154.
52. Johnson, I.S. (1982). Authenticity and purity of human insulin (recombinant DNA). Diabetes Care, 5(Suppl. 2), 4-12.
53. Frank, B.H. and Chance, R.E. (1983). Two routes for producing human insulin utilizing recombinant DNA technology. Münch med Wschr, 125(Suppl. 1), S14-S20.
54. Frank, B.H. and Chance, R.E. (1986). The preparation and characterization of human insulin of recombinant DNA origin. In: Joyeaux, A., Leygue, G., Morre, M., Roncucci, R. and Schmelck, P.H. (eds.). Therapeutic Agents Produced by Genetic Engineering Quo Vadis? Symposium, Sanofi Group. Toulouse-Labège. Sanofi Recherche, Montpellier. pp. 137-146.
55. Prouty, W.F. (1991). Production-scale purification processes. In: Chiu, Y-y. H. and Gueriguian, J.L. (eds.). Drug Biotechnology Regulation. Scientific Basis and Practices. Marcel Dekker, Inc., New York. pp. 221-262.
56. Kroeff, E.P. et al. (1989). Production scale purification of biosynthetic human insulin by reversed-phase high-performance liquid chromatography. Journal of Chromatography, 461, 45-61.
57. Slieker, L.J. et al. (1994). Insulin and IGF-I analogs: novel approaches to improved insulin pharmacokinetics. In: LeRoith, D. and Raizada, M.K., (eds.). Current Directions in Insulin-Like Growth Factor Research. Plenum Press, New York. pp. 25-32.
58. Atkins, L.M. et al. (1987). Recommendations for establishment of reference standards for recombinant-DNA-derived proteins and polypeptides. Journal Association of Official Analytical Chemists, 70(4), 610-617.
59. Farid, N.A. et al. (1989). Liquid chromatographic control of the identity, purity and "potency" of biomolecules used as drugs. Journal of Pharmaceutical and Biomedical Analysis, 7(2), 185-188.
60. Kroeff, E.P. and Chance, R.E. (1982). Applications of high-performance liquid chromatography for analysis of insulins. In: Gueriguian, J.L., Bransome, E.D., Jr. and Outschoorn, A.S. (Workshop Organizers) (eds.). Hormone Drugs. Proceedings of the FDA-USP Workshop on Drug and Reference Standards for Insulins, Somatropins, and Thyroid-axis Hormones. United States Pharmacopeial Convention, Inc., Rockville. pp. 148-162.
61. Weiss, M.A. et al. (1991). Heteronuclear 2D NMR studies of an engineered insulin monomer: assignment and characterization of the receptor-binding surface by selective ^2H and ^{13}C labeling with application to protein design. Biochemistry, 30, 7373-7389.

62. Riggin, A. et al. (1997). A non-isotopic probe-hybridization assay for residual DNA in biopharmaceuticals. Journal of Pharmaceutical and Biomedical Analysis, 16(4), 561-572.
63. Slieker, L.J. and Sundell, K. (1991). Modifications in the 28-29 position of the insulin B-chain alter binding to the IGF-I receptor with minimal effect on insulin receptor binding. Diabetes, 40(Suppl. 1), 168A.
64. Slieker, L.J. et al. (1993). Insulin and IGF-I analogs: novel approaches to improved insulin pharmacokinetics. In: Du, Y-C., Tam, J.P. and Zhang, Y-S. (eds.). Peptides- Biology and Chemistry. Proceedings of the 1992 Chinese Peptide Symposium. ESCOM, Leiden. pp. 7-10.
65. Schwartz, G.P. et al. (1987). A superactive insulin [B10-aspartic acid] insulin (human). Proceedings of the National Academy of Sciences of the United States of America, 84, 6408-6411.
66. Drejer, K. (1992). The bioactivity of insulin analogues from in vitro receptor binding to in vivo glucose uptake. Diabetes/Metabolism Reviews, 8(3), 259-286.
67. Dideriksen, L.H. et al. Carcinogenic effect on female rats after 12 months administration of insulin analogue B10 Asp. Diabetes, 41(Suppl. 1), 143A.
68. Jørgensen, L.N. et al. (1992). Carcinogenic effect of the human insulin analogue B10 Asp in female rats. Diabetologia, 35(Suppl. 1), A3.
69. Jørgensen, L.N. and Dideriksen, L.H. (1993). Preclinical studies of rapid-acting insulin analogues. In: Berger, M. and Gries, F.A. (eds.). Frontiers in insulin pharmacology. Thieme Medical Publishers, Inc., New York. pp. 110-117.
70. DeMeyts, P. et al. (1993). Enhanced mitogenic potency of insulin analogues in a cell line devoid of IGF-I receptors correlates with slow dissociation from insulin receptors. Diabetes, 42(Suppl. 1), 163A.
71. DeMeyts, P. (1994). The structural basis of insulin and insulin-like growth factor-I receptor binding and negative co-operativity, and its relevance to mitogenic versus metabolic signalling. Diabetologia, 37(Suppl. 2), S135-S148.
72. Danielsen, G. et al. (1995). Early signalling events of insulin analogs. European Journal of Endocrinology, 132(Suppl. 1), 8.
73. Lundemose, A.G. et al. (1995). Molecular actions of insulin analogues. In: Baba, S., Kaneko, T. (Eds.) Diabetes, 1994. Elsevier Science BV, Amsterdam 469-472.
74. Hansen, B.F. et al. (1996). Sustained signalling from the insulin receptor after stimulation with insulin analogues exhibiting increased mitogenic potency. Biochemical Journal, 315, 271-279.
75. Liu, L. et al. (1997). IGF-I receptor-mediated signalling of the human insulin analogue HOE 901. Diabetologi,a 40(Suppl. 1), A355.
76. Berti, L. et al. (1998). The long acting human insulin analog HOE 901: characteristics of insulin signalling in comparison to ASP(B10) and regular insulin. Hormone and Metabolic Research, 30, 123-129.
77. Llewelyn, J. et al. (1998). Preclinical studies on insulin lispro. Drugs of Today, 34(Suppl. C), 11-21.
78. Drejer, K. et al. (1991). Receptor binding and tyrosine kinase activation by insulin analogues with extreme affinities studied in human hepatoma HepG2 cells. Diabetes, 40, 1488-1495.
79. Hamel, F.G. et al. (1997). B10-Asp insulin (B10), but not B28-Lys, B29-Pro insulin (LYSPRO), is resistant to metabolism by hepatocytes and insulin degrading enzyme (IDE). Diabetes, 46(Suppl. 1), 204A.

80. Somwar, R. et al. (1998). Stimulation of glucose and amino acid transport and activation of the insulin signalling pathways by insulin lispro in L6 skeletal muscle cells. Clinical Therapeutics, 20(1), 125-140.
81. Slieker, L.J. and Sundell, K.L. (1994). In vitro analysis of Lys(B28), Pro(B29) human insulin (LY275585): comparison to human insulin in terms of insulin and IGF-I receptor binding, glucose uptake into adipocytes and thymidine incorporation into smooth muscle cells. Unpublished report on file Lilly Research Laboratories, Preclinical Pharmacology Report No. 7.
82. Fawcett, J. et al. (1998). Effect of insulin analogs on DNA synthesis in cultured rat liver cells. Diabetes, 47(Suppl. 1), A410.
83. Lilly Research Laboratories. Data on file.
84. Helton, D.R. et al. (1996). General pharmacology of insulin lispro in animals. Arzneimittel-Forschung/Drug Research, 46(I), 91-97.
85. Zimmermann, J. (1994). Subchronic and chronic toxicity, and mutagenicity studies conducted with LysPro [Lys(B28), Pro(B29)] human insulin analog, LY275585. Fifteenth International Diabetes Federation Congress (Nov 6-11, Kobe, Japan). p. 123.
86. Zimmermann, J. (1994). A 12-month chronic toxicity study of LY275585 (human insulin analog) administered subcutaneously to Fischer 344 rats. Diabetes, 43(Suppl. 1), 166A.
87. Zimmermann, J.L. and Truex, L.L. (1997). 12-month chronic toxicity study of LY275585 (human insulin analog) administered subcutaneously to Fischer 344 rats. International Journal of Toxicology, 16, 639-657.
88. Buelke-Sam, J. et al. (1994). A reproductive and developmental toxicity study in CD rats of LY275585, [Lys(B28), Pro(B29)]-human insulin. Journal of the American College of Toxicology, 13(4), 247-260.
89. Zwickl, C.M. et al. (1995). Immunogenicity of biosynthetic human LysPro insulin compared to native-sequence human and purified porcine insulins in rhesus monkeys immunised over a 6-week period. Arzneimittel-Forschung/Drug Research, 45(I), 524-528.
90. Howey, D.C. et al. (1994). [Lys(B28), Pro(B29)]-human insulin. A rapidly absorbed analogue of human insulin. Diabetes, 43, 396-402.
91. Howey, D.C. et al. (1995). [Lys(B28), Pro(B29)]-human insulin: effect of injection time on postprandial glycemia. Clinical Pharmacology and Therapeutics 58, 459-469.
92. Heinemann, L. and Woodworth, J. (1998). Pharmacokinetics and glucodynamics of insulin lispro. Drugs of Today, 34(Suppl. C), 23-36.
93. Heinemann, L. et al. (1996). Prandial glycaemia after a carbohydrate-rich meal in type 1 diabetic patients: using the rapid acting insulin analogue [Lys(B28), Pro(B29)] human insulin. Diabetic Medicine, 13, 625-629.
94. Anderson, J.H., Jr. and Koivisto, V.A. (1998). Clinical studies on insulin lispro. Drugs of Today, 34(Suppl. C), 37-50.
95. Anderson, J.H. et al. and the Multicenter Insulin Lispro Study Group. (1997). Reduction of postprandial hyperglycemia and frequency of hypoglycemia in IDDM patients on insulin-analog treatment. Diabetes, 46, 265-270.
96. Anderson, J.H. et al. (1997). Mealtime treatment with insulin analog improves postprandial hyperglycemia and hypoglycemia in patients with non-insulin-dependent diabetes mellitus. Archives of Internal Medicine, 157(11), 1249-1255.
97. Anderson, J.H. et al. and the Multicenter Insulin Lispro Study Group.(1997). Improved mealtime treatment of diabetes mellitus using an insulin analogue. Clinical Therapeutics, 19(1), 62-72.

98. Lean, M.E. et al. (1985). Interval between insulin injection and eating in relation to blood glucose control in adult diabetics. British Medical Journa,1 290, 105-108.
99. American Diabetes Association. (1997). Standards of medical care for patients with diabetes mellitus. Diabetes Care, 20(Suppl. 1), S5-S13.
100. Dimitriadis, G.D. and Gerich, J.E. (1983). Importance of timing of preprandial subcutaneous insulin administration in the management of diabetes mellitus. Diabetes Care, 6(4), 374-377.
101. Vignati, L. et al. (1997). Efficacy of insulin lispro in combination with NPH human insulin twice per day in patients with insulin-dependent or non-insulin-dependent diabetes mellitus. Clinical Therapeutics, 19(6), 1408-1421.
102. Zinman, B. et al. (1997). Insulin Lispro in CSII. Results of a double-blind crossover study. Diabetes, 46, 440-443.
103. Melki, V. et al. (1998). Improvement of HbA_{1c} and blood glucose stability in IDDM patients treated with lispro insulin analog in external pumps. Diabetes Care, 21(6), 977-982.
104. Pfützner, A., Renner, R. and The German Humalog CSII Study Group. (1997). CSII therapy with insulin pumps using insulin lispro. Diabetes, 46(Suppl. 1), 34A.
105. Schmauss, S. et al. (1998). Human insulin analogue [LYS(B28), PRO(B29)]: the ideal pump insulin? Diabetic Medicine, 15(3), 247-249.
106. Campbell, R.K. et al. (1998). Impact on clinical status and quality of life of switching from regular insulin to insulin lispro among patients using insulin pumps. The Diabetes Educator, 24(1), 95-99.
107. Schernthaner, G. et al. (1998). Postprandial insulin lispro: a new therapeutic option for type-1 diabetic patients. Diabetes Care, 21(4), 570-573.
108. Holcombe, J.H. et al. (1998). Comparative study of insulin lispro and regular insulin in prepubertal children with type 1 diabetes. Diabetes, 47(Suppl. 1), A96.
109. Rutledge, K.S. et al. (1997). Effectiveness of postprandial Humalog in toddlers with diabetes. Pediatrics, 100(6), 968-972.
110. Holcombe, J. et al. (1997). Insulin lispro (LP) results in less nocturnal hypoglycemia compared with regular human insulin in adolescents with type 1 diabetes. Diabetes, 46(Suppl. 1), 103A.
111. Holcombe, J. et al. (1997). Patient preference for insulin lispro versus Humulin R in adolescents with type 1 diabetes. Diabetologia, 40(Suppl. 1), A343.
112. Holcombe, J. et al. (1997). Comparative study of insulin lispro and regular insulin in 481 adolescents with type 1 diabetes. Diabetologia, 40(Suppl. 1), A344.
113. Rami, B. and Schober, E. (1997). Postprandial glycaemia after regular and lispro insulin in children and adolescents with diabetes. European Journal of Pediatrics, 156, 838-840.
114. Kotsanos, J.G. et al. (1997). Health-related quality-of-life results from multinational clinical trials of insulin lispro. Assessing benefits of a new diabetes therapy. Diabetes Care, 20(6), 948-958.
115. Desmet, M. et al. (1994). [Lys(B28),Pro(B29)] human insulin (LysPro): patients treated with LysPro versus human regular insulin: quality of life assessment (QOL). Diabetes, 43(Suppl. 1), 167A.
116. Anderson, J.H., Jr. et al. (1996). Insulin analogues: designer insulins with improved characteristics for better patient care. Diabetes News, 17, 5-7.

Chapter 7

Interferon beta-1b - the first long-term effective treatment of relapsing-remitting and secondary progressive multiple sclerosis (MS)

R. Horowski, J.-F. Kapp, M. Steinmayr, St. Stuerzebecher
Schering AG, SBU Therapeutics, D-13342, Berlin

Key words: Interferon Beta-, multiple sclerosis

Abstract: Beta-interferons like other type I interferons are produced in response to viral infections by various mammalian cells. These include macrophages, dendritic cells and fibroblasts. Type I interferons have non-specific antiviral and anti-proliferative effects, as well as a broad spectrum of immunomodulatory activities. On this basis, type I interferons are used successfully in the treatment of viral infections such as hepatitis C, papilloma and HIV-1 virus. They are also used (mostly in combination with other drugs) to treat some forms of cancer, including leukemia. Whilst these effects could be anticipated from the biological function of type I interferons, it came as a surprise to most when a group of neurologists presented convincing clinical and laboratory evidence of a relevant and important effect of interferon beta-1b in multiple sclerosis (MS). These effects were first noted with regard to its earlier relapsing-remitting form, which is marked by reversible exacerbation, and subsequently also in secondary progressive MS with its increasing physical and especially motor disability. We describe the development, clinical effects and side effects as well as the mode of action of this new therapeutic approach to MS.

1. MULTIPLE SCLEROSIS AND INTERFERON BETA-1B

The effectiveness of interferon beta-1b in relapsing-remitting MS (1, 2) has been confirmed by various groups using three different brands of interferon beta-1a (3, 4). With both types of beta-interferons, other studies in secondary progressive MS are on-going or have already been finished (5). The results obtained with interferon beta-1b have led to the first approval of any long-term treatment of MS by the American and European drug authorities. This has changed the general perception of MS, an insidious, unpredictable and progressively disabling demyelinating disease, which can at any time hit any CNS function. MS is no longer a prototype orphan disease where doctors only too often did not dare to tell the patients their diagnosis (which was often disguised by nouns such as encephalomyelitis disseminata or Charcot's disease). Obviously, medical education and self-perception did not permit doctors to state frankly: "We know very little, and we can do virtually nothing." Needless to say, patients given a diagnosis of MS but then deprived of effective medical treatment turned to all kinds of paramedical, 'wonder drugs', 'miracle cures' or 'magic bullets' which were promoted not only by quacks but also by some serious doctors (and indeed, some patients and their relatives). These repeated endeavors were facilitated by great inter-patient variability, not only in the symptomatology but also in the natural course of the disease. Some 10% or more of patients exhibit an extremely benign development of the disease. (If some of these had been recruited by chance, with other patients with more aggressive forms dropping out quite rapidly, *bona fide* belief into a new 'effective' treatment could arise quite easily). All of these magic cures ultimately turned out to be ineffective. Only a few dedicated scientists (many supported by patients' organizations), did not give up but developed new strategies for testing supposed anti-MS therapies. This task was facilitated by the great historical progress in brain imaging, culminating in various MRI (magnetic resonance imaging) technologies used in MS (6), including enhancement by gadolinium (7). In the meantime, less spectacular but extremely important progress in symptomatic treatments, (i.e. palliative care) greatly improved life quality and expectation in the MS population. Finally, improved understanding of immune mechanisms and the availability of various animal models of MS (the so-called 'experimental allergic encephalomyelitis', based on the concept of an autoimmune-triggered demyelination) had prepared the route to a better situation for the patients. Indeed, the success finally obtained with the positive results of the interferon beta-1b trial in MS had three major consequences for the patients and their disease:

1. For the first time, a significant sustained and relevant reduction (30-50%) in the clinical manifestations of the disease (i.e. frequency and severity of attacks, disease related hospitalization frequency and duration, etc.) was achieved. Furthermore, a dramatic reduction in the number, activity and

area of MRI lesions as an indicator of an impact on the pathogenetic manifestations of MS, has been noted.

2. For the first time an option for patients for actively fighting against their disease was made available.

3. A sudden increase in medical, therapeutic and scientific interest in MS by clinicians, scientists and drug companies was recorded. This resulted in an almost immediate exponential growth in publications, as well as in research and therapy projects. This ended the previous status of MS as an orphan disease.

The history of the use of interferons in MS has been reviewed by two leading pioneers in this field (8). In a chronic disease of unknown origin, but with environmental factors involved (and, in its epidemiology, some similarities with poliomyelitis), a viral etiology had been suspected for a long time. One after another candidate virus, however, failed to become accepted as a cause of MS. It was not unreasonable to test the non-specific antiviral interferons in this condition, once enough material had been made available by recombinant technologies. A surprise observation was that γ-interferon (the first interferon type made available in large quantities), not only did not alleviate MS, but resulted in acute exacerbations of the condition (9). As it was known, however, at this time that beta-interferons can antagonize some of the effects of γ-interferon, this result gave direction to a small group of American and Canadian clinicians to study interferon beta-1b.

2. CLINICAL TRIALS

In a large double-blind multicentre placebo-controlled trial in patients with relapsing-remitting MS, two dosages of interferon beta-1b were tested (1.5 and 8 Million International Units (MIU) given s.c. e.o.d). This was designed to achieve maximum plasma levels and efficacy, as shown in an earlier small dose-finding study (10). An interim analysis of this 2-year study by independent experts came to the conclusion that the drug had obvious, significant and dose-dependent effects in the relevant parameters and virtually all patients from the study volunteered to have a third year of therapy. As a final result, it became clear (11) that 8 MIU of IFNb1b administered by s.c. injection every other day reduced MS attacks by 35%, major attacks by 50% and MS-related hospitalizations by 40%. Duration of attacks averaged 44 days in the placebo arm and 19 days in the 8 MIU arm over 3 years. Median time from entry until first attack was 153 days for the placebo arm and 295 days for the 8 MIU arm, with similar delays noted for subsequent attacks. All who had more than three attacks requiring glucocorticoid treatment in any 12-month period were labelled treatment failures. There were 10 failures in the placebo arm and 1 in the 8 MIU arm over the first 2 years. Efficacy has held for 5 years. IFNb is well tolerated and appears safe.

Mean disability score for the placebo group was 2.8 (Kurtzke scale) at entry and 3.2 after 3 years. For the 8 MIU-treated group, values were 3.0 at entry and 3.1 after 3 years. The differences were not significant. For patients who lost at least 1 point on the Kurtzke scale over 3 years, there was a difference in favor of the treated group ($p < 0.043$).

Yearly magnetic resonance imaging (MRI) scans showed a 17% increase in disease burden (median) over 3 years in the placebo-treated group and a 6% decrease in the 8 MIU IFNb1b-treated patients. Serial 6-weekly MRI scans were performed over 2 years on patients at the University of British Columbia. Active scans were reduced by 83% (median) in the 8 MIU IFNb-treated population; mean reduction was 62%. Scans showing active lesions were reduced by 80% (median), with a mean reduction of 55%.

3. PRODUCT APPROVAL

On this basis, the FDA approved the use of the drug in relapsing-remitting MS. The product was subsequently approved within the EU by the EMEA. Interferon beta-1b, which had been investigated under 'orphan drug' status in the US, thus became available to tens of thousands of patients with the relapsing-remitting form of MS and, indeed, is now the 'gold-standard' for any subsequent new development.

In the following years more than 70 000 patients with MS have been treated with interferon beta-1b, and with this increasing long term experience some consensus became possible on how to most effectively use this new therapeutic instrument in MS (12): The decision to treat a patient with interferon beta-1b should be individualized; that is based on each patient's clinical presentation and course of MS. The most common adverse effects noted include (a) injection-side-reactions and (b) flu-like symptoms, which are generally manageable and usually abate after the first few months of treatment. Spasticity may increase. Patients with severe depression or suicidal tendencies should be monitored carefully, and symptomatic treatment should be pursued. Interferon beta-1b is contraindicated in pregnant and nursing women. Interferon beta-1b is effective in reducing the progression of total disease burden as seen on MRI in patients with MS. Its use is relatively straightforward and generally does not require alteration in the symptomatic treatment of MS. Patient education and support remain the mainstays of maintaining compliance through the early phases of therapy.

It is of utmost importance that patients to be treated with beta-interferons have adequate information and realistic expectations and are taught how to self-administer this drug. As with all protein- and peptide-based biological drugs, parenteral application is necessary, but local side-effects and reactions are frequent. Fortunately, as a rule, patients do not quit treatment as a consequence of these difficulties, and even in the rare occurrence of necrotic lessons patients have continued or restarted therapy with no further incident. It is not known what causes such local necrotic reactions. S.C. injection is the preferred way of administration; it is much more convenient to patients in terms of self-administration.

4. ADVERSE EFFECTS

Flu-like reactions, sometimes severe with fever, chills, myalgia and general malaise, are frequent initial side effects of beta-interferon therapies. They usually subside after a few months of therapy but can also be treated by concomitant administration of non-steroidal anti-inflammatory drugs such as ibuprofen (in the pivotal clinical study with interferon beta-1b, NSAID's were not accepted). Another possibility is to use only half of the dose during the initial weeks of therapy. American neurologists also report very good initial tolerability when treatment with interferon beta-1b is started at the same time as a clinical exacerbation is treated with glucocorticosteroids. Depression is another potential side effect. In the pivotal study one patient committed suicide 1 month after she stopped therapy with the low dose of interferon beta-1b. This highlights the problem of depression in people with MS. In addition to depression triggered by a relapse, MS lesions may directly affect neuronal connections in the brain which control mood. Furthermore, it may indirectly produce cytokines which cause depression and fatigue. The failure of a new therapy to effectively treat the disease might also cause another form of reactive depression. All these forms of depression cannot be distinguished with certainty from a drug-induced depressive state. Thus it is advisable to keep close contact with the patients, and to stop interferon therapy whenever there are signs of depression, especially when suicidal tendancies become an issue. Only when antidepressant drugs have stabilized the mood of the patient a cautious restarting of interferon-beta treatment is advised.

Pregnancy is a contraindication as type-I-interferons in animals have been shown to induce abortions. In the limited human experience there have been abortions but also some full-term pregnancies (interferon beta-1b treatment is stopped once pregnancy had been diagnosed).

An issue which has received significant attention is the occurrence of neutralizing antibodies against beta-interferons in MS patients. Again, this is a potential problem with all parenteral protein drugs, and it is also known that MS patients have a strong tendency to develop auto-antibodies. Neutralizing antibodies are described by their inhibitory effect on some interferon beta-1b action (antiviral effect or induction of the MXA proteins) but their clinical significance is still unclear. Whilst there is a tendency that patients, who have developed neutralizing antibodies, have a higher relapse rate and MRI burden of disease than others, the statistical association does not permit individual predictions (13, 14). Furthermore, antibody-positive patients subsequently tend to become negative again (possibly as a result of intermittent glucocorticosteroid therapy of relapses) and in long-term observations no patient has retained neutralizing antibodies after 102 weeks of treatment (15). Only in rare instances, and in patients with very high titers, is there an impression of reduced therapeutic activity. Indeed, lower concentrations of neutralizing antibodies may bind circulating interferon beta-1b in a reversible way, thus even prolonging the effects of this drug.

Side effects and effects of interferon beta-1b appear to be dose-dependent as has been shown in the first pivotal trial. This has been confirmed by subsequent studies. Unfortunately there seems to be a ceiling effect which prevents the drug from becoming 100% effective, and so far, the s.c. application every other day seems to be an optimum, as also indicated by peripheral markers (β2-microglobulin, neopterin, 3'5' oligoadenylate synthetase). Another point of discussion is whether this new therapy has a long-lasting impact on the course of the disease. Whilst the strong effects of interferon beta-1b on MRI lesions predict a strong impact of this therapy on the course of the disease, there are not yet sufficient clinical long-term observations to prove this point. Assessment of disability is quite difficult, and the so-called expanded disability status scale (EDSS), unfortunately, mixes disability with neurological deficits and motor problems, whilst neglecting fatigue and other major early symptoms. It is thus of little value in early disease characterized by little progression of disability in the placebo groups. For this reason, another large double-blind multi-centre trial was started in secondary progressive MS with 718 patients participating. In this population with more severe MS there was a highly significant and relevant effect of interferon beta-1b (8 MIU s.c. e.o.d.) on all disability-related parameters. These include progression in EDSS score and time to wheelchair. A delay of 9-12 months in disease progression resulted from only 2 years' therapy with this drug. Furthermore there was again a clear reduction in the frequency and severity of relapses, MS-related hospitalisations and glucocorticosteroid use. A very strong positive effect on the numbers of new lesions in the MRI was also noted (5).

5. MODE OF ACTION

There can be no doubt that interferon beta-1b has a strong, relevant and lasting impact on all major aspects of MS. This effect is achieved without jeopardizing the immune status of the patients and without any bone marrow depression or other relevant toxicity. How can this be achieved and what is the mechanism of action of this drug? This has not been elucidated and, given the lack of detailed knowledge of MS pathogenesis, the answer will be hard to find.

It is generally assumed that the pathology of MS is related to an activation and proliferation of T cells (Th1 type CD4+) specific for epitopes of myelin antigens (MOG, MBP, MOG, PLP, β-crystalline, etc.). These T cells subsequently enter the brain. By interacting with MHC II molecules on microglia and macrophages (and possibly other cells) and with their specific antigen, they form a so-called tri-molecular complex. This is part of an inflammatory reaction, with production and release of pro-inflammatory cytokines such as γ-interferon, IL-1, IL-2 and TNFα. Probably facilitated by antibodies and complement, these acute inflammation products (via intermediates such as NO and other free radicals), destroy myelin around the axons. Also, either directly or in a secondary way, the oligodendrocytes of which myelin is a part are also destroyed. In addition to demyelination, these mechanism might also interfere with axonal and neuronal function, and this reaction is also correlated with the disruption of the blood-brain barrier and local oedema. All these events may contribute to acute symptomatology, not just the demyelination. Recovery follows by the subsequent activation of anti-inflammatory mechanisms or remyelination. If these fail, persistent local damage with scar formation and lasting axonal and neurological deficit symptoms will result (for review see 15). These symptoms depend on the localisation of the lesions, and are due to the local damage but lesions can also remain 'silent' due to redundancy and plasticity in the brain. Acute symptoms, on the other hand, may also be caused by the same mechanisms or by other effects, such as opening of the blood-brain barrier (which can be visualized by gadolinium-enhanced MRI). Local oedema and inflammation and spreading of cytokines and their products (such as NO) may also be underlining factors. This could explain why in early MS we very frequently see severe fatigue. Finally these humoral factors might also functionally reactivate old lesions and thus cause a relapse of previous symptomatology.

Interferon beta-1b counteracts acute symptoms as well as general disease processes such as the frequency of attacks and the occurrence of new MRI lesions. Therefore, its mechanism of action should explain all these effects. Unfortunately beta-interferons have been shown to have many different effects related to their immunomodulatory action *in vitro* as well as *in vivo*. However, there are already a number of proposed mechanisms for the therapeutic effect of interferon beta-1b in MS (for review see 16). There seems to be some equilibrium between pro- and anti-inflammatory T cell populations and immune mechanisms, which apparently is disturbed in MS. A reduced suppressor cell function, and possibly low endogenous levels of beta-interferon, is likely associated with the disease. On this basis, it can be assumed that interferon beta-1b antagonizes proinflammatory mechanisms, likely by its known inhibitory effects on the production of γ-interferon, TNFα and TNFβ as well as on NO (17-19). At the same time, this drug restores or enhances anti-inflammatory mechanisms by activating so-called suppressor T cells with increased production of IL-10 and TGFb (20,21). Special attention should be given to the action of interferon beta-1b on the opening of the blood-brain barrier where its effects have been described as immediate and dramatic (22). Here, the effects of interferon beta-1b on adhesion molecules such as ICAM may be important (23-25). Effects on cell trafficking, on metalloprotenase (26, 27) and NO radicals (28) may also be important. Last but not least, one should keep in mind that beta-interferon can induce its own synthesis ('cascade-effect') and that its therapeutic effects may also be propagated by migration and trafficking of T cells activated in a paracrine way. From the results so far, there is no evidence that the other classical actions of interferons, i.e. antiviral and anti-proliferative effects, play any prominent role in its therapeutic use in MS. A simplified general schedule of events during MS attacks and possible effects of interferon–beta 1b is given in Table 16 (from 16, modified).

Interferon beta-1b - the first long-term effective treatment of relapsing-remitting and secondary progressive multiple sclerosis (MS)

Table 16. Schedule of likely events occurring during an MS attack and the possible mechanisms by which interferon beta-1b counteracts some of these events

Steps in the cascade of an MS attack and pathogenetic mechanisms	Effects of interferon beta-1b
Etiology unknown	—
Activation (by non-specific stimuli) of peripheral Th1 cells specific for MOG, MAG, MBP, PLP, β-cystalline etc. epitopes.	Restoration of deficient suppressor cell function (16,17) (increased/variable MHC I/MHC II expression, decreased B7-1 expression, increased B 7-2 expression (16))
Recruitment of peripheral specific Th1 cells and cell entry into lymph nodes	Apparent fall in circulating lymphocytes (1,2)
Expansion and T cell proliferation.	Inhibition of T cell proliferation (16)
Cell trafficking, increased adhesion molecule expression, interferon-γ production	Reduced adhesion molecule production (but variable serum levels, 18, 19, 20) antagonism of γ-interferon effects (16) and down-regulation of interferon-γ-induced MHC II expression (21)
Penetration of blood-brain barrier by NO, and metalloproteases	Closing of blood-brain barrier disruption (22) inhibition of metalloproteases and NO production *in vitro* (23, 24)
Local inflammation and tissue damage by trimolecular complex (MS antigens) with increased interferon-γ, TNFα, TNFβ, NO production and demyelination	Inhibition of pro-inflammatory cytokine production and effects of NO production (23-25), activation of anti-inflammatory mechanisms (increased IL-10 (26, 27) TGFβ (27), IL-6 (28)
Recovery (with or without scar formation and persistent axonal loss, T cell apoptosis, macrophage emigration until next attack	

6.   CONCLUSION

It is not believed likely that higher dosages of interferon beta-1b or related drugs will further enhance efficacy against MS. For a complete suppression of MS activity, combinations of this main therapy with other drugs displaying new and complementary actions may be the best option in the future. Other attempts such as desensitization strategies, induction of tolerance, immunization or promotion of remyelination are, at this moment, quite unlikely to have an impact on the patients' situation within the next 10 years.

In conclusion, general scientific progress, increased knowledge of immune mechanisms, gene technology and especially improved diagnostic methods such as MRI and new therapeutic strategies, have all contributed to the development of interferon beta-1b as a major breakthrough in the therapy of MS. This endeavour has been supported by favourable conditions such as orphan drug legislation and other public support (e.g. by NIH, patients and their organisations).

BIOGRAPHY

R. Horowski is an M.D. and pharmacologist who has contributed to the development of a new anti-Parkinson drug and a new hypnotic. J.-F. Kapp is an M.D. and head of the Therapeutics Group of Schering A.G. M. Steinmayr is a Ph.D. and in charge of marketing in the Therapeutic Group and St. Stürzebecher is an M.D. and head of clinical development in Therapeutics.

Schering AG is a pharmaceutical company based in Berlin/Germany for more than 125 years (it is not related to the U.S. Schering Corporation). Schering AG has a strong tradition as a research-oriented company, with a strong record in the field of sexual steroids (with the first discovery resp. isolation of estrogens and androgens and with ethinylestradiol, gestodene, cyproterone acetate, norethisterone acetate, etc. all originating here). This has made Schering the world's largest producer of this type of drug and a leader in the field of contraception and hormonal therapies. Furthermore, Schering AG is a leader in the field of contrast agents used for imaging (X-ray, MRI, ultrasound) after the first iodinated compound useful for selective urography had been developed in Berlin in the 1920s. More recently, Schering and its US affiliations, Berlex/Berlex Biosciences, have entered the fields of oncology, cardiology and CNS diseases with interferon beta-1b (Betaseron®) as one of the last and most successful new developments. Thus, Schering is now also a market leader in the field of multiple sclerosis.

REFERENCES

1. The IFNB Multiple Sclerosis Study Group. (1993). Interferon beta-1b is effective in relapsing-remitting multiple sclerosis. I. Clinical results of a multicenter, randomized, double-blind, placebo-controlled trial. Neurology, 43, 655-661.
2. Paty, D.W., Li, D.K.B., the UBC MS/MRI Study Group and the IFNB Multiple Sclerosis Study Group. (1993). Interferon beta-1b is effective in relapsing-remitting multiple sclerosis. II. MRI analysis results of a multicenter, randomized, double-blind, placebo-controlled trial. Neurology, 43, 662-667.
3. Jacobs, L.D. et al. (1996). Intramuscular Interferon Beta-1a for Disease Progression in Relapsing Multiple Sclerosis. Ann. Neurol., 39, 285-294.
4. Ebers, G.C. (1997). The multiple sclerosis PRIMS study: prevention of relapses and disability by interferon-beta-1a subcutaneously in multiple sclerosis, Paper presented at the 122. ANA Meeting, San Diego, Sept. 30, 1997.
5. Kappos, L. et al. (1998). Interferon Beta-1b (IFN Beta-1b) Delays Progression of Disability in Secondary Progressive Multiple Sclerosis: Results of the European Multicentre Study. J. Neurol., 245, 357.
6. Willoughby, E.W. et al. (1989). Serial magnetic resonance scanning in multiple sclerosis: a second prospective study in relapsing patients. Ann. Neurol., 25, 43-49.
7. Gonzales-Scasano F. et al. (1987). Multiple Sclerosis disease activity correlates with gadolinium-enhanced MRI. Ann. Neurol., 21, 300-306.
8. Jacobs, L. and Johnson, K.P. (1994). A Brief History of the Use of Interferons as Treatment of Multiple Sclerosis. Arch. Neurol., 51, 1245-1252.
9. Panitch, H.S. et al. (1987). Treatment of MS with gamma interferon: exacerbations associated with activation of the immune system. Neurology, 37, 1097-1102.
10. Johnson, K.P. et al. (1990). Rec. Interferon beta treatment of RR MS: pilot study results Neurology, 40, Suppl. 1, 261.
11. Arnason, B.G.W. and Reder, A.T. (1994). Interferons and Multiple Sclerosis. Clin. Neuropharmacol., 17 (6), 495-547.
12. Lublin, F.D. et al. (1996). Management of patients receiving interferon beta-1b for multiple sclerosis: Report of a consensus conference. Neurology, 46, 12-18.
13. The IFNB Multiple Sclerosis Study Group and the UBC MS/MRI Study Group. (1996). Neutralizing antibodies during treatment of multiple sclerosis with interferon beta-1b: Experience during the first three years. Neurology, 47, 889-894.
14. Pachner, A.R. (1997). Anticytokine antibodies in beta-interferon-treated MS patients and the need for testing. Neurology, 49, 647-650.
15. Rice, G. (1997). The evolution of neutralizing antibodies in patients taking beta interferon 1b. Multiple Sclerosis, 3(5), 344.
16. Arnason, B.G.W. et al. (1996). Mechanisms of action of interferon-b in multiple sclerosis. Springer Semin Immunopathol, 18, 125-148.
17. Wandinger, K.-P. et al. (1997). Diminished production of type-I interferons and interleukin-2 in patients with multiple sclerosis. J. Neurol. Sci., 149, 87-93.
18. Dhib-Jalbut, S. et al. (1996). The effect of interferon β-1b on lymphocyte-endothelial cell adhesion. J. Neuroimmun., 71, 215-222.

19. Miller, A. et al. (1996). Immunoregulatory effects of interferon-β and interacting cytokines on human vascular endothelial cells. Implications for multiple sclerosis and other autoimmune diseases. J. Neuroimmun., 64, 151-161.
20. Defazio, G. et al. (1998). ICAM 1 expression and fluid phase endocytosis of cultured brain microvascular endothelial cells following exposure to interferon β-1a and TNFα. J. Neuroimmun. 88, 13-20.
21. Huynh, H.K. et al. (1995). Interferon-β downregulates interferon-γ-induced class II MHC molecule expression and morphological changes in primary cultures of human brain microvessel endothelial cells. J. Neuroimmun., 60, 63-73.
22. Stone, L.A. et al. (1995). The Effect of Interferon-β on Blood-Brain Barrier Disruptions Demonstrated by Contrast-enhanced Magnetic Resonance Imaging in Relapsing-Remitting Multiple Sclerosis. Ann Neurol., 37, 611-619.
23. Defazio, G. et al. (1996). Interferon β-1b effects on cytokine mRNA in peripheral mononuclear cells in multiple sclerosis. Multiple Sclerosis, 1, 262-269.
24. Dayal, A.S. et al. (1995). Interferon-gamma.secreting cells in multiple sclerosis patients treated with interferon beta-1b. Neurology, 45, 2173-2177.
25. Genc, K. et al. (1997). Increased CD80 B Cells in Active Multiple Sclerosis and Reversal by Interferon β-1b Therapy. Clin. Invest., 99 (11), 2664-2671.
26. Porrini, A.M. et al. (1995). Interferon effects on interleukin-10 secretion. Mononuclear cell response to interleukin-10 is normal in multiple sclerosis patients. J. Neuroimmun. 61, 27-34.
27. Rep, M.H.G. et al. (1996). Recombinant interferon-b blocks proliferation but enhances interleukin-10 secretion by activated human T-cells. J. Neuroimmun. 67, 111-118.
28. Brod, S.A. et al. (1996). Interferon-β 1b treatment decreases tumor necrosis factor-α and increases interleukin-6 production in multiple sclerosis. Neurology, 46, 1633-1638.
29. Corsini, E. et al. (1997). Effects of β-IFN-1b treatment in MS patients on adhesion between PBMNCs, HUVECs and MS-HBECs: an *in vivo* and *in vitro* study. J. Neuroimmun. 79, 76-83.
30. Calabresi, P.A. et al. (1997). Increases in Soluble VCAM-1 Correlate with a Decrease in MRI Lesions in Multiple Sclerosis Treated with Interferon β-1b. Ann. Neurol., 41, 669-674.
31. Billiau, A. (1995). Interferons in multiple sclerosis: Warning from experiences. Neurology, 45, 50-53.

Chapter 8

Reteplase, a recombinant plasminogen activator

Dr Michael Waller and Dr Ulrich Kohnert
Boehringer Mannheim Therapeutics, Mannheim and Penzberg, Germany

Key words: Acute myocardial infarction, thrombolytic therapy, recombinant plasminogen activator, reocclusion, reteplase, alteplase, streptokinase.

Abstract: Thrombolysis is now standard therapy for the treatment of acute myocardial infarction. However, evidence suggests that earlier treatment and more rapid and complete recanalization of infarct-related coronary arteries may lead to greater survival benefits than achieved by current therapies. Furthermore, first- and second-generation thrombolytics can be associated with side effects, such as bleeding complications, and often require prolonged infusion or complex dosing regimens to optimize clinical outcome. Reteplase, a novel recombinant plasminogen activator (thrombolytic), is a deletion variant of native, human t-PA that has been designed to provide a longer half-life plus more specific and rapid lysis of coronary thrombi, using a bolus dosing regimen. Clinical studies comparing reteplase with the current standard thrombolytic agents have demonstrated that reteplase has a highly favourable pharmacological profile with prolonged half-life, low bleeding risk and low potential for antigenicity.

1. INTRODUCTION

Thrombolytic therapy is now established as standard emergency treatment for acute myocardial infarction (AMI) in industrialized countries and its use has been proven to reduce mortality rates and enhance recovery of left ventricular function (1-7). As experience with thrombolytic therapy has increased, it has been confirmed that the faster and more complete the restoration of blood flow in the infarct-related coronary artery, the greater the potential reduction in mortality and residual dysfunction (8).

Until recently, four agents were approved for use in this indication: streptokinase, urokinase, recombinant tissue-type plasminogen activator (rt-PA, alteplase) and acetylated plasminogen streptokinase complex (APSAC, anistreplase) (9). Even using optimal regimens, these first- and second-generation thrombolytics only achieve complete reperfusion (Thrombolysis in Myocardial Infarction [TIMI] grade 3 flow) within 90 minutes in about half of patients, with even poorer patency rates at earlier time points (10). The therapeutic benefit of thrombolysis is often further limited by early reocclusion, the risk of haemorrhage, especially intracranial bleeding, hypotension and allergic reactions (with streptokinase), and the inconvenience and delay involved in having to administer drugs by continuous intravenous infusion, due to their short half-lives.

Reteplase was developed in response to the need to improve the risk/benefit ratio of thrombolytic therapy for patients. Using molecular biological techniques, reteplase was designed to eliminate the undesirable features of native t-PA to produce a therapeutic thrombolytic agent with a superior clinical profile (11).

2. BIOCHEMICAL CHARACTERISTICS

Reteplase is genetically engineered. Its expression in *Escherichia coli* (*E. coli*) results in a non-glycosylated protein, which accumulates (inside the cells) as inactive inclusion bodies. The reteplase gene lacks the complementary DNA sequences for the three N-terminal domains found in t-PA (finger, epidermal growth factor [EGF] and kringle-1 domains), but retains the kringle-2 and the serine protease domains (12). Like natural t-PA, reteplase is a single-chain molecule, which can be converted to the two-chain form by cleavage at the Arg_{275}-Ile_{276} bond during fibrinolysis (12,13).

2.1 Molecular structure

Reteplase contains 355 amino acids, which comprise amino acids 1-3 and 176-527 of human t-PA (12). The molecular structure of reteplase is shown in Figure 19. The arrowhead indicates the site at which the single-chain is cleaved to the two-chain form; active site residues are marked by black circles; disulphide bonds are represented by connecting bars; and K_2 and P denote the kringle-2 and protease domains, respectively.

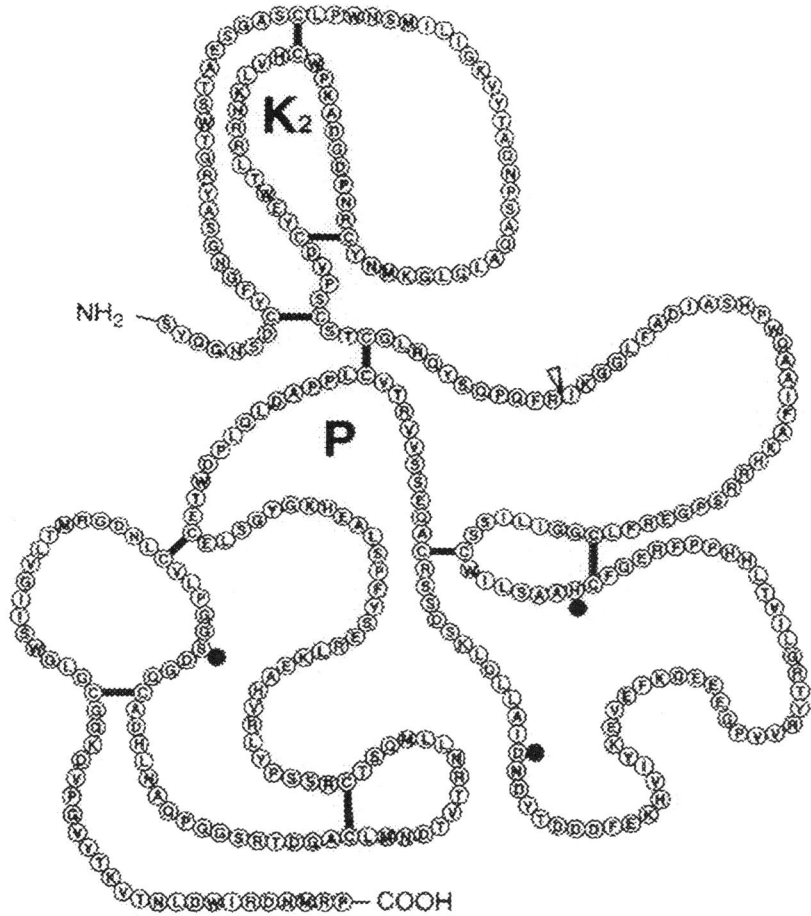

Figure 19. Molecular structure of reteplase

2.2 Chemical name

The chemical name for reteplase, assigned by the World Health Organization, is 173-L-serine-174-L-tyrosine-175-L-glutamine-176-527-plasminogen activ-ator (human tissue type).

2.3 Molecular weight

Reteplase has a molecular weight of 39571.5 Daltons.

2.4 Molecular formula

The molecular formula of reteplase is: $C_{1736}H_{2653}N_{499}O_{522}S_{22}$

2.5 Pharmacological class

Reteplase is a plasminogen activator, and belongs to the same pharmacological class as alteplase, anistreplase, streptokinase and urokinase.

3. INDICATION AND RATIONALE

Reteplase is indicated for the thrombolytic treatment of AMI, within 12 hours after the onset of symptoms.

3.1 AMI overview

The acute formation of an occlusive thrombus within a coronary artery causes ischaemia in the area of the myocardium supplied by the occluded vessel and abnormal contraction of the left ventricle. Prolonged ischaemia may progress to myocardial necrosis, or infarction, causing death or permanent left ventricular dysfunction, the severity of which is proportional to infarct size (14). Use of thrombolytic therapy to dissolve the fibrin clot in an occluded coronary vessel and restore blood flow allows salvage of jeopardized (ischaemic, but not yet necrosed) myocardium, thereby minimizing infarct size and reducing post-AMI morbidity and mortality (1-7,15,16).

3.2 The open-artery theory and thrombolytic therapy

According to the open-artery theory, first proposed in 1941, myocardial damage following AMI may be minimized by restoring blood flow in an occluded infarct-related artery as rapidly as possible (17). This proposition was confirmed by the landmark GUSTO angiographic study (8), and the dangers of early reocclusion recognized (10). Thus, the primary goal of thrombolytic therapy has been to achieve stable reperfusion (TIMI grade 3 flow) of the coronary artery as rapidly as possible after the onset of AMI, without causing adverse effects.

3.3 Problems with first- and second-generation thrombolytics

All available therapeutic thrombolytic agents act either by directly or indirectly cleaving plasminogen at the peptide bond Arg_{560}-Val_{561} to generate plasmin (9), which then lyses fibrin (Figure 20).

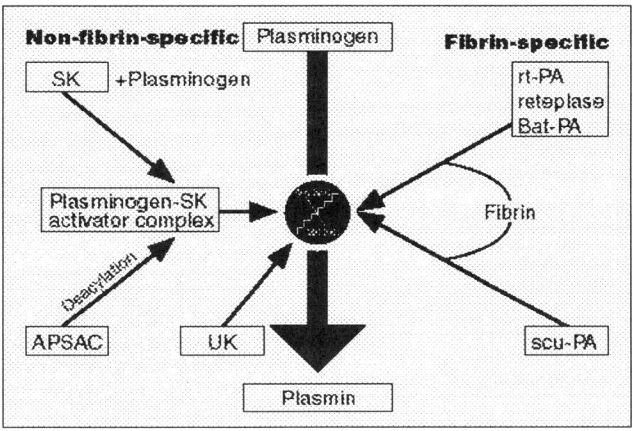

Figure 20. Action of thrombolytic agents according to fibrin specificity (9). SK = streptokinase; UK = urokinase; APSAC = anistreplase; rt-PA = alteplase; scu-PA = saruplase; Bat-PA = vampire bat plasminogen activator

However, these agents differ in their specificity for fibrin and exhibit other important differences in their pharmacological properties, creating significant variation in their clinical performance (9,18) (Table 17).

Streptokinase, urokinase and anistreplase are not fibrin-specific; all convert plasminogen to plasmin whether or not it is bound to fibrin, causing systemic activation of plasminogen (18). This not only produces slower recanalization than with fibrin-specific alteplase, but also depletes α^2-antiplasmin, the primary plasmin inhibitor, causing high levels of circulating plasmin, systemic fibrinolysis and an elevated risk of bleeding (18). Of the first- and second-generation agents, alteplase achieves the highest rate of early patency, but provides complete reperfusion after 90 minutes in only 54% of patients (8). Despite its fibrin specificity, alteplase is associated with an even higher risk of cerebral bleeding complications than streptokinase (19). As a result of their short half-lives, streptokinase, urokinase and alteplase must be administered by intravenous infusion (9), limiting their use outside the hospital environment and thus, probably, delaying the initiation of thrombolysis.

Table 17. Characteristics of first- and second-generation thrombolytic agents [Adapted from (9,18)]

Agent	Streptokinase	Urokinase	Alteplase	Anistreplase
Source	gp C streptococci	Recombinant human foetal kidney	Recombinant human	gp C streptococci plasminogen anisoylated
Molecular weight	47 kD	35–55 kD	63–70 kD	131 kD
Fibrin specific?	No	No	Yes	No
Metabolism	Hepatic	Hepatic	Hepatic	Hepatic
Mode of action	Activator complex	Direct	Direct	Activator complex
Antigenicity	Yes	No	No	Yes
Peak effect	6–24 h	20 minutes	45 minutes	45 minutes
Half-life (minutes)	18–23	14–20	3–4	70–120
Coronary patency rate (TIMI 2+3 flow) at 90 minutes	50–70%	50–70%	70–75%	55–75%
Complete reperfusion (TIMI 3 flow) at 90 minutes	40–50%	35–45%	49–54%	40%

3.4   Profile of the ideal thrombolytic

Experience with the first- and second-generation agents, and identification of key requirements for optimizing the outcome of thrombolytic therapy, suggested that improved outcomes could be obtained by eliminating the drawbacks associated with these earlier agents. The aim in developing reteplase was, therefore, to produce an agent with a profile that closely matches that for the ideal thrombolytic (20) (Table 18).

These characteristics were simplified into four key clinical objectives for a new thrombolytic:

1. Quick, easy, convenient administration
   Preferably by fast, intravenous bolus injection of the entire dose to ensure early reperfusion, and enable emergency therapy to start prior to hospital admission (21).
2. Fast action
   Minimizing time to reperfusion, thereby maximizing mortality benefits (2,8,19,22).
3. Efficient lysis
   Achieving complete patency in as many patients as possible (8,19,22).

4. Good safety profile
   Including a low risk of bleeding (especially intracranial bleeding) and a low risk of antigenicity.

Table 18. Characteristics of the ideal thrombolytic agent

Characteristics
Rapid recanalization after administration
Approach 100% efficacy for recanalization
Can be given as rapid intravenous bolus injection
Specific for recent thrombi
Assists in preventing reocclusion
Should result in sustained patency over the first 24 hours
Appropriate half-life for bolus injection
Targets thrombus induced by plaque rupture
No effect on circulatory haemodynamics
No negative interactions with adjunctive therapies
No antigenicity
No significant side effects

4. THE MOLECULAR DEVELOPMENT OF RETEPLASE

4.1 Analysing structure-function relationships within the t-PA molecule

Natural t-PA is a single-chain polypeptide serine protease with a molecular weight of approximately 70 kD. The chain contains 527 amino acid residues, and can be cleaved at the Arg_{275}-Ile_{276} bond into a two-chain form (26). Recombinant t-PA for therapeutic use (alteplase) is produced in Chinese hamster ovary (CHO) cells. It has the same amino acid sequence and domain composition as natural t-PA (26). The structure-function relationships of t-PA have been analysed by testing mutant forms of the molecule in *in vitro* and *in vivo* models (27) (Table 19).

Table 19. Structure-function relationships of t-PA domains (27)

Domain	Function
Finger (F domain)	High-affinity fibrin binding
Epidermal growth factor (EGF domain)	Receptor binding (liver)
Kringle-1 (K_1 domain)	Receptor binding
Kringle-2 (K_2 domain)	Fibrin specificity (stimulation)
Protease (P domain)	Plasminogen-specific protease, PAI-1 binding site
Carbohydrates	Mediators of plasma clearance

The heavy A chain of t-PA contains the fibronectin finger-like domain, which is associated with high-affinity fibrin binding, the EGF domain, which binds hepatic receptors, accelerating plasma clearance, the kringle-1 domain associated with receptor binding, and the kringle-2 domain which allows stimulation of the activity by fibrin.

The light chain contains the plasminogen-specific protease domain. The t-PA molecule contains four potential glycosylation sites (26), three of which are actually glycosylated. The site on kringle-1, ASN_{117}, is a high-mannose-type side chain, which is associated with rapid plasma clearance (27). The two other glycosylation sites, ASN_{184} in kringle-2 and ASN_{448} in the protease domain, carry complex side chains (28).

4.2   Molecular modifications to produce reteplase

Extensive work on t-PA variants in our laboratories demonstrated that it is possible to produce new molecules with different properties. Reflecting the requirements for an improved thrombolytic, reteplase was designed to retain the plasmin forming activity, the fibrin specificity and to have a longer plasma-half-life which allows bolus injection instead of infusion.

The complementary DNA sequences coding for the finger domain, responsible for high affinity fibrin binding, the EGF- and the kringle-1 domains, which contribute for plasma clearance, were deleted (12, 23). The deletion of domains was carried out according to the intron-exon-transitions of the t-PA gene in order to reduce the risk of producing a modified molecule that may be recognized as a foreign protein by the human immune system. The remaining sequence coding for the kringle 2 and the protease domain was inserted into a plasmid and introduced into *E. coli* for protein synthesis.

In contrast to mammalian cells, the production of reteplase in *E. coli* results in a non-glycosylated protein without introduction of specific point mutations because bacteria lack the enzyme systems necessary for the glycosylation of proteins. The absence of carbohydrate side chains prevents the clearance of reteplase by hepatic carbohydrate receptors and allows a maximum prolongation of the plasma half-life.

5.   COMPARISON OF THE BIOCHEMICAL PROPERTIES OF RETEPLASE AND ALTEPLASE

Reteplase has undergone extensive biochemical analysis, confirming that the kringle-2 and protease domains are refolded into the correct structure, and that their biochemical properties are independent of the three N-terminal domains of native t-PA (12).

SDS-polyacrylamide electrophoresis (SDS-PAGE) analysis, amino acid sequence analysis and analysis by mass spectroscopy have revealed that the molecular weight of the single-chain form and the N-terminal amino acid sequence of reteplase are as expected from the DNA sequence (12).

5.1 Cleavage by plasmin

Incubation of reteplase with plasmin gives only one additional N-terminal sequence starting with Ile_{276}, indicating that reteplase, like native t-PA, is cleaved by plasmin specifically at the Arg_{275} –Ile_{276} bond. This was demonstrated by incubating both reteplase and rt-PA with Sepharose-bound plasmin, to generate the two-chain forms of both enzymes. Both the single- and two-chain forms of reteplase and rt-PA underwent SDS-PAGE analysis under reducing conditions (12). A comparison of the SDS-PAGE analysis is shown in Figure 21.

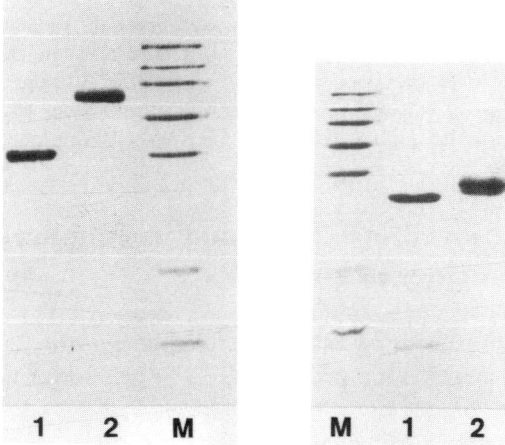

Figure 21. SDS-PAGE analysis of (a) the single-chain forms and (b) the two-chain forms of reteplase and rt-PA under reducing conditions (12). The marker proteins (Lane M) are: α_2-macroglobulin (170 kDa); β-galactosidase (116 kDa); fructose-6-phosphate-kinase (85 kDa); glutamate dehydrogenase (56 kDa); aldolase (39 kDa); carbonic anhydrase (30 kDa, only visible in b); soybean trypsin inhibitor (20 kDa); lysozyme (14 kDa). Lane 1 is reteplase and Lane 2 is rt-PA. Electrophoresis was carried out on 12.5% and 18% gels for the single-chain and two-chain forms, respectively.

5.2 Amidolytic activity

The amidolytic activity of reteplase, indicated by kinetic constants K_m and k_{cat}, is similar to that of rt-PA, both for the single- and two-chain forms. Incubation with plasmin (i.e. generation of the two-chain forms) increases the catalytic efficiency (k_{cat}/K_m) of both reteplase and rt-PA by factors of 12 and 17, respectively. These results were obtained in an assay using S-2288 as the substrate (12). Data is provided in Table 20.

Table 20. Kinetic constants for the amidolytic activity of reteplase and rt-PA on S-2288 (12)

Enzyme	K_m (mmol/l)	k_{cat} (s^{-1})	k_{cat}/K_m (s^{-1} l/mmol)
Single-chain forms			
BM 06.222	2.5	13.9	5.6
CHO-t-PA	2.1	11.4	5.4
Two-chain forms			
BM 06.222	0.5	33.9	67.8
CHO-t-PA	0.3	27.1	90.3

The amidolytic activity of BM 06.222 and CHO-t-PA on S-2288 was determined in 0.1 M Tris-HCl, pH 8.5, 0.15% Tween 20 at enzyme concentrations of 1 and 1.5 µg/ml, respectively. The substrate concentrations were varied between 0.1 and 5 mmol/l for the single-chain and the two-chain forms of the enzymes. The reaction rate was determined within 2.5 minutes. Due to the poor solubility of S-2288, K_m and k_{cat} could not be determined from the plots of velocity versus substrate concentration. Therefore both values were only calculated from the double reciprocal Lineweaver-Burk plots. All values are the mean of two experiments.

5.3 Plasmin-forming activity and its enhancement by fibrin, fibrinogen and FDPs

The rate of plasmin-forming activity of reteplase in the presence of fibrin monomer and fibrin degradation products (FDPs) was lower by factors of 2.0 and 4.3, respectively, as compared to alteplase. Like t-PA, reteplase was only marginally stimulated by fibrinogen (29). This data demonstrates that comparable plasminogenolytic activities and fibrin specificities of reteplase and alteplase can be expected *in vivo*.

5.4 Inhibition by PAI-1

Plasminogen activator inhibitor (PAI-1) rapidly inhibits t-PA in human plasma by forming a tightly bound 1:1 complex with the plasminogen activator (30). Inhibition of reteplase by PAI-1 is identical to that of rt-PA, indicating that the PAI-1 binding sites within the protease and the kringle-2 domains of reteplase are functionally and spatially preserved (12) (Figure 22).

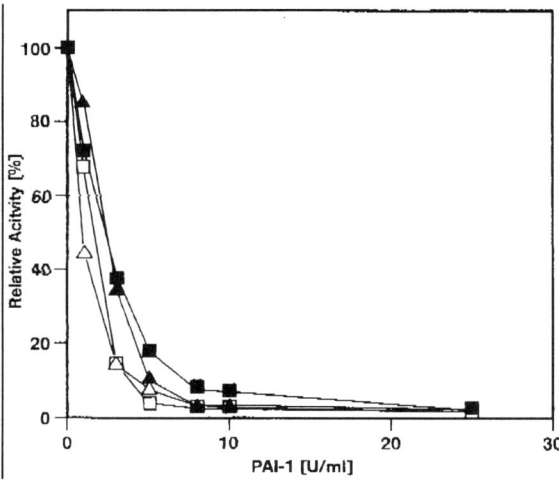

Figure 22. Inhibition of reteplase and rt-PA by PAI-1 (12). 5 ng of the respective single- and two-chain forms of reteplase and rt-PA were incubated with increasing amounts of PAI-1. The relative activity was calculated as (activity in the presence of PAI-1 divided by activity in the absence of PAI-1) x 100. (The closed squares = reteplase [single-chain form]; closed triangles = rt-PA; open squares = reteplase [two-chain form]).

5.5 Affinity for fibrin and lysine

The comparison of the fibrin binding of reteplase and alteplase reveals significant differences. Alteplase binds completely to a fibrin clot, whereas 65-70% of reteplase is found in the supernatant of the clot. This was confirmed in a clot-binding assay in which 200 ng each of reteplase and alteplase were mixed in increasing concentrations of fibrinogen, to which thrombin was added to induce clot formation. As shown in Figure 23, alteplase almost completely binds to a fibrin clot formed from approximately 100 µg of fibrinogen, whereas less than 40% of reteplase is bound at levels of 100 and 150 µg (25).

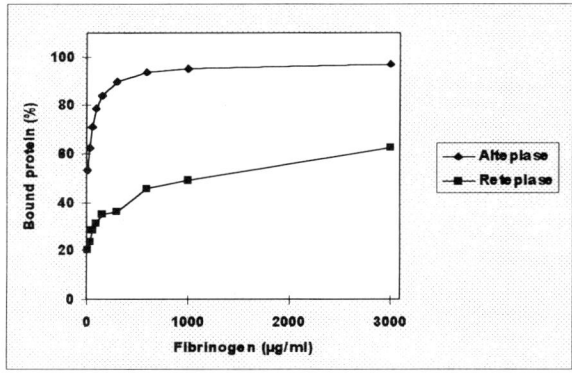

Figure 23. Comparison of the fibrin binding of 200 ng of alteplase and reteplase. The plasminogen activators were mixed with increasing concentrations of fibrinogen. The formation of the clot was induced by the addition of thrombin. The amount of bound plasminogen activators was calculated as the difference between the plasminogen activator added and the activator in the supernatant of the clot after centrifugation. Reteplase and alteplase were determined by an ELISA using polyclonal anti-t-PA-antibodies and a standard curve for each plasminogen activator (25).

The non-specific binding of reteplase to fibrin in this *in vitro* model is almost completely suppressed by 0.3 mM ε-aminocaproic acid (EACA) (a lysine analogue) whereas rt-PA has a residual affinity of 55% (12). These results arise from the different structures of t-PA and reteplase. Besides the binding of t-PA via the high-affinity fibrin binding site on the finger domain, the *in vitro* model also allows binding via the non-specific lysine binding site on kringle-2. The lysine binding site is maintained in reteplase. Reteplase and rt-PA possess the same affinity for lysine and lysine analogues (12).

5.6 Expression of reteplase in *E. coli*

A complementary DNA library from a Bowes melanoma cell line was screened with a mixture of three oligodeoxynucleotides which were designed on the basis of the published t-PA sequence (26). A full-length t-PA complementary DNA clone was reconstituted from several overlapping clones. The coding sequence for the finger, EGF and kringle domains of the t-PA (nucleotides 199-714) was removed according to the intron-exon organization of the t-PA gene (31). The coding sequence of reteplase was introduced into the vector plasmid pKK223-3 as previously described (32). The resulting plasmid pA27 fd was introduced into *E. coli* K12 C600+ by transformation (33). The production level was improved by cotransformation with the pUBS520 plasmid containing the DNA Y gene (34).

In this assay, the plasminogen activator is pressed into the clot by a peristaltic pump, mimicking the *in vivo* situation more closely than the static model. At low concentrations, both reteplase and alteplase have a similar activity. At high concentrations, as achieved during the treatment of AMI, reteplase is more potent than alteplase (Figure 25 (25)).

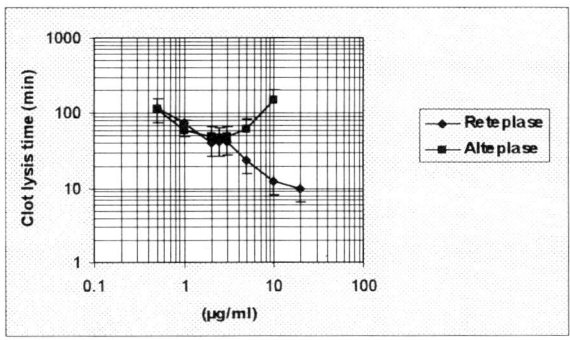

Figure 25. Comparison of the plasma clot lysis activity of alteplase and reteplase in the dynamic plasma model. Increasing concentrations of both plasminogen activators were added to 1 ml plasma on top of a preformed clot. All values are the mean ± S.D. of five experiments.

6.3 Clot penetration studies *in vitro*

The ability of reteplase and alteplase to penetrate into plasma clots was analysed by adding the inhibited plasminogen activators to the surface of the clot and determining their location by immunostaining before and after washing the clot. These experiments demonstrated that alteplase was tightly bound to the fibrin matrix and accumulated at the surface of the clot. As a consequence, activation of plasminogen and the subsequent degradation of the fibrin matrix is supposed to occur from the surface to the interior of the clot, depending on the permanent supply of plasminogen from the plasma. In contrast, reteplase penetrated into the clot due to its lack of fibrin binding, which in turn allowed the activation of plasminogen inside the clot (25). The ability of plasma clot penetration might be a prerequisite for the achievement of a high *in vivo* efficacy of fibrin-specific thrombolytic agents, especially when applied by bolus injection.

6.4 Clot lysis *in vivo*

In contrast to the static *in vitro* clot lysis model, in a rabbit jugular vein model, reteplase was 5.3 times more effective than rt-PA in lysing venous thrombi following intravenous bolus injection (13). This efficacy benefit may have been due to slower plasma clearance of reteplase: pharmacokinetic analysis revealed a much longer half-life for reteplase (18.9 ± 1.5 minutes) compared with rt-PA (2.1 ± 0.1 minutes), and a 4.3-fold lower clearance rate, following the same bolus dose (13). The improved clot lysis *in vivo* may also reflect the lower fibrin binding of reteplase compared with rt-PA, discussed above, which may allow better clot penetration, particularly following a bolus dose which produces high peak plasma levels (25).

6.5 Canine model of coronary artery thrombosis

Reteplase provided faster reperfusion than rt-PA and all other thrombolytics in a canine model of coronary artery thrombosis (42). Of six dogs treated with an intravenous bolus injection of reteplase (140 kU/kg (0.24 mg/kg)) four achieved reperfusion at 18.3 ± 6 minutes, compared with a mean reperfusion time of 76.5 ± 16.1 minutes in four out of six dogs treated with rt-PA; 1.33 mg/kg as an initial bolus, then by infusion (0.66 mg/kg over 1 hour and 0.53 mg/kg over 2 hours) ($p < 0.05$ vs reteplase). Residual fibrinogen was comparable in dogs treated with reteplase and rt-PA, as was bleeding time measured at 90 minutes. These results demonstrate that, in the animal model, reteplase fulfils its developmental goals by providing rapid reperfusion in a high proportion of subjects without inducing a systemic lytic state (42).

A double bolus of reteplase (140 kU/kg + 140 kU/kg, 45 minutes apart) in another canine study (13) significantly prolonged the duration of arterial patency, and this effect was further enhanced by co-administration of platelet and thrombin inhibitors (43). These results suggest that double bolus dosing optimizes the clinical performance of reteplase.

6.6 Animal pharmacokinetics

The goal of prolonging the half-life of reteplase relative to rt-PA has been achieved. Following a single intravenous bolus dose of 200 kU/kg in different mammals, reteplase exhibits a half-life of between 7.2 ± 0.5 minutes (non-human primate) and 15.4 ± 2.6 minutes (rabbit), which is up to 10-fold longer than rt-PA. Plasma clearance of reteplase is significantly lower than that of rt-

PA in all species (44,45), presumably as a result of the lack of carbohydrate side chains and deletion of the kringle-1 and EGF domains.

6.7 Toxicology

Reteplase exhibits low potential for acute and long-term toxicity, demonstrated by studies in rabbits, rats and dogs (13). In rats, the minimum lethal dose exceeds 8400 kU/kg. Following dosing in dogs for 14 days, there was a dose-dependent reduction of fibrinogen, plasminogen and α_2-antiplasmin 2 hours post-injection. Mutagenicity, assessed in *Salmonella typhimurium* and rat bone marrow erythrocytes, revealed no mutagenic activity (13).

7. CLINICAL PHARMACOLOGY

Reteplase exhibits a highly favourable pharmacological profile in humans, consistent with its developmental goals of a prolonged half-life compared with rt-PA, low bleeding risk, and low potential for antigenicity.

7.1 Pharmacokinetics

7.1.1 Healthy subjects

In healthy volunteers, the area under the activity concentration-time curve (AUC) for reteplase increased in a linear and dose-dependent fashion. Following a single bolus dose of 5.5 U reteplase, the total plasma clearance was 306 ± 40 ml/minute and plasma half-life was 14.4 ± 1.1 minutes (46). In a second study, following a single bolus dose of 6 U of reteplase injected over 2 minutes, the half-life of reteplase activity (determined with a plasminogenolytic assay) was 11.2 ± 0.4 minutes and that of antigen (assessed using an enzyme-linked immunosorbent assay [ELISA]) was 13.9 ± 0.7 minutes, followed by a terminal half-life for antigen of 173 ± 33 minutes. Plasma clearance was 371 ± 13 ml/minute for activity and 183 ± 15 ml/minute for antigen. Compared with previously published values for rt-PA (47), the activity half-life for reteplase was 3.3-fold higher and the clearance 3.3-fold lower.

7.1.2 Following AMI

Reteplase exhibits similar pharmacokinetics in patients with AMI. Following administration of a single bolus dose of 10 U or 15 U reteplase, the plasma half-life, assessed using ELISA, was approximately 19 minutes (48). This is almost four times as long as the half-life reported for rt-PA in AMI patients, measured after a 50 mg single bolus dose (49).

7.2 Haemostatic effects

7.2.1 Healthy subjects

Reteplase has modest effects on haemostatic variables, indicating a low potential for causing bleeding (46,50). Intravenous bolus doses of reteplase ranging from 0.11 U to 5.5 U in 18 subjects had no effect on plasma fibrinogen, and reduced plasminogen levels only at higher doses. Fibrin D-dimers and α_2-antiplasmin were reduced in a dose-dependent fashion (46). In a separate randomized, single-blind, placebo-controlled, crossover study, seven healthy volunteers received placebo or 6 U reteplase as a bolus injection over 2 minutes (50). Fibrinogen levels remained unchanged and plasminogen and α_2-antiplasmin levels fell to $83 \pm 1\%$ and $64 \pm 3\%$, respectively, of their baseline levels. Reteplase was well-tolerated in both studies and no antibodies were detected up to 1 year after dosing.

7.2.2 Following AMI

Transient and dose-dependent reductions in fibrinogen, plasminogen and α_2-antiplasmin have been observed in patients following AMI.

Two hours after a single bolus dose of reteplase, fibrinogen, plasminogen and α_2-antiplasmin levels fell to 60%, 55% and 30% of their respective baseline values following a 10 U dose, and to 44%, 41% and 26%, respectively, following a 15 U dose (51). After a double bolus of 10 + 5 U reteplase, given 30 minutes apart, levels of the same three variables fell to 45%, 42% and 19%, respectively, of their baseline values (52).

8. CLINICAL DATA

Clinical trials with reteplase, summarized in Tables 21 and 22, have confirmed the profile of reteplase suggested by biochemical, pharmacological and preclinical studies.

Table 21. Dose-finding clinical trials using reteplase in patients with AMI

	GRECO (51)	GRECO-DB (52)	MF4292 (53)
Location	Germany	Germany	Germany
Study design	Open-label	Open-label	Randomized, open-label
Population	AMI	AMI	AMI
Dose strength and form	10 U RP (n=42) 15 U RP (n=100)	10 + 5 U RP (n=52)	15 U RP (n=9) 10 + 5 U RP (n=8) 10 + 10 U RP (n=8)
Frequency duration	Single bolus	Double bolus	Single bolus Double bolus
Primary evaluation	90-minute patency and TIMI 3 rates	90-minute patency and TIMI 3 rates	90-minute patency and TIMI 3 rates, pharmacokinetics, haemostasis parameters
Key entry criteria	ST-segment elevation Onset of ischaemic pain within 6 hours 18–75 years old	ST-segment elevation Onset of ischaemic pain within 6 hours 18–75 years old	ST-segment elevation Onset of ischaemic pain within 6 hours 18–75 years old

GRECO = German Recombinant Plasminogen Activator Study; GRECO-DB = German Recombinant Plasminogen Activator Double Bolus Study; RP = Reteplase.

The trials demonstrate that reteplase can achieve rapid reperfusion of occluded coronary arteries in a high proportion of patients, while maintaining a favourable safety profile. Through dose-finding studies, the optimal dosing regimen has been identified as a double-bolus dose of 10 + 10 U reteplase, administered 30 minutes apart. The TIMI flow grades, used in most trials of thrombolytics, are summarized (Table 23).

Table 22. Comparative studies using reteplase in patients with AMI

	RAPID 1 (phase II) (54)	RAPID 2 (55)	INJECT (56)	GUSTO III (57)
Location	USA, Germany, UK, Austria	USA, Germany	UK, Germany, Poland, Sweden, Hungary, Finland, Spain, Lithuania, Austria	USA, Canada, Europe, Australia, New Zealand, South Africa, Argentina
Study design	Randomized, Open-label (angiograms read blinded)	Randomized, Open-label (angiograms read blinded)	Randomized, double-blind	Randomized, Open-label
Population	AMI	AMI	AMI	AMI
Dose strength and form	15 U RP (n=146) 10 + 5 U RP (n=152) 10 + 10 U RP (n=154) 100 mg AP (standard dose) (n=154)	10 + 10 U RP (n=169) 100 mg AP (accelerated dose) (n=155)	10 + 10 U RP* (n=3004) 1.5 MU SK* (n=3006)	10 + 10 U RP (n=15059)
Frequency duration	Single bolus Double bolus Double bolus Bolus + 3-hour infusion	Double bolus Bolus + 1.5-hour infusion	Double bolus 1-hour infusion	Double bolus
Primary evaluation	90-minute patency and TIMI 3	90-minute patency and TIMI 3	35-day mortality	30-day mortality
Key entry criteria	ST-segment elevation Onset of ischaemic pain within 6 hours 18–75 years old	ST-segment elevation or bundle branch block Onset of ischaemic pain within 12 hours ≥ 18 years old	ST-segment elevation or bundle branch block Onset of ischaemic pain within 12 hours ≥ 18 years old	Within 6 hours of symptom onset, ST-segment elevation or bundle branch block ≥ 18 years old

*Seventy-four patients were randomized but were not treated. Overall, 2965 reteplase patients and 2971 streptokinase patients received treatment. AP = alteplase; DB = double-blind; GUSTO III = Global Use of Strategies to Open Occluded Coronary Arteries; INJECT = International Joint Efficacy Comparison of Thrombolytics; RAPID I = Reteplase Angiographic Phase II International Dose-finding Study; RAPID 2 = Reteplase versus Alteplase Patency Investigation During Acute Myocardial Infarction Study; RP = reteplase; SK = streptokinase.

Table 23. TIMI flow grades (58)

TIMI grade	Angiographic features of coronary artery flow
0	No penetration of contrast beyond the point of obstruction
1	Contrast penetrates the point of obstruction but does not completely opacify the entire distal vessel
2	Complete contrast opacification of the infarct-related artery but neither contrast opacification nor washout is delayed
3	Brisk, 'normal' flow

8.1 GRECO study

The GRECO study (51) demonstrated that a single bolus dose of reteplase achieves rapid reperfusion and high rates of early patency.

Forty-two patients with AMI received 5000 IU heparin and a single bolus of reteplase, either 10 U or 15 MU within 6 hours of symptom onset. The group receiving the 10 U dose reached the lower preset efficacy limit (90-minute patency of 70%) and in accordance with the study protocol, the higher dose of 15 U was given to a further 100 patients. All patients also received oral aspirin.

At 30, 60 and 90 minutes after injection of 10 U reteplase, angiography revealed TIMI grade 2 or 3 patency in 65%, 73% and 66%, respectively. In the 15 MU group, these values were 66%, 74% and 75%, respectively. Very early reocclusion (prior to 90 minutes) occurred in 5 of 30 (17%) patients in the 10 U group, and 10 of 78 (13%) of those given 15 MU reteplase. Frequency of bleeding complications was as would be expected for standard thrombolytic therapy.

8.2 GRECO-DB

A second dose-finding study, GRECO-DB (52), demonstrated that a double bolus dose of reteplase is well tolerated and may offer efficacy advantages over a single bolus regimen by extending the period of high plasma levels of reteplase.

Fifty-one patients with AMI received 10 U of reteplase as a bolus dose within 6 hours of symptom onset, and a further 5 U bolus 30 minutes later. Angiography after 30, 60 and 90 minutes demonstrated TIMI grade 2 or 3 flow in 50%, 72% and 78% of patients, respectively. Early reocclusion (prior to 90 minutes) occurred in 10% of patients, and reocclusion between 90 minutes and 24 hours occurred in 2% (1 patient). Bleeding complications were no higher than for standard thrombolysis. This study showed that the double-bolus regimen reduces but does not completely eliminate incomplete initial lysis or reocclusion. A subsequent study of 24 patients (53) showed that a double bolus regimen of 10 + 10 U reteplase, given 30 minutes apart, prolongs the thrombolytic activity compared with the 10 + 5 U regimen.

8.3   RAPID-1 study

The RAPID-1 study (54) provides evidence that reteplase fulfils the developmental goal of offering clinical benefits over standard treatment with rt-PA in the treatment of AMI.

Six hundred and six patients with AMI were randomized to receive either rt-PA, 100 mg intravenously over 3 hours; reteplase as a single 15 U bolus; reteplase as a 10 U bolus followed by 5 U 30 minutes later; or reteplase as a 10 U bolus, followed by another 10 U bolus 30 minutes later. The 10 + 10 U reteplase regimen led to the highest early complete patency rates, providing TIMI 3 flow in 63% of patients at 90 minutes, compared with 49% for rt-PA ($p = 0.019$). The TIMI 3 flow in the 10 + 10 U reteplase group at 60 minutes was equivalent to that in the rt-PA group at 90 minutes (51% versus 49%). At hospital discharge, the global ejection fraction and regional wall motion in the 10 + 10 U reteplase group were superior to those in the rt-PA group, and the other reteplase groups. Bleeding complications were similar in all groups.

8.4   RAPID-2 study

The GUSTO angiographic study (8) established an accelerated rt-PA regimen as the gold standard in the management of AMI. The RAPID-2 study (55) demonstrates that a double bolus dose of reteplase, 10 + 10 U 30 minutes apart, achieves faster and more complete recanalization than accelerated rt-PA, without an increased risk of bleeding.

Three hundred and twenty-four patients with AMI were randomized to receive either reteplase, 10 + 10 U or rt-PA, as a 15 mg bolus, followed by 0.75 mg/kg over 30 minutes, then 0.5 mg/kg over 60 minutes. After 90 minutes, 83% of patients treated with reteplase achieved TIMI grade 2 or 3 flow, compared with 73.3% of those given rt-PA ($p = 0.03$). The rate of TIMI grade 3 flow was 59.9% in the reteplase group compared with 45.2% in the rt-PA group ($p = 0.01$). A clear advantage for reteplase was also observable at 60 minutes (TIMI grade 2 or 3: 81.8% versus 66.1%, $p = 0.01$; TIMI grade 3: 51.2% versus 37.4%, $p < 0.03$). Reteplase-treated patients required fewer additional interventions (13.6% versus 26.5%, $p < 0.01$) and had a lower 35-day mortality (4.1% versus 8.4%) although this difference was not significant. There were no significant differences between the groups in terms of bleeding requiring a transfusion (12.4% for reteplase; 9.7% for rt-PA) or haemorrhagic stroke (reteplase 1.2%; rt-PA 1.9%).

8.5   INJECT trial

Streptokinase is still the most widely used thrombolytic agent. The International Joint Efficacy Comparison of Thrombolysis (INJECT), a phase III study, compared the efficacy of reteplase and streptokinase in preventing mortality following AMI (56).

A total of 6010 patients in nine European countries were randomized to receive reteplase as a double bolus of 10 + 10 U, 30 minutes apart, or streptokinase, 1.5 MU, infused over 60 minutes. Treatment was initiated within 12 hours of symptom onset, and all patients received adjunctive aspirin and heparin.

After 35 days, the mortality rates in the two groups were equivalent (reteplase, 9.02%; streptokinase, 9.53%). There were no significant differences between the groups in terms of incidence of stroke (reteplase, 1.23%; streptokinase, 1.0%), stroke disablement at 6 months (reteplase 0.17%; streptokinase 0.27%) or requirement for blood transfusions (reteplase, 0.7%; streptokinase, 1.0%). The rate of recurrent AMI was similar in both groups, but the incidence of atrial fibrillation, asystole, cardiac shock, heart failure and hypotension was significantly lower in the reteplase group.

The INJECT study included insufficient patient numbers to demonstrate a statistically significant difference in mortality between the therapeutic regimens. In addition, the trial protocol allowed randomization to thrombolysis up to 12 hours after symptom onset, possibly masking any potential benefit of reteplase.

In contrast to the mortality results, a substudy of the INJECT trial which involved 1398 patients revealed a significant benefit for reteplase over streptokinase in the extent of resolution of ST-segment elevation (59). The results were classified as complete resolution ($\geq$ 70%), partial resolution (70 to 30%) and no resolution (< 30%) (59,60). Although the 35-day mortality rates did not differ significantly between the groups in the substudy (reteplase, 5.2%; streptokinase, 7.2%; p = 0.12), more patients in the reteplase group had complete resolution and fewer had no resolution, compared with the streptokinase group (p = 0.006).

8.6   GUSTO-III trial

The GUSTO trial (19) demonstrated a 14% relative reduction in mortality from AMI for the accelerated alteplase regimen compared with streptokinase which, the data suggest (8), resulted from a more rapid and complete restoration of coronary flow through the infarct-related artery. The RAPID-2 study revealed that double-bolus reteplase in the 10 + 10 U regimen provides more rapid and complete recanalization of an occluded coronary artery than accelerated rt-PA.

A new mega-trial, GUSTO-III (57), was therefore designed to compare the impact of these two regimens on 30-day mortality in 15,059 patients from 20 countries, presenting with AMI within 6 hours of symptom onset. Randomization was in a 2:1 allocation reteplase (10 + 10 U, 30 minutes apart), or accelerated rt-PA (15 mg bolus followed by 0.75 mg/kg over 30 minutes, then 0.50 mg/kg over 60 minutes).

Reteplase and rt-PA showed close and consistent similarity across all primary and secondary end-points. The 30-day mortality rates were equivalent (7.47% for reteplase; 7.24% for rt-PA), as was the incidence of stroke, including haemorrhagic and non-haemorrhagic stroke (overall stroke rate 1.64% for reteplase; 1.79% for rt-PA). The combined rate of death or disabling stroke was almost identical for the two groups (7.89% for reteplase; 7.91% for rt-PA) as were the rates of serious and moderate bleeding, and need for transfusions (0.95%, 6.92%, 5.90%, respectively, for reteplase; 1.2%, 6.82%, 6.2% for rt-PA). The incidence of reinfarction, congestive heart failure and arrhythmias were also similar in the two groups.

The GUSTO III results reveal no efficacy differences between the two thrombolytic regimens. However, the double-bolus dosing of reteplase is simpler and more convenient than the accelerated rt-PA regimen, which involves intravenous infusion. In practice, this may enable thrombolysis to begin before hospital admission, and reduce staff time in administering therapy.

8.7 Safety profile of reteplase

The safety of reteplase has been assessed in 45 healthy volunteers and almost 14,000 patients, most of whom received the 10 + 10 U double-bolus regimen. Reteplase has been shown to be similar to other available thrombolytic agents with respect to bleeding complications, including stroke, cardiac events, allergic events and other adverse events. Risk of bleeding, including stroke, for the 10 + 10 U double-bolus regimen of reteplase has been shown to be equivalent to streptokinase (56) and rt-PA, whether as a standard (54) or accelerated regimen (55, 57). The INJECT trial showed a benefit of reteplase over streptokinase in terms of lower incidence of new or worsening congestive heart failure, cardiogenic shock, hypotension, pulmonary oedema, atrial fibrillation or flutter, and asystole (56). The RAPID-1 (54) and RAPID-2 (55) trials demonstrated equivalent rates of cardiac adverse events for rt-PA and reteplase. The overall stroke rates in RAPID-1 and RAPID-2 were higher for patients treated with rt-PA than those treated with reteplase ($p = 0.05$) (54, 55). In GUSTO III (57), overall stroke rates were similar for reteplase and rt-PA. Reteplase is associated with a lower incidence of allergic reactions and other serious adverse events than streptokinase (56). Reteplase and rt-PA have an equivalent rate of allergic reactions.

The dosing of reteplase is independent of body weight. The safety profile of reteplase in low-weight ($\leq$ 65 kg) and higher-weight (> 65 kg) patients is similar to that of the weight-adjusted, accelerated dosing regimen of rt-PA.

9. CONTRAINDICATIONS

9.1 Absolute contraindications (61)

Reteplase must not be used in the following situations:
1. known bleeding tendency
2. concomitant medication with oral anticoagulants (e.g. phenprocoumon)
3. brain tumours, arteriovenous malformations or aneurysms
4. tumours associated with increased risk of bleeding
5. history of cerebrovascular events
6. recent (within the last 10 days) prolonged, intensive (traumatic) external cardiac massage
7. severe uncontrolled hypertension
8. active peptic ulcers
9. portal hypertension (oesophageal varices)
10. acute pancreatitis, pericarditis, bacterial endocarditis
11. haemorrhagic retinopathy in diabetes mellitus or other conditions of the eye associated with a bleeding tendency
12. less than 3 months after any of the following: severe bleeding, severe trauma, major surgery (e.g. coronary artery bypass surgery, intracranial or intraspinal surgery or trauma), childbirth, organ biopsy or preceding puncture of non-compressible vessels.

9.2 Relative contraindications (61)

Careful risk-benefit assessment is necessary in the following cases:
1. cerebrovascular disease
2. systolic blood pressure > 160 mm Hg before treatment
3. recent (within last 10 days) gastrointestinal or urogenital bleeding
4. conditions with high probability of left ventricular thrombosis (e.g. mitral valve stenosis with atrial fibrillation)
5. septic thrombophlebitis or occluded arteriovenous fistula

6. advanced age (i.e. patients > 75 years)
7. any other condition in which bleeding constitutes a significant hazard, or would be difficult to control due to its location

At present there are insufficient data on the use of reteplase in patients with a diastolic blood pressure > 100 mm Hg before initiation of thrombolytic therapy.

9.3　Drug interactions (61)

No specific studies of interactions of reteplase with other drugs normally used to the treatment of AMI have been conducted. Retrospective analyses of clinical studies have revealed no clinically relevant interactions of reteplase with drugs normally used in patients with acute coronary thrombosis. The risk of bleeding may be increased if the following are given prior to, during or after reteplase, especially when plasma fibrinogen levels are reduced (e.g. up to 2 days following fibrinolytic therapy):
– heparin;
– vitamin K antagonists;
– drugs that alter platelet function (such as ASA, dipyridamole).

10.　RETEPLASE – FORMULATION AND FORMAT (61)

10.1　Name of finished medicinal product

10.1.1　International non-proprietary name

Reteplase

10.1.2　Proprietary names

Rapilysin® 10 U
Retavase™ (United States and Canada only)

10.2　Description of active substances and pharmaceutical formulation

10.2.1　Organoleptic characteristics

The active solution of reteplase is clear and odourless.

10.2.2 Purity

The purity of reteplase, determined by high resolution chromatographic methods (RP-HPLC and SE-HPLC) is as follows:

Reteplase protein makes up > 99.99% of the total protein.
Impurities:

- *E. coli* protein < 70 ppm
- *Erythrina* trypsin inhibitor < 2 ppm
- DNA < 25 pg/10 U
- Endotoxins < 5 EU/ml

10.2.3 Solubility

Reteplase is poorly soluble in water and the usual buffers, and requires the addition of stabilizers (e.g. amino acids such as arginine).

10.2.4 Absorption spectra (visible, UV, IR)

The protein concentration is determined by measurement of the absorption at 280 nm against the respective buffer as reference. With a light path of 1 cm and a protein concentration of 1 mg/ml, the absorbance is 1.69.

10.2.5 Isoelectric point (IEP)

The IEP of reteplase calculated from the amino acid composition is 7.23.

10.2.6 pH

The pH of the reconstituted medicinal product of reteplase is in the range pH 7.0 to pH 7.4

10.2.7 Stability

- If stored in the original package, the injection vials of reteplase 10 U are stable for two years at temperatures of 2°C to 25°C. During prolonged storage the powder should be protected from excessive exposure to light.
- The lyophilisate is also stable if deep-frozen. Exposure to temperatures above 25°C shortens the shelf-life; temperatures greater than 30°C should be avoided.
- After reconstitution as directed, the solution should be used immediately. The reconstituted solution is chemically stable for 4 hours.

10.2.8 Nature and risk of the decomposition products

Decomposition results in transition to the two-chain form of reteplase.

10.3 Product presentation

10.3.1 Description

- Reteplase is presented as a sterile, dry, white, lyophilized powder in a glass vial, mostly adhering to the glass wall of the vial.
- Each vial contains 1.16 g powder, equivalent to 10 U (a single dose) of reteplase.
- Each dose of the medicinal product contains 1.0 mg polysorbate 20, plus 871.0 mg arginine and 268.6 mg phosphoric acid as stabilizers.
- The powder is reconstituted with 10 ml diluent (water) for intravenous injection.

10.3.2 Potency units

The potency of reteplase is given in units (U) based on a reference standard which is reteplase-specific and is not comparable with the units used for other thrombolytic agents. The designation MU (1 MU $\approx$ 1 U) is also sometimes used in publications.

BIOGRAPHY

Ulrich Kohnert was the Project Leader Biotechnology for the development of reteplase. He is a chemist by training and has special expertise in *in vitro* folding, purification and *in vitro* characterization of proteins, as well as the development of pharmaceutical formulations for therapeutic proteins. Boehringer Mannheim is a high technology company in the healthcare market. It holds a leading position in research development and production of diagnostics and new drugs. Biotechnology has a prominent position and long tradition within the company profile. Boehringer Mannheim operates one of Europe's largest biotechnology facilities in Penzberg, Bavaria. Boehringer Mannheim GmbH is part of the Roche Deutschland Holding GmbH.

REFERENCES

1. Gruppo Italiano per lo Studio della Streptochinasi nell'Infarto miocardico (GISSI). (1986). Effectiveness of intravenous thrombolytic treatment in acute myocardial infarction. Lancet, 1, 871-874.
2. ISIS-2 (second International Study of Infarct Survival) Collaborative Group. (1988). Randomised trial of intravenous streptokinase, oral aspirin, both or neither among 17,187 cases of suspected acute myocardial infarction: ISIS-2. Lancet, 2, 349-360.
3. ISAM Study Group. (1986). A prospective trial of intravenous streptokinase in acute myocardial infarction (ISAM). N. Engl. J. Med., 314, 1465-1471.
4. AIMS Trial Study Group. (1988). Effect of intravenous APSAC on mortality after acute myocardial infarction: Preliminary report of a placebo-controlled clinical trial. Lancet, 1, 545-549.
5. Wilcox, R.G. et al. (1988). Trial of tissue plasminogen activator for mortality reduction in acute myocardial infarction: Anglo-Scandinavian Study of Early Thrombolysis (ASSET). Lancet, 2, 525-530.
6. White, H.D. et al. (1987). Effect of intravenous streptokinase on left ventricular function and early survival after acute myocardial infarction. N. Engl. J. Med., 317, 850-855.
7. Van De Werf, F. and Arnold, A.E.R. (1988). Intravenous tissue plasminogen activator and size of infarct, left ventricular function, and survival in acute myocardial infarction. Br. Med. J., 297, 1374-1379.
8. GUSTO-Angiographic Investigators. (1993). The effects of tissue plasminogen activator, streptokinase or both on coronary-artery patency, ventricular function and survival after acute myocardial infarction. N. Engl. J. Med., 1329, 1615-1622.
9. Granger, C.B. et al. (1992). Thrombolytic therapy for acute myocardial infarction - a review. Drugs, 44, 293-325.
10. Sobel, B.E. and Collen, D. (1992). Questions unresolved by the Third International Study of Infarct Survival. Am. J. Cardiol., 70, 385-389.
11. Bode, C. et al. (1997). Clinical trial results with a new plasminogen activator. European Heart Journal, 18 (suppl F), F17-F21.
12. Kohnert, U. et al. (1992). Biochemical properties of the kringle 2 and protease domains are maintained in the refolded t-PA deletion variant BM 06.022. Protein Eng., 5, 93-100.
13. Martin, U. et al. (1993). BM 06.022: A novel recombinant plasminogen activator. Cardiovasc. Drug Rev., 11, 299-311.
14. Braunwald, E. (1987). The path to myocardial salvage by thrombolytic therapy. Circulation, 76(suppl II), II-2 - II7.
15. Chazov, E.I. et al. (1976). Intracoronary administration of fibrinolysis in acute myocardial infarction. Ter Arkh, 48, 8.
16. Rentrop, K.T. et al. (1979). Initial experience with transluminal recanalization of the recently occluded infarct-related coronary artery in acute myocardial infarction. Comparison with conventionally treated patients. Clin. Cardiol., 2, 92.
17. Blumgart, H.L. et al. (1941). Experimental studies on the effect of temporary occlusion of coronary arteries. Am. J. Heart., 22, 374-389.
18. Smalling, R.W. and Hanna, G.P. (1966). Clinical pharmacology and mechanisms of action of thrombolytic agents. In: Califf RM (ed.). Thrombolytic Therapy: New Standards of Care, Part I: the Science of Plasminogen Activators. American Journal of Cardiology. Belle Mead, NJ: Excerpta Medica, 9-15.
19. The GUSTO Investigators (1993). An international randomized trial comparing four thrombolytic strategies for acute myocardial infarction. N. Engl. J. Med., 329, 673-682.

20. Rapaport, E. (1992). The ideal thrombolytic agent. In: Sleight P, Tavazzi L (eds). The Major Clinical Trials on Thrombolysis for Acute Myocardial Infarction. New York: Raven Press, 1-5.
21. Vahanian, A. (1996). Thrombolytic therapy in Europe: current status. Eur. Heart J., 17(suppl E), 21-27.
22. Simes, R.J. et al. (1995). Link between the angiographic substudy and mortality outcomes in a large randomized trial of myocardial reperfusion: Importance of early and complete artery reperfusion. Circulation, 91, 1923-1928.
23. Fischer, S. (1998). The molecular design of reteplase. Drugs of Today, 33, 641-648.
24. Bode, C. et al. (1996). The future of thrombolysis in the treatment of acute myocardial infarction. Eur. Heart J., 17(suppl E), 55-60.
25. Fischer, S. and Kohnert, U. (1997). Major mechanistic differences explain the higher clot lysis potency of reteplase over alteplase: lack of fibrin binding is an advantage for bolus application of fibrin-specific thrombolytics. Fibrinolysis and Proteolysis, 11(3), 129-135.
26. Pennica, D. et al. (1983). Cloning and expression of human tissue-type plasminogen activator cDNA in E. coli. Nature, 301, 214-221.
27. Smalling, R.W. (1996). Molecular biology of plasminogen activators: What are the clinical implications of drug design? Am. J. Cardiol., 78 (suppl 12A), 2-7.
28. Hotchkiss, A. et al. (1988). The influence of carbohydrate structure on the clearance of recombinant tissue-type plasminogen activator. Thromb. Haemost., 60, 255-261.
29. Kohnert, U. et al. (1993). A variant tissue plasminogen activator (t-PA) comprised of the kringle 2 and protease domain shows a significant difference in the in vitro rate of plasmin formation as compared to the recombinant human t-PA from transformed Chinese hamster ovary cells. Fibrinolysis, 7, 365-372.
30. Sprengers, E.D. and Kluft, C. (1987). Plasminogen activator inhibitors. Blood, 69, 381-387.
31. Ny, T. et al. (1984). The structure of the human tissue-type plasminogen activator gene: correlation of intron and exon structures to functional and structural domains. Proc. Natl. Acad. Sci., 81, 5355-5359.
32. Stern, A. et al. (1990). Gewebs-Plasminogenaktivator-Derivat. European Patent Application 382174.
33. Maniatis, T. et al. (1992). Molecular Cloning: A Laboratory Manual. Cold Spring Harbor, NY: Cold Spring Harbor Laboratory Press.
34. Brinkmann, U. et al. (1989). High-level expression of recombinant genes in Escherichia coli is dependent on the availability of the dnaY gene product. Gene, 85, 109-114.
35. Rudolph, R. et al. (1987). Verfahren zur Aktivierung von gentechnologisch hergestellten, heterologen, Disulfidbrücken aufweisenden eukaryontischen Proteinen nach Expression in Prokaryonten. European Patent Application 219874.
36. Rudolph, R. (1990). In: Tschesche H (ed.). Modern Methods in Protein and Nucleic Acid Research. Berlin: Walter de Gruyter, 149-172.
37. Rudolph, R. and Fischer, S. (1990). Process for obtaining renatured proteins. United States Patent, 4933434.
38. Heussen, C. et al. (1984). Purification of human tissue plasminogen activator with Erythrina trypsin inhibitor. J. Biol. Chem., 259, 11635-11638.
39. Kruithof, E.K. et al. (1985). Human tissue-type plasminogen activator. Production in continuous serum-free cell culture and rapid purification. Biochem. J., 226, 631-636.

40. Verheijen, J.H. et al. (1982). A simple spectrophotometric assay for extrinsic (tissue-type) plasminogen activator applicable to measurements in plasma. Thromb. Haemostas., 48(3), 266-269.
41. Kohnert, U. et al. (1996). The recombinant *Escherichia coli*-derived protease-domain of tissue-type plasminogen activator is a potent and fibrin specific fibrinolytic agent. Fibrinolysis, 10, 83-102.
42. Martin, U. et al. (1992). Evaluation of thrombolytic and systemic effects of novel recombinant plasminogen activator BM 06.022 compared with alteplase, anistreplase, streptokinase and urokinase in a canine model of coronary artery thrombosis. J. Am. Coll. Cardiol., 19, 433-440.
43. Martin, U. et al. (1992). Hirudin and sulotroban improve coronary blood flow after reperfusion induced by the novel recombinant plasminogen activator BM 06.022 in a canine model of coronary artery thrombosis. Int. J. Hematol., 56, 143-153.
44. Martin, U. et al. (1991). Pharmacokinetic properties of an *Escherichia coli*-produced recombinant plasminogen activator (BM 06.022) in rabbits. Thromb. Res., 62, 137-146.
45. Martin, U. et al. (1992). Pharmacokinetics of the novel recombinant plasminogen activator BM 06.022 in rats, dogs and non-human primates. Fibrinolysis, 6, 39-43.
46. Martin, U. et al. (1991). Dose-ranging study of the novel recombinant plasminogen activator BM 06.022 in healthy volunteers. Clin. Pharmacol. Ther., 50, 429-436.
47. Seifried, E. et al. (1988). Pharmacokinetics of antigen and activity of recombinant tissue-type plasminogen activator after infusion in healthy volunteers. Drug Res., 38, 418-422.
48. Seifried, E. et al. (1992). Bolus application of a novel recombinant plasminogen activator in acute myocardial infarction patients: Pharmacokinetics and effects on the hemostatic system. Ann. N.Y. Acad. Sci., 667, 417-420.
49. Tebbe, U. et al. (1989). Single bolus injection of recombinant tissue-type plasminogen activator in acute myocardial infarction. Am. J. Cardiol., 64, 448-453.
50. Martin, U. et al. (1991). Pharmacokinetic and hemostatic properties of the recombinant plasminogen activator BM 06.022 in healthy volunteers. Thromb. Haemost., 66(5), 569-574.
51. Neuhau,s K-L. et al. (1994). Dose finding with a novel recombinant plasminogen activator (BM 06.022) in patients with acute myocardial infarction: Results of the German recombinant plasminogen activator study. J. Am. Coll. Cardiol., 24, 55-60.
52. Tebbe, U. et al. (1993). Open, noncontrolled dose-finding study with a novel recombinant plasminogen activator (BM 06.022) given as a double bolus in patients with acute myocardial infarction. Am. J. Cardiol.,72, 518-524.
53. Grünewald, M. et al. (1995). Single vs double bolus thrombolysis with the recombinant plasminogen activator BM 06.022 in patients with acute myocardial infarction - pharmacokinetics and hemostatic changes (abstract). Thromb. Haemost., 73, 1328.
54. Smalling, R.W. et al. (1995). More rapid, complete, and stable coronary thrombolysis with bolus administration of reteplase compared with alteplase infusion in acute myocardial infarction. Circulation, 91, 2725-2732.
55. Bode, C. et al. (1996). Randomized comparison of coronary thrombolysis achieved with a double-bolus reteplase (recombinant plasminogen activator) and front-loaded, accelerated alteplase (recombinant tissue plasminogen activator) in patients with acute myocardial infarction. Circulation, 94, 891-898.
56. International Joint Efficacy Comparison of Thrombolytics. (1995). Randomised, double-blind comparison of reteplase double-bolus administration with streptokinase in acute myocardial infarction (INJECT): trial to investigate equivalence. Lancet, 346, 329-336.

57. The Global Use of Strategies to Open Occluded Coronary Arteries (GUSTO III) Investigators. (1997). A comparison of reteplase with alteplase for acute myocardial infarction N. Eng.l J. Med., 337, 1118-1123.
58. Chesebro, J.H. *et al.* (1987). Thrombolysis in Myocardial Infarction (TIMI) Trial, Phase I: A comparison between tissue plasminogen activator and intravenous streptokinase. Clinical findings through hospital discharge. Circulation, 76, 142-154.
59. Schröder, R. *et al.* (1995). Extent of early ST segment elevation resolution: A strong predictor of outcome in patients with acute myocardial infarction and a sensitive measure to compare thrombolytic regimens. A substudy of the international joint efficacy comparison of thrombolytics (INJECT) trial. J. Am. Coll. Cardiol., 26, 1657-1664.
60. Schröder, R. *et al.* (1994). Extent of early ST segment elevation resolution: A simple but strong predictor of outcome in patients with acute myocardial infarction. J. Am. Coll. Cardiol., 24, 384-391.
61. Rapilysin: Reteplase. (1998). Standard Information for Hospital Pharmacists. Boehringer Mannheim.

Chapter 9

Stabilisation of biopharmaceutical products and finished product formulations

Maninder S. Hora and Bao-lu Chen
Department of Formulation Development, Chiron Corporation, 4560 Horton Street, Emeryville, Ca 94608, USA

Key words: stabilisation, protein, formulation, stability, drug product, drug substance

Abstract: This chapter overviews the stabilization of protein-based biopharmaceuticals. Finished biopharmaceutical products must have the required stability from a clinical and regulatory point of view. Furthermore, the stability profile should cover the manufacturing and marketing cycles of the product to minimise costs. The major routes by which proteins may be degraded are outlined. A brief description of biopharmaceutical production processes is provided to acquaint the reader with issues relating to stabilisation during processing. The basic principles of how in-process material and finished product may be stabilized are presented. Finally, a table of major constituents in approved biopharmaceuticals is included at the end of the chapter to illustrate how stabilisation principles have been put to practice to date.

1. INTRODUCTION TO BIOPHARMACEUTICALS

Over the past 15-20 years, we have witnessed the approval for clinical use of a number of biotechnology-derived therapeutic products, i.e. biopharmaceuticals. Such products include recombinant insulin, interferons, interleukins, erythropoietin, etc.

In contrast to more conventional drugs, biopharmaceuticals are generally large macromolecules. Although many nucleic acid-based products are currently in development, only protein products have reached the marketplace so far. Adequate stability of therapeutic and antigenic proteins is a critical requirement for their successful entry into the marketplace. In this chapter, we wish to focus our attention upon the formulation and stability of such biopharmaceutical proteins.

2. WHY IS STABILISATION IMPORTANT?

Stability may be defined as the absence of a change in the properties of a product over a given period of time. This all-encompassing statement presents an almost impossible task to achieve in practice. A more realistic expectation for stability of pharmaceutical products is provided by the International Conference for Harmonisation (ICH) guidelines (1) in the following statement:

> The purpose of stability testing is to provide evidence on how the quality of a drug substance or drug product varies with time under the influence of a variety of environmental factors such as temperature, humidity, and light, and enables recommended storage conditions, retest periods, and shelf lives to be established.

Here, the term 'drug substance' for a biopharmaceutical product refers to the purified bulk of the active therapeutic or vaccine antigen. The term 'drug product' describes the finished formulated product, packaged in its final container-closure system.

With this backdrop, let us consider why stabilisation is important for a biopharmaceutical product. The first and foremost reason is that the product must maintain its properties of safety and efficacy throughout its declared shelf life. In the case of products still under clinical testing - where the efficacy is unknown - the stability characteristics must ensure that the product is safe to be administered to patients. In addition to such ethical and regulatory issues, there are obvious commercial advantages associated with developing a product whose shelf-life is maximised. An efficient biopharmaceutical production plant is often a multi-product manufacturing facility requiring flexibility in its operation. Such plants usually manufacture different products in "campaigns" in which several lots of a given product are manufactured consecutively before switching to a different product. Therefore, stabilisation of a biopharmaceutical at various stages of its production process is important to create several key in-process hold points. This strategy allows batch processing of several partially purified lots, which could, in turn, be further purified as separate or combined batches.

3. STRATEGIES FOR STABILITY EVALUATIONS

Proteins are large, complex molecules that are subject to a variety of degradative influences during storage and use. The complexity of their structure necessitates the use of an array of analytical techniques for their characterisation. Stability of a biopharmaceutical cannot be defined by a single analytical method, but by a panel of techniques.

Stability issues are of critical importance at two stages of drug development. Firstly, stabilisation of biomolecules is a key issue during process and formulation development. Secondly, stability of the finished product must be monitored to assure maintenance of identity, quality, potency and purity during its shelf life. Due to the aggressive pace of the drug development process, a pharmaceutical scientist relies heavily on methods of stability prediction to make decisions during the formulation selection process. Data generated during formulation development forms the basis for predicting the shelf life of the product used in clinical trials. Generally, enough real time (i.e., storage under the intended conditions) data is generated by the time the product reaches the market.

The fundamental strategy for understanding product stability and identifying suitable stabilizers is by studying its degradation. Since proteins are subject to many types of degradation reactions, a "degradation profile" needs to be constructed using many assays to determine a rank order relationship among the various stabilising conditions. After evaluation of the degradation reactions, the major degradation pathways are identified and used as inverse markers for stability during formulation development. Since changes do not occur in a rapid enough time frame for development studies, the scientist has to rely upon "accelerated" stability evaluations. Elevated temperatures are often used to expedite changes in products. Unlike with small molecules, traditional reaction rate analyses using the Arrhenius equation do not always apply to proteins. In general, the chemical degradation reactions (cleavage of a peptide bond, oxidation, deamidation, etc.) obey the classical rate equations and have been used to develop preformulation profiles of proteins. For example, such analyses have been used for relaxin (2), tumour necrosis factor (3) and DNase (4). However, unlike with small molecules, these rates are not frequently used to estimate shelf life of a protein product. Instead, relative rates are generally employed for selecting a stable formulation from the available choices.

4. DEGRADATION-INDUCED CHANGES IN BIOPHARMACEUTICAL PROTEINS

Typical degradation reactions which proteins may undergo are often caused by either denaturation (non-covalent modification) or chemical bond formation/cleavage (covalent modification). For a more comprehensive overview of degradation pathways, the reader is referred to additional publications (5-8).

Since proteins are complex macromolecules, limited changes at one or more sites sometimes do not alter their biological activity. For example, oxidation of interleukin-2 (IL-2) (9) and deamidation of hGH (10) has little effect on their biological activity. Many of these altered species are "handed over" as part of the bulk drug substance to the formulator by the process scientist, and the formulator's job is to preserve the species composition as much as possible during finished product manufacturing, storage shelf life and use by the patient.

4.1 Physical Inactivation

4.1.1 Unfolding

Proteins are composed of amino acids linked together via peptide bonds. The hierarchy of protein structure consists of several levels. The primary structure represents the protein amino acid sequence while the secondary structure describes the organisation of the polypeptide chain into α-helices, β-sheets and β-turns. The tertiary structure characterises the overall folding of the primary and secondary structures into a unique three-dimensional structure that brings distant residues on the polypeptide chain in close vicinity. Some native proteins exist in an oligomeric form as an assembly of subunit polypeptides; such a high level assembly is known as quaternary structure.

Protein unfolding describes the loss of its unique three-dimensional structure (secondary and tertiary structure). This is normally accompanied by loss of its biological activity. Unfolding is often observed as a highly co-operative process between the native (N) and unfolded (U) forms with rarely detected intermediates as described in Eq. 1. Here K_d equals [U]/[N], the unfolding equilibrium constant.

$$N \text{ (native)} \xrightarrow{K_d} U \text{ (unfolded)} \tag{1}$$

The stability of a protein can be defined as the difference in free energy between the unfolded (G_U) and the native (G_N) forms, which can be related to the equilibrium constant by the following equation (11):

$$\Delta G = G_U - G_N = -RT\ln K_d \tag{2}$$

Here, R is the gas constant and T is the temperature in degree Kelvin. The relationship between the free energy, enthalpy (ΔH) and entropy (ΔS) is given by the Gibbs equation:

$$\Delta G = \Delta H - T\Delta S \tag{3}$$

The protein native conformation is marginally stabilised by the enthalpy gain, (realised from hydrophobic interactions, hydrogen bonding, van der Waal bonding, electrostatic and disulphide bond formation), over the disfavoured entropy loss due to an ordered matrix structure (12). The value of free energy difference between the native and unfolded form is around 5 to 20 kcal/mol, which is equivalent to the contribution of several hydrogen bonds (13). The marginally stabilised native conformation is thus sensitive to environmental parameter changes such as temperature, pH, salt concentration, contact with surfaces, etc.

4.1.2 Aggregation and precipitation

Protein aggregation frequently occurs subsequent to the exposure of interior hydrophobic residues during unfolding. The exposed hydrophobic residues from different molecules interact together directly. A simple aggregation pathway is described in Eq. 4, although aggregation can also occur from partially unfolded intermediates (14):

$$N \leftrightarrow U \rightarrow \text{aggregation} \tag{4}$$

Heating, cooling, pH shift, solvent condition changes, or contact with surfaces can trigger unfolding of a protein and result in aggregation. Therefore, aggregation is often observed during accelerated stability evaluations at elevated temperatures, reconstitution of freeze-dried products and manipulations during manufacturing processes.

Protein aggregates can also be formed via intermolecular covalent interactions, mainly through disulphide linkages. Formation of disulphide bonds from free cysteines is known to occur via deprotonated thiol intermediates. The pKa of the thiol group of cysteine is about 8. Thus, covalent aggregation through disulphide bonding is promoted under alkaline pH conditions.

Aggregation through hydrophobic interactions eventually leads to formation of large visible precipitates, probably due to the non-specific nature of the forces involved. In contrast, due to its specific nature, covalent aggregation may invoke smaller oligomers, as the number of free cysteines in a protein limits this process. In the evaluation of protein aggregation, one needs to account for soluble aggregates in addition to the visible precipitated protein. Soluble aggregates are defined operationally as the fraction of oligomers remaining in solution without generating a visible turbidity in the solution. Insoluble aggregates are visible and can be removed from solution by low speed centrifugation or filtration (15).

Aggregation often follows protein unfolding. Unlike unfolding, which is usually a unimolecular reaction and is independent of protein concentration, aggregation depends strongly on the initial protein concentration. A higher initial protein concentration results in aggregate formation in greater amounts. The kinetics of aggregation may involve a nucleation step, which accumulates denatured chains to a critical concentration, followed by a rapid growth phase.

4.1.3 Adsorption

Proteins are sometimes lost from solution due to adsorption onto contacting surfaces. The interaction is believed to be due to non-specific forces, as it occurs regardless of the nature of the surface. The adsorbed layer consists of unfolded and partially unfolded protein molecules bound to the surface. In most instances, a protein monolayer is formed which saturates the surface, generally preventing further adsorption (16). The adsorption follows the Langmuir isotherm, which is expressed as

$$C_s = M/AN_A \, (KC_b/(1+KC_b)) \tag{5}$$

C_s is the amount of protein adsorbed per unit area of surface, M is the molecular weight of the adsorbing protein, K is an equilibrium adsorption constant, C_b is the bulk protein concentration in solution, A is the surface area per site, and N_A is Avogadro's number. A characteristic of this isotherm is that a plot of C_s versus C_b yields a curve that starts at zero (at the origin) and rises monotonically with increasing bulk concentration, until a plateau is reached. The plateau surface concentration is the monolayer concentration, C_m, and is equal to M/AN_A. In the case of proteins that are prone to aggregation, a multilayer aggregate may be formed.

4.2 Chemical Degradation

Proteins chemically degrade via side chain modification or via hydrolysis. Although there are a number of reactions in addition to those discussed herein, deamidation, oxidation, covalent aggregation (already discussed within the physical degradation section) and hydrolysis represent the most commonly encountered degradation pathways for biopharmaceutical proteins.

4.2.1 Deamidation

Deamidation reactions involve chemical modification of the amide side chain of either Asn residues or Gln residues in the protein. Under general acid and base catalysis mechanisms, formation of an oxyanion transition state through protonation of the amide leaving group and hydroxide nucleophilic attack of the amide carbonyl carbon is believed to be the rate controlling step (8). The eventual product is obtained from the hydrolysis of the amide group to aspartate or glutamate residues with a charge change. The oxyanion transition state can be stabilised by hydrogen bonding with the amide group and carbonyl group on the main chain. Since glutamine has an extra CH_2 group in its side chain, the larger distance from the oxyanion transition state to the main chain nitrogen disfavours hydrogen bond formation. Therefore, the oxyanion transition state for glutamine is less stabilised compared to that of asparagine. The deamidation rate, therefore, is usually faster for asparagine than for glutamine.

At neutral and alkaline pH, deamidation of aspargine can also go through a five-member ring cyclic imide intermediate. The main chain peptide nitrogen on the adjacent C-terminal side residue can cause a nucleophilic attack on the carbonyl carbon of the asparagine side chain to form this cyclic imide intermediate. The resulting product can be either isoAsp or Asp depending upon which C-N bond is hydrolysed. The deamidation rate is greatly affected by the bulkiness of the adjacent residue and the local conformation of Asn and the adjacent residues. Glutamine can form a six member ring cyclic imide intermediate, but this is less favoured.

Deamidation is affected by solvent conditions such as pH, ionic strength, buffer anions and temperature. Because of the dependence on a general acid-

and base-catalysed mechanism, the deamidation rate has a pH of minimum reactivity. For peptides, this pH is around 6. The minimum pH for protein deamidation varies from protein to protein. High ionic strength and elevated temperature all accelerate the deamidation rate. Phosphate ions are also known to promote deamidation.

4.2.2 Oxidation

Residues susceptible to oxidation are cysteines, methionine, histidine, tryptophan and tyrosine. Cysteines are easily oxidised at neutral to alkaline pHs to form intra- or intermolecular disulphide bonds. Cysteine forms disulphide bridge through a deprotonated thiol intermediate (Eq. 6).

$$R\text{-SH} + OH^- \leftrightarrow R\text{-}S^- + H_2O$$

$$R\text{-}S^- + R'\text{SH} + OH^- \rightarrow R\text{-}S\text{-}S\text{-}R' + H_2O \tag{6}$$

Under conditions unfavourable to disulphide formation (e.g. a lack of nearby thiols or when the protein is in dilute solution), cysteine can also be oxidised to a series of cystic acids (Eq. 7).

$$R\text{-SH} \xrightarrow{[O]} R\text{-SOH} \xrightarrow{[O]} R\text{-}SO_2H \xrightarrow{[O]} R\text{-}SO_3H \tag{7}$$

Oxidation of methionine, on the other hand, is acid catalysed. Methionine can be oxidised to sulphoxide and further to sulphone (Eq. 8).

$$R\text{-}CH_2\text{-}CH_2\text{-}S\text{-}CH_3 \xrightarrow{[O]} R\text{-}CH_2\text{-}CH_2\text{-}CH_2\text{-}S(O)\text{-}CH_3 \xrightarrow{[O]} R\text{-}CH_2\text{-}CH_2\text{-}(O_2)\text{-}CH_3 \tag{8}$$

Oxidative reactions can be catalysed by oxygen from air, dissolved molecular oxygen, trace amount of metal ions remaining in solution, hydrogen peroxide and many other oxidant residuals.

Tryptophan, histidine, cysteine, methionine and tyrosine are susceptible to photooxidation in the presence of oxygen and sensitising dyes (17). The oxidative rate depends upon the solution pH. At acidic pH, Trp and Met are readily oxidised. At neutral pH, oxidation of histidine is rapid and only the neutral form is susceptible. At alkaline pH, Tyr is most reactive.

4.2.3 Hydrolysis

The peptide bond in proteins is susceptible to acid hydrolysis resulting in the generation of fragmented peptides. Peptide bonds adjacent to Asp residues (particularly Asp-Pro and Asp-Gly), are labile under acidic conditions. The reaction is thought to occur via a six-member ring intermediate formed by the carboxyl group on the Asp side chain with the carbonyl group on the main chain on either the C-terminal or N-terminal side. The hydrolysis rate is accelerated by an increase in temperature and lowering of pH. Peptide bonds X-Ser and X-Thr are also labile but scission of these occur only under strongly acidic conditions.

5. OVERVIEW OF PRODUCTION PROCESSES FOR PROTEINS

A production process for a recombinant protein generally consists of a series of steps as shown in Figure 26. The protein is expressed in an appropriate host system, either bacterial such as *Escherichia coli,* or yeast, or in mammalian cells. After the fermentation step, a concentration step is employed to collect the protein of interest. In the case of a secreted protein, the cell supernatant is collected and concentrated. When the protein is expressed intracellularly, the cells are collected, disrupted and the protein extracted. At this point, a crude concentrate is generated for further purification. Purification steps usually consist of one or more chromatographic steps (e.g., reverse phase, size exclusion or ion exchange), coupled with other steps such as precipitation and diafiltration. The final purification step generally yields a bulk protein solution, which is designated as the "bulk drug substance".

The bulk drug substance is formulated to the desired final composition by a combination of diafiltration, dilution or addition steps. At this stage, the product is known as the "final formulated bulk". For products, which cannot be sterilised by a final filtration, the final formulated bulk is prepared by combining prefiltered components aseptically. Otherwise the final formulated bulk is subjected to a final 0.2 μm filtration. The sterile liquid is dispensed into the desired final containers (vials, syringes, *etc.*), and freeze-dried if appropriate. The finished product is closed, sealed and labelled. For non-parenteral products, the fill-finish procedures are modified to fit the need.

In general, a biopharmaceutical production process is a complex procedure requiring a variety of skills, facilities and equipment. Tight process monitoring and controls are included at each stage to assure the successful completion of various steps in the process train.

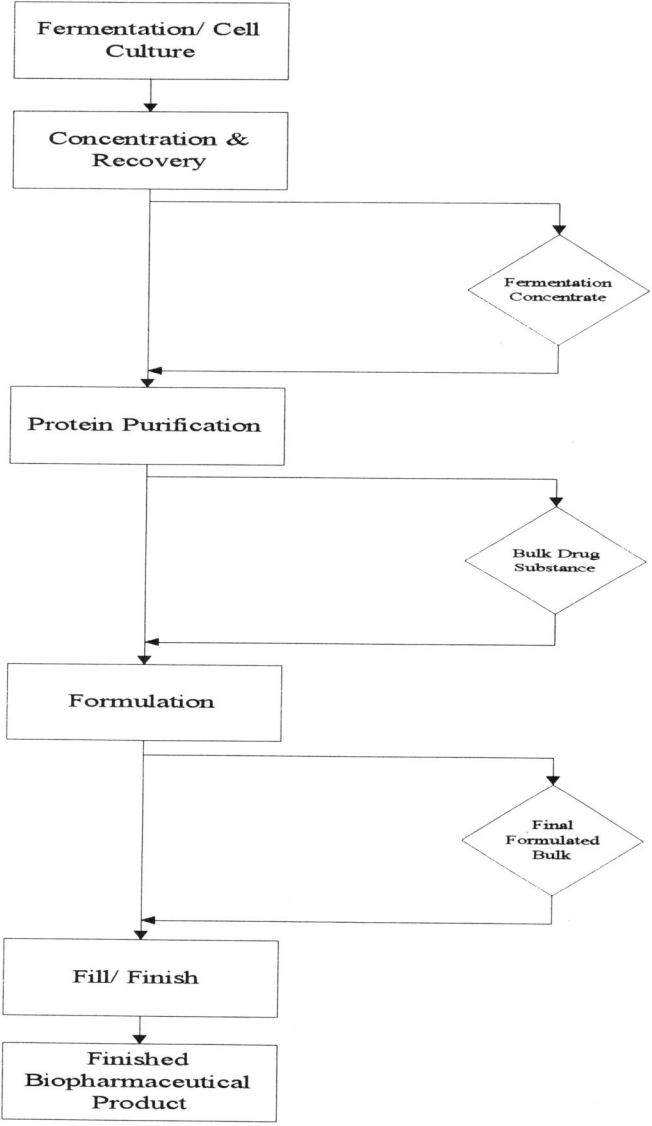

Figure 26. An overview of a generalized biopharmaceutical production process

6. STABILIZATION OF KEY INTERMEDIATES

As already pointed out, key intermediates in a biopharmaceutical production process are the fermentation concentrate, the bulk drug substance and final formulated bulk. Other intermediate steps may also be important as dictated by the nature and length of a particular process.

As biopharmaceuticals are relatively expensive to produce, one way to reduce costs is by sharing a facility that is utilised to produce a number of products. In this manner, the overhead of the infrastructure of the plant is distributed across different products. Stabilisation of key intermediates is a critical requirement for operating a given process in a multi-product facility. In addition, since fermentation, purification and finished product operation require different types of facilities, it is desirable to break the process at these logical points. Finally, the specialised nature of these processes may require the use of facilities at different locations, necessitating shipping of stable intermediates.

We shall now discuss two important intermediates that are most frequently stored for prolonged periods to manage a biopharmaceutical manufacturing process. These are the fermentation concentrate and bulk drug substance. Stability issues for other intermediates such as certain column eluates from a purification step or the final formulated bulk are similar to those discussed for these two key intermediates.

6.1 Fermentation Concentrate

The objective at this stage of the process is to stabilise the protein of interest for reasonable periods of time (usually months to years) to allow manufacturing flexibility. At this stage, the protein of interest is present in the mixture with other constituents of the host system and fermentation medium. These may include proteolytic enzymes and high concentrations of other solutes (e.g., salts and residual metal ions).

6.2 Bulk Drug Substance

For traditional pharmaceuticals, the chemically synthesised bulk drug (known as the "drug substance") is usually supplied as a solid. In contrast, proteins are derived from aqueous solution-based processes. Therefore, protein bulk drug substances are supplied as bulk, concentrated buffered solutions. The bulk protein is extensively characterised for integrity, purity, potency and other quality parameters. This stage represents a practical hold point for multi-formulation, multi-dosage strength products and for products which are filled at a remote location. The drug substance is also the most logical point for detailed characterisation of finished products, which may not be fully analysed in their final formulations (e.g., human serum albumin (HSA) containing formulations and vaccine antigens adsorbed onto alum).

6.3　Stability Issues for Key Intermediates

6.3.1　Storage conditions

At the fermentation concentrate stage, the product is present in a very complex mixture and dissection of stability effects due to each component of the mixture is at best difficult to accomplish. In terms of product stability, the most prudent course usually is to store the fermentation concentrate as frozen material until used by the next step in the process. In contrast to the fermentation concentrate, the protein is present in its purest state as the bulk drug substance. The composition of the buffer and the pH are usually derived from careful experimental studies similar to those of the finished drug product. Nonetheless, relatively inconvenient storage conditions (such as freezing) are quite acceptable, as the goals at this point are to prevent microbial contamination of the bulk during storage, and not compromise the stability of the subsequent finished product. In case of stable biomolecules, storage under refrigeration conditions can also be used.

Freezing is considered innocuous by many, as it is felt that lowering of the temperature would reduce or eliminate the rate of deleterious reactions in a system. However, Franks (18) has provided ample evidence regarding the injurious effects of freezing on biological systems in an excellent treatise on the subject. Briefly, freezing of water in a biological system leads to concentration of the biomolecule and other solutes within an ice matrix. For example, an initially isotonic salt solution (0.154 M NaCl) undergoes a 20-fold concentration when it is frozen and the temperature reduced to -10°C. At its eutectic temperature of -23°C, the salt concentration increases to a value of 4.7 M. Similar concentration effects are seen with other solutes such as buffer ions, chelators and biomolecules present in a freezing mixture. Such freeze concentration processes often affect the buffering ability of standard buffer mixtures. Shifts in pH occur due to temperature, concentration or, in extreme cases, by the actual crystallisation of one or more of the buffer components under eutectic conditions.

Based on the discussion above, it is clear that care must be taken to map the phase transitions of a solution before recommending a subzero temperature of storage. Presence of phase transition temperatures can be detected by a differential scanning calorimetry (DSC) technique. In general, storage temperatures significantly different from phase transition temperatures should be chosen to avoid exposure to temperature cycling effects. Temperature cycling refers to the temperature variation (usually ±5°C) around the freezer set point that occurs as the freezer mechanism attempts to maintain the nominal freezer temperature by repeated heating or cooling bursts. Such cycling may inactivate the protein of interest by exposing it to high salt concentrations, extreme pHs or other concentration effects. After a temperature of storage away from phase transition events is chosen, stability of the protein must be confirmed by other biochemical analyses. Assays are usually limited to enzyme linked immunosorbent assays (ELISA) or HPLC at the fermentation concentrate stage, but involve a full battery of techniques at the drug substance stage.

6.3.2 Storage containers

Large volumes of aqueous solutions need to be stored at this stage. An ideal container should be compatible with the fermentation concentrate or the bulk protein being stored. In addition, the container should have adequate thermal characteristics that it can withstand both freezing and autoclaving conditions, and the rigours of shipping and handling. To facilitate their use for freezing applications, their brittleness temperature should be lower than their storage temperature. Table 24 lists typical characteristics for some common plastics.

Table 24. Physical characteristics of some common plastics

Resin	Appearance	Maximum Use Temp. °C	Brittleness Temp. °C	Autoclave-ability	Water Absorption %
High density polyethylene	Translucent	120	-100	No	<0.01
Polycarbonate	Transparent	135	-135	Yes	0.35
Polypropylene	Translucent	135	0	Yes	<0.02
Polytetra-fluoroethylene	Translucent	260	-267	Yes	0.01

Glass and polypropylene containers have been used in the past. As the science of large scale biotechnology progressed, these containers have given way to stainless steel containers (primarily for refrigerated or ambient storage) and polycarbonate or polytetrafluoroethylene containers for storage under freezing conditions.

Large scale freezing of liquids is not simple, as the outer layer of frozen mass first formed can inhibit the subsequent bulk freezing process. To minimise these effects, one alternative is to use small containers. Controlled freezing in large-scale containers (19) is another option for storage of bulk drug substance and the final formulated bulk. Finally, a relatively new technique called "cryogranulation" has recently been described as an alternative means for storage of bulk drug substances (20). Cryogranules of proteins are prepared by exposing solutions or slurries of bulk proteins to a cryogenic material such as liquid nitrogen.

7. STABILISATION OF PROTEINS IN FINISHED PRODUCT FORMULATIONS

7.1 Key Objectives and Considerations

In an earlier section, we noted the importance of having an adequate shelf life for a biopharmaceutical finished product. Common sense dictates that the finished product formulation should be suitable for use in the patient population selected for the drug. The adequacy of the formulation, and length and storage temperature of a product are subjects for many debates within the pharmaceutical, clinical, regulatory and marketing departments of a biopharmaceutical organisation. A detailed discussion of these issues is beyond the scope of this article. Briefly, however, the formulation and stability of a product should be consistent with its intended clinical, regulatory and marketing settings after commercialisation.

7.1.1 Clinical considerations

The finished product should be formulated such that it is amenable to normal manipulations and easy use in a clinic or by the patient. In contrast to traditional drugs, biopharmaceutical products can not be administered by the most convenient (oral) route. Practically all biopharmaceuticals presently on the market are administered by intravenous, intramuscular or subcutaneous routes. In addition, some products are also given by nebulizers (e.g., DNAse) or topically (e.g., platelet derived growth factor-BB). Finally, much research is under way to develop delivery mechanisms by the pulmonary (21) or nasal routes. Development of slow release dosage forms is also continuing as certain disease states may be most effectively treated by continuous delivery of a biopharmaceutical over prolonged periods.

Formulation components should also be selected such that the formulation is biocompatible when administered. The buffer chosen should be sufficient to provide pH stability to the protein during storage, but be incapable of altering the local pH of the biological microenvironment for prolonged periods after administration. Certain buffers have been found to cause pain when used at high buffer strengths (22-23). The buffer components should not alter the pharmacological and toxicological attributes of the drug unless specifically intended.

7.1.2 Regulatory considerations

Components of finished formulations are subject to thorough scrutiny by regulatory agencies. These are obviously directly administered to the patient and often in quantities that are much greater than the active substance itself. A small number of excipients used in finished formulations already on the market normally form the 'first list' for the formulator. These excipients are not automatically "approved" for use in new products but are approved for use in the specific product(s) in question. However, there is usually a fair amount of safety information on their chronic use. Proper toxicological experiments can thus be designed to test the excipients in combination with the new active drug. If a new excipient is used for the first time in a finished product for injectable use, regulatory authorities expect a thorough physicochemical and biological characterisation, essentially similar to that expected from a new active ingredient (24).

Recently, concerns regarding the use of human and animal derived substances in pharmaceutical products have attracted much regulatory interest. For example, guidelines are now in place to restrict or eliminate the use of substances of animal origin in such preparations (25). These considerations affect many common excipients such as human serum albumin, polysorbate surfactants, amino acids and carbohydrates. When a formulation is finalised and introduced to pilot manufacturing, the formulator must make sure that acceptable excipients are used.

Some regulatory authorities test every batch of the finished product by some key release tests before the product is released for sale by the manufacturer. This adds 1-3 months to the manufacturing and release cycle of the product, thus decreasing its useable shelf life. The Food & Drug Administration (FDA) has now instituted a policy for certain specified biologicals (previously known as well characterised biologicals), for which sufficient characterisation data are made available during or after their licensure, which exempts them from every lot testing by the agency.

7.1.3 Marketing considerations

An ideal formulation from a marketing standpoint is one that is convenient for the health practitioner or patient (ready to use, can withstand extreme temperatures for short periods during shipping and handling, compatible with a majority of diluents, devices and accessories prevalent in medical practice). To gain a competitive edge, the product may be packaged in a unique presentation such as self-injecting pens, prefilled syringes, prediluted infusion solution bags, etc. Each of these systems represent a separate challenge as unique manufacturing and stability considerations apply. The pharmaceutical scientist should work closely with their marketing colleagues to introduce these programs early to ensure the best presentation reaches the market and allows the patient to enjoy the ease and convenience of modern technology without adversely affecting the product quality.

7.2 Stabilisation to Extend Long Term Storage

For many years, it was believed that most proteins were too unstable to be formulated as solutions. As a result, stabilisation of a protein by lyophilisation attracted much of the attention up until the 1980s. Since then, however, much progress with regard to stabilisation of proteins in solution has been made, and several solution formulations with long storage shelf lives are on the market. In this section, we wish to discuss protein stabilisation in solution and lyophilised states. Strategies for stabilisation of proteins in these two forms are quite different.

7.2.1 Solution stabilisation

Stabilisation of a protein in solution is achieved by overcoming its conformational instability and chemical lability. A protein's native conformation is stabilised by hydrophobic forces, electrostatic interactions, hydrogen bonding, van der Waal forces and disulphide bonds. Changes in the microenvironment of the protein, such as extremes of cold, heat, pH, and high concentrations of salt or denaturing solvents, disfavour these stabilising interactions. Stabilisation of protein conformation in solution is the key to its overall stability, if unfolding is the rate-limiting step in the major degradation pathway. In these situations, stabilisation of the protein at the unfolding step can be most effective in preserving its function.

Chemical degradation involves specific protein side chains interacting directly with solute components or among themselves. Stabilisation of protein conformation may protect interior residues but not the surface residues. Thus, chemical stabilisation depends upon protectinng the surface reactive

Stabilisation of biopharmaceutical products and finished product formulations

groups and preventing the formation of key intermediate in the degradation pathway. Elimination of certain solvent components such as dissolved metal ions, and changes in bulk solvent properties such as red-ox potential, may also be necessary to achieve the desired stability.

7.2.1.1 pH stabilisation

Optimisation of the stability of a protein through pH adjustment is a simple but an extremely effective approach. Proteins are macro-ions and charge distribution on the protein surface depends upon the degree of protonation or deprotonation of ionisable side chains. The conformation is affected by acidic and basic conditions because of changes in the electrostatic repulsive or attractive forces between protonated and deprotonated residues. A change in pH of the medium influences charge distribution on the protein surface thus affecting protein conformational stability.

Physical and chemical degradation reactions are also affected by the pH. Proteins may precipitate out of solution at a pH close to their pI. The effects of pH on typical chemical degradation pathways are summarised in Table 25. Rates of these chemical reactions are strongly influenced by the solution pH. A simple rule of thumb is to avoid the use of basic pH conditions as side chain modifications and cysteine oxidation are promoted under these conditions. Similarly, avoidance of very low pH values is recommended as these usually facilitate hydrolysis and methionine oxidation. The range 5 to 7.5 is the recommended working range for pH optimisation of most therapeutic proteins.

Table 25. Effect of pH on chemical degradation

Process catalyzed by acidic pH values:
Hydrolysis of peptide bonds
Methionine oxidation
Deamidation through direct hydrolysis of amide side chain
Maillard reaction
Processes catalyzed by basic pH values:
Cysteine oxidised to disulphide or sulphonic acid
β-elimination
Disulphide scrambling
Deamidation through cyclic imide intermediate

The presence of additional components in the formulation may alter the pH of optimum stability of the protein. The optimum pH may change if there are other stabilisers or additives present, since charge distribution on the protein surface depends not only on the pH but also on its ionic strength. Additionally, the chemical degradation pathway may be altered in the presence of other stabilisers. Buffer species used for pH adjustment may also affect the physical or chemical inactivation rate.

7.2.1.2 Stabilisation through specific interactions

Certain compounds which exhibit preferential binding to the native protein can also enhance protein conformational stability. Proteins are known to be stabilised by ligand binding, as this shifts the unfolding equilibrium (Eq. 1) towards the native side (26). Similarly, compounds which have specific affinity to the protein's native conformation, will also stabilise proteins (27). For example, detailed studies reveal that anionic polymers have a great stabilisation effect on several members of the fibroblast growth factor (FGF) family such as acidic FGF (28), basic FGF (29) and keratinocyte growth factor (15). Arginine or lysine are known to interact with Kunitz domains and therefore are employed in the processing and final formulation of tPA (30). Zinc is often detected in insulin crystals and promotes formation of its hexamer. Zinc-stabilised insulin provides prolonged insulin levels in the body and such preparations have long been used in commercially available formulations of insulin (31).

7.2.1.3 Stabilisation through non-specific interactions

Timashiff and colleagues (32) have proposed the preferential exclusion principle to account for protein stabilisation by compounds that interact with proteins in a non-specific manner. These are compounds originally found to increase osmotic pressure of living cells under stressed environment conditions. They are thus called osmolytes (33). Equilibrium dialysis results revealed that these compounds are excluded from the protein surface. This results in an increase in the chemical potential of the protein, therefore this process is not favoured thermodynamically. However, since denatured protein molecules have a much greater surface area than that of their native counterparts, the increase in chemical potential is greater for the denatured protein than for the native protein. Therefore, overall exclusion favours the native conformation and stabilised it.

This class of compound includes carbohydrates, amino acids, salts, glycerol and polyethylene glycol. Thermal unfolding results confirms that these osmolytes increase thermal melting temperature of proteins (34) and the storage stability (35). However, the energy gain from the difference in the exclusion between the native and the denatured forms from each solute molecule or ion is small. As a consequence, high solute concentrations (e.g., 500 mM to 1 M) are needed for the desired stabilisation, thus limiting the practical application of this theory. Nonetheless, this stabilisation method can be used in combination with others to achieve an overall positive stabilisation effect.

7.2.1.4 Other stabilisers and additives for solution formulations

Proteins can also be stabilised by solutes that interact indirectly with the protein molecules. Compounds, which modify bulk solvent properties or act as a scavenger of certain solutes, can also stabilise proteins from certain chemical degradation pathways. The so-called antioxidants, which are scavengers of radicals, can protect cysteine residues from oxidising to disulphide bonds. Commonly used metal chelators such as EDTA chelate transition metal ions to prevent oxidative degradation.

Surfactants are often used in a protein preparation to prevent surface denaturation during processing (36), protein loss due to adsorption to surfaces (37) and damage due to freeze-thaws (38). Human serum albumin (HSA) is a blood-abundant protein and has been used in many commercial formulation to protect proteins from loss due to adsorption or storage. The use of HSA is gradually decreasing due to the regulatory issues surrounding the use of blood-derived products.

For sparingly soluble proteins, solubilisers need to be added to maintain an adequate amount of protein in solution for prolonged periods of time. A detailed discussion of solubility is beyond the scope of this article and the reader is referred to the relevant literature (39). In general, some of the same strategies that are applicable to stabilisation are used for solubilisation (pH, ionic strength, and specific solubilisers). Solubilisers include surfactants such as polysorbate-20 and -80, sodium dodecyl sulphate, cyclodextrins, polyanions and amino acids.

Antibacterial agents are included in formulations to allow the finished formulation to be used as a multidose product. These agents normally interact with proteins through hydrophobic or electrostatic forces and are therefore capable of perturbing their fine structure. Careful studies should be performed to ensure long term stability of the finished formulation in the presence of these agents. Common antibacterial agents include alcohols (benzyl alcohol, methyl- and propyl parabens, phenol or m-cresol) and quaternary amino compounds (benzalkonium chloride and benzethonium chloride). The preservative efficacy of the finished multi-dose formulation should be evaluated by the European, Japanese or United States Pharmacopoeia depending upon the marketing territory of interest. These tests must be carried out during development to demonstrate preservative efficacy for the formulation immediately after manufacturing and after the intended storage shelf life.

Finally, tonicity agents are added to make the formulation isotonic with physiological fluids. Normally, one of the stabilisers or solubilisers is increased to a concentration to render the formulation isotonic. If the cost, toxicity or other concerns limit the use of stabilisers/solubiliser to a low value, common substances such as salts or sugars may be added (if compatible with the rest of the formulation) to balance the tonicity.

7.2.2 Lyophilised product stabilisation

The stability of a lyophilised protein depends upon many factors such as the formulation components, state of the solid during freezing and drying processes, level of residual water in the finished product and storage temperature.

7.2.2.1 Stabilisation within a glassy matrix

It is now well accepted that immobilisation of a protein within a glassy matrix is essential for achieving good stability during storage. Franks (40) has theorised that stabilisation is realised as proteins and other molecules lose mobility due to the high viscosity of the glassy state. Thus, transitional movement is prohibited and only small vibrational and rotational motions are allowed. Physical and chemical reactions are hindered because there is little diffusion of molecules.

Primary criteria for selection of a glass-forming stabiliser are: a) the compound itself is chemically inert, b) it is compatible with the protein so that protein molecules can be dispersed in the glass without any phase separation during lyophilisation, and c) the glass transition temperature ($Tg`$) for the solid is substantially higher than the storage temperature. Among commonly used pharmaceutical formulation excipients, the disaccharide sucrose seems to meet all these requirements. Sucrose has a $Tg`$ of $-32°C$ in the frozen state. Glass transition temperature of solid sucrose is well above refrigerated or ambient storage temperature (40). Sucrose is chemically inert and it can form amorphous solids with protein molecules without phase separations. Another disaccharide, trehalose, although not in an approved pharmaceutical formulation yet, also deserves attention (41). Trehalose has a similar $Tg`$ to sucrose but has a much higher glass transition temperature in the amorphous form, which may enhance the storage stability of a protein to a greater extent than sucrose.

Many monosaccharides such as glucose and sorbitol also form a glassy matrix. However, their $Tg`$ values are much lower than that of sucrose and this increases freeze-drying time, making the process less economical. In addition, glucose and lactose are reducing sugars, which may interact with the lysine groups in protein side chains resulting in modified protein molecules via the "Maillard reaction" (42). Since glass transition temperature has an inverse relationship with molecular weight, macromolecular compounds have a higher $Tg`$ value than sucrose and could be useful for stabilisation with the advantage of having a short freeze-drying cycle. For example, HSA has been used to stabilise proteins in many commercial lyophilised formulations. However, compounds with higher molecular weight are not always useful for increasing protein stability. For example, both sucrose and dextran were lyophilised with hGH and formed amorphous solids. While sucrose stabilised the protein, dextran failed to do so. The authors speculate that dextran may be too bulky to form hydrogen bonds with proteins and may phase separate during freeze-drying (43).

Another proposal for protein stabilisation in the lyophilised state is the "water substitution" hypothesis (44). According to this hypothesis, good stabilisers can replace water to interact with protein surface groups through hydrogen bonding in the lyophilised state. FTIR studies of proteins lyophilised with mono- and disaccharides show evidence of hydrogen bonding between proteins and sugar molecules (45). This mechanism seems to work well for stabilising sugars, which have good glass formation and hydrogen bonding properties. However, this hypothesis fails to describe the protein stabilisation effect of other compounds, such as tetramethylglucose and polyvinyl pyrrolidone, which form glass after freeze-drying but can not form hydrogen bonds with protein molecules (46). In addition, there is a discrepancy in this logic as these compounds are known to stabilise proteins in solutions by the exclusion principle. This would mean that these molecules are excluded from the protein surface during freeze-drying thus contradicting the hypothesis.

Drying to a low residual water level is considered as another requirement for stabilisation. One obvious advantage of freeze-drying preservation is that water-mediated reactions are retarded. Chemical degradation reactions such as hydrolysis and deamidation all need water as a reaction component.

The stability of a protein in a glass is dependant upon the difference between the storage temperature (Ts) and the glass transition temperature (Tg`) of the solid (46). The temperature dependence of protein stability obeys the Williams-Landel-Ferry relation in the glassy state and not the Arrhenius equation, which is applicable in solutions. Therefore, the lower the storage temperature of the protein glass compared to its glass transition temperature, the greater is the stability of the protein. Since water is a plasticiser for the glass and the glass transition temperature of a solid depends upon its water content, a low moisture level in the glass is desirable for achieving greater protein stability.

Interestingly, the concept "as dry as possible" may not always benefit protein stability. DSC results show that the protein unfolding temperature in solids increases at lower residual moisture levels (47). However, it is thought that protein molecules need at least a monolayer of water molecules to retain their essential conformational structure in the dried state, probably by satisfying hydrogen bonding of the surface groups (48). Thus, adequate residual moisture for a particular protein formulation needs to be determined to maximise its stability.

One must also consider preservation of a protein's native conformation throughout the freeze-drying process for improving the protein stability. If more of the protein native structure in the solid is retained (as determined by, e.g., FTIR) more of the active protein is obtained after rehydration (49). Presumably, storage of unfolded protein in the solid form leads to irreversible aggregation.

7.2.2.2   Other additives for lyophilised formulations

During storage, the product should not undergo recrystallisation events. Crystallising solutes, e.g., mannitol or glycine, are often included as bulking agents in lyophilised formulations. The freeze-drying cycle should be designed such that the crystallising excipients reach their crystalline state after the process. Otherwise, the remaining, uncrystallised bulking agent dried into the amorphous state along with the protein and glassy protectant may recrystallise during product storage. The recrystallisation event may release water, which might adversely affect the glass transition temperature of the remaining glass thus potentially reducing stability of the product.

Other additives such as solubilisers, antibacterial agents, antioxidants and tonicity modifiers are not specific to lyophilised products. Guidelines similar to those proposed for products in solution (see section 7.2.1.4) apply here, as long as these substances do not disturb glass formation during lyophilisation. For some of these reasons and often to overcome limited stability in the presence of some of these agents, they are included in the diluent provided for reconstitution of the product.

7.2.3   Stabilisation in novel drug delivery systems

As the use of new delivery routes and systems progresses, stabilisation of proteins under some rather unfriendly environments becomes a new challenge for the pharmaceutical scientist. In the case of implantable pumps or depots and injectable microspheres or liposomes, the protein is exposed to prolonged and intimate contact with device or carrier surfaces and biological tissues and fluids at 37°C. These conditions typically lead to aggregation and degradation. For example, denaturation of interleukin-2 (IL-2) at a pump surface has been reported to occur within a day of the implantation (50). However, this effect is prevented by inclusion of HSA in the diluting medium, and the product is stable at 37°C for at least six days under certain, well-defined conditions (51).

The use of poly-lactide-co-glycolide (PLG) microspheres is an area of active interest for *in vivo* delivery of proteins for prolonged periods. In such a system, the protein is exposed not only to a hydrophobic, degrading surface at 37°C but also to acidic pH due to lactic and glycolic acid formation during degradation. The protein is first subjected to stresses such as exposure to solvents (ethyl acetate or methylene chloride) or ultra-low temperatures used during manufacturing. Hora *et al.* demonstrated that inclusion of HSA but not of polymers, such as polyethylene glycol and polyvinyl pyrrolidone, or carbohydrates, such as mannitol and hydroxyethyl starch, prevented IL-2 from associating with the PLG polymer and allowed it to continuously release for 30 days *in vitro*. (52-53). Recently, Cleland and Jones have stabilised hGH by coencapsulation with trehalose and mannitol (54). In addition, encapsulation of a co-precipitate of hGH and zinc exhibited enhanced stability within PLG microspheres. The hGH-PLG system has been shown to be successful for *in vivo* delivery of the protein in clinical trials (55).

In the case of liposomes, stability of proteins may be perturbed by peroxy lipid free radicals, as peroxides are known to oxidise proteins during prolonged contact (56-58). Lipids also form complexes with proteins through hydrophobic or ionic binding via the fatty acid tail or the charged headgroup, respectively, thus altering their conformation and activity upon release. These effects need to be characterised and resolved for these systems to be applied generally for protein delivery.

For pulmonary delivery of proteins, scientists have adopted the stabilisation techniques from solution formulations, freeze-drying and spray-drying. In general, for powder inhalation dosage forms, proteins are included within a glassy matrix and spray dried as extremely small particles. Stabilisers for solutions are used for solution finished products for inhalation. Important parameters to consider are the viscosity and density of the solution.

7.3   Stabilisation to Prevent Acute Damage or Loss

As noted in sections 2, 3 and 7.1, the finished product formulation must be capable of withstanding the stresses encountered during shipping, handling and use of the product to ensure its safety and efficacy.

7.3.1   Shear damage

Acute damage to proteins can occur during product handling, transfer and usage, as proteins are unstable to agitation stresses. Mechanical agitation and container movements result in exposure of proteins to shearing forces. These perturbations can cause unfolding of the protein, thereby exposing the inner hydrophobic core to the solvent. This sequence of events eventually leads to aggregation and precipitate formation. Such instability has previously been reported for enzymes long before the biotechnology era. For example, Charm and Wong demonstrated that shearing of catalase, rennet and carboxypeptidase resulted in their partial inactivation in the 1970s(59). More recently, Sluzky *et al.* (1991) studied the kinetics of aggregation of insulin upon agitation (60). Exposure to an air-water interface was accomplished by agitating a partially filled glass vial containing the insulin solution. Exposure to a hydrophobic surface was achieved by agitating insulin solution after addition of tetrafluoroethylene (Teflon®) spheres in a glass tube. It was reported that aggregation occurred only when the insulin solution was agitated in the presence of hydrophobic surfaces.

Proteins experience inactivation during pumping and filtration steps associated with the manufacturing process, and also during shipping. This is probably due not to the mechanical shear *per se,* but to the concomitant rapid generation of air-liquid surface. For example, hGH is sensitive to shearing perturbations such as those encountered during vortexing. It was observed that 67% of hGH was converted to insoluble aggregates by a one-minute vortex mixing (36). Pikal tested shear-induced damage to hGH by pumping solutions through capillary tubes in the absence of air-liquid interfaces and found no shear-induced protein aggregation (43). These results are consistent with similar data obtained in our laboratories. In our hands, a shear-sensitive protein such as IL-2, exhibited aggregation only when a partially filled vial was agitated, but not when a completely filled vial with no headspace was shaken vigorously. These results indicate that exposure of the protein to the hydrophobic, air-liquid interface is responsible for the inactivation, as opposed to the shearing force itself.

Proteins degrading at air-liquid interfaces often produce fibre-like aggregates. The use of a non-ionic surfactant at a concentration above its critical micelle concentration (CMC) can prevent this phenomenon. In this case, the air-liquid interface is covered by a layer of surfactant molecules which block the access of protein molecules to the surface and thus prevents the protein from undergoing surface denaturation. Surfactants such as polaxomers, which can form multiple layers on a surface, are more effective in impeding surface exposure than the polysorbate surfactants.

Susceptibility of proteins to agitation needs to be evaluated during early development of a biopharmaceutical product, because appropriate measures need to be taken to control or eliminate this phenomenon in the eventual product.

Proteins can also precipitate during the buffer exchange process which is often used to bring the purified protein into the finished product buffer medium. Usually an ultrafiltration step is used for this purpose. Protein precipitation occurring at this stage can be explained by at least two mechanisms. The protein may undergo damage upon contacting the membrane surface under pressure. Alternatively, the protein may precipitate due to insufficient solubility at an intermediate buffer composition in the midst of the exchange process. Precipitation due to the first mechanism would cause a continuous decrease in the protein concentration throughout the ultrafiltration process, and is preventable by addition of a non-ionic surfactant. If the precipitation were due to a solubility limitation, the concentration of soluble protein would decrease at a certain intermediate point of the buffer exchange step, and may cause turbidity in the solution. Often, this type of loss at an intermediate step is resolvable as the protein redissolves in the final buffer.

7.3.2 Freeze damage

Some proteins are susceptible to damage due to exposure to freezing. Considerations outlined in section 6.3.1 for freezing of process intermediates also apply for the finished product. Protein inactivation due to freezing can be accounted for by one or more of the following mechanisms. The protein may be thermodynamically unstable at low temperatures as it undergoes cold denaturation (61); or it may be denatured at the ice-water surface (62); or it may be affected by an increased salt concentration or a pH shift (18, 63).

Slow cooling combined with the use of a surfactant may alleviate cold protein damage if denaturation at the ice-water surface is the cause. If the protein is intrinsically unstable to cold temperatures, the use of cryoprotectants may enhance stability of the protein (64). When a pH change is the cause, a proper buffer selection may prevent a pH shift during freezing. For example, potassium phosphate has a much smaller pH shift during freezing than its corresponding sodium salt. Certain pH indicating agents can be used to test pH of a buffer upon freezing (65).

7.3.3 Adsorptive losses

Adsorption of proteins occurs when a protein solution comes in contact with a surface. This process has been studied most in the biomaterials field, where this phenomenon is known to take place when an artificial organ is implanted in the body (66). A protein is thought to unfold as selected domains interact with the surface upon contact, thus forming an adsorbed protein layer and reducing the protein concentration in solution. These adsorptive processes can be described by the Langmuir isotherm, which assumes that surface saturation takes place after a monolayer of protein molecules is formed on the surface. In some cases, adsorption does not stop at a monolayer as adsorbed molecules further aggregate with protein molecules from the bulk solution. This leads to formation of multilayers.

Protein loss due to adsorption can be important as commercially a protein solution is filled into final product containers using long tubing lines. Subsequently, solution formulations are in constant contact with components of the container-closure system during its shelf life. For protein formulations at concentrations below approximately 20 µg/ml, these processes could remove a substantial percentage of the drug by adsorption (67). If the adsorption process follows a well behaved monolayer pattern, an overage can be estimated (or determined experimentally) and added during filling to compensate for losses during processing and storage. Alternatively, carrier proteins (such as HSA) or surfactants can be included in the formulation, which prevent the adsorption of the protein from occurring.

Protein adsorption is a significant potential problem in the case of infusion solutions as the protein concentration for protein drugs is usually <50 µg/ml. These are exposed to large areas comprising the infusion bag and administration tubing surfaces. In these cases, inclusion of an excess of a carrier protein in the solution prior to the drug addition minimises adsorption of the biopharmaceutical. Figure 27 illustrates this point for an experimental drug: macrophage colony-stimulating factor (M-CSF). In this example, we evaluated adsorption of M-CSF from a variety of protein concentrations and obtained a monolayer saturation value of 0.2 µg/cm^2 (5.7x10^{-6} µmoles/cm^2). The system followed the Langmuir adsorption isotherm and M-CSF adsorption could be eliminated by inclusion of 0.1% HSA in the dilution medium (68). Recently, Johnston has reported similar results for GM-CSF (37).

Stabilisation of biopharmaceutical products and finished product formulations

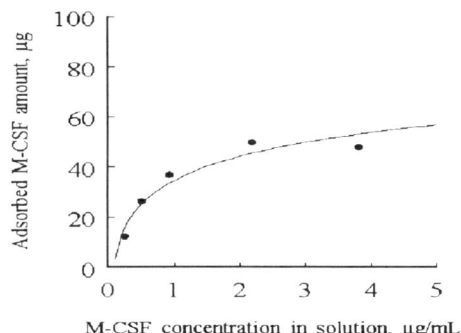

Figure 27. Adsorption of M-CSF to polyvinyl bag surfaces at 23°C from solutions at various protein concentrations. Adsorbed protein was estimated by bioactivity measurements before and after the 2-hour equilibration

Table 26. Selected components of biopharmaceutical finished product formulations

Component Class	Component	Example Recombinant Protein Finished Product(s)
Buffers	acetate	G-CSF (Neupogen, Amgen), PDGF-BB (Regranex, Johnson & Johnson), IFN-α-2a (Roferon, Roche)
	arginine phosphate	tPA deletion mutein (Retavase, Centecor), tPA (Activase, Genentech)
	citrate	erythropoeitin (Epotein-α, Amgen), hGH (Nutropin AQ, Genentech), rituximab (Rituxan, Genentech/ Idec), ß-glucerebrosidase (Cerezyme, Genzyme),
	histidine	coagulation factor IX (BeneFix, Genetics Institute)
	phosphate	IFN-alfacon-1 (Infergen, Amgen), IFN-ß-1a (Avonex, Biogen), abciximab (ReoPro, Lilly/ Centecor), imciromab (Myoscint, Centecor), IL-2 (Proleukin, Chiron), hGH (Protropin and Nutropin, Genentech), IL-11 (Neumega, Genetics Institute), insulin lispro (Humalog, Lilly), hGH (Humatrope, Lilly), orthoclone OKT3 (Johnson & Johnson), IFN-α-2b (Intron, Schering Plough)
	succinate	interferon-γ (Actimmune, Genentech)
	tris	GM-CSF (Leukine, Immunex)
Lyophilisation stabilisers	arginine	as phosphate in tPA deletion mutein (Retavase, Centecor), tPA (Activase, Genentech)
	glycine	factor VIII (Kogenate, Bayer), coagulation factor IX (Benefix) and IL-11 (Neumega, Genetics Institute), hGH

Component Class	Component	Example Recombinant Protein Finished Product(s)
Lyophilisation Bulking Agents	HSA	(Humatrope, Lilly and Nuropin, Genentech), IFN-α-2b (Intron, Schering Plough) factor VIII (Kogenate, Bayer), IFN-ß-1b (Betaseron, Berlex/ Chiron), IFN-ß-1a (Avonex, Biogen), IFN-α-2b (Intron, Schering Plough)
	sucrose	coagulation factor IX (BeneFix, Genetics Institute), GM-CSF (Leukine, Immunex),
	mannitol	IL-2 (Proleukin, Chiron), hGH (Protropin and Nutropin, Genentech, Humatrope, Lilly), ß-glucerebrosidase (Ceredase, Genzyme), GM-CSF (Leukine, Immunex)
Solution Stabilisers	sucrose	GM-CSF (Leukine, Immunex),
	mannitol	G-CSF (Neupogen, Amgen), IFN-γ (Actimmune, Genentech), GM-CSF (Leukine, Immunex)
Surfactants	polysorbate 20	IFN-γ (Actimmune, Genentech), hGH (Nutropin AQ, Genentech), tPA mutein (Retavase, Centecor)
	polysorbate 80	G-CSF (Neupogen, Amgen), tPA (Activase, Genentech), rituximab (Rituxan, Genentech/ Idec), coagulation factor IX (BeneFix, Genetics Institute), ß-glucerebrosidase (Ceredase, Genzyme), murmonab CD3 (Othoclone OKT3, Johnson & Johnson), IFN-∝-2b (Intron, Schering Plough), IFN-α-2a (Roferon, Roche)
	Sodium dodecyl sulphate	IL-2 (Proleukin, Chiron)

8. EXAMPLES OF FINISHED PRODUCT FORMULATIONS IN MARKETED PRODUCTS

To illustrate the application of key principles discussed in this chapter, we thought it useful to include information on some key components used in biopharmaceutical finished product formulations currently on the market. Such information for all parenteral products must be disclosed to the public once a product gains a marketing approval in the USA. Information in Table 26 has been gathered from a variety of sources such as the Physician Desk Reference (69), the FDA freedom of information website (70) or the manufacturer's website.

9. CONCLUSION

Stabilisation of biopharmaceutical products and finished product formulations is a critical aspect of the biopharmaceutical industry. One must take into account manufacturing, clinical, regulatory, marketing and other requirements in designing the best practical strategy for stabilisation of intermediate bulks, purified drug substance and the finished product. The ultimate goal is to ensure that economically feasible, safe and efficacious finished products are made available to the clinician.

BIOGRAPHY

Dr. Maninder Hora is the Senior Director and Dr. Bao-lu Chen a Senior Scientist in the Department of Formulation Development at Chiron with research interests in formulation and delivery of therapeutic proteins and vaccines. Chiron Corporation, headquartered in Emeryville, California, USA, is a leading biotechnology company that participates in three global healthcare markets: blood testing, therapeutics and vaccines. Chiron is involved with research and development in the fields of biological proteins, novel antigens and adjuvants, gene therapy and combinatorial chemistry.

REFERENCES

1. Federal Register (1996). International conference on harmonisation; final guidelines on stability testing of biotechnological/ biological products; availability; notice. 61,36466-36469.
2. Nguyen, T.H. and Shire, S.J. (1996). Stability and characterization of recombinant human relaxin. In: Pearlman, R. and Wang, Y.J. (Eds.) Formulation, characterization and stability of protein drugs. Plenum. New York.
3. Hora, M.S. et al. (1992). Lyophilized formulations of recombinant tumor necrosis factor. Pharm. Res., 9, 33-36.
4. Shire, S.J. Stability, characterization and formulation development of recombinant human deoxyribonuclease I [Pulmozyme® (dornase alpha)] In: Pearlman, R. and Wang, Y.J. (Eds.) Formulation, characterization and stability of protein drugs. Plenum. New York.
5. Wang, Y.-C. J. and Hanson. M. A. (1988). Parenteral Formulations of proteins and peptides: stability and stabilisers. J. Parenteral Sci. & Tech. 42, S4-S26.
6. Manning, M. C. et al. (1989). Stability of protein pharmaceuticals. Pharm. Res. 6, 903-918.
7. Chen, T. (1992). Formulation concerns of protein drugs. Drug Dev. and Industrial Pharmacy. 18, 1311-1354.
8. Cleland, J. L. et al. (1993). The development of stable protein formulations: a close look at protein aggregation, deamidation, and oxidation. Critical Reviews in Therapeutic Drug Carrier Systems. 10, 307-377.
9. Kunitani, M. et al. (1986). Reversed-phase chromatography of interleukin-2 muteins. J. Chromatogr. 359, 391-402.

10. Prescribing information for Nutropin AQ (1997), In: Physician desk reference, Medical Economics Company, Inc., Montville, NJ.
11. Becktel, W. J. and Schellman, J. A. (1987). Protein stability curves. Biopolymers. 26, 1859-1877.
12. Kauzmann, W. (1959). Some factors in the interpretation of protein denaturation. Adv. Protein Chem. 14, 1-64.
13. Pace, C. N. (1975). The stability of globular proteins. CRC Crit. Rev. Biochem. 5, 1-43.
14. King, J. (1989). Deciphering the role of protein folding. Chem. Eng. News. 67, 32-54.
15. Chen, B.-L. et al. (1994). Aggregation pathway of recombinant human keratinocyte growth factor and its stabilisation. Pharm. Res. 11, 1581-1587.
16. Young, B.R. et al. (1988). Protein adsorption on polymeric biomaterials I. Adsorption isotherms. J. Colloid Int. Sci. 124, 28-43.
17. Foote, C. S. (1968). Mechanisms of photosensitized oxidation. Science, 162, 963-970.
18. Franks, F. (1985). Biophysics and biochemistry at low temperatures. Cambridge University Press, London.
19. Wisniewski, R. and Wu, V. (1996) Large scale freezing and thawing of biopharmaceutical products. In: Avis, K. and Wu, V. (Eds.) Biotechnology and biopharmaceutical manufacturing, processing and preservation. Interpharm. Buffalo Grove, IL.
20. Akers, M.J. and Schmidt, D.J. (1997). Cryogranulation: A potential new final process for bulk drug substances. Biopharm, 10, 28-32.
21. Patton, J. (1998). Breathing life into protein drugs. Nature Biotechnology, 16, 141-143.
22. Fransson, J. and Espander-Jansson, A. (1996). Local tolerance of subcutanous injections. J. Pharm. Pharmacol., 48, 1012-1015.
23. Franken, L.A.M. et al. (1994). Analysis of the efficacy of measures to reduce pain after subcutaneous administration of epotein alpha. Nephrol. Dial. Transplant., 9, 1295-1298.
24. CPMP working party on quality of medicinal products note for guidance (1994). Excipients in the dossier for application for marketing authorization of medicinal product. Commission of the European Communities, Brussels.
25. Commission decision on the prohibition of the use of material presenting risks as regards transmissable spongiform encephalopathies (1997). Commission of the European Communities, Brussels.
26. Schellman, J. A. (1975). Macromolecular binding. Biopolymers. 14, 999-1018.
27. Schellman, J. A. (1987). The thermodynamic stability of proteins. Ann. Rev. Biophys. Biophys. Chem. 16, 115-137.
28. Volkin, D.B. and Middaugh, C.R. (1996). The characterization, stabilization and formulation of acidic fibroblast growth factor. In: Pearlman, R. and Wang, Y.J. (Eds.) Formulation, characterization and stability of protein drugs. Plenum. New York.
29. Wang, Y.J. et al.(1996). In: Pearlman, R. and Wang, Y.J. (Eds.) Formulation, characterization and stability of protein drugs. Plenum. New York.
30. Cleary, S. et al. (1989). Purification and characterization of tissue plasminogen activator Kringle-2 domain expressed in *Escherichia coli*. Biochemistry, 28, 1884-1891.
31. Chien, Y.-W. (1996). Human insulin: basic sciences to therpeutic uses. Drug Dev. and Industrial Pharmacy. 22, 753-789.
32. Timasheff, S. N. and Arakawa, T. (1989). Stabilisation of protein structure by solvents. In: Creighton, T. E. (Ed.) Protein Structure, a practical approach. IRI Press, Oxford, pp. 301-345.
33. Yancey, P. H. et al. (1982). Living with water stress: evolution of osmolyte systems. Science, 217, 1214-1222.

34. Santoro, M. M. et al. (1992). Increased thermal stability of proteins in the presence of naturally occurring osmolytes. Biochemistry, 31, 5278-5283.
35. Chen, B.-L., and Arakawa, T. (1996). Stabilisation of recombinant human keratinocyte growth factor by osmolytes and salts. J. Pharm. Sci. 85, 419-422.
36. Katakam, M. et al. (1995). Effect of surfactants on the physical stability of recombinant human growth hormone. J. Pharm. Sci. 84, 713-716.
37. Johnston, T. P. (1996). Adsorption of recombinant human granulocyte colony stimulating factor (rhG-CSF) to polyvinyl chloride, polypropylene, and glass: effect of solvent additives. PDA J. Pharm. Tech. 50, 238-245.
38. Chang, B. S. et al. (1996). Surface-induced denaturation of proteins during freezing and its inhibition by surfactants. J. Pharm. Sci. 85, 1325-1330.
39. Schein, C.H. (1990). Solubility as a function of protein structure and solvent components. Bio/Technology, 8, 308-315.
40. Franks, F. (1990). Freeze drying: from empiricism to predictability. Cryo-Letters 11, 93-110.
41. Roser, B. (1991). Trehalose Drying: a novel replacement for freeze-drying. BioPharm. 4, 47-53.
42. Wolfe, S.P. et al. (1991). Protein glycation and oxidative stress in diabetes mellitus and ageing. Free Radical Biology & Medicine, 10, 339-352.
43. Pikal, M. J. et al. (1991). The effects of formulation variables on the stability of freeze-dried human growth hormone. Pharm. Res. 8, 427-436.
44. Carpenter, J. F. et al. (1990). Comparison of solute-induced protein stabilisation in aqueous solution and in the frozen and dried states. J Dairy Sci. 73, 3627-3636.
45. Carpenter, J. F. and Crowe, J. H. (1989). An infrared soectroscopic study of the interactions of carbohydrates with dried proteins. Biochemistry, 28, 3916-3922.
46. Pikal, M. J. (1994). Freeze-drying of proteins. In: Cleland, J. L. and Langer, R. (Eds.) Formulation and delivery of proteins and peptides. ACS Symposium Series. 567, 120-133.
47. Bell, L. N. et al. (1995). Thermally induced denaturation of lyophilised bovine somatotropin and lysozyme as impacted by moisture and excipients. J. Pharm. Sci. 84, 707-712.
48. Rupley, J. A. and Careri, G. (1991). Protein hydration and function. Adv. Prot. Chem. 41, 37-172.
49. Prestrelski, S.J. et al. (1995) Optimization of lyophilization for recombinant human interleukin-2 by dried-state conformational analysis using Fourier-transform infrared spectroscopy. Pharm. Res. 12, 1250-1259.
50. Tzannis, S.T. et al. (1996). Irreversible inactivation of interleukin-2 in a pump-based delivery environment. Proc. Natl. Acad. Sci. USA, 93, 5460-5465.
51. Hora, M. and Rana R.K. (1991). Delivery of Proleukin from a Deltec CADD pump, Data on file at Chiron Corporation, Emeryville, CA.
52. Hora, M.S. et al. (1990). Controlled release of interleukin-2 from biodegradable microspheres. Bio/Technology, 8, 755-758.
53. Hora, M.S. et al. (1989). Biodegradable polymeric microspheres for controlled release of interleukin-2. Pacific Polymer Preprints of the First Pacific Polymer Conference, Maui, HI, 1, 519-520.
54. Cleland, J.L. and Jones, A.J.L. (1996). Stable formulations of recombinant human growth hormone and interferon-γ for microencapsulation in biodegradable microspheres. Pharm. Res., 13, 1464-1475.

55. Cleland, J.L. *et al.* (1997). Recombinant human growth hormone poly(lactic-co-glycolic acid) microsphere formulation development. Adv. Drug Del. Rev., 28, 71-84.
56. Hunt, J.V. *et al.* (1988). Hydroperoxide-mediated fragmentation of proteins. Biochem. J., 250, 87-93.
57. Hora, M.S. *et al.* (1991) Development of a lyophilized formulation of interleukin-2, In: Develop. Biol. Standard., Karger, Basel.
58. Nguyen, T.H. *et al.* (1993). The kinetics of relaxin oxidation by hydrogen peroxide. Pharm. Res., 10, 1563-1571.
59. Charm, S.E. and Wong, B.L. (1970) Enzyme inactivation with shearing, Biotech. Bioeng., 12, 1103-1109.
60. Sluzky, V. *et al.* (1991) Kinetics of insulin aggregation in aqueous solutions upon agitation in the presence of hydrophobic surfaces. Proc. Natl. Acad. Sci. USA, 88, 9377-9381.
61. Privalov, P. L. (1990). Cold denaturation of proteins. Crit. Rev. Biochem. Molec. Biol. 25, 281-305.
62. Strambini, G. B. and Gabellieri, (1996). Proteins in frozen solutions: evidence of ice-induced partial unfolding. Biophys. J. 70, 971-976.
63. Murase, N. and Franks, F. (1989). Salt precipitation during the freeze-concentration of phosphate buffer solutions. Biophysical Chemistry, 34, 293-300.
64. Carpenter, J. F. and Crowe, J. H. (1988). The mechanism of cryoprotection of proteins by solutes. Cryobiology, 25, 244-255.
65. Hill, J. P. and Buckley, P. D. (1991). The use of pH indicators to identify suitable environments for freezing samples in aqueous and mixed aqueous/nonaqueous solution. Analytical Biochem. 192, 358-361.
66. Horbett, T. A. and Brash, J.L. (1987) Proteins at interfaces: Current issues and future prospects. In: Brash, J.L. and Horbett T.A. (Eds.), Proteins at interfaces: Physicochemical and biochemical studies, American Chemical Society, Washington, D.C.
67. Burke, C.J. *et al.* (1992). The adsorption of proteins to pharmaceutical container surfaces. Int. J. Pharm., 86, 89-93.
68. Hora, M. *et al.* (1991). Compatibility of macrophage colony stimulating factor (M-CSF) with plastic components of intravenous administration systems. Pharm. Res., 8, S-59.
69. Physicians' Desk Reference. (1996). Medical Economics Company, Monvale, NJ.
70. The Food and Drug Administration. (1998). http://www.fda.gov/cber/efoi/approve.htm

Chapter 10

Patent Law for Biopharmaceuticals

R. Stephen Crespi
European Patent Attorney, West Sussex, UK

Key words: Patent, biotechnology, biopharmaceutical, litigation

Abstract: The patenting of all technological innovations has to be accommodated within one and the same patent law. This is true for engineering, physics, chemistry, and biotechnology in all their many sub-categories. To be patentable, biopharmaceuticals and their processes of manufacture have to meet the same three legal tests of novelty, inventiveness, and practical utility as do the simpler compounds and preparative methods of synthetic chemistry. As for all such other inventions biopharmaceuticals must also be capable of adequate description and definition in the patent specification, the all-important document upon which rest all questions of patent scope (how much the patent covers) and patent validity (whether it is enforceable).

It is in this last of these legal requirements that biopharmaceuticals present a special case, because of their exceptionally complex structure and the complexity of the methods by which they are prepared.

For the first generation of biopharmaceuticals the requirement of novelty had the important consequence that the recombinant forms of the blood proteins and other valuable naturally-occurring proteins could not be patented as products. This will be explained in some of the case studies to be outlined below, but first it may be helpful to summarise some basic distinctions between product and process patents. These distinctions are concerned entirely with the 'claims' of the patent i.e. the legal clauses which come after the technical description and form the last and most important section of the specification.

1. TYPES OF PATENT

1. A product patent is one which claims (i.e. protects) the product per se without limitation as to the process used to make it. This is the claim normally granted where the product is new and can be defined per se in physical, chemical, or other structural terms.
2. A product-by-process patent is one which claims the product in terms of a particular process of preparation described in the patent. This is frequently used when the product is new but cannot be defined in any of the usually accepted ways, e.g. because its constitution is unknown.
3. A process patent covers the procedural act of preparing a product by the steps defined in the process claim.

The unauthorised act of making or in any way dealing in a product or carrying out a process covered by the patent claims is termed 'infringement' of the patent. It follows that a product patent protects against unauthorised manufacture in, or importation into, the 'patent-country' no matter by what process the product has been made. Before the other types of patent can be exercised, it will be crucial to determine what process has been used either locally or abroad. One might think that one who imports a product made outside the jurisdiction of the 'patent-country' might be immune from the effect of a process patent. However, in Europe a process patent automatically also covers the product made by that process, with the qualification that it is the direct product of the process and not one resulting from downstream processing of the direct product.

1.1 Natural product patents

Naturally occurring substances might appear to pose a problem for the concept of novelty (and inventiveness) since it may be asked how something can be patented that already exists in nature, whether it be an inanimate substance, an enzyme, a micro-organism or higher life form, or a gene? This question has added force when one recalls that the patent laws of most countries make a distinction between "discovery" and "invention" and exclude mere discoveries from patentability. The 'product of Nature' question has a long history in patent law but in most industrially developed countries the extraction of valuable substances from natural sources, so as to make them available for the first time in a form in which their properties can be utilised in medicine, agriculture, or any other useful art, is recognised as worthy of patent protection. In terms of what belongs to the state-of-the-art i.e. what is already available to the public (the true test of novelty) these materials can be just as new and inventive as the synthetic materials created by the chemist and biochemist. In this vein, The US, European, and Japanese Patent Offices have declared their joint agreement to the patenting of purified natural products in appropriate circumstances (1).

1.2 Procuring and enforcing biopharmaceutical patents

Applying for a patent is an adversarial process. The applicant has first to overcome objections from the patent office examiner. When this has been done, and the patent granted, the patent owner (patentee) can take action to enforce the patent against infringers. However, the patent is open to attack by others in opposition proceedings or in Court. As all these procedures take time to complete, the litigation, in which the courts establish principles which become legal precedents, is always some years behind the fast-developing technology. With this reservation in mind case studies help to illuminate these principles.

2. OUTLINE CASE STUDIES

Two important modern examples of patents for purified natural products can be cited (Scripps Clinic and Research Foundation v Genentech, and Amgen Inc. v Genetics Institute) The first of these involved Factor VIII, the blood clotting factor important in the treatment of hemophilia. The second example is that of erythropoietin, the hormone which stimulates the formation of red blood cells. These patents disclose particular chromatographic methods of purifying or isolating the naturally-occurring protein. However the claims are not limited to products produced by these methods but also cover the proteins defined solely in terms of units of activity, a parameter connected with the degree of purity which these methods have made possible.

2.1 Factor VIII litigation

In the Scripps case (2), the crucial claim of their US Re-issue patent 32,011 was:-

"A human VIII:C preparation having a potency in the range of 134 to 1172 units per ml and being substantially free of VIII:RP."

Scripps asserted that this claim was infringed by the defendants manufacture of recombinant Factor VIII since the latter would fall within its purity criteria. The defendant argued that this claim should be interpreted in the context of the Scripps method of purification and therefore limited to material purified from natural sources.

The case went before the District court of Northern California on a motion for Summary Judgement (a procedure for obtaining a speedy decision preliminary to a full trial) and in 1987 the court decided (in favour of Scripps) that 'human VIII:C' described the substance in terms not of derivation from plasma but in terms of its fundamental characteristics peculiar to the human species. However, on appeal to the Court of Appeals for the Federal Circuit (CAFC) it was decided in 1991 that this and the other issues in the case were too complex to be resolved in preliminary proceedings. The final outcome of this case is difficult to determine. There was probably an out of court settlement.

2.2 Erythropoietin litigation

In the Amgen case (3), Genetics Institute Inc.(GI) had obtained US patent 4,677,195 for a method of purification of epo from human urine by reverse phase high performance liquid chromatography. But the patent also has broad product claims, including the following:-

"Homogeneous erythropoietin characterised by a molecular weight of about 34,000 Daltons on SDS PAGE, movement as a single peak on reverse phase high performance liquid chromatography and a specific activity of at least 160,000 IU per absorbance unit at 290 nanometers".

The dispute in this case involves issues similar to those in the Factor VIII case. For similar reasons the US court of first instance held that Amgen's recombinant product infringed the GI claim to the purified protein. However this holding was later suspended on appeal.

The two instances outlined above are examples of conflict between the respective proprietors of a natural product patent and a recombinant DNA patent in which it has to be decided who is free to operate commercially with the recombinant protein. In these two examples one question for the court is whether a patent based entirely on the purification of the natural product can be permitted to cover the protein as made by recombinant methods. Unfortunately, single issues are rare in patent lawsuits and the answers are often complicated by other factors. It seems possible that the US courts could still affirm in principle the dominance of the natural product patent.

3. RECOMBINANT PROTEINS AND DNA SEQUENCES

The most common types of claim found in recombinant DNA patents are the following

a) Recombinant protein products (and alleles, variants, derivatives)
b) DNA sequences coding for the products of (a)
c) Vectors containing the DNA sequences of (b)
d) Micro-organisms, cell lines, and other organisms transformed with vectors (c)
e) Processes for constructing the micro-organisms etc. of (d)
f) Processes for producing products (i) by expression of DNA sequences; (ii) in a recombinant host organism (d)
g) Products of processes (f)
h) Pharmaceutical (or other) compositions containing products of (a) or (g)

The first recombinant DNA inventions to be patented were directed to the manufacture of known proteins. In both US and European patent practice the product in these circumstances could not be claimed under type (a) or even in product-by-process claims of type (g). One example of the refusal of a product claim by the US Patent Office is the following:

3.1 Nerve growth factor case

This was an application dealing with a recombinant nerve growth factor protein(4), claimed as : 'Human beta-NGF comprising the amino acid sequence Ser-Ser-Ser-His .....(full sequence given)...... Lys-Ala-Val-Arg and which is free of other proteins of human origin.'

This claim does not use the word "recombinant" but it was assumed that this was implied in view of the method described in the specification. The claim was rejected because there was no evidence that the recombinant product was significantly different from the natural material or possessed any unexpected properties.

The more usual types of claim allowed in these circumstances, especially in the US, were those directed to DNA sequences, recombinant vectors, and transformed cells. These products can be fairly described as the tool kit or as intermediates for producing the ultimate commercial product. The more generous UK practice does allow a product-by-process claim in these circumstances. Thus a claim to 'recombinant human tissue plasminogen activator' was allowed by the UK Patent Office, as discussed below.

3.2 Genentech's UK patent 2,119,904 on tissue plasminogen activator (t-PA)

In this patent the principal claims were product claims directed to the recombinant product (5), e.g.-

"Human tissue plasminogen activator as produced by recombinant DNA technology".

This claim covers human t-PA prepared by any recombinant DNA technique and it is a product-by-process claim of broad scope. The claim would appear to cover only the recombinant version of the natural product but a much broader interpretation was intended by the patentee, as was made clear from certain "Definitions" given in the specification. Thus, in addition to allelic variations of t-PA it was stated that all modifications or derivatives of the natural product were to be included, such as those obtained by mutagenesis e.g. having amino-acid deletions, substitutions, additions, or replacements, provided the essential activator function was retained.

The natural t-PA protein was a known substance having been previously isolated from the Bowes melanoma cell line. The amino-acid sequence of t-PA had not previously been determined, but one cannot patent a known substance just by being the first to determine its structure. In holding the patent invalid the court of first instance said that had Genentech been the first to discover t-PA, its decision would have been different. Since most of the initial applications of this technology have been aimed at producing naturally occurring proteins of known therapeutic value this observation was cold comfort for those using standard techniques in this field. The only consolation that this statement offers is that novel derivatives or analogues of t-PA or other proteins which show some advantage over the natural product cannot be so readily dismissed as unpatentable. In the Court of Appeal the recombinant product was held obvious first, because it was a known desirable objective on which a number of groups were working and secondly, because the court was persuaded that the product was obtained by using conventional text-book methods of gene cloning.

Genentech's British patent was invalidated in late 1988 but, in addition to filing applications in various countries in Europe under their national patent systems, Genentech had also filed a duplicate European patent application, thus providing an alternative route to protection throughout Europe.

3.3 The European patent on t-PA

The corresponding European patent 93619 issued in September 1989 (6), had a claim structure very different from that of the British patent. It relied mainly on process claims, such as "A process which comprises preparing cDNA from mRNA extracted from the Bowes melanoma cell line and isolating from it a DNA sequence having the restriction pattern shown in Figure 4 hereof for the putative mature tissue plasminogen activator sequence and which encodes a 527 amino-acid polypeptide having human tissue plasminogen activator function."

This was followed by process claims to the expression of the DNA to produce the protein. There were no product claims to the recombinant form of natural t-PA but various related substances were claimed in the following way :-

"A protein having human tissue plasminogen activator function and which comprises an allele or derivative by way of amino-acid deletion, substitution, insertion, inversion, addition or replacement of the 527 amino-acid sequence as encoded by the DNA product of claim 1 or 2"

This patent was formally opposed by seven opponents. The EPO Appeal Board at first commented adversely on the claim to derivatives, which they considered bad for want of a clear definition of the term "human tissue plasminogen activator function". No specific examples of any such derivatives were given in the patent. Eventually the Board settled on a definition in terms of catalyzing the conversion of plasminogen to plasmin, binding to fibrin, and by immunological properties.

As to the broad claim to derivatives, the board took the view that, the basic molecular structure of t-PA being given, it would not require inventive skill or undue experimentation to prepare such functional derivatives. Some commentators argue that the tendency for broad dominating claims to be granted in recombinant DNA patents, which "reach through" to all conceivable second-generation derivatives, is bad for the industry as a whole because it is inhibitory to further research. There are however many instances of patents of such breadth. For reasons given earlier the claims in the US patent on recombinant TPA are directed to the DNA coding sequences. (see later)

3.4 The Erythropoietin litigation (Further developments)

As noted above, one of the parties to this dispute has a patent for a purified form of erythropoietin while the other has a patent covering recombinant techniques for producing the protein. The case throws further light on the question of mutual impact between two patents of this kind (7).

The plaintiff Amgen Inc. holds US patent 4,703,008 (the '008 patent) which describes the production of recombinant erythropoietin. The patent states that:

"The present invention provides, for the first time, novel purified and isolated polypeptide products having part or all of the primary structural conformation (i.e. continuous sequence of amino acid residues) and one or more of the biological properties (e.g. immunological properties and *in vivo* and *in vitro* biological activity) of naturally occurring erythropoietin, including allelic variants thereof."

However, for reasons explained above, there are no claims to the final products. and no claims to methods of producing the final products. The claims are of the following types:-

1. A purified and isolated DNA sequence encoding erythropoietin, said DNA sequence selected from the group consisting of:
    (a) the DNA sequences set out in FIGS 5 and 6 or their complementary strands; and
    (b) DNA sequences which hybridize under stringent conditions to the DNA sequences defined in (a).
2. A purified and isolated DNA sequence consisting essentially of a DNA sequence encoding a polypeptide having an amino acid sequence sufficiently duplicative of that of erythropoietin to allow possession of the biological property of causing bone marrow cells to increase production of reticulocytes and red blood cells, and to increase hemoglobin synthesis or iron uptake.

The defendants in this case were Chugai Pharmaceutical Co., the licensee of US patent 4,677,195 (the '195 patent) granted to Genetics Institute Inc. for a method of purifying erythropoietin (see above). Although the '195 patent mentions cultures of recombinant cells as one possible source of erythropoietin to which the method can be applied, it does not describe the use of such source material in detail in the patent examples. The latter are restricted to the preparation of homogeneous erythropoietin from a mixture of several polypeptides, of molecular masses ranging from 30,000 to 70,000 Daltons, derived from human urine. The process is based on the use of reverse-phase high-performance liquid chromatography and is applicable to crude erythropoietin "no matter what the source of the epo (erythropoietin). The product claim in this patent has been set out above.

Amgen failed to persuade the court that its recombinant erythropoietin did not meet the characteristics specified in this claim. The court observed that it makes no difference to infringement liability that Amgen's erythropoietin is produced by genetic engineering. The '195 patent includes product claims, the court explained, and it makes no difference by what path or process the infringing material is manufactured. If Amgen's material produced by recombinant technology falls within the claims of Chugai's asserted product claims, the court reasoned, then Chugai's asserted patent claims are infringed.

Moreover, Amgen's counterclaim that Chugai's recombinant erythropoietin imported from Japan infringes the '008 patent did not succeed in the summary judgement phase of the trial. The Amgen patent contains no claims to the recombinant erythropoietin product and no process claims for making such a product (the latter having been refused by the US Patent Office). The court then had to decide the question of infringement on the basis of the claims to the recombinant DNA starting materials and the corresponding genetically manipulated host cells, all of which would have been used abroad and therefore outside the jurisdiction of the US court. This unsatisfactory state of the law has now changed, fortunately for those in a similar position to Amgen. First, importation into the US of the product of a US process patent is now an infringement. Secondly, process claims for producing a recombinant product will not now be routinely refused (as they once were) by the US Patent Office.

In a further phase of this US litigation the validity of the Amgen patent was attacked. The Massachusetts District court reviewed the course of the research by Amgen and others from 1991 onward into the cloning of the erythropoietin gene. This was not a large gene and for this reason was seen in 1981 as a good candidate for cloning. But it took two years of painstaking work by the Amgen inventor before the cloning strategy succeeded. This entailed determining the correct amino-acid sequence of sufficient portions of the protein structure to enable oligonucleotide probes to be designed which would extract the gene from a genomic library. A large number of probes were constructed and it was necessary to use a combination of probes which would hybridise to different parts of the gene. The great magnitude of the task cannot be denied. Genetics Institute, Cal-Tech, and Biogen had also embarked on the quest for this gene.

Here there seem to be some similarities to the t-PA case. The gene was a good candidate for cloning and a number of competitive groups were pursuing it along their own trails. But apparently there were important differences from the t-PA development. First, the protein was an exceptionally scarce material when the Amgen inventor started work, a fact which made sequence information that much more difficult to obtain. Secondly, and of greater significance, was the fact that by 1981 and even 1982 there had been little or no precedent in the screening of a genomic library using two sets of probes. The court therefore concluded that it was not within the power of one of ordinary skill in the art to reduce this invention to practice at that time.

In approaching the question of the inventive step the US court did not put the partial sequencing of epo into a separate category from the steps which followed. The correct sequencing and the consequential design of two sets of fully degenerate probes from different regions of the sequence to explore a genomic library amounted to an inspirational combination in the eyes of the court. The court concluded that "the unique probing and screening method employed in isolating the epo gene was what distinguished the invention from the prior art"

In the decision on appeal the Court of Appeals for the Federal Circuit (CAFC) noted that it was the product that was claimed and not the process, but decided to follow the approach of the lower court and the parties to the litigation. The point was therefore by-passed.

Before describing the proceedings in the European patent for recombinant erythropoietin corresponding to US 4,703,008, it will be helpful to discuss the fourth of the basic requirements for patentability mentioned earlier, namely, the need for a patent to contain an adequate disclosure of the claimed invention.

4. THE ENABLING DISCLOSURE REQUIREMENT

A patent can be conceived in terms of a 'bargain' between an inventor and the public. Rather than using an invention as an exclusive secret, the inventor discloses the invention for the public benefit. In return for this gesture, the inventor is protected for a limited time against copying by others. When this time comes to an end, use of the invention is free for all. It follows that there must be some correlation between the extent of the inventor's patent disclosure and the scope of the inventor's patent protection. When the patent application is officially examined, sufficiency of disclosure is a critical issue considered by the patent examiner, along with those of novelty and inventiveness.

For biotechnology patents, this question has assumed special importance at the present time. The patent description, based on data provided by the inventor, must enable the skilled person to perform the invention (the "enabling disclosure"). The patent description must also "support" the claims of the patent. In Europe these two criteria are dealt with as separate Articles both in the European law (EPC) and in the national patent laws of individual countries.

Why these points were seen as separate requirements in the EPC has previously not been at all clear and this has given rise to some difficulty in the courts. In his own sphere, the research scientist can well understand the distinction between a paper which provides enough information to enable the author's results to be reproduced by his peers and one which provides enough data to support the author's conclusions. But for European patent law, sufficiency and support are now coming to be considered as two sides of the same coin, since both have to be assessed in relation to the patent claims, which determine the scope of the legal protection. Both investigations converge on the effectiveness of the patent description to teach the skilled person how to put the invention to use. What the inventor has actually done in the laboratory is, in a sense, less important than the ability to write a recipe for others to follow.

For biotechnology, patent law first grappled with this problem in relation to inventions in classical microbiology, especially those involving the use of newly isolated or developed strains of micro-organism to produce some useful product. The repeatability problem was solved by using culture collections as official patent depositories of such strains as a substitute for the well-nigh impossible task of providing an adequate written description both of the new organism and how to obtain it. This became formalised in an international convention (the Budapest Convention) whereby deposit in any one such depository was recognised by all member countries as complying with their own national requirements. This convention now extends beyond micro-organisms to any biological material that cannot be adequately described by the written word. It follows that this principle must apply to many of the materials referred to in gene cloning and monoclonal antibody protocols which are not available 'off the shelf' to the skilled person wishing to repeat the prescribed procedure.

4.1 Erythropoietin in the European Patent Office (Appeal case T412/93)

European patent No.148 605 was opposed by six Opponents. The case has assumed considerable legal and technical complexity, including consideration of over 500 cited documents, and is not concluded at the time of this writing. The following remarks are therefore limited to what can be determined from published accounts.

This patent was issued with claims to certain specified DNA sequences coding for expression of a polypeptide product having at least part of the primary structural conformation of erythropoietin. The claim covers (i) specific tabulated sequences, (ii) sequences which hybridize to them, and (iii) those "which but for the degeneracy of the genetic code would hybridize to any of (i) or (ii)". There are also claims to the polypeptide expression product of an exogenous DNA sequence. The polypeptide claims are skilfully worded product-by-process claims which do not cover the polypeptide as formed in nature.

A considerable part of the technical arguments of expert witnesses revolved around whether the description was adequate to enable the preparation of erythropoietin cDNA. This particular enquiry always involves deciding who is the typical "skilled person" who has to repeat the described method. The EPO Appeal Board decided that the notional skilled person is "a Ph.D researcher with several years of experience and two laboratory assistants having the necessary manual dexterity and lack of fatigue". The matter was heavily contested by the scientific experts for all sides. The end result is that the cDNA claim has not survived but the broader DNA claim is still intact.

As to whether the claims to the recombinant polypeptide (r-Epo) were permissible, the question was whether the product was new in the sense of being different from erythropoietin obtained from urine (u-Epo). As indicated previously this is crucial to the allowability of product claims. The scientific experts were much exercised on this point also. The claim was finally amended to specify that the product has greater molecular weight than u-Epo.

Another decision of the Appeal Board on this issue is awaited on certain procedural points following which the matter will be concluded in the EPO. It may of course re-surface in national courts.

4.2 Recombinant Factor VII patent

In view of the inventiveness issues that appeared in the t-PA and epo cases, it is instructive to note how these problems were anticipated by the writer of the Factor VII patents (US 4,784,950 and EP 0 200,421).

The patent points out that Factor VII is a trace plasma protein and the mRNA encoding it is rare. Purification without degradation was also problematical. Consequently Factor VII was poorly characterised and was difficult to obtain in quantities sufficient for sequence analysis.

It was nevertheless necessary to expand on this argument to persuade the EPO examiner to withdraw her objection that, at the priority date of this patent application (April 1985), it was obvious to clone this particular gene and how it could be achieved. The applicant pointed out in detailed manner the uncertainty of making a cDNA library, the inadequacies of sequence information for oligonucleotide screening, the uncertainty of antibody tools then available, and the lack of success of other groups working on the problem. The EPO examiners are open to this kind of reasonable persuasion and will concede in the face of such impressive arguments.

For reasons mentioned above, these patents do not have a product claim to the naked recombinant protein. The claims are mostly directed to the DNA constructs, recombinant plasmids, transfected cells and methods of production. However, the European Factor VII patent has a commercially very valuable claim to "a pharmaceutical preparation for the treatment of bleeding disorders containing a protein having an amino acid sequence as shown in Figure 1b and free of contaminating human proteins".

4.3 Recombinant Factor IX

In some countries, the Factor IX gene and the protein are covered in separate patents. The recombinant Factor IX is covered in US 5,171,569 wherein it is claimed as a plasma-free preparation containing the recombinantly-derived Factor IX protein defined, inter alia, as derived from a single individual, free of pox viruses and other plasma constituents, and having a specific activity (as defined in a certain way) of at least 90% of average normal human plasma.

In Europe the gene for the Factor IX precursor polypeptide (convertible *in vivo* into human Factor IX) is the subject of EP 107,278 wherein it is claimed as a specified 129 nucleotide sequence.

4.4 Hepatitis B patent (Biogen v Medeva)

This case is an even more striking example of different outcomes obtained on equivalent patents in different jurisdictions. This patent covers recombinant DNA molecules having sequences coding for polypeptides displaying HBV antigen specificity (claim 1) and HBV antigenicity (claim 2). There are specific claims to the coding for HBV core antigen and HBV surface antigen and product-by-process claims to the expression products of these. Biogen's European patent 182,442 survived opposition but its British counterpart underwent the whole rigour of litigation in the UK courts. Questions of adequacy of description and support for the broad claims figured prominently throughout this case, the most remarkable feature of which was the diverging opinions expressed on these issues by the various judges en route to the final decision. Finally the UK House of Lords decided that its claims were too broad. The complexities of this case have been described in detail elsewhere (8).

5. THERAPEUTIC ANTIBODIES

The patent literature is replete with inventions in the field of monoclonal antibodies. Most of the early development of this technology, as reflected in patents, has been directed to the use of monoclonals for the separation and purification of bio-molecules and for diagnostic applications. One example of an early patent to be granted is US patent 4,361,549 which describes an OKT3 hybridoma which produces a complement fixing monoclonal antibody to an antigen found on normal human T cells. The claims are restricted to mouse monoclonals but the broadest is not restricted to any one T cell antigen. Diagnostic and therapeutic applications are indicated. This product is now on the market for use in controlling transplant rejection.

The use of mouse-derived antibodies in humans can induce a human anti-mouse antibody (HAMA) response which might be serious if repeated use is necessary. The full therapeutic potential of monoclonal antibodies may therefore depend on techniques for reducing or eliminating the anti-globulin response to mouse and other non-human antibodies. The development of chimeric and 'humanised' antibodies formed by recombinant DNA methods as hybrid immunoglobulins now offers the promise of reducing the immune response to such antibody products. For reasons touched on earlier, (e.g. for the t-PA and Epo patent situations), protein engineering technology gives rise to the opportunity for new final product patents i.e. those containing product-per-se claims for the new polypeptides as well as for DNA sequences that result from these techniques. Patents now being granted in this field include some for new principles of immunoglobulin reshaping but most are for specific products with partially or fully defined amino-acid sequences. A selection of these is given below.

The basic principle of antibody reshaping is covered by US patent 5,225,539 and its European equivalent EP 239,400. The claims granted on this invention cover any 'altered' antibody in which the variable domain framework regions and the complementarity-determining regions (CDRs) are derived from different immunoglobulins. The US claims specify this difference as one of antigen binding specificity or affinity, species, class or subclass. Inventions which are the first to open a new field are often described as "pioneering", especially if they are followed by a stream of later developments in the same field. The first patent usually issues with claims of very broad scope because the patent examiner has not been able to cite seriously damaging prior art relevant to either novelty or inventiveness. Such patents are often called "master patents" because they dominate later inventions. In these situations the requirement for adequate supporting disclosure has often been met by providing a relatively small number of 'proof of principle' model examples. This is the case with the patent numbered above in which one of the specific worked examples describes the humanisation of one chain of a mouse antibody to a small chemical hapten.

The first example of an antibody humanised to a complex antigen in both the light and heavy chains is the subject of European patent 328,404. This antibody (known as Campath 1-H) binds to the CD52 antigen.

The essential feature of this patent is the choice of CDRs of specified amino-acid sequence derived from a particular rat antibody. This patent seems also to have been the first to include an amino-acid change in the originally fully human framework region of one of the immunoglobulin chains. This favoured the packing of the CDRs and improved the binding power of the humanised antibody.

Framework changes designed to achieve the above effect are covered in US patent 5,585,089. This is a more than usually complex document and one of its European counterparts, EP451,216, has itself provoked a response from others working in this field of research and in industry.

This patent covers a number of principles involved in selecting pairings of CDRs from a donor Ig and frameworks from a human acceptor Ig showing a high degree of homology to the framework of the donor Ig. These principles include certain defined criteria for the replacement (if necessary) of amino-acids in the heavy or light chain human frameworks by the corresponding amino-acids in the donor Ig sequence.

Some other patents on products undergoing clinical evaluation are US 5,585,097 which covers an aglycosyl anti-CD3 antibody claimed in terms of fully sequence-defined light and heavy chains and aglycosylated constant region. Aglycosylation reduces the first-dose (cytokine) response which may be encountered with the parent antibody. As with many similar US patents, there is also a claim in 5,585,097 to a method of treating a patient with this antibody to prevent renal allograft rejection. Method claims of this type are not allowed under European patent law. The reason for this is not primarily an ethical one, as is sometimes thought, but because the procedural act of treating humans is not an industrial process and therefore does not meet the 'susceptibility of industrial application' requirement for patentability.

Another example of method claims is US patent 5,656,272 which describes the preparation of an anti-tumour necrosis factor-á chimeric antibody for treating Crohn's disease. This product has received US FDA approval. The patent claims are all directed to the method of treatment. A parallel patent application for the antibody itself may also exist.

6. PATENT INFRINGEMENT

The writing of patents for nucleic acids and proteins presents an acute dilemma. One knows that the composition of these molecules may be modified in various ways, leading to mutants, variants and derivative forms which may either retain, enhance or reduce, or totally lose the original biological activity. The patent draftsman attempts to guard against third party avoidance of claims tied too closely to the limited range of specific products made by the inventors and presented as the patent examples. But in the absence of data on the effect of compositional variation on activity there is nothing to guide him. The patent examiners will normally stress the uncertain effect of variation and will insist that the claims are limited to what has been disclosed.

The patent draftsman will usually prepare the ground for a broad interpretation of the claims by the use of a skilfully drawn "Definitions" section. A comprehensive model of this tactic is to be found in the human tissue plasminogen activator patents, especially US patent 4,766,075 which is the equivalent of UK patent 2,119,804 discussed above. These definitions of t-PA embrace natural allelic variations and derivatives modified by single or multiple amino-acid substitutions, deletions, additions or replacements in the t-PA molecule so long as the essential biological function of t-PA is retained.

An infringement suit on the US t-PA patents provides an instructive example of how these matters are treated by the courts.

6.1 Genentech v Wellcome Foundation and Genetics Institute (1990)

Genentech sued these defendants for infringement of the following three US patents relevant to t-PA (9) :-

- 4,752,603 (the '603 patent) is directed to Human plasminogen activator derived from the Bowes melanoma cell line and it covers the original work carried out by Leuven Research and Development. The claims are limited to material of specific activity of 500,000 IU/mg against a specified reference standard. This limitation was necessary because of an earlier publication by one of the inventors describing material purified to an activity of 266,000 units.

- 4,766,075 (the '075 patent) covers "A DNA isolate consisting essentially of a DNA sequence encoding human tissue plasminogen activator" and the corresponding recombinant expression vectors.

- 4,853,330 covers the process of expressing the DNA of the '075 patent to produce t-PA.

Wellcome's product, made in the UK and exported to the US, differs by only one amino-acid from human t-PA,"the product of the patent". The Wellcome product (met-t-PA) contained methionine in place of valine at position 245.

The GI product (FE1X) was a product having 81 amino acid deletions from t-PA. It lacked the finger region and most of the epidermal growth region and had other differences in the kringle region of the native protein.

The District Court of Delaware (in March 1990) held that, on their literal interpretation, the claims in both the '603 and '075 patents were limited to the full-length amino-acid sequences of naturally occurring human t-PA and its naturally occurring allelic variants. The court also decided that the specific activity limitation in the '603 patent should also apply to the definition of t-PA in the context of the '075 patent. This was particularly surprising because the claims of the '075 patent were to the DNA coding and not to the protein, either per se or at any level of purity.

It being admitted that neither met-t-PA nor FE1X naturally occur in humans and that their specific activities were lower than that specified (or assumed) in the claims, these products were held not to infringe the patent on the literal interpretation. Wellcome also argued successfully that importation of met-t-PA into the US, which involved no use of the claimed DNA or recombinant cell lines in the US, was non-infringing for this reason also.

But US law also has a "doctrine of equivalents" which the Delaware court described as "an equitable doctrine permitting a more expansive interpretation of patent claims than the literal scope thereof". In view of the material issues of law involved, the court declined to rule on this issue and reserved it for further trial. Genentech then applied for a jury trial on the equivalents issue. This commenced 7 days later and resulted within 15 days in verdicts of infringement by equivalents for both products. One might well marvel at the idea that a jury could be expected competently and fairly to assess technical and legal issues as complex as those involved in this case. However, this victory was relatively short-lived (as legal processes go) because it was reversed by the Court of Appeals for the Federal Circuit (CAFC) in June 1994.

While the Appeal to CAFC was pending, Wellcome announced their decision to discontinue development of a t-PA product. Genentech had also decided not to cross-appeal on the issue of the literal interpretation of the claims. Therefore the only issue for the CAFC was whether FE1X infringed under the doctrine of equivalents.

The court identified three key issues. The first was whether the specific activity limitation in the '603 patent applied to the '075 and '330 patents, to which the court gave a negative answer because these were in every way independent patents. The second was the basis of measurement of the specific activity figure. The court decided that this was to be measured by the bovine fibrin plate assay. The third and much the most important question for the court was the meaning of "human tissue plasminogen activator".

The essential nature and properties of human t-PA had been described in various ways in the patent with the result that the court had to choose from at least the following four possible definitions of the substance :

1. recombinant t-PA having the structure (composition) of native t-PA.
2. products containing the kringle and serine protease regions.
3. products containing just the enzymatically active portion i.e the serine protease region.
4. products which convert plasminogen to plasmin, bind to fibrin, and are classified as t-PA on the basis of immunological properties.

On the evidence of witnesses, the CAFC opted for the first of these definitions because "it is the most consistent with the limited form in which the claims are drafted, and the others are hopelessly over-broad." The court concluded that the jury's finding of equivalents was not supported by the evidence on any of the key issues mentioned above. FE1X was also shown to behave significantly differently from human t-PA in the body (e.g. ten-fold increased half-life, decreased binding affinity). Therefore, FE1X did not infringe any of the asserted patents.

7. ISSUES FOR THE FUTURE

As indicated previously, many of the recombinant DNA patents applied for in the early 1980s have been justified on the basis that gene cloning involved considerable difficulties at this early stage of the science and technology. It follows that, as some scientists now say that gene cloning has become largely routine, the case for inventiveness may have to be buttressed with other arguments. As they contend with one another in legal disputes, the biotechnology companies must be careful how they use the 'largely routine' argument against their competitors because it can easily return to haunt them when their own patents are under challenge. Patent attorneys usually find that inventors can often supply good reasons why not everything is as easy as it might seem in particular cases and they will hold these in readiness for use when difficulties arise with the patent authorities.

Finally one development is worthy of mention here although the materials are not themselves biopharmaceuticals. An increasing number of patent applications are being published in the field of receptors and the corresponding genes for use in screening for biologically active substances, including many kinds of agonists and antagonists of known agents. Examples of some pending patent applications are:

- WO/9413799 Human GABA receptor cell line
- WO/9509872 Prostaglandin receptor IP cDNA
- WO/94/09828 Human serotonin receptor DNA
- EP.514207A Human neurokinin-1 receptor DNA

Applications of this general type will usually claim the DNA coding, the protein sequence, the receptor system, and the method of use in screening for particular activity. Some even claim all (unspecified) products which might be discovered in this way. How patents of this kind will be viewed by the authorities is yet one more of the continuing supply of questions with which biotechnology exercises patent law.

BIOGRAPHY

R.Stephen Crespi is a British and European patent attorney, a former Head of Patents of the British Technology Group (BTG) and now practising as an independent consultant to industry, Universities, and patent firms. He is author of 'Patenting in the Biological Sciences' (John Wiley 1982) and 'Patents - a Basic Guide to Patenting in Biotechnology' (Cambridge University Press 1988) and has contributed to a number of OECD Reports on aspects of biotechnology and intellectual property.

REFERENCES

1. Comparative Studies of Patent Practice in the Field of Biotechnology (1988). 7 Biotechnology Law Report, (Mar/April), 153-193, Mary Ann Liebert Inc
2. Scripps Clinic and Research Foundation v Genentech (1989). 11 United States Patent Quarterly, 2d 1187
3. Amgen Inc v Genetics Institute, (1989). 10 United States Patent Quarterly 2d, 1906
4. Ex Parte Gray, (1989), 10 United States patent Quarterly, 2d, 1922
5. Genentech UK tpa, (1989). UK Reports of Patent cases 1473, reviewed in Nature, 337, 317.
6. European Patent Office Reports (1996). EPOR 275.
7. Amgen Inc v Chugai (1989), 13 United States Patent Quarterly, 2d, 1737.
8. Crespi, R.S. (1989). Patenting in Biotechnology—the saga continues, Biotechnology and Genetic Engineering Reviews, 15, 229.
9. Genentech v Wellcome Foundation and Genetics Institute (1990). 14 United States Patent Quarterly 2d, 1363.

Chapter 11

The development of new medicines: an overview

Dr. John C. Stinson
Medical Director, Leo Laboratories Ltd., Crumlin, Dublin 12, Ireland

Key words: Medicines, discovery, toxicity, clinical trials, regulatory approval

Abstract: Natural remedies for disease have been used for thousands of years, though their discovery was almost always serendipitous and the formulations and applications crude. Many modern medicines were initially derived from natural substances, such as digitalis from the Foxglove plant, penicillin and fusidic acid from fungi. Recent medicines originating from plants include the anti neoplastic agents paclitaxel and docetaxel which come from the Yew tree. Luck, scientific observation and reasoning frequently combined to yield successes. Gold salts such as sodium aurothiomalate were originally developed as anti-tuberculus therapy. They were found to be ineffective but a chance finding was that patients treated with these salts who had concomitant rheumatoid arthritis obtained great relief and now gold salts are commonly used by rheumatologists.

However, it was unacceptable that chance should play such an important role in research. Today the drug development process has become a disciplined, modern science and this has occurred, almost entirely, in the last 40 years and at an ever increasing rate. If one looks back to just before the second world war, there were no antibiotics, no anti-hypertensive agents, no medicines for asthma, peptic ulcer disease -indeed there were very few effective medicines at all. Taking hypertension as one example, effective therapy came along with diuretics in the '50s, beta receptor antagonists in the '60s, calcium channel antagonists in the '70s, ACE inhibitors in the '80s and now AII receptor antagonists in the '90s. This series of therapeutic advances for one disease has been mirrored by advances in many others.

The process of developing new medicines has evolved over the past half century and this chapter aims to give an overview of the current process.

1. DISCOVERY OF NEW MEDICINES

With notable exceptions, such as the discovery of Insulin by Banting and Best in 1922 (1), early discovery of medicines was more due to random screening than scientific reasoning. However, as advances occurred in physiology and as biochemistry developed, better strategies and methodologies for medicinal chemists arose. The development of receptor science was followed by advances in protein and peptide chemistry. These were assisted greatly by advances in NMR and X-ray crystallography. As these advances occurred, so the drug discovery process became more selective and less dependent on chance. The last decade has seen the advancement of rational screening of molecules based on 3D databases and automated molecule design. Molecular biology has introduced the ability to clone receptors, to produce mononclonal antibodies and recombinant products, to study transgenic animals, to select gene therapies and to develop antisense oligonucleotides. These advances have brought us to the possible future where doctors will not only treat patients with disease, but will treat individuals on the basis of altering their genetic makeup because of its propensity to cause future disease. This treatment of the genotype rather than phenotype has many implications, but is not so far away.

Currently when a pharmaceutical company considers the development of new medicines, it focuses in certain therapeutic areas only. It is still considered very difficult for even the biggest companies to be active in all therapeutic areas. Then within the therapeutic area, the company will select a disease target. Therefore if it is decided to concentrate on respiratory medicine, a target disease might be asthma or cystic fibrosis or chronic bronchitis. A company will consider many factors before embarking on a particular search for a new medicine. For example, the company will look at diseases which are common but for which there are no, or only unsatisfactory, treatments available. The company will consider their internal expertise, they will consider the commercial return, the competitor activity in this therapeutic area, the current therapies available, the epidemiology of the disease, the cost of the research program etc., before starting.

Once the target disease has been selected then a mechanism to attack the disease process is chosen. As scientific advances increase the understanding of the pathophysiology of a disease then new routes to attack the disease become apparent. For example as it was discovered that asthma was primarily caused by inflammation, then the search for new anti-asthma compounds shifted from β receptor agonists to anti-inflammatory agents such as leukotriene antagonists. It is through better understanding of conditions such as Alzheimer's disease or Multiple Sclerosis that possible pharmacological approaches to treatment will arise.

The development of new medicines: an overview 271

1.1 New Chemical Entities

When a potential therapeutic approach is investigated, molecules will be synthesised which are considered likely to have the required activity. For example if one wanted to suppress an enzyme involved in a harmful biological process, then if one understands the structure, 3-D shape and actions of the enzyme, it may be possible to design chemicals that will inhibit this activity. The discovery that high levels of cholesterol were associated with premature death from coronary heart disease was soon followed by the discovery that the enzyme HMG CoA reductase was the rate limiting enzyme in the endogenous synthesis of cholesterol. Several pharmaceutical companies then invested many millions of dollars looking for agents that would inhibit this enzyme. Through intelligent screening of many molecules, it was discovered that molecules of a particular class had the desired activity. These molecules would be further divided to find those with most activity. This class of compounds are now referred to as the statins and 4 have reached the market.

When *in vitro* activity of a molecule has been shown it is considered a new chemical entity (NCE). It is not a medicine, and will remain an NCE as it hopefully passes more and more hurdles on the way to becoming a medicine. An NCE does not become a medicine until it has passed through all the developmental stages, through toxicology testing, through trials in healthy humans and hence into trials in patients and then shown some benefit. A medicine may be defined as: "*Any treatment or preparation used for the treatment or prevention of disease*".

When an NCE has demonstrated a significant effect in an *in vitro* model, (and perhaps even before), it is likely to be patented, usually along with several analogues which may be "backup" compounds should the initial NCE fail one of the following developmental hurdles.

1.2 Toxicology testing

Before an NCE can be given to human volunteers its toxicological potential must be assessed in animals. Here the term toxicology is used in its broadest sense as it includes safety pharmacology and genotoxicity testing. The safety pharmacology testing involves giving the NCE at levels much higher than those to be used in patients (but lower than that used in toxicology testing) and looking for any serious effects primarily on the "life support organs" i.e. the cardiovascular, respiratory and central nervous systems. This is usually performed in two animal species (e.g. rat and dog) in parallel to the toxicology testing.

Genotoxicity always includes the Ames' test which involves exposing bacteria to high concentrations of the NCE and looking for any gene mutations. The mouse lymphoma cell line is used to test the potential for mammalian gene mutation. Chromosomal damage in mammals is tested in Chinese hamster ovary cell lines. *In vivo* testing can be performed by exposing rats to the NCE and then looking for chromosomal damage in the blood stem cells of the bone marrow. It is also possible to assess DNA synthesis repair mechanisms in the rat liver.

Acute toxicity testing involves administering single doses to two species of animals, usually rat and dog. In different groups of animals increasing doses are given to define doses such as the maximum non-lethal dose, the maximum well tolerated dose and the no observable effect dose. The tests are repeated by two routes of administration, intravenous and oral, this ensures that the animals are exposed systemically to non-absorbed NCEs.

Repeat dose toxicity testing involves administering regular doses to two species (rodent and mammal) for a certain period of time. The longer one intends giving the NCE to humans, the longer the repeat dosing studies in animals must last. For example if the NCE is a new antibiotic with an intended 5 day course in humans, then shorter repeat dose studies in animals are needed than if one intends the NCE to be given to humans for say 3 months. In general such studies start with a study to estimate the maximum repeatable dose (MRD) and these are followed by 1 month, 3 or 6 month and 12 month studies (2).

Reproductive toxicology is a subspeciality, and only tends to be necessary when women of child bearing potential are to be included in the clinical trial programme. Including women in the early clinical trial programme used to be a very rare occurrence but in 1993 the FDA changed policy to insist that pharmaceutical companies ensure that fertile women are adequately represented in clinical trials and to state that lack of data on gender differences may lead to their refusal to accept an NDA (3).

In addition, if there is no data on toxicological effects of an NCE in pregnant animals, then it will never be given to pregnant humans, even if the risk/benefit ratio is expected to weigh heavily toward using the new medicine. Teratogenesis is essentially defined as the abnormal growth of a foetus. The thalidomide disaster which occurred 40 years ago horrifically proved that the placenta was not a perfect protective barrier. With only one exception, every drug that has been since shown to be teratogenic in humans has been shown to be teratogenic in animals (4). Adequate reproductive toxicology testing in animals cannot guarantee safety in pregnant women, but it is the best available and with the anti acne treatment isotretinoin it has prevented a disaster of the magnitude of thalidomide where a foetus had a 20-30% risk of malformation.

Carcinogenicity (oncogenicity) studies involve administering the NCE to a rodent species for the lifetime of the animals and looking for any development of any tumours. These studies are only necessary when the NCE will be given for continuous periods in excess of 6 months, when the genotoxicity studies or the chemical structure of the compound suggest oncogenic potential.

Major concerns from the results of the toxicology programme will usually permanently stop any further development of the NCE.

1.3 Animal Pharmacokinetic studies.

These studies are performed usually in parallel with the toxicology programme. The purpose is to determine how the NCE is absorbed, distributed, metabolized and excreted in two animal species, one a rodent and one a mammal. These studies are often referred to as ADME studies and are mainly performed with radioactive versions of the NCE. Hence radioactive metabolites discovered in the urine, faeces and blood can be identified as breakdown products of the NCE. It is important to examine major metabolites for toxicological and pharmacological effects as some NCEs will have pharmacologically active metabolites, indeed some marketed medicines are not effective until they have undergone a metabolic step and are known as prodrugs (e.g. levodopa).

Animal pharmacokinetic studies are also an important milestone in the evolution of an NCE. In certain cases the results may suggest that there is no point in any further development of this NCE. For example, if the NCE is being developed as a once daily treatment for urinary tract infections, and dog studies show it has a half life of 1 hour and is exclusively excreted by the biliary tract, then there is really no point proceeding.

1.4 Formulation Development

From early in the development of an NCE, it will have been decided what formulation(s) would be desirable for the marketed product. Whilst most medicines are in oral formulations, many are in topical, injectable, inhalable, rectal, sublingual etc. forms. Even amongst oral preparations there are possibilities for tablet, capsule, sustained release, solution and suspension formulations. Preformulation consultations with chemists will define the physical and chemical properties of the compound. If all its salts are insoluble then a tablet will be useless and a solution impossible to produce. Different salts of the NCE may have different potencies, for example plain hydrocortisone is a mild steroid and hydrocortisone butyrate is a potent steroid. Excipient compatibility and method of synthesis of active ingredient will also be studied carefully.

At the end of this process the company should have a very good idea of what the final formulation for trials in patients and for the marketed product will be. However for the very first studies in humans a solution is usually the preferred formulation, as this best allows for many incremental increases in dose in what are essentially dose finding and safety/tolerability studies.

The overall aim of the complete pharmaceutical development programme is to retain the pharmacological activity of the NCE in a stable form, whilst adding predictability and ensuring that this formulation will be easily administered and manufactured. In addition the manufacturing process should be as simple and economic as possible and the end product as easy to transport and store as possible.

2. FROM ANIMALS TO HUMANS.

At the end of the toxicology and pharmacokinetic studies in animals, if no safety issues have arisen then the decision to move into trials of the NCE in humans can be made. It is important to realise that whilst animal toxicology studies assist greatly and are the best tests available, they do not guarantee safety in introducing an NCE to man. However, at least with studies in mammals we can be reasonably confident that we can predict by extrapolation the findings from these higher organisms to man (5).

Care needs to be taken as there are many examples of exceptions to the predictive nature of such animal studies. The rat has no gallbladder and hence intrahepatic metabolism is altered. The dog lacks the enzyme N-acetyl transferase and thus can't acetylate, whilst the cat lacks the enzyme glucuronyl transferase and thus cannot glucuronate. If one considers the metabolism of amphetamine, parahydoxylation is the major route in the rat, a moderate route in the mouse, a minor route in man, yet the rabbit deaminates amphetamine. With the anti-inflammatory agent phenylbutazone, a dose of 300mg/kg in rabbits is needed to obtain the desired effect compared to only 5mg/kg in man. These different doses give the same plasma concentrations in the two species. If one tried to predict the half life of phenylbutazone in humans from the animal pharmacokinetic data, then as in both the dog and rat give the same value of 6 hours, one would be surprised to find that it is 72 hours in man. Nonetheless, there are many more similarities between higher species and man than there are differences

In most cases, the first studies of an NCE in man are performed in healthy volunteers. The FDA have defined a "normal volunteer" as those who are free from abnormalities which would complicate the interpretation of the experiment or which might increase the sensitivity of the subject to the toxic potential of the drug. It would be important to add that the volunteer must be able to give valid, informed consent.

Studies in healthy volunteers are termed phase 1 trials and are extremely carefully designed and performed. The design and rationale for the study must be approved by an ethics committee, which is made up of medical and non-medical professional people, who act completely independently of the pharmaceutical company. The volunteers (almost always males) are extensively screened to assure they are healthy. The first such study usually involves giving single doses to about 4-8 volunteers in a very controlled environment. The first dose is usually about 1% of the "no observable effect level" (NOEL) dose in the most toxicologically sensitive animal species studied. It is usually given as a solution in a weight adjusted (per kg) dose. Ideally some of the volunteers will be given a "dummy" or placebo (from Latin: "I will please") in a "double blinded" manner. This means that neither the volunteers nor those interpreting the results know which volunteer received active and which received the placebo. During this study the volunteers will be monitored extremely closely and blood samples will be taken frequently and these samples will be assayed to determine the concentration of the NCE in the blood. After the first group of volunteers have received the NCE, medical examinations, safety blood tests and assays for the NCE in the blood and urine will be performed. When the results are available and if they show no cause for concern, then another group of volunteers will receive another, larger dose of the NCE. This will then be repeated until a certain, preset dose or blood concentration of NCE is reached or until some safety issue is noted. Usually the dose is doubled at each step, but sometimes smaller increments are used as one gets towards the preset limits.

This first in man study will supply pharmacokinetic (PK) data and should give much guidance as to whether the NCE is metabolised differently in man as compared to animals. The PK data gathered will usually include the following: time to maximum concentration (T_{max}), the maximum concentration achieved (C_{max}) and the time taken for the concentration to fall from its maximum to 50% of this figure which is also known as the half life ($t_{1/2}$). Although phase 1 studies are performed in healthy volunteers, who by definition, do not have the disease the NCE is aimed at, it may be possible to get some idea of possible efficacy from studies in healthy volunteers. Often this involves using a "surrogate marker" to assess possible pharmacodynamic (PD) effects. A surrogate marker is a validated, measurable effect which should be statistically and mechanistically related to the clinical effect one eventually wants the NCE to demonstrate in patients (see below).

If there has been no safety or tolerability problems in the first study in man then a series of trials in healthy volunteers will follow. The phase 1 trial programme for an NCE will be specific for that NCE, but in general single dose studies are followed by multiple dose studies and then by special trials to assess bioavailability (the % of the dose of NCE given that is absorbed). Depending on what is known from the animal studies and what is predicted to occur in humans, certain specific trials may be designed to assess whether food affects absorption, whether alcohol may affect the metabolism etc.

Traditionally, phase 1 trials were seen simply as gathering data on pharmacokinetics, safety and tolerability. However these "clinical pharmacology" studies have become more pivotal as the cost of taking an NCE into a full clinical trial programme has escalated exponentially. Pharmaceutical companies now wish to see some evidence of efficacy or pharmacological effect in the phase 1 trials. Therefore they want to be as confident as possible that the NCE will be both safe and effective in advance of it proceeding to trials in patients. From the time that the first safety data in healthy humans is obtained, the animal toxicology data becomes very much of secondary importance.

Phase 1 trials also involve studies to assess if the NCE will have interactions when taken with other medicines. There is likely to be some assessment of potential interactions with other medicines which are likely to be co-prescribed. This is particularly true if the NCE is likely to be co-prescribed with medicines such as digoxin or oral anticoagulants which have many interactions. If the NCE is shown to be metabolised by the cytochrome P450 enzyme system or to be extensively metabolised in the liver, then interaction studies of metabolism are probably necessary.

Phase 1 clinical trials are not usually performed on NCEs which are designed as anti-cancer medicines. These usually have many noxious effects and whilst it may be acceptable to study these very experimental products in patients with life threatening cancer, who have little other therapeutic option, it is not ethical to give them to healthy volunteers.

2.1 Clinical Trial Design

It is important to discuss the concept of a clinical trial. Many human diseases are not completely progressive and hence tend to have fluctuations in severity. Hence patients with rheumatoid arthritis might have months of severe pain and disability followed by a natural improvement in their symptoms, without any medical intervention. In addition, there can be psychological as well as pharmacological effects of taking medicines. The responses of patients symptoms to dummy or placebo "medicines" are well known. When a patient is given a potential new medicine, any improvement may therefore be due to a pharmacological effect of the medicine, a psychological effect of the medicine or the natural course of the disease. It is obviously important to be able to prove if an NCE has a beneficial effect or not in treating disease.

If, as a true scientist should, one wishes to get as close to the absolute truth as possible, then many features can be incorporated into the design of a clinical trial to minimise the risk of an incorrect result. For example, if we consider that an NCE, "A", is better than giving the patients no treatment, it can be decided to study this effect in 200 patients. The first 100 might be given "A" and the second 100 patients given no treatment. However it is better to randomly assign the patients to treatment or no treatment to exclude a possible investigator bias in assignment. Randomisation will also help exclude any "period" effect bias, for example if "A" was a new treatment to prevent 'flu and the first 100 patients were recruited in the summer months and the second 100 patients recruited in the winter months one can see an obvious flaw.

The measurement of any data is extremely important also, especially where the patient or physician know which treatment the patient had been given. In the above example the patients who received the anti 'flu treatment A might report less symptoms because they expect the treatment to work. This sort of bias can also be expected of the physician. To exclude these possible errors, those randomised not to receive "A" may receive a dummy form of it (called placebo). If neither the patient nor doctor know whether treatment or placebo has been given (as placebo will be identical in every way other than active ingredient to "A") then the study is said to be "double blind".

Companies will always make sure all the data is collected and the statistical data base "locked" before the code is broken to allow for a statistical assessment of which treatment was best.

A pharmaceutical company will frequently come up with a hypothesis that their new drug is better than a comparator. They will wish to test this hypothesis in a clinical trial. Any difference between the two treatments may be due to one (or more) of 4 possible reasons: they could be real, could be due to chance, or could be due to either an allocation bias or an assessment bias.

If one ensures that the number of patients in the trial is as large as possible and are randomised to the treatment this goes a long way to excluding allocation bias. Other specialist methods that may be appropriate to exclude allocation bias include minimisation, stratification and a crossover design.

By using a prospective design with objective measures and having both patient and doctor "blind" to treatment will minimise allocation bias. Thus the description of a clinical trial as being "a prospective, randomised, double blind trial" means that both assessment and allocation bias have been excluded.

The final issue that needs to be excluded before one can confidently state that any effect seen is real is the possibility that the results are purely due to chance. If this can be excluded (in addition to allocation and assessment biases) then the results must be real.

Chance is traditionally excluded by the use of statistical analyses of the data. Although statistics is a major discipline in itself (5), it is fair to say that most statistical analyses aim at minimising the risk of accepting a result as real when it was due to chance. Most clinical research physicians wish to see that the chance of a false positive result (type I error) is less than 5%, although for practical purposes they will accept a false negative rate (type II error) of 10-20%.

Clinical trials are extremely expensive to perform and require extensive resources from a pharmaceutical company. The company, therefore, wishes to find out as quickly as possible if their NCE displays any efficacy. They also want to know if there are any safety concerns that might halt its development. Hence they direct a lot of effort and finance to supervising and monitoring their clinical trials. The quality of this, very expensively obtained, data is crucial to decision making about the NCE's future.

2.2 First Clinical Trials in Patients

The first clinical trials in patients are known as phase 2 clinical trials and involve giving the NCE to relatively small numbers of patients (usually 15-100) with the disease(s) the NCE is aimed at treating. The rationale of the phase 2 trial programme is to extend the safety and tolerability knowledge gained in phase 1 into a patient population. In addition the phase 2 studies will determine dose ranges, dose frequency and efficacy data that will be used in the large phase 3 clinical trials. These trials will hopefully give an idea of the dose response curve which relates clinical effect to tissue concentration of the NCE. Pharmacokinetic data may be gathered from these studies, as it is unwise to assume that patients will metabolise an NCE in the same way as healthy volunteers patients (who are usually younger). Overall the data gathered should give an initial idea of the risk/benefit ratio (7).

Phase 2 studies must be aimed at preventing mistakes in the design of the phase 3 studies which would be extremely expensive and would seriously delay the time to market. Hence phase 2 trials are extremely important as the transitional step between healthy volunteer studies and the large multinational studies in patients which will be pivotal and form the bulk of the efficacy and safety data to be submitted as part of the application for a marketing authorisation (8).

The first phase 2 studies are performed with low doses of the NCE in a small number of patients who are intensively monitored for both safety and for any evidence of efficacy. If no safety issues are discovered then the studies will advance to higher doses and to multiple dose studies. As long as there is adequate long term safety toxicology data in animals then the duration of these studies are also increased. These early phase 2 studies are usually performed by very experienced physicians who have a specialist knowledge of the therapeutic area and preferably of clinical pharmacology as well. Hence trials with new anti-asthma compounds will be carried out by respiratory physicians usually in a large university linked teaching hospital or "centre of excellence".

There is usually an overlap between phases 1 and 2 of NCE development, i.e. one does not complete the phase 1 programme totally before entering phase 2. Indeed phase 1 trials may sometimes be needed long after an NCE has reached the market as a medicine. Any trial in healthy volunteers is considered phase 1, therefore if an alcohol interaction safety issue arises after the product reaches the market, it may be necessary to go back and investigate this issue in a study in healthy volunteers.

As phase 2 trials are, in part at least, dose finding, they (and also the phase 1 studies) reach the highest doses that will ever be given to humans and the safety data gathered from these trials is extremely important.

2.3 Surrogate Markers

Surrogate markers are often used as the endpoints in clinical trials. A surrogate marker is a measurable entity which is directly related to the actual disease process. For example when we treat patients with hypertension one of the main aims is to prevent stroke. However, it is necessary to treat patients for many years to prevent a stroke occurring. If one used stroke as the endpoint of a 200 patient trial of a new antihypertensive medicine, then one might have to follow these patients for 10 years or more to demonstrate an effect. However as it is known that high blood pressure is directly related to risk of stroke, the blood pressure reduction is used as the surrogate marker of the effectiveness of the medicine. The drop in blood pressure will occur quickly and certainly maximal effect can be measured within weeks rather than waiting for 10 years to discover the effect on stroke incidence. A surrogate marker must have a statistical relationship to the actual endpoint one is trying to alter and it should have a pathophysiological relationship. The surrogate marker should be measurable by a validated methodology.

Examples of surrogate markers that might be used in phase 2 trials would include measurement of gastric pH in patients with peptic ulcer disease, or measurement of cholesterol levels in patients with hyperlipidaemia. The benefits to the clinical research programme is that surrogate endpoints are easier and cheaper to measure, the response rate may occur more rapidly and that a significant response rate may be seen with fewer subjects. Another advantage of surrogate endpoints is that they may allow an assessment of pharmacological effect in the phase 1 healthy volunteer studies. For example when ACE inhibitors were being developed for treating high blood pressure, it was not expected that they would have much effect in healthy volunteers who had normal blood pressure. However as ACE inhibitors work by suppressing the hormone renin, then as measurements of renin levels decreased in these volunteers (although blood pressure did not) the companies had some evidence of efficacy from their phase 1 studies.

2.4 Phase 3 Clinical trials.

The end of the phase 2 clinical trial programme represents a major milestone or decision making point in the development of an NCE. At this stage there should be some concept of the efficacy of the NCE, there should be knowledge of its pharmacokinetics and a preliminary assessment of its tolerability and safety in patients. The company should have established the final formulation and method of synthesis, they will have analysed the commercial opportunities and will know the competitor products in the therapeutic area. A major decision has to be taken, for a full phase 3 trial programme could cost $100,000,000, although there are opportunities to halt the programme should the need arise.

In the phase 3 programme the number of patients is increased as the company becomes more confident that the NCE will be a safe and effective medicine. The initial studies, sometimes called phase 3(a), may not be that much larger than those in phase 2, but later phase 3(b) trials may involve 1000 patients or more. The ultimate aim of the phase 3 programme is to gather sufficient data to enable the company to apply successfully to the regulatory authorities for authorisation to market the NCE as a medicine. This will require adequate efficacy and safety data, and sufficient information to allow a doctor to appropriately prescribe it to his patients. The doctor will need to know what diseases can be treated with the product, at what dose, for how long and what effect is likely.

The prescriber will also want to know what restrictions apply to the use of the product -- can it be used in children or patients with liver or kidney failure? The doctor will also want to know the common side effects (see pharmacovigilance below), and whether it interacts with other prescribed medicines. Many of these questions cannot be adequately addressed even by large phase 3 trials and so frequently the regulatory authorities place multiple restrictions on the use of a new medicine, and the company attempts to remove these restrictions by post marketing (phase 4) trials which address each specific restriction.

Phase 3 trials usually will compare the NCE with an appropriate medicine which is already on the market and is currently used to treat the same disease. If no such product exists then a placebo is used as comparator. It is important for these studies to be prospective, randomised and double blind if possible. Some regulatory authorities will request at least 2 studies showing the safety and efficacy of a new medicine before granting approval. These studies may be designed to show statistical equivalence or superiority to the comparator. Although larger numbers of patients may be needed to show superiority, the expense may be justified as this data, especially when published, will be of great value to the Sales and Marketing departments in the promotion of the new medicine

3. REGULATORY APPROVAL

When a company is ready to apply for a marketing authorisation, the product has normally been in development for about 10 years. Although medicines must be approved in any country before they are marketed, individual countries had their own individual system of approval. There has been extensive harmonisation between European Union countries, and in the last 7 years the ICH (International Congress on Harmonisation) has attempted to standardise the data requirements needed for approval in the USA, Japan and the E.U.

In 1965, the (then) EEC passed a directive (65/65EEC) which required each member state to assess and authorise all medicines before they could be marketed in that state. The E.U. have further legislated and refined the process of harmonisation and a new "EU-wide" system came into place on January 1st 1995 with a transitional phase which ended on 31st December 1997. The new procedures are to ensure uniform standards of safety, efficacy and quality for all marketed products.

As part of this process, the EMEA (European Medicines Evaluation Agency) was established which acts as a co-ordinating centre and hosts the CPMP (Committee for Proprietary Medicinal Products). The CPMP is the body that advises the European Licensing authority, the European Commission. The CPMP consists of 2 representatives of each national authority.

Since 1st January 1998, there are three methods of applying for a marketing authorisation. A company may apply directly to the EMEA via a "centralised procedure" for any new product (and must use this method for medicines derived by recombinant DNA, monoclonal antibodies and hybridoma technology or biotechnology compounds). The second method is to apply by the "decentralised (mutual recognition) procedure" whereby a licence granted in one state is recognised in any other state if no objections are raised. The final system allows for an application to the regulatory authority of one individual state, if that is the only state in which the product will be authorised.

The dossier that must be submitted is in standardised form. Part 1 is a summary of the dossier and contains the Summary of Product Characteristics (SmPC) and expert reports. Part 2 contains the data on the chemical, pharmaceutical and biological testing of the product. The third part contains the toxicological and pharmacological data and part 4 contains the clinical studies.

The dossier is carefully scrutinised by the medical, pharmaceutical and toxicological experts of the regulatory authority, who may also seek the opinion of additional, independent experts in specific cases, especially where the medicine has a novel mode of action or represents a new class of compound.

Assuming the authorities are satisfied, and this may be after certain queries have been adequately addressed, then an authorisation to market the medicine will be granted and the company can launch the product.

3.1 Marketing

A pharmaceutical company must make a profit if it is to continue to research and develop new medicines. Obviously to make a profit the company will need to more than recoup the costs of developing each new medicine. This requires successful marketing of the product. One might expect after the arduous path the product has so far followed that this would be the easiest stage. However healthcare costs are rising rapidly throughout the world and are under intense scrutiny. Although the mean expenditure on prescription medicines in Europe is only about 11% of healthcare costs, it is a growing fraction. Furthermore, as it is politically difficult to cut back on the healthcare employees' payroll or on hospital building programmes, restricting the medicines bill is regarded as one of the easiest areas to limit health expenditure. Therefore very significant efforts are being made by governments to, at least, slow the rise in the "medicines bill".

This is occurring at the very time that the costs of new drug development are greatly increasing. Modern medicines are expensive - in 1980 the average cost of bringing a new prescription medicine to the market was $230 million. That figure is almost 20 years old and the cost has risen since as regulatory authorities seek larger, higher quality databases before licensing new medicines. The companies only have a limited period of patent protection (10-

15 years) in which to recoup the costs, before "generic companies" launch cheaper versions.

New medicines are therefore expensive, although in most countries the cost to the patient is heavily reduced by government reimbursements. When a new medicine gets a marketing authorisation in a country, it does not automatically get "reimbursement status". Indeed limiting new products' reimbursement and removing this status from older products are methods employed by various countries to slow the rise in the national medicines bill.

In many ways the marketing of medicines is very similar to that of any other product. Pharmaceutical companies will have a commercial division with sales and marketing departments and sales' representatives in each country. They will advertise, they will "sell" the product to their customer (the prescriber, or health authority or insurer) and they will seek competitive advantage over other companies' products.

However, in other ways the marketing of medicines differs from that of other products. In most countries prescription medicines cannot be advertised to patients who are, after all, the end-users, but only to healthcare workers. However in the US prescription medicines can be advertised directly to the public and pressure is building to extend this to other countries (9). In most countries the promotional activity of the pharmaceutical industry is either self regulated for example by a "code of practice" or by the national regulatory authority. This type of regulation in essence aims at preventing excessive promotion to prescribers as medicines are a necessity and not a luxury. It is in the interests of the industry that it is not seen to be making excessive profits on the back of patient's illnesses and using these profits for over indulgent promotional activity to doctors.

3.2 Phase 4 Clinical Trials

Phase 4 clinical trials are almost always interventional and by definition occur after the medicine has been licenced. In addition these trials involve a licenced formulation used within the terms of its licence. In the vast majority of cases the product will be compared to another licenced product and all the clinical trial material is supplied by the sponsor. The aim of these studies is to extend the knowledge about the effectiveness of the product. Thus the study may aim to show that a new medicine is more effective than another which is the market leader for that indication. Phase 4 trials are extremely important to the marketing personnel of a pharmaceutical company, who will be actively involved in the design of these studies. Nonetheless these trials must have a scientific rationale, must be assessed by an independent ethics' committee, data must be objectively collected and the patients must give informed consent before participating.

In the past, trials were occasionally performed which were purely commercial and these involved paying doctors to gather information on patients prescribed a new medicine. A comparator was rarely employed, little data was collected and the patients (or state) paid for the medicines. The sponsoring company paid the prescriber per patient enrolled. These trials were called "seeding" trials as the concept was to get a certain number of patients prescribed a medicine, who would then continue with this therapy long after the study was complete. These trials are now seen as being unethical and should no longer be performed.

4. PHARMACOECONOMICS.

Pharmacoeconomics is the relatively new discipline which attempts to assess the financial cost of medicines relative to the clinical outcome. In addition it allows the cost-outcome of one medicine to be compared with another and also with non-pharmacological intervention such as surgery etc.

Whilst traditionally a pharmaceutical company had to prove that a new medicine was effective, safe and manufactured to high quality, now many governments are requesting pharmacoeconomic data. However even if the state does not require such economic assessments, the ultimate payer, be it the patient, the prescriber, the hospital or the insurance agency will want such data.

The types of outcomes and costs used are very important in addressing these issues. One needs to consider the direct costs, the indirect costs and even the intangible costs compared to the outcome. For example if a new medicine costs £10 per day and is 20% more effective than the current treatment which costs £1 per day, then is it better from a pharmacoeconomic point of view? What if this new medicine also allows for the patient to be treated at home rather than in hospital? The direct costs in this example might be the cost of the medicine and the cost of a hospital stay. The indirect costs would include the loss of earnings by the patient. Intangible costs would include the degree of suffering or pain experienced by the patient.

Different types of evaluations have developed. These include cost effectiveness analysis (CEA), cost-minimisation analysis (CMA), cost-benefit analysis (CBA) and cost-utility analysis (CUA). Cost-effectiveness analyses are the most commonly performed by pharmaceutical companies as they allow direct comparisons with other medicines or treatments

5. PHARMACOVIGILANCE.

Pharmacovigilance consists of the continual collection, review and analysis of adverse reactions (ADRs) to a medicinal product. This involves the spontaneous reports received by a pharmaceutical company from doctors, healthcare workers and patients and also may involve formal studies of a medicines adverse reactions. ADRs are notoriously under reported by healthcare workers and hence pharmacovigilance studies may be necessary.

Pharmacovigilance studies are sometimes classified as phase 4 trials, but are more appropriately called post-marketing surveillance studies. These are non-interventional and essentially observational with the aim of gathering more safety data about the newly licenced medicine. Various techniques can be used to gather such data such as cohort studies, case-control studies and computerised data bases which link prescriptions to ADRs. On average when a new medicine is licenced about 1500-3000 humans will have been exposed to it. If a particular adverse reaction to this new medicine only occurs in 1 in 5000 patients, then it is obvious that the pre-licencing data has little chance of detecting this. Hence pharmacovigilance is only beginning when a medicine reaches the market.

There are well known examples of medicines which were withdrawn from the market place when previously unknown adverse reactions became apparent. It is in the best interests of any company that they should learn of any safety issues as soon as possible so they may react accordingly. For example it may be discovered that the product interacts with another medicine or that the dose needs to be carefully monitored in a certain group of patients (the elderly, those with liver failure etc.). The company will want to protect patients from any harm, will want to further investigate the problem and will want to issue any warnings that are appropriate.

There is a legal responsibility in EU countries for any pharmaceutical company to report any serious and unexpected ADR to the national regulatory authority within 15 days of learning about the incident. In addition, (within the EU) for the first 2 years after licencing, the company must submit, every 6 months, detailed records of _all_ other ADRs reported (worldwide) to the EMEA. Furthermore, this must be performed annually from years 3 to 5 after licencing and every 5 years thereafter. It is also a legal requirement to report ADRs which occur in phase 2 and 3 clinical trials to the regulatory authorities and to the ethics committees.

6. PRODUCT DEVELOPMENT

When a new medicine reaches the market, it will have a limited period of time under patent protection. During that period of time the company will hope to recoup its developmental costs and make a profit. The company will try to develop different formulations and combinations of a product in an attempt to satisfy the clinical needs of patients, but also to prolong patent protection over some of the formulations.

For example, if a new medicine comes to the market as a tablet, a solution or suspension formulation may allow much smaller doses to be given accurately to children. A capsule might be developed which will give controlled, sustained release which may allow once daily dosing. An injectable form might allow intravenous or intramuscular administration. A topical formulation might allow a patch application or an aerosol preparation an inhaled product.

The product might be combined with another commonly prescribed medicine in one product for ease of administration, for example several versions of a combined diuretic and ACE inhibitor are available. Combining another agent to the initial medicine may improve potency, for example the addition of clavulanic acid (a beta lactamase inhibitor) to amoxycillin helped prevent bacteria which produced beta lactamase from becoming resistant to the product.

Another approach to product development is to seek new therapeutic indications for the medicine. For example when the ACE inhibitors were first licenced they were indicated only for the treatment of hypertension. Further studies showed that these products could also be used successfully to treat heart failure, even if no hypertension was present. Another example might be a new antibiotic, initially shown to be effective in treating chest infections, may later be shown to be effective in treating urinary tract infections and thus get an extension to its licenced indications.

7. CONCLUSION

The processes by which new medicines are developed are continually evolving in parallel with new technologies, medical discoveries and new regulatory requirements. Many discoveries in the pathophysiology of diseases are made by researchers in, or collaboration with, the pharmaceutical industry. The ever spiralling costs of developing new medicines gives great cause for concern, whilst the advances in molecular biology and gene therapy give great hopes for therapeutic advances. As man has a finite lifespan, there will always be therapeutic challenges which demand the development of new medicines.

BIOGRAPHY

John Stinson is medical director with Leo Laboratories Ltd., Ireland. Leo is an independent, research based pharmaceutical company, globally renowned for its R & D in dermatology, coagulation and bone turnover. Dr. Stinson also maintains a small clinical commitment in St. James's Hospital and is a part-time lecturer in Clinical Pharmacology in Trinity College, Dublin.

REFERENCES

1. Banting, F.G. and Best, C.H. (1922). The internal secretion of the pancreas. J. Lab. Clinical Med., 7, 256-71.
2. Scales, M.D.C. (1994). Toxicity Testing. In: Griffin, J.P., O'Grady, J., Wells, F.O. and D'Arcy, P.F. (Eds.) The textbook of Pharmaceutical Medicine. Queens University, Belfast.
3. Merkatz, R.B. et al. (1993). Working Group on Women in clinical trials. Women in Clinical Trials of new drugs—a change in Food and Drug Administration policy. N. Engl. J. Med , 329, 292-6.
4. Koren, G, et al. (1998). Drugs in pregnancy. N. Engl. J.Med, 338, 1128-1137.
5. Zbinden, G. (1987). The predictive value of animal studies in toxicology. Annual CMR lecture.
6. Altman, D.G. (1991). Practical Statistics for Medical Research. Chapman and Hill. London.
7. Burley, D.M. and Glynn, A. (1985). Clinical Trials. In: Burley, D.M. and Binns, T.B. (Eds.). Pharmaceutical Medicine. Edward Arnold, London.
8. Colburn, W.A. (1996). Decision making during NME development. Applied Clinical Trials, October 44-55.
9. Editorial. (1998). Pushing ethical pharmaceuticals direct to the public. Lancet, 351, 921.

Chapter 12

The EMEA and regulatory control of (bio)pharmaceuticals within the European Union

Dr. Gary Walsh
Lecturer, Industrial Biochemistry Programme, University of Limerick, Limerick, Ireland

Key words: Pharmaceutical regulation, European Union, EMEA

Abstract: Over the past 10-15 years a substantial body of harmonizing pharmaceutical legislation has been adopted by all constituent countries of the European Union. This has facilitated the creation of a common, European-wide system for the authorization and subsequent supervision of medicinal products. Central to this was the creation of the European Agency for the Evaluation of Medicinal Products (the European Medicines Evaluation Agency, EMEA). The EMEA, which became operational in February 1995, is charged with managing and coordinating the new drug approval system within the European Union. This new approval system provides for the evaluation of any new product marketing application via one of two routes; a 'centralized' route mandatory for all products of biotechnology, and a 'decentralized' route. Despite the inevitable teething problems, the EMEA has proven itself efficient. Within its first three years of operation, it has facilitated the granting of 52 European-wide marketing authorizations, of which 23 were products of biotechnology. By and large, the EMEA has now gained the confidence of both the European pharmaceutical industry and the various national regulatory authorities within each EU member state.

1. GENERAL INTRODUCTION

This chapter aims to overview the regulatory control of (bio) pharmaceutical products within the European Union. Prior to discussing these core issues, it may be beneficial to acquaint the reader with some basic information regarding the EU and, particularly, to provide a summary overview of European pharmaceutical law. These issues are therefore discussed over the first few pages of this Chapter. The reader is then acquainted with the operational aspects of the EMEA and the new drug approval system, particularly as it applies to products of biotechnology. Subsequently, the performance of the EMEA to date is evaluated, and a summary overview of the actual biopharmaceutical products approved thus far under the new system is presented.

1.1 The European Union

The founding principles of the European Union are enshrined in the Treaty of Rome, initially adopted in 1957 by six countries (Germany, France, Italy, Netherlands, Belgium, Luxembourg). Subsequently, Ireland, Denmark and the UK joined in 1973 and today the EU is comprised of 15 member states and boasts a total population of 371 million citizens (1). Along with the USA and Japan, the EU forms one of the three major global markets for ethical pharmaceutical sales, with a market value estimated to be in excess of £40 billion (Table 27). In addition, upwards of 13 other countries (mainly newly formed eastern European states) have expressed an interest in joining the EU. Should they be successful in doing so, the total EU population would increase by almost a further 200 million, making it the most powerful commercial block in the world.

Table 27. Profile of annual sales value of prescription drugs within the major world pharmaceutical markets

World region	Population (millions)	Market size ($ billions)	% world sales	Average cost per capita ($)
E.U.	371	68	32	183
USA	249	66	31	265
Japan	124	47	21	379

The main official bodies of the European Union are listed in Table 28. In the context of pharmaceutical regulatory affairs, the European Commission is probably the single most important body.

1.2 The European Commission and European Law

The European Commission consists of 20 commissioners (at least one of whom is nominated by each member state, with larger states appointing more than one), and a supporting civil service of several thousand. It is organized into 23 directorates general (DGs). The Commission has numerous responsibilities, including the proposal of new laws (including pharmaceutical law) and other measures required to ensure implementation of EU policy. The

Commission is also charged with ensuring that these laws are actually being enforced in all Member States.

Table 28. Some of the major institutions of the EU, and their responsibilities. Additional information available on the Internet may be accessed at http://europa.eu.int/inst-en.htm

Institution	Role within the EU
The European Commision	Proposes new EU legislation and ensures that EU laws are upheld in individual member countries
The Council of Ministers ('The Council' or 'The Council of the European Union')	Consisting of ministers from each Member State, the Council legislates for the Union. It also sets the Union's political objectives and helps co-ordinate their national policies
The European Parliament	Consisting of 626 directly elected members from all Member States, the Parliament has a number of functions. It has some legislative power, along with the Council of Ministers. It approves the European budget. It is the President of the Parliament who signs the budget into law. It also functions to supervise and scrutinize the use of executive power divested in the main EU institutions.
The Court of Justice	The Court of Justice comprises of 15 judges and 9 advocates general. The Court bears responsibility to ensure that the law is observed with regard to the various treaties signed by EU Member States.
The Economic and Social Committee	The 222 members of the Economic and Social Committee are nominated by national governments. They may be categorized into one of 3 groups: employers, workers and special interest groups. The Committee acts in a consultative manner, issuing opinions on various matters referred to them by the Commission or the Council of Ministers.

The Commission can issue opinions of different legal standing. These are termed 'regulations' and 'directives', respectively. A regulation is a strong legal instrument which, when approved at EU level, must be enforced immediately and without alteration in all Member States. A directive is a looser legal term, with Member States being given a period of time (usually up to 18 months) to integrate the 'essence' of the law into their national laws.

In addition to regulations and directives, the Commission will occasionally issue non-binding decisions which include 'recommendations' and 'opinions'.

The drafting and eventual adoption of EU laws is a relatively complex process. Once the Commission identifies a need for a new law, it seeks advice from appropriate experts, and usually accepts submissions from interest groups. After evaluating the advice/information provided, the Commission draws up a draft of the proposed directive/regulation. This is then made publicly available by its publication in one of the official journals of the European Communities. The draft law is then considered by other EU bodies (e.g. the Council of Ministers, the Economic and Social Committee and the European Parliament), which often recommend alterations to the proposed legislation. The final draft of the legislation becomes law upon its adoption and approval by the Council of Ministers.

1.3   Pharmaceutical law within the EU

While the Treaty of Rome committed all signatories to a range of co-operation and harmonization measures, it largely deferred the organization of healthcare and related systems to individual Member States. As a consequence, each Member State developed its own set of pharmaceutical regulations, enforced by its own national regulatory agency (Table 29). Thus, in contrast to the approach taken in the USA, any company wishing to market a product within the EU was obliged to apply independently to each individual Member State for marketing authorization. Although many of the basic requirements for gaining marketing authorization in the various countries were very similar, different national regulatory authorities could develop somewhat different opinions regarding the relevant weighting of various dossier elements when assessing benefit : risk ratio. Thus, conformance of regulatory authority response to any single dossier across Europe was not guaranteed. Additionally, each country enforced its own language requirements, scale of fees, appeals mechanism, processing times, etc. Such national disharmony, in addition to being contrary to the European ideal, created an enormous duplication of effort, both for the pharmaceutical companies and the regulatory authorities. The need for the introduction of a substantial body of harmonizing European pharmaceutical legislation was obvious. The first European Pharmaceutical Directive was adopted in 1965 (Council Directive 65/65/EEC, January 26, 1965). Although subsequent directives were introduced in the mid-1970s, it was not until the 1980s that a determined attempt was made by the Commission to introduce harmonizing European pharmaceutical legislation. Since the mid-1980s, a total of 18 directives and 8 regulations have been adopted which relate specifically to medicinal products for human or veterinary use. In introducing this body of harmonizing legislation, the EU has constantly aimed to pursue two objectives: the protection of public health and the promotion of the free movement of pharmaceutical products within the EU.

Table 29. The National Drug Regulatory Authorities (Human Medicines) of the 15 EU Members States

Austria Bundesministerium fur Arbeit Gesundheit und Soziales (Federal Ministry for Labour, Health and Social Affairs) Stubenring 1 1030 Wien Tel. + 43 1711 72 Fax + 43 1714 9222	**Belgium** Ministere de la Sante Publique Vesaliusgebouw, Quartier Vesale Rijjksadministratief Centrum Cite Administrative de L'Etat B-1010, Brussels Tel. + 32 2210 49 24 Fax + 32 2210 49 35
Denmark Danish Medicines Agency Frederikssundsveg 378 DK-2700 Bronshoj Tel. + 45 4488 9111 Fax + 45 4494 0237	**Finland** National Agency for Medicines Siltasaarenkatu 18A P.O. Box 278 Fin-00531 Helsinki Tel. + 358 0 396 2115 Fax + 358 0 714 469
France Agence du Medicament 143/147, Boulevard Anatole France 93285 Saint-denis Cedex Tel. + 33 1 48 13 20 00 Fax + 33 1 48 13 20 98	**Germany** Bundesinstitut fur Arzneimittel und Medizinprodukte (BfArM; Federal Institute for drugs and Medical Devices) Seestrasse 10, D-13353, Berlin Tel. + 49 30 4548 3200 Fax + 49 30 4548 3332
Greece National Drug Organization (E.O.F.) 284 Mesogion Ave GR-15562 Cholargos Athens. Tel. + 30 1 6549 500 Fax + 30 1 6545 535	**Ireland** Irish Medicines Board Earlsfort Centre Earlsfort Terrace Dublin 2 Tel. + 353 1 676 4971 Fax + 353 1 676 7836
Italy Dipartimento per la Valutazione dei Mediciniali e la Farmacovigllanza Ministero della Sanita Viale Civilita Romana 7 I-00144 Rome Tel. : 39 6 5994 3666 Fax : 39 6 5994 3365	**Luxembourg** Laboratoire Nationale de Sante 1A, rue A Lumiere B.P. 1102 Tel. : 352 478 5590 Fax : 352 22 4458

The Netherlands
Medicines Evaluation Board
P.O. Box 5811
2280 HV Rijswijk
Tel. : 31 70 3407 152
Fax : 31 70 3405 155

Spain
Direccion General de Farmacia y Productos Sanitarios
Ministerio de Sanidad y Consumo
Paseo del Prado 18-20
Planta 9, 28071, Madrid
Tel. : 34 1 596 40 68
Fax : 34 1 596 40 69

U.K.
Medicines Control Agency (MCA)
Market Towers
1 Nine Elms Lane
London SW8 5NQ
Tel. : 44 171 273 0100
Fax : 44 171 273 0548

Portugal
Instituto Nacional Da Farmacia E Do Medicamento (INFARMED)
Parque de Saude de Lisboa
Avenida do Brasil, 53
1700 Lisboa
Tel : 351 1790 8500
Fax : 351 1795 9116

Sweden
Medical Products Agency
P.O. Box 26
S-751 03, Uppsala
Tel. : 46 18 17 46 00
Fax : 46 18 54 85 66

1.4 The rules governing medicinal products in the European Union

In addition to publishing the full text of the actual binding legislation (regulations and directives), the European Commission has facilitated the preparation and publication of several guides, largely designed to assist the pharmaceutical industry and other interested parties interpret/meet the requirements of the legislation. The legislation, along with these guides/guidelines are published in a series of volumes entitled "The rules governing medicinal products in the European Union" (Table 30)(2). Although these volumes were originally published over a period of several years, they have recently been revised and updated, and a 1998 edition of each is now available. They represent an essential source of reference material for all personnel who wish to manufacture and/or market pharmaceutical products within the EU, and they may be purchased from the Office for Official Publications of the European Union in Luxembourg. Sales outlets are also present in all EU Member States. Within the USA, these publications may be purchased from Bernan Associates, MD. (Toll free telephone: 800 274 44 47). In Japan, they may be purchased from PSI Japan, Tokyo (Telephone 813 32 34 69 21). As is evident from Table 30, volumes 1 and 5 contain the text of all European directives and regulations relating to medicinal products destined for human and veterinary use, respectively.

Table 30. The 9 volumes which comprise the rules governing medicinal products in the European Union. Refer to text for specific detail

Volume Number	Details
Volume 1	Pharmaceutical legislation: medicinal products for human use
Volume 2	Notice to applicants: medicinal products for human use
Volume 3	Guidelines: medicinal products for human use
Volume 4	Good manufacturing practices: medicinal products for human and veterinary use
Volume 5	Pharmaceutical legislation: veterinary medicinal products
Volume 6	Notice to applicants: veterinary medicinal products
Volume 7	Guidelines: veterinary medicinal products
Volume 8	Maximum residue limits: veterinary medicinal products
Volume 9	Pharmacovigilance: medicinal products for human use and veterinary medicinal products

Volume 2 represents a guide to those seeking an EU marketing authorization for a medicinal product for human use. As such, it holds particular significance in the context of this chapter. The volume is presented in two parts. Volume 2A also contains information relating to the procedures that apply when seeking variations to a marketing authorization. Volume 2B seeks to provide practical guidance for the compilation of dossiers submitted when applying for European marketing authorizations.

Volume 3 aims to provide practical guidance as to how EU regulatory authorities are likely to interpret and apply EU directives concerning the demonstration of safety, quality and efficacy of any medicinal product. This volume facilitates the preparation of applications for marketing authorization within the EU. The volume is presented as 3 parts. Volume 3A provides quality and biotechnology guidelines. The biotechnology guidelines (Table 31) represent an important source of regulatory guidance to those seeking marketing authorization for any biopharmaceutical.

Table 31. EU guidelines of specific relevance to biopharmaceutical products, as presented in Volume 3A of the rules governing medicinal products in the European Union

- Production and quality control of medicinal products derived by recombinant DNA technology
- Quality of biotechnological products: analysis of the expression construct in cells used for production of rDNA derived protein products
- Production and quality control of cytokine products derived by biotechnological processes
- Production and quality control of monoclonal antibodies
- Quality of biotechnological products: stability testing of biotechnological/biological products
- Gene therapy; product quality aspects in the production of vectors and genetically modified somatic cells
- Use of transgenic animals in the manufacture of biological medicinal products for human use
- Virus validation studies: the design contribution and interpretation of studies validating the inactivation and removal of viruses
- Validation of virus removal/inactivation procedures: choice of viruses
- Minimising the risk of transmitting agents causing spongiform encephalopathy via medicinal products
- Tests on samples of biological origin
- Plasma-derived medicinal products
- Plasma pool testing
- Harmonization of requirements for influenza vaccines
- Allergen products
- Assessing the efficacy and safety of human plasma-derived factor VIII:C and factor IX:C products in clinical trials in haemophiliacs, before and after authorization.
- Assessing the efficacy and safety of normal intravenous immunoglobulin products for marketing authorizations

Volume 3B ('Safety, Environment and Information') provides pharmaco-toxicological and environmental guidelines, as well as guidelines relating to information provided on medicinal products (e.g. summary of product characteristics, some information on user/package leaflets and information detailing the rapid alert systems (RAS) in pharmacovigilance). Volume 3C relates to efficacy guidelines and thus largely concerns itself with guidance on conduct and interpretation of clinical trials. The other volumes provide a guide to good manufacturing practice of medicinal products (Volume 4), guidelines relating to veterinary medicinal products (Volumes 6-8) and pharmacovigilance (Volume 9).

2. THE EMEA AND THE EU'S NEW DRUG REGISTRATION SYSTEM

By the early 1990s, sufficient pharmaceutical directives and regulations had been implemented to render the harmonization of pharmaceutical legislation in all EU states a practical reality. This facilitated the implementation of a common, EU-wide system for the authorization (and subsequent supervision) of medicinal products. The European Medicines Evaluation Agency (EMEA) was set up to co-ordinate and manage this new system.(3-6) Essentially, there are now two routes by which EU-wide marketing authorization can be obtained. The centralized route entails direct submission of the application to the EMEA, which then arranges to have the application evaluated. Under the 'decentralized' procedure ('mutual recognition'), the marketing authorization application is normally forwarded to the national regulatory authority of one Member State (Table 29). If, after evaluation, national authorization is granted in that Member State, the mutual recognition process allows for the extension of the marketing authorization to one or more additional Member States. If disputes arise (i.e. mutual recognition breaks down), then the EMEA will arbitrate. Purely national authorizations are still available for medicinal products to be marketed in a single Member State.

2.1 Structure and role of the EMEA

The EMEA was established (on paper) in 1993 by EU Council Regulation, EEC 2309/93. It is located in Canary Wharf, London; was inaugurated on January 25th, 1995 and became operational on February 1st of that year. Contact details for the agency are as follows: European Medicines Evaluation Agency, 7 Westferry Circus, Canary Wharf, London E14 4HB. Tel: +44 171 418 8400. Fax: +44 171 418 8416. E-mail: mail@emea.eudra.org. The Agency is composed of 4 main elements:

- A Management Board, which serves as the EMEA's governing body. It appoints the Executive Director, is responsible for budgetary matters, approves EMEA work programmes and monitors the performance of the agency. The Management Board generally meets 4-5 times a year and it consists of two representatives per EU Member State, two representatives of the European Commission and two representatives of the European Parliament.

- Two scientific committees: the Committee for Proprietary Medicinal Products (CPMP) and the Committee for Veterinary Medicinal Products (CVMP). These are responsible for formulating the EMEA's scientific opinion on marketing authorizaton applications relating to human and veterinary medicines, respectively. In addition to providing scientific assessments/opinions on such marketing applications, the members of the committees also facilitate technical co-ordination between the EMEA and the national regulatory authorities.

  Each committee consists of 30 members (two representatives from each EU State), each of which serves a (renewable) term of 3 years. All members are expert in medicinal product evaluation and most are actually employees of national regulatory authorities of the Member States. The committees each convene once a month at the EMEA offices, and a typical meeting lasts 3-5 days.

- The EMEA Secretariat largely consists of the personnel based permanently at EMEA headquarters of which there are approximately 150 (Table 32). It is headed by the EMEA's Executive Director, Fernand Sauer. The Secretariat provides technical and administrative support for the two scientific committees and their working parties (discussed later), and ensures appropriate co-ordination between them.

Table 32. Budgetary and staff profile of the EMEA, 1995-1997
While the European Commission directly contributes to the financing of the EMEA, the level of self-funding in the form of evaluation fees is continually increasing. Currently, the fee charged for assessment of a biotechnology-derived product (for a single indication) via the EMEA's centralized procedure is 140,000 ECU.

Year	Staff numbers	Budget (ECUs). Amount raised via evaluation fees given in brackets
1995	50	14,412,000 (4,000,000)
1996	100	22,550,000 (8,600,000)
1997	143	28,530,000 (14,000,000)

The target objectives of the EMEA are clearly defined by Council regulation EEC No. 2309/93. These responsibilities are reflected in the EMEA Mission Statement which states that the EMEA's goal is to contribute to the protection and promotion of public and animal health by:

- Mobilizing scientific resources from throughout the European Union to provide high quality evaluation of medicinal products, to advise on research and development programmes and to provide useful and clear information to users and health professionals.

- Developing efficient and transparent procedures to allow timely access by users to innovative medicines through a single European Marketing Authorization.

- Controlling the safety of medicines for humans and animals, particularly through a pharmacovigilance network and the establishment of safe limits for residues in food-producing animals.

Much of the EMEA's work is thus concerned with facilitating assessment of Marketing Authorization applications via either the centralized or decentralized procedures, both of which are now described in detail.

2.2 Centralized procedure for Marketing Authorization applications

EU Marketing Authorization applications for all biopharmaceutical products must be assessed via the centralized procedure (Figure 28). Other innovative medicines (mainly New Chemical Entities, NCEs) can also be assessed via this route, although it is not mandatory for such products. The centralized evaluation procedure entails direct submission of the application to the EMEA. For centralized assessment purposes, products of biotechnology are classified as 'List A', while other innovative products are classified as 'List B'.

Upon receipt of any such Marketing Authorization application, a Project Manager from the EMEA Secretariat is appointed. The Project Manager will help co-ordinate the assessment procedure of the application, will liaise with the CPMP and will serve as a contact point for the sponsoring company. The full application will initially be scanned to ensure that it is complete and presented in the proper format (as, for example, laid down in Volume 2 of the Rules Governing Medicinal Products in the European Union). This validation procedure will take in the order of 7-10 days. The appropriate fee must also accompany the application.

Once validation has been satisfactorily completed, the application will be presented at the next meeting of the CPMP (or CVMP, if the product is a veterinary one). The CPMP then appoints one of its members to be a 'rapporteur' for that application. The rapporteur serves to organize and champion technical evaluation of the application (i.e. evaluation of safety, efficacy and quality of the product). The choice of rapporteur is made by taking into account any preferences listed by the submitting company, as well as the availability and technical expertise of the CPMP members. The Committee can also appoint a second member to act as co-rapporteur. The rapporteur (and co-rapporteur, if appointed) then arranges to have the authorization assessed in detail). Usually, this is undertaken in the (co)rapportueurs home national regulatory authority. The EMEA has entered into partnership agreements with the various national regulatory authorities of the EU Member States, and contracts have been signed for the provision of such evaluation services.

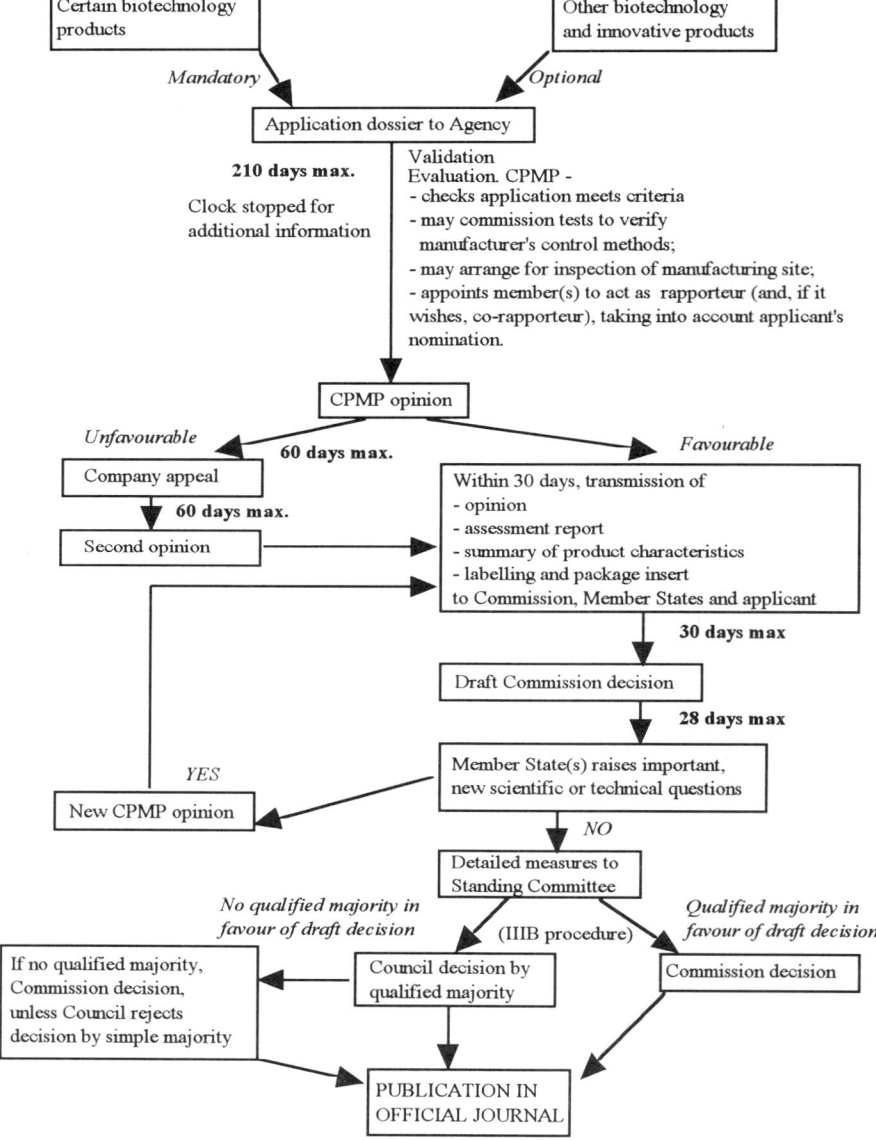

Figure 28. The EMEA's centralized procedure for approval of medicinal products within the EU. (Reproduced from Eur. J. Clin. Pharmacol. (1995), 47: p. 474, with kind permission of Springer-Verlag)

Upon completion of evaluation, the rapporteur (and co-rapporteur) each draw up a report and recommendation and present these at the next relevant CPMP (or CVMP) meeting. After discussion, the CPMP will issue a decision (an 'opinion'). Under EMEA operational procedures, the 'opinion' should be issued within 210 days of receipt of the Marketing Authorization application. However, if questions are raised during evaluation which require attention by the company, this 210 day 'clock' stops until the questions are resolved to the evaluators' satisfaction.

The EMEA's scientific 'opinion' is then conveyed to the European Commission (the Commission, not the EMEA is the body with the authority to allow issue of a marketing licence). The Commission officials can raise any supplementary questions they have at this stage, and the EU Member States are afforded a final opportunity to raise any scientific/technical misgivings they may have. The Commission then issues a final decision (this is invariably in agreement with the EMEA 'opinion').

The Commisson decision is then published in an official EU journal and is transmitted to the sponsoring company. The maximum time-scale allowed from issue of EMEA opinion to issue of the final Commission decision is 90 days. Thus (excluding 'clock' stops), the entire centralized evaluation procedure should be completed within 300 days of receipt of the application.

2.3 Mutual recognition

In addition to the centralized route, Market Authorization applications may also be assessed via the mutual recognition procedure. This is also the route that other medicinal products (e.g. generics, new line extensions, etc.) must take. (Biopharmaceuticals are amongst the only product type excluded from being evaluated by this decentralized route).

The mutual recognition route entails the initial submission of a Marketing Authorization application to a single national (EU) regulatory authority (Figure 29). That national regulatory authority then assesses the application within a period of 210 days. If successful, natonal authorization is granted in that state and the company can apply via 'mutual recognition' to extend the Marketing Authorization to additional Member States. During this 'bilateral' phase, the marketing application (along with the assessment report compiled by the national regulatory authority of the Member State where the application was first filed) is conveyed to the national regulatory authorities of other Member States. If, after consideration of the data provided, the additional Member States concur with the positive decision made by the first Member State, then they too will issue national marketing licences. Although it can take longer, this bilaterial phase is usually undertaken within 60 days.

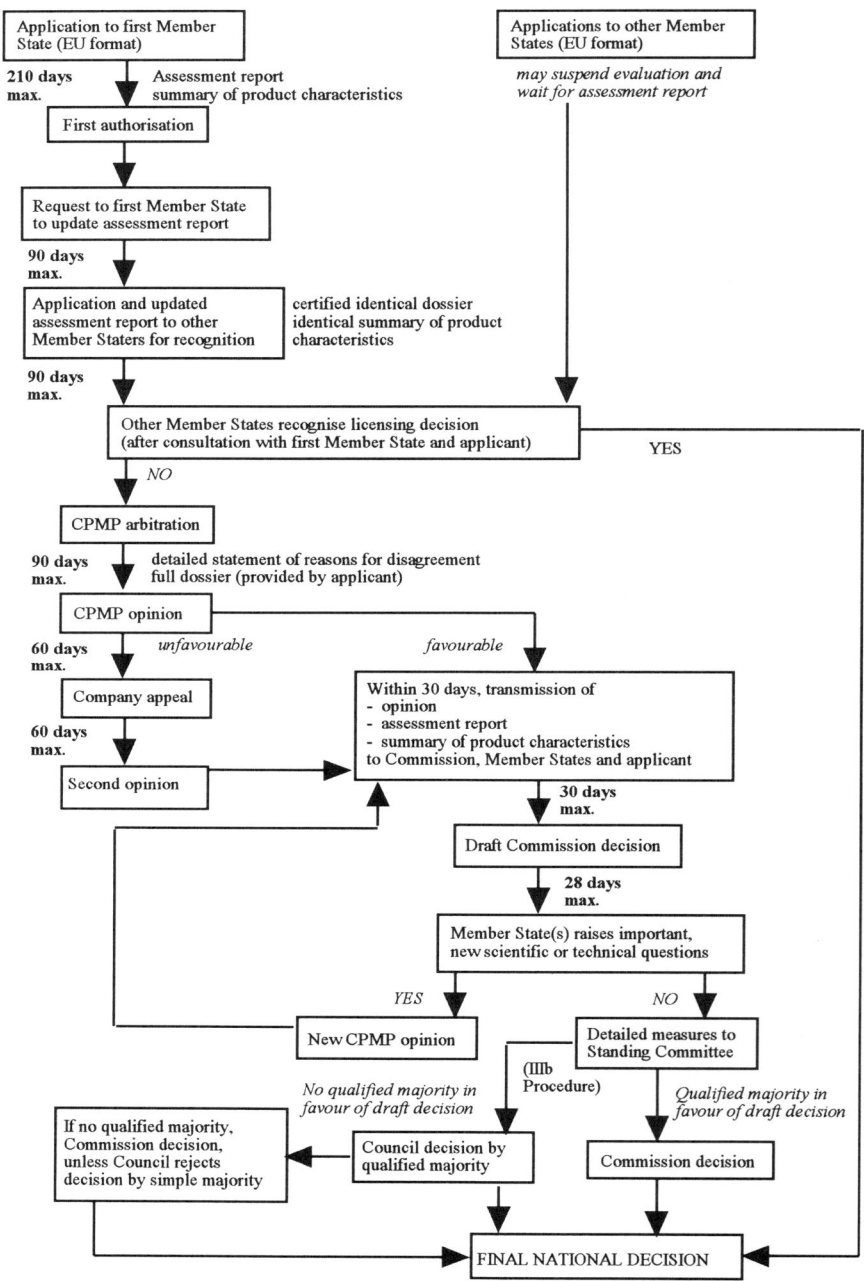

Figure 29. The EMEA's decentralized procedure for approval of medicinal products within the EU. (Reproduced from Eur. J. Clin. Pharmacol. (1995), 47: p.475, with kind permission of Springer-Verlag

The basis of such mutual recognition, of course, lies in the fact that pharmaceutical law - and hence Marketing Authorization requirements - are harmonized throughout the EU.

If, however, one or more states do not automatically accept the decision made by the original national regulatory authority, the dispute is brought to EMEA arbitration. Taking the EMEA view very much into account, the European Commission will then issue a final, binding decision.

2.4 The EMEA as an information provider

A key goal of the EMEA is that of transparency and the provision of information detailing its operations to interested parties. Information packs containing an EMEA directory and its most recent annual report are obtainable by contacting the agency directly. The EMEA internet home page (http://www.eudra.org/emea.html), however, provides the most comprehensive information. The home page is clearly and well laid out and houses in excess of 550 documents. It received over 4 million hits last year alone, and over 120,000 documents were downloaded from the site.

One of the most useful document types available via the EMEA's home page is the European Public Assessment Report (EPAR). Once any product evaluated via the centralized procedure gains Marketing Authorization, the EMEA will make available an EPAR detailing that product. The aim of the EPAR (which is generally about 35 pages long) is essentially to outline the reasons why the CPMP issued its positive opinion. In addition to providing background information on, for example, the submission of the dossier, how it was assessed and the general conditions of the Marketing Authorization granted, the EPAR houses a major section, termed 'scientific discussion'. As part of this section, an overview of the product's characteristics and how it is manufactured is presented. In addition, an overview of dossier content (e.g. the product's chemical/pharmaceutical/biological characteristics, its toxico-pharmacological characteristics, as well as clinical performance) is provided. The EPAR thus represents a valuable source of technical information detailing a product.

2.5 Additional EMEA activities

In addition to its role in co-ordinating the drug approval process, the EMEA also plays a central role in other facets of drug registration within the EU (7). It is, for example, responsible for co-ordinating national efforts with respect to pharmacovigilance. It is also involved in co-ordinating issues relating to the inspection of pharmaceutical manufacturing facilities in

Member States, although such inspections are still conducted by the individual home national regulatory authorities.

In conjunction with the European Union, the EMEA is also responsible for drafting 'soft' laws - non-binding guides designed to provide advice to the European pharmaceutical industry in the form of, for example, guidelines, SOPs, etc. Such additional EMEA activities are driven mainly by its two scientific committees, the CPMP and CVMP. Each committee is also supported by a number of specialist working parties which serve to provide additional advice on specific matters relating to the quality, efficacy and safety of medicinal products. There are presently 4 CPMP and 3 CVMP working parties, as well as a further joint CPMP/CVMP one (Table 33). The biotechnology working party advises the CPMP on specialist issues relating to the manufacture and control of products of biotechnology and other biological medicinal products, including immunological products, as well as products derived from blood or plasma. It normally meets about 10 times a year and plays a particularly prominent role in assisting the CPMP in drawing up the various biotechnology product guidelines published (Table 34).

Table 33. Working parties of the CPMP and CVMP. In addition to those listed, a joint CPMP/CVMP quality working party also exists

CPMP working parties	CVMP working parties
Pharmacovigilance working party	Efficacy working party
Biotechnology working party	Immunologicals working party
Efficacy working party	Pharmacovigilance working party
Safety working party	

In addition to the technical expertise of its own members, the CPMP and CVMP can draw upon the technical knowledge of a bank of European experts as required. A European list of over 2,100 experts with proven experience in the assessment of medicinal products has been compiled by the EMEA. Most of these are employees of national regulatory authorities. Such experts may be called upon to attend CPMP or CVMP meetings, or may be invited to sit on specific working parties.

Table 34. Guidelines relating to biotech-derived products which have been adopted (or released for consultation) by the CPMP during 1996/1997

Guideline	Status	Guideline	Status
Virus validation studies: the design, contribution and interpretation of studies validating the inactivation and removal of viruses	Adopted (Feb. 1996)	Quality of biotech products: viral safety evaluation of biotech products derived from cell lines of human or animal origin.	Adopted (March 1997)
Guidelines to assess efficiency and safety of normal intravenous immunoglobulin products for marketing authorizations	Adopted (Feb. 1996)	Quality of biotech/biological products: derivation and character-ization of cell substrates used for production of biotech/biological products	Adopted (Sept. 1997)
Efficiency and safety of human plasma-derived factor VIII.c and IX.c products in clinical trials in haemophiliacs before and after authorization	Adopted (Feb. 1996)	Pre-clinical safety evaluation of biotech-derived pharmaceuticals	Adopted (Sept. 1997)
Plasma-derived medicinal products	Adopted (March 1996)	Minimizing the risk of transmitting animal spongiform encephalapathy agents via medicinal products	Adoption pending
Allergen Products	Adopted (March 1996)	Pharmaceutical and biological aspects of combined vaccines	Consultation stage
Harmonization of requirements for influenza vaccines	Adopted (March 1997)	Plasma-derived medicinal products	Consultation stage
Core SPC for human immuno-globulin	Adopted (March 1997)		

3. NEW CENTRALIZED SYSTEM

December 31st, 1997 marked the end of the initial (3 year) introductory phase of the EMEA's existance. By that date, the CPMP had issued scientific opinions for 27 products of biotechnology, 23 of which had also been granted Marketing Authorization by the European Commission (Table 35). The first product of biotechnology to gain approval under the new system was Serono's Gonal F (recombinant follicle stimulating hormone). This product was evaluated by the CPMP in an active time of only 107 days. With only two exceptions, the CPMP completed evaluation of biotechnology products within the 210 day guideline. The average duration of evaluation of all such products during this period was 175 days. Lilly's insulin product, Liprolog, was evaluated in the shortest time (48 days). The duration of 'stop clocks' recorded during evaluation of biotech products varied widely. No 'clock stop' was recorded in the case of Liprolog while a clock stop of 398 days was recorded in the case of the Ciba Novartis Anticoagulant, Revasc.

The CPMP has proven very efficient in assessing Marketing Authorization applications for products of biotechnology. However, the total time (excluding clock stops) elapsing, between dossier submission and product approval, can be disappointingly long. After issue of the CPMP opinion, it generally takes the EMEA in the region of 40 days to complete the paperwork and transmit the decision to the European Commission. Furthermore, it usually takes the Commission in excess of 80 days to issue a final decision. A reduction in these time scales is desirable.

Table 35. Biopharmaceutical products which have succeeded in securing marketing approval within the EU via the EMEAs centralized procedure (1995-1997)

Product Brandname (non-proprietary name listed in brackets)	Therapeutic Indication	Company	a) Duration of active evaluation by CPMP b) Clock Stop	a) Total time; from commencement of EMEA evaluation to Commission decision (excludes clock stops) b) Date of final approval by commission
Gonal F (Follitropin-alpha) (Follicle Stimulating Hormone)	Treatment of Infertility	Serono Laboratoiries	a) 107 days b) 30 days	a) 262 days b) 20.10.95
Betaferon (Interferon β-1b)	Immunostimulation, multiple sclerosis	Schering AG	a) 138 days b) 55 days	a) 278 days b) 30.11.95

The EMEA and regulatory control of (bio)pharmaceuticals within the European Union

Product Brandname (non-proprietary name listed in brackets)	Therapeutic Indication	Company	a) Duration of active evaluation by CPMP b) Clock Stop	a) Total time; from commencement of EMEA evaluation to Commission decision (excludes clock stops) b) Date of final approval by commission
Novo Seven (Factor VII$_a$)	Treatment of some forms of haemophilia	Novo-Nordisk	a) 210 days b) 80 days	a) 338 days b) 23.02.96
Humalog (Insulin Lispro)	Treatment of diabetes mellitus	Lilly Industries	a) 245 days b) 81 days	a) 403 days b) 30.04.96
Puregon (Follitropin-beta) (Follicle stimulating hormone)	Treatment of infertility	Organon	a) 203 days b) 151 days	a) 336 days b) 03.05.96
Tritanrix - HB (A combined vaccine)	Vaccine against: Hepatitis B, Diphtheria, Tetanus, and Pertussis	SmithKline Beecham	a) 180 days b) 240 days	a) 324 days b) 19.07.96
CEA - Scan (Arcitumomab; a monoclonal antibody)	Diagnosis of colonic and rectal carcinoma	Immunomedics	a) 110 days b) 386 days	a) 255 days b) 04.10.96
Tecnemab K1 (Anti-melanoma mouse monoclonal fragments)	Diagnosis of cutaneous melanoma lesions	Sorin	a) 187 days b) 320 days	a) 292 days b) 05.09.96
Rapilysin (Retephase, i.e. tPA)	Treatment of acute myocardial infarction	Boehringer Mannheim	a) 204 days b) 83 days	a) 310 days b) 29.08.96
Ecokinase (Reteplase, i.e. tPA)	Treatment of acute myocardial infarction	Galenus Mannheim	a) 204 days b) 83 days	a) 310 days b) 29.08.96
Twinrix Adult (Combination vaccine)	Vaccine against hepatitis A and B	SmithKline Beecham	a) 197 days b) 83 days	a) 317 days b) 20.09.96
Indimacis 125 (Igovomab; i.e. a murine monoclonal antibody fragment)	Diagnosis of ovarian adeno-carcinoma	CIS-Bio International	a) 154 days b) 363 days	a) 278 days b) 04.10.96

Product Brandname (non-proprietary name listed in brackets)	Therapeutic Indication	Company	a) Duration of active evaluation by CPMP b) Clock Stop	a) Total time; from commencement of EMEA evaluation to Commission decision (excludes clock stops) b) Date of final approval by commission
LeukoScan (Sulesomab; i.e. a murine monoclonal antibody fragment)	Diagnostic imaging for infection/inflammation in bone of patients with suspected osteomyelitis	Immunomedics	a) 210 days b) 183 days	a) 337 days b) 14.02.97
Insuman (Human insulin)	Treatment of diabetes mellitus	Hoechst AG	a) 158 days b) 182 days	a) 260 days b) 21.02.97
Twinrix Paediatric (Combined vaccine)	Immunization against hepatitis A & B in children	SmithKline Beecham	a) 132 days b) 35 days	a) 230 days b) 10.02.97
Avonex (Interferon β)	Treatment of multiple sclerosis	Biogen	a) 216 days b) 307 days	a) 329 days b) 13.03.97
Refludan (Lepirudin; i.e. hirudin)	Anti-coagulation therapy for heparin associated thrombocytopenia	Behringwerke AG	a) 200 days b) 112 days	a) 310 days b) 13.03.97
Liprolog Bio Lysprol (Insulin lispro)	Treatment of diabetes mellitus	Lilly	a) 48 days b) 0 days	a) 191 days b) 07.05.97
Revasc (Desirudin; i.e. hirudin)	Prevention of deep venous thrombosis	Ciba Novartis Europharm Ltd	a) 181 days b) 398 days	a) 332 days b) 09.07.97
Neo Recormon (Epoetin beta; i.e. erythropoietin)	Treatment/prevention of anaemia under various conditions	Boehringer-Mannheim	a) 209 days b) 140 days	a) 482 days b) 16.07.97
Infanrix-HepB (Combined vaccine)	Active immunization of infants against diphtheria, tetanus, pertussis & hepatitis B	SmithKline Beecham	a) 199 days b) 217 days	a) 313 days b) 30.07.97

Product Brandname (non-proprietary name listed in brackets)	Therapeutic Indication	Company	a) Duration of active evaluation by CPMP b) Clock Stop	a) Total time; from commencement of EMEA evaluation to Commission decision (excludes clock stops) b) Date of final approval by commission
Benefix (Nonacog-α; i.e. coagulation factor IX)	Haemophilia B	Genetics Institute	a) 162 days b) 55 days	a) 283 days b) 27.08.97
Cerezyme (imiglucerase, i.e. recombinant macrophage-targeted beta-glucocerebrosidase)	Type 1 Gaucher's disease	Genzyme B.V.	a) 175 days b) 30 days	a) 274 days b) 17.11.97

4. CONCLUSION

Thus far, the EU's new drug registration system appears to be working well. While the EMEA co-ordinates this sytem, it is obviously dependent upon the co-operation and good will of the national regulatory authorities. A recent EMEA/EFPIA (European Federation of Pharmaceutical Industries Association) survey found that all the national regulatory authorities were satisfied with the operation of the centralized procedure. The same survey also found that the majority of companies who submitted applications evaluated via the centralized procedure, were also satisfied with this new system and the EMEA's performance. Although some teething problems remain, the EMEA has proven receptive to taking on board criticisms and have been found willing to re-evaluate aspects of their operation in order to further improve their service. The EMEA is also making its mark on a global scale. It continues to provide technical support to the activities of ICH, and participates in on-going discussions concerning the development of a common technical dossier (the so-called global application dossier). Thus far, Europe's new drug regulatory system - and the agency set up to manage this sytem - appears to be functioning quite effectively.

REFERENCES

1. http://europa.eu.int (Internet home page of Europa, the European Union's server).
2. The rules governing medicinal products in the European Union, Volumes 1-9. 1998 ed. European Commission, Directorate General III (Industry, Pharmaceuticals and Cosmetics).
3. Jeffries, D. & Jones, K. (1995) EMEA and the new pharmaceutical procedures for Europe. Eur. J. Clin. Pharmacol. 47, 471-476.
4. Sauer, F. (1997) A new and fast drug approval system in Europe. Drug Inform. J. 31, 1-6.
5. http://www.eudra.org/emea.html (Internet home page of the EMEA).
6. http://www.eudra.org/home.html (Internet home page of network services for the European Union Pharmaceutical Regulatory sector).
7. The European agency for the evaluation of medicinal products. First, second and third general reports (1995, 1996 and 1997, respectively). Office for Official Publications of the European Communities, L-2985, Luxembourg.

Chapter 13

Biopharmaceutical Validation: an overview

Stephen Slater
Raytheon Engineers and Constructors

Key words: Validation, Qualification, Quality, GMP

Abstract: General validation principles are explained. Documentation requirements and the practical organisation of a validation project are described. The specific considerations for qualifying biopharmaceutical equipment and processes are described.

1. INTRODUCTION

In this chapter we will examine the fundamentals of validation within a biopharmaceutical facility. Areas covered will include validation definitions and concepts, the documentation associated with validation and organising for a validation project.

Validation is defined as "the process of providing documented evidence that provides a high degree of assurance that a process will consistently produce a product meeting its pre-determined specifications and quality attributes." It is performed in order that a manufacturing company can be assured of the consistent quality of their products, improve its performance and meet its regulatory obligations. Validation is required in order to meet the regulatory requirement embodied in the concept Good Manufacturing Practice (GMP) (1-3). GMP is continuously evolving and is not only manifest in written guidelines but also comprises of best industry practice and regulatory expectations that may not be formally published. When a company intends to market a product in a specific country, the requirements of that countries regulatory body must be met. A validated process is a requirement of almost all regulatory agencies. The Food and Drugs Administration (FDA) of the United States and the European Authorities (amongst others) enforce these requirements by physical inspection and the review of documentation. If these agencies are not satisfied with the validation status of the facility then various punitive and restrictive measures can be put in place cumulating in a refusal to allow the marketing of product in that country.

Validation should always be considered as part of the company's overall quality structure. One of the major responsibilities of the Quality Assurance (QA) department is to authorise product release. They should only do this if they are confident that anything that could affect the quality of the product is working properly and under control. In many facilities, the Validation Department reports into the QA department. Even in instances where this is not the organisational structure, QA must still have the final approval on all validation documents. However, because of the nature of validation work, validation personnel will need to be drawn from many disciplines.

The validation of biopharmaceutical facilities is no different in concept and philosophy than the validation of any other pharmaceutical plant. The main differences are due to the type of equipment and processes that are typically employed.

The initial part of this chapter describes the basic fundamentals of validation that are generally applicable, before moving onto the more specific requirements of the biopharmaceutical industry.

As with other specialist areas, some specific words and phrases have come into common usage. Validation personnel can come from engineering or scientific backgrounds and some of the terms used originate from within the engineering field. These are discussed below to facilitate understanding of the remainder of this chapter.

2. TERMINOLOGY AND DOCUMENTATION

The following three items are the key documents that control and describe the validation activities. These are followed by some definitions of common terms used in the validation field.

2.1 The Validation Master Plan (VMP)

This is the most significant single document in any validation programme. It serves as the principal guide for the validation programme, defines the responsibilities of those engaged in validation activities, identifies all items which are subject to validation and describes the nature and extent of testing on each item.

2.2 A Validation Protocol

A document that describes the item to be qualified, the tests and checks to be performed and the results that are expected to be obtained. Often the protocol will incorporate test sheets or sections for recording data. In this way once the protocol has been executed the document constitutes the record of results and conclusions.

2.3 A Summary Report

A document produced after testing has been completed summarising the validation activities, the validation data and the conclusions drawn.

2.4 Acceptance Criteria

Specifications assigned, before undertaking testing, to allow evaluation of test results to demonstrate compliance with a test phase or delivery requirement. Acceptance criteria should be relevant to the impact on product quality. They may or may not be the same as design specifications. However, they cannot ever exceed the design specifications, as this would imply that the required criteria exceed the capability of the equipment.

2.5 Approval

A critical activity that applies to all validation documents. All protocols must be "pre-approved" (approved prior to use). The completed protocol should again be approved. Review and approval must be carried out by competent individuals normally from Manufacturing, Engineering and Quality Assurance.

2.6 Change Control

A formal system by which qualified representatives of appropriate disciplines review proposed or actual changes that might affect a validated status. The intent is to determine the need for action that would ensure and document that the system is maintained in a validated state.

2.7 Commissioning

An engineering term that covers all aspects of bringing a system or sub-system to a position where it is regarded as adjusted for operational use and ready for validation.

2.8 Concurrent Validation

Validation carried out while routine manufacture of products is also taking place.

2.9 Mechanical Completion

An engineering term that covers the basic activities that are performed after installation is complete. This usually includes inspection of pipes and ductwork, flushing and blowing lines clean, reinstatement of instruments, installation of final filters, chemical cleaning and passivation, general cleaning and debris removal.

2.10 Piping & Instrument Diagram (P & ID)

Engineering schematic drawings that provide details of the interrelationship of equipment, utilities, material flows, plant controls and alarms. The P & ID also provide the reference for each tag or label used for identification.

2.11 Prospective Validation

Validation carried out prior to the routine manufacture of products.

2.12 Qualification

The performance of individual elements of a validation programme when one element is complete, the item is qualified, when all elements are satisfactorily completed the equipment and process is validated.

2.13 Retrospective Validation

Validation carried out by a review of historical records. Not recommended unless the alternatives are impractical.

2.14 Re-Validation

The repeat of some of the initial validation to provide an assurance that changes in the process/process environment whether introduced intentionally or unintentionally do not adversely affect process characteristics and product quality.

3. VALIDATION ACTIVITIES

Not all items within a pharmaceutical facility will require validation. Only items that may impact on the product quality need to be validated. Only systems and processes that are used to manufacture or test products for sale or for clinical trial require validation. In addition the requirements of the different regulatory authorities vary to some extent. This will impact the depth and type of testing. Where validation is required, it is necessary to examine the process for producing the product, the equipment on which the process will run, the utilities or services that may come into contact with the product and the environment in which the product is manufactured. To simplify and rationalise the task of validation, it is important to start at the basic building blocks before tackling the complexities of the process. This will prevent the situation where a product fails to meet the required quality without a clear understanding of the likely causes. For this reason, validation is generally split into five distinct and sequential activity types:

3.1 Design Qualification (DQ)

This refers to the auditing of the design of a facility, equipment and utilities to ensure that the items are compliant with the specifications, and capable of meeting GMP requirements.

This activity will only apply to new items that are in the design or procurement stage of the project. It may consist of several audits carried out at critical points in the design phase. It should address specific and detailed requirements that may have an effect on product quality for each item. The requirements will be different depending on the item and the type of product and the stage in the process. For example a parenteral product that uses water in the final formulation will require water of a different quality from an oral product that uses water for washing equipment. The type of equipment used to produce the water may also be different, as may the specifications for such equipment. The DQ process should challenge the design to ensure that the system is capable of producing product meeting the stated quality requirements. For water systems the objective may be to ensure that the system is designed to be able to produce water meeting United States Pharmacopoeia and European Pharmacopoeia standards with particular emphasis on ensuring that the design does not encourage microbial proliferation (4, 5). In this case DQ should include evaluating the technology used, examining the proposed specifications for materials, surface finishes, type of valves, monitoring instruments, checking the overall system design, operating temperature and sanitisation methodology. It can be carried out in a number of ways. It often comprises of individual reviews of equipment specifications combined with assembling a team to audit aspects of design within a formal meeting where the designers are present. As with all validation activities it is important that records are kept to prove that the design was subject to qualified review. These records do not have to be lengthy or complex. Often a brief audit report or a validation/QA approval signature on a specification is all that is required.

3.2   Installation Qualification (IQ)

This refers to the performance and documentation of checks and tests to ensure that equipment for use has been appropriately selected, and correctly installed in accordance with established specifications.

To perform IQ properly, it is important to examine purchasing documentation, as well as documents for operating and maintaining equipment, As-built drawings, certificates of calibration and certificates of material compliance. Examination of the equipment should be carried out to verify in particular that items that may compromise product quality are correctly installed, have the specified surface finish and are made of the specified materials. Instruments should be calibrated and tagged. All services should be of the correct capacity and connected in the specified manner. Any control systems hardware and software should be checked to ensure compliance with the Functional Design Specifications (FDS). IQ cannot be carried out without detailed information on the acceptance criteria for each critical component, each critical instrument and all product contact materials. Physical checks should be made to confirm that the item meets the acceptance criteria. For example, a simple vessel may include an agitator, a vent filter, several valves and pressure and temperature monitoring instruments. In this instance the acceptance criteria should include the material specifications of the vessel, the valves (including seals), the agitator and filter (including filter housing), the finish for the internal vessel surface, the type of valve, the filter rating, the pressure rating for the vessel, the specified lubricant (if potentially in product contact) and so on. Instruments should be classified as critical (e.g. used to monitor or control the process) or non-critical (e.g. "guide" instruments used by maintenance personnel). The instrument requirements with respect to the process should be known. Therefore instrument acceptance criteria should include range, tolerance and requirement or otherwise for hygienic design.

The actual process of IQ will consist of obtaining documentation (specifications, data sheets, user requirements, P & IDs etc.) in order to write the protocol and then writing and approving the protocol and finally executing the protocol. Field execution can usually commence for new items after mechanical completion and concurrently with commissioning. It involves:
- Checking the physical installation against the P & ID and marking up the P & ID to indicate any discrepancies
- Reviewing documentation (including mill certificates, calibration certificates, raw data) to ensure that the specifications for material, calibration and pressure testing have been met
- Checking the nameplates on items to ensure that the requirements with regard to size, capacity, rating etc. are consistent with the order documentation
- Recording manufacturers' name and serial numbers,
- Checking that services (compressed air, gases, power, water) are connected to the item and that the connection size, connection material and specifications (temperature, pressure) are as specified
- Checking the position and type of instrument
- Checking that the equipment has been entered into a Planned Maintenance and routine calibration system.

3.3   Operational Qualification (OQ)

This refers to the performance and documentation of tests to verify that the system or sub-system performs as intended throughout all anticipated operating ranges. In general, OQ does not check the functioning of equipment with a load unless it is the only sensible way to evaluate operation. For example, OQ would check that an agitator rotates in the correct direction, at the specified speed and in compliance with the noise specifications. It would not check that the agitator was capable of mixing a specified product to a defined homogeneity. This is carried out at the next qualification stage. The objective of OQ is to check that each component works reproducibly and in accordance with the user requirements. It is, therefore, paramount that OQ only takes place when the system has been satisfactorily commissioned. Attempting to qualify equipment that has not been commissioned will inevitably result in a series of failures and adjustments having to be made. It is possible to integrate commissioning and validation (by using the same resources and using adequately controlled and documented commissioning procedures) but this does not mean that commissioning should not be performed. For example, in a classified environment, it is important to ensure the air balance has been completed (a commissioning exercise) before checking for compliance with pressure differential acceptance criteria (a validation requirement). Some typical OQ expectations for a classified environment would include acceptance criteria for air change rates, differential pressure, airflow patterns, particle count at rest, temperature, humidity, induction leak testing and recovery rate. It is important to establish the acceptance criteria with respect to the process and specific regulatory guidelines.

In order to demonstrate reproducibility it is normal practice to perform a number of the more critical tests several times.

Whilst, for most items, it is preferable for practical reasons to keep IQ and OQ separate, for simpler items, such as incubators, refrigerators and freezers, it may be desirable to combine the installation and operational qualification. For most new items of process equipment and utilities, the time interval between the earliest possible date for IQ and the earliest possible date for OQ makes it sensible to keep the activities distinct.

3.4 Performance Qualification (PQ)

This refers to the documented demonstration that equipment and processes operate as required at the normal operating limits of critical parameters using actual product or placebo. PQ is the interface between equipment qualification and the process. The focus is on proving that the equipment is suitable for a specific manufacturing process. The emphasis therefore shifts to observing the effect of the equipment on the product. A great deal of PQ work will look at the compliance of product (or product contact utility) to regulatory or in house specifications. PQ tests on product contact compressed air, for example, usually include checking for particulates and microbes. It should also include other "quality of product tests" such as the presence of hydrocarbons and the moisture level unless these tests have been included in the OQ. For items that require thermal mapping (autoclaves, steam-in-place systems, temperature-controlled storage), it is the normal practice at OQ to determine temperature distribution and cold spots without a load and then at PQ to evaluate the effect of standard loads on the equipment performance.

The nature of PQ demands that many specialists become involved at this stage. This includes laboratory expertise, operators trained in sampling techniques and specialists trained in the use of thermal mapping equipment. It is also critical at this point (if not before), that standard operating procedures on the operation of the equipment, the sampling procedures and the supporting laboratory procedures are available and approved. This is to ensure that all tests and equipment operations are performed in a standard and reproducible manner.

3.5 Process Validation (PV)

This refers to the documented demonstration that processes operate as required using normal operating conditions during the actual manufacture of product. This is typically performed as the manufacture of three consecutive batches of product; all meeting the product release specifications and other pre determined acceptance criteria. In practical terms the PV is performed after all systems involved in the process have been satisfactorily qualified. The batch records should be used to document the steps carried out. Inevitably more extensive monitoring of product and environments will be carried out than would be the normal practice during routine operations. It is important to include all parts of the production process in PV including equipment preparation, manufacture, sampling and filling, packaging and sterilisation activities. It is not necessary to include raw laboratory data in the process validation documentation, although of course they should be made available for regulatory review.

Other validation activities that are distinct from the above are described below.

3.6 Methods Validation

The validation of laboratory methods is a regulatory requirement for those methods that are non-compendial. That is, if the method is not an official procedure stated in the United States Pharmacopoeia or the European Pharmacopoeia (3, 4), then validation should be performed. Naturally most companies would choose to use a compendial method where available. If an alternative method is used then equivalence or superiority of the preferred method must be demonstrated. For compendial methods the company must still demonstrate that the method works under the specific conditions of use. The International Conference on Harmonisation (ICH) has published guidelines on analytical method validation (6). In general, depending on the specific test, method validation seeks to verify the test's accuracy, precision, specificity, detection and quantitation limits, linearity and range in order to demonstrate that the procedure is suitable for its intended purpose. It is important to ensure that all tests used to support PQ and PV have been validated where required.

3.7 Cleaning Validation

Cleaning validation should be distinguished from the demonstration that the cleaning equipment works properly. This demonstration is part of OQ and typically comprises of verifying spray-ball coverage and operation of clean-in-place (CIP) systems. Cleaning validation is the specific verification of the removal of product down to a pre-determined level. As it is product specific, it cannot be performed until the process and product requirements are defined. However data to support the cleaning regime must be in place before product is released for sale. Therefore, cleaning validation should ideally be performed between PQ and PV. Because of the importance of validating the cleaning process, it is not unusual to start consideration of this aspect very early in the validation life-cycle. It would of course be opportune to consider the cleaning process during product development. In practice, a cleaning philosophy is often written during the design and equipment procurement stages of the project. The cleaning philosophy should identify the requirements for cleaning of process equipment, the issues to be considered for product changeover (in multi-product facilities), the prevention of contamination with degradation products and the control of microbial and endotoxin contamination. Such a document will describe the main objectives of the design with respect to cleaning. For example:
- Adequate design to control cross contamination (closed systems, classified areas, and local containment)
- Water systems capable of producing water quality appropriate to the product and process
- Systems and procedures designed to control bioburden and endotoxin for sterile products prior to final sterilisation (equipment design, sterilisation and cleaning procedures)
- Demonstrable cleanable design for multipurpose equipment and facilities (materials and surface finishes, use of CIP)

Like most other validation activities, cleaning validation should be conducted in accordance with approved protocols. Potential contaminants should be identified. These may be residues from previous batches, degradation products, cleaning agents, reaction by-products and microbes. The method of sampling to prove reduction of residues to acceptable levels should be justified, as should the type of testing employed. Finally it is important to justify an "acceptable" level based on scientific reasoning. The requirement to be below the limit of detection is nowadays hard to justify with the increasing sensitivity of many assay methods. From a practical standpoint some of the key areas to consider before embarking on a cleaning validation programme are:
- What are the products?

- Is the equipment dedicated or multipurpose?
- Are there assay methods for detergent and formulation ingredients?
- Is there a sampling plan?
- What is the percentage recovery from swabs (if used)?
- Should a specific assay be carried out or is a non-specific assay (e.g. total organic carbon) acceptable?

3.8 Computerised Systems Validation

Computer systems are used in almost every pharmaceutical facility in small or large ways. From the simplest Programmable Logic Controller (PLC) fitted to a piece of process equipment to the complex systems used to monitor and control manufacturing and warehousing systems, computers have become part of pharmaceutical manufacturing. Like other aspects of the manufacturing process where product quality may be affected, validation is required. In principle the validation of computer systems is no different than the validation of any other item. In practice, it is recognised that computer systems are so complex that a relatively simple series of functional tests is insufficient to assure that the system will work reproducibly to produce product meeting its quality criteria. Therefore, in addition to these functional tests, it is important that documented proof is obtained that the software was developed in a structured and controlled way within an overall quality system and that structural testing was performed on the source code. Because of these requirements it is extremely difficult to perform satisfactory validation of complex computerised systems unless the requirements have been stated prior to software design. A number of "life cycle" models have been proposed to explain the steps required to validate a computer system. All include the requirement for a user input at the beginning of a project, the production of a FDS by the supplier that details how the User Requirements will be addressed. This is followed by production of design specifications. All this should happen prior to any coding starting. Finally test specifications are developed, the software is structurally tested, and the system is constructed and installed and then functional testing starts.

As with all systems, the level of complexity, whether the system is mass produced or one off special, and the degree with which product may be impacted will determine the extent and depth of testing required. A number of publications are available to guide both suppliers and users through this area. In Europe the Good Automated Manufacturing Practice (GAMP) guidelines (7) are widely accepted and provide an excellent basis to establish a validation programme for new computerised systems.

4. A SYSTEMATIC APPROACH TO VALIDATION.

Any significant validation project must be organised as any other complex task. This means that to assure success the fundamentals of planning, scheduling, cost control and resource allocation must be established. All anticipated activities should be budgeted and a multi-disciplinary team assembled to execute the work. As it is expected that validation requirements for a GMP compliant facility must be built into the design, construction and procurement phases of a project, some validation projects extend far beyond the traditional activities of writing and executing protocols. Although that is the cornerstone of validation, validation personnel should and do become involved in other activities. This can best be illustrated by the following examples that describes some of the activities that may be project requirements.

4.1 Procurement

When purchasing items of process equipment or asking a subcontractor to bid for the construction and installation of pharmaceutical systems, validation personnel should be part of the team. This is to ensure suppliers are able to comply with GMP and to ensure that the equipment is designed to permit practical validation activities. This includes, for example, the provision of sampling valves in appropriate locations and access for test instrumentation. In order for a validation specialist to assess each vendor's capability to produce equipment and documentation meeting GMP standards, it may be necessary to attend vendor interviews, to audit vendor quality systems, to review bid packages and to review samples of vendor documents such as quality plans, commissioning procedures and operating manuals. In addition to assessing the vendors, validation should also have an input into the documents sent to potential suppliers to ensure that the requirements of GMP compliance and validation are carefully explained. Vendors cover the whole spectrum when it comes to GMP compliance. Some are familiar with GMP, as they are specialists focusing on the pharmaceutical industry, others provide products for many market sectors and may not be familiar with the regulatory expectations. All of them have a business to run and are unlikely to offer extra testing or documentation for free if these requirements are not made plain in the bid process. If the vendor is expected to assemble the turnover package in a specific format or with specific documents, this should be clear in the invitation to bid package. It is particularly important if a vendor claims capability to perform validation on his own equipment to establish clearly the scope of what is provided. Should a vendor deviate from the specification, a validation specialist should be part of the team ensuring that the deviation does not compromise GMP requirements.

4.2 Preparing Strategic Documentation

This should include a Validation Master Plan, which in complex projects may have a series of sub plans dealing with specific validation topics or items of equipment. It may include a cleaning philosophy and a project specific quality plan. For the VMP, provision should be made to keep this up to date as the project progresses.

4.3 Reviewing computer system documents for compliance with life cycle requirements

For computer systems in particular, the documentation produced by the vendor is critical in the verification of its fitness for purpose. A computer validation specialist is required to review these items. Additionally the specialist should work with the user to produce the user requirements. This is a critical document, upon which all further work is based. Never underestimate the amount of literature required to adequately underpin the validation of a complex computerised system. The vendor should have a quality system in place and the standard operating procedures, which are the visible expression of that system, may require review. Project specific documentation may include:
- The vendor's Quality Plan,
- The FDS,
- Software and hardware design specifications,
- Software module design specifications,
- Test specifications for software modules and for the integrated system,
- Hardware test procedures,
- Software structural test results,
- Factory acceptance test procedures and results,
- Site acceptance test procedures and results,
- Software listings,
- Pseudo code and flow charts,
- Operating manuals.

For large systems some of these items may be split into sections related to specific functionality or area of control.

4.4 Standard Operating Procedures (SOPs)

A number of SOPs are required to document validation methodology. If not already written, these should be prepared in the early stages of the project. This includes documenting the format and contents required in each type of protocol, the method for producing and revising the VMP and the distribution and control of validation documentation. Many other SOPs will also be required; some of these will need the input of a validation specialist. These include SOPs for vendor audits, for performing factory acceptance tests, for sampling, for operating various types of test equipment. SOPs will also be required for the future routine operations of the facility. Writing these are not the direct province of a validation specialist, but they are essential to the successful completion of qualification and are therefore required to be present at OQ and PQ.

4.5 Turnover Package

The documentation that is provided by vendors, system constructors and installers and the designers of the facility should be assembled into a carefully structured package for hand-over to the facility user at the conclusion of a project. This package is sometimes referred to as a Turnover Packages (TOP) or Validation Turnover Packages (VTOP). The TOP should include complete documentation for all pharmaceutical systems and will become the basis of the GMP library that the user will manage and keep current. Although the TOP will include information relevant to many departments (e.g.. manufacturing and maintenance), it will in particular contain documentation supporting validation. Validation personnel will utilise these documents particularly when executing installation qualification. In addition, as a key user, it is useful to involve validation specialists when determining the structure and contents of the TOP. Ideally the TOP should be assembled by system, concurrently with the progress of the project. Routine auditing of the TOP is valuable during assembly in order to highlight non-conformance prior to qualification starting. Typical documents relevant to validation that should be found in the TOP include:
- As-built drawings
- Equipment specifications and data sheets
- Method statements for commissioning
- Factory acceptance tests and site acceptance tests
- Results of tests with supporting raw data
- Welding documentation
- Life-cycle documents for computer systems

- Test methods for passivation, flushing, cleaning and pressure testing as applicable
- Certificates of conformity and material certificates for items in contact with product or clean utilities
- Surface finish certificates for clean utilities and product contact metal surfaces
- Certificates for passivation, flushing, and cleaning as appropriate
- Pressure/leak test certificates
- Equipment alignment and level documentation
- Pump performance documentation
- System isometrics where applicable
- Operating and maintenance manuals

4.6 Factory Acceptance Tests (FATs)

FATs may be performed on items of equipment that are constructed at suppliers premises prior to shipment for installation in the pharmaceutical facility. Items that may be subject to FAT are skid-mounted equipment, discrete Process items and the generating equipment for utility systems. Whilst FAT cannot substitute for properly performed tests on the installed item, it may if performed and documented correctly, permit significant reduction in some qualification activities. If results are used to support validation, it is critical that the tests are approved by the user and witnessed. FATs cannot be used to support equipment qualification where the conditions in the factory do not match the conditions as installed. This applies particularly to tests that rely on the provision of specific services such as steam and water. However, failure at the FAT usually indicates that problems are likely to be encountered later. As logistically and financially it is invariably easier to correct deviations at the factory, FAT is an important stage leading up to qualification and a validation specialist should be part of the team evaluating FAT procedures, witnessing FATs and reviewing FAT reports.

4.7 Installation and construction

As part of the overall quality system, checks should be made to ensure that construction and installation activities that have a significant impact on GMP are in compliance to approved procedures. Examples of critical activities include storage of stainless steel, welding of sanitary systems, hydrostatic testing, flushing and passivation. Validation specialists may be involved in auditing subcontractors work and documentation, reviewing non-conformity reports and commenting on materials, equipment and quality documentation problems.

4.8 Calibration

Calibration is an essential precursor of OQ and normal operation. It is important to establish the policies and responsibilities with regard to calibration. Instruments are usually supplied individually or batch calibrated. They should be recalibrated when the equipment is installed on site particularly if they have been removed and refitted after flushing and blowing of pipework. It must be established who has the responsibility for recalibration, who will label items as calibrated and who will enter all instruments into a calibration database. Validation specialists may be required to review and comment on calibration procedures and to approve calibrating instruments and review certification in order to ensure that instrumentation meets required standards of accuracy and precision.

4.9 Validation Protocols

Validation protocols are required to describe the objective, methodology and acceptance criteria for installation qualification, operational qualification and performance qualification. They are written to ensure test methods and acceptance criteria are reviewed and approved before practical qualification commences. In practical terms there are several stages to the production of protocols. Firstly an acceptable format needs to be agreed. There is no universal format for protocols and to some extent, the type of equipment, the size of the project and personal preferences will dictate the protocol style. However, some norms have been established. Like other controlled documents, protocols are assigned unique reference numbers and revision numbers. They are titled and numbered on every page. In addition, they have spaces for approval signatures. Other common elements in protocols tend to be brief descriptions of the item being qualified and clear statement of responsibilities.

Next the list of items to be qualified needs to be produced and the approach agreed. This will include answering questions such as:
- Is it preferable to group identical items together under one protocol or to have a protocol per item?
- For a simple system, is it preferable to write a single protocol covering IQ, OQ and PQ?
- For complex items such as large utilities, is it acceptable to address the generating system and the distribution system in separate protocols?

- Should computer systems validation on control systems for process equipment be documented within the mechanical validation protocol or should a separate protocol be written?
- Is it acceptable to have different boundaries for PQ than for IQ? (An IQ protocol may group all vessels together, the PQ protocol may group the vessel with the system it serves e.g. harvest vessel with bioreactors).

There is no single right answer for every project. However, it is important that the approach is agreed and adhered to, to ensure internal consistency and to prevent items being inadvertently being forgotten.

Once the approach and format is agreed the protocol preparation can start. Preparing protocols requires individuals who are trained to extract information from a variety of sources and synthesise it into a coherent whole. Typically, sources of information include, material requisitions, specifications, data sheets, P & IDs, manufacturers literature and equipment operating manuals. Meaningful protocols cannot be written until sufficient sources of information are available. Once the first draft is written, protocols require review. Normally, individuals from manufacturing, engineering and quality assurance will perform the review. It is quite normal for first draft protocols to require considerable modification, as this will be the first time that other interested parties have seen them. Rather than check everything three times, it is preferable that the review disciplines concentrate on their specialist areas. For example, manufacturing should evaluate if the proposed acceptance criteria are consistent with the process requirements, engineering should ensure that the list of instruments, major components and utilities requirements are accurate and QA should ensure that compendial requirements have been met and that the protocol meets normal quality expectations.

In a large project, the co-ordination of protocol review, keeping track of the revision status and the storage and retrieval of protocols, is a task that requires carefully planning. At some points it is probable that a number of protocols at different draft stages are circulating for review, some are being modified, new protocols are being prepared and approved protocols are stored prior to execution starting. A tracking system should be implemented that can very quickly identify where a protocol is and provide a snapshot of the current protocol production status. The method and time allowed for review must also be agreed. Review methods range from the traditional circulation of a single document to each individual for comment to assembling a team for a joint review to electronically accessing the document and revising or commenting on it. The method chosen must reflect the available time, resource and technology. In some cases the resource for protocol review is a limiting factor that can critically affect the schedule.

4.10 Commissioning

The various checking and testing activities that occur after equipment installation may be divided up into mechanical completion, pre-commissioning, site acceptance tests and commissioning. At each of these stages, some of the tests performed may be critical to validation. For example, the HEPA filter integrity tests performed to check the installation of filters installed in clean rooms, need not be repeated during qualification as long as there is documented evidence that the tests have been performed properly. In practice, this means that, as with FATs, validation personnel should review the procedure, witness the tests and check the documentation supporting the results. Whilst it is normal practice to utilise certain commissioning test results in the qualification of a system, in recent years there has been a move towards attempting to more fully integrate commissioning and equipment qualification. The obvious advantages in this are the potentially shortened time between installation and PV and the potential financial savings. There are a number of pitfalls to be aware of. The skills and training of a commissioning engineer are fundamentally different from a validation specialist. Commissioning involves repeated checks and adjustments until the items function correctly. Validation verifies that the items can function correctly immediately and reproducibly. Commissioning will not include many of the checks routinely carried out by validation specialists.

4.11 Change Control Monitoring

In any major construction and installation activity, there will be a requirement to perform changes in the field to the way systems are constructed, installed and tested. These changes must be controlled by a formal change request system. Changes must be reviewed for validation and GMP implications.

4.12 Gap Analysis

Where a validation project involves items that have already been subject to validation, it may be necessary to carry out gap analysis. This is the process of determining if there are significant validation requirements that have not yet been addressed. Gap analysis may also be required when the validation scope is divided amongst different contractors and vendors. This is to determine if any elements have been unintentionally omitted. It is a review process that compares what is planned or has been executed against what is required.

4.13 Protocol execution

This refers to the execution of checks and tests specified in pre-approved protocols and completing the protocols. That is, test sheets and blank spaces within the protocols will be filled in with results obtained and supporting data and documentation attached or referenced. Unless the documentation is organised properly, the execution of IQ can be particularly difficult and time consuming. The documentation package that supports IQ may be extensive. If it is inadequate, the entire validation of that system may be at risk. OQ testing needs to be scheduled sensibly, with an appreciation that services supplying a system should be qualified before the system is qualified. For example the steam supply to the autoclave should be qualified prior to thermally mapping the autoclave. Consideration should also be given to access. In some cases access to ductwork is impossible once ceilings have been sealed. If floors are being laid, no other work may be possible in the area at all. There is often a relationship between other items as well. For example the working of a fume hood will have an impact on the room air pressure differential.

The practical execution of OQ will usually be a combined effort between the equipment operator and the validation specialist. While some equipment can be tested without external testing and measuring apparatus, it is quite common to utilise a number of specialist tools for OQ. These include:
- Data-loggers and thermocouples for thermal mapping activities
- Particle counters, smoke generators and differential pressure gauges for evaluating environments
- Flow meters for measuring velocity of fluids in pipework
- Tachometers for measuring rotational speed

4.14 Summary Reports

Summary reports are required to summarise the results of tests and analysis after qualification of GMP critical systems. These may be written at the conclusion of each phase of qualification or after all phases are complete. There should be at least one report per GMP system. These are important documents that must enable regulators and other interested parties to very quickly assess the status of validation

4.15 Action Plans

All deviations identified at gap analysis, or during field execution, should be documented as having had a satisfactory resolution. In some instances, a punch list supported by a punch list procedure is adequate. Where resolution is more complex, an action plan should be produced detailing the deviation, the proposed action to correct the deviation, the responsible individuals and the due date. Not only will this assist in expediting the corrective action, it will also demonstrate to the regulators that the company is in control.

4.16 Training

All key personnel should be fully aware of GMP and validation requirements. Team building activities can include elements of GMP awareness in order to increase understanding and build cohesiveness amongst all team members. It is important that contractors and subcontractors, that have a responsibility for GMP critical systems, are trained in the correct way in which to produce and complete documentation. They should also be made aware of the importance of their work in relation to product quality.

4.17 Project Management

Validation Project Managers are required to ensure project resources are used efficiently, to provide up to date information on validation progress and budget and to ensure activities are consistent with the agreed strategy. They should have technical knowledge and team management skills. On large projects, the Validation Manager should be supported by expert planners and cost control personnel. The schedule is an important tool in complex validation projects. Validation is often required to fit into the time between completion of commissioning and manufacture of product. Inevitably, the commissioning dates will slip and the date for product manufacture will remain immovable. In this situation the ability to perform "what if" analysis is important. Many decisions impact on validation. These include the type of equipment, the use of the equipment and the product. Sometimes these are subject to change as the validation project progresses. For this reason it is recommended that a steering group be established to assist in communication and as a forum to agree changes in strategy.

5. BIOPHARMACEUTICAL FACILITY CHARACTERISTICS

The sections above describe validation concepts, approaches and activities that are broadly applicable to most pharmaceutical facilities. The following section considers the specific characteristics of the biopharmaceutical plant and how this impacts on validation.

5.1 Containment

Biopharmaceutical facilities manufacture product typically either by mammalian cell culture, microbial cell culture or extraction from natural sources. One of the major concerns in these facilities is operator and environmental safety. To protect the operator and the external environment, containment measures must be taken. These measures generally consist of facility designs and equipment to ensure that potentially contaminated air and waste cannot escape from operating areas into the surrounding environment. Containment requirements can conflict with product protection requirements. Where product is potentially exposed to the environment, the air quality is expected to meet certain particulate standards. To achieve these standards, it is normal practice to design a facility where the pressure differentials between rooms ensures that air flows from the most critical areas into the least critical areas. This is in direct contrast to the design for containment. Both requirements can be met by the use of airlocks or corridors as air sinks or air curtains. The validation specialist must be aware of the design intent and ensure during GMP review and at subsequent testing that both operator and product protection have been achieved.

5.2 Sterility

Even if a sterile product is not being produced, the upstream stages of Biopharmaceutical production may require a sterile environment in order for the process to work. Mammalian cell culture, in particular, requires protection from contamination. This protection is usually provided by using a closed system. A critical aspect of validation is verifying the system's integrity. This is performed at the OQ stage by leak testing and at the PQ stage by media holds trials. In the latter test, the bioreactor is filled with sterile microbial media for at least seven days and subject to normal operational interventions. In fact, it is efficient to execute the other PQ tests (such as pH, dissolved oxygen and temperature control) at this time. Daily samples should be taken and examined for contamination. If the product is expressed under conditions of limited oxygen, then it is also important to look for anaerobes. Sterile filtration is often employed in the biopharmaceutical industry for product, raw materials or media components. This is usually because these items are not heat tolerant and cannot be sterilised easily in any other way. Depending on the material, mycoplasma may be a potential contaminant. If this is the case, filtration of $0.1\mu m$ is required. For all sterile filtration, filter validation should be undertaken. This is a complex area and assistance can usually be obtained from the filter manufacturer. The critical elements are that a microbiological challenge should be performed using the specific process conditions (viscosity, flow rate, pressure and temperature) and a reproducible standard challenge methodology. If the solution to be filtered is toxic to the test organism (*Brevundimonas dimunuta* is the usual organism of choice), then a substitute fluid of similar characteristics should be selected.

5.3 Purpose built systems rather than packaged items

Most process equipment in a biopharmaceutical facility is not purchased off-the-shelf. Systems often consist of purpose-designed skids with associated vessels and pipework. A typical list of process equipment could include bioreactors; filtration, ultrafiltration and chromatography skids; centrifuges; harvest vessels and media vessels. As each system is built for a specific purpose, the validation requirements must be conceived from first principles, rather than from a set menu of activities which is so often the case when validating an off-the-shelf package.

5.4 The "Family" Approach

Biopharmaceutical plants will often utilise a number of items that are identical in configuration and intended to be used for the same purpose. A good example is the number of portable vessels used as feed or harvest containers. The practicality of performing full validation on perhaps 30 identical vessels is daunting. Consideration should be given to the use of the "family approach". This is where full IQ checks are carried out to prove that each item is identical. Full OQ tests are carried out on a representative item and a smaller number of OQ tests are carried out on the remaining items to confirm that the performance is identical to the representative item. This then qualifies each item to be used interchangeably within a complete system during PQ.

5.5 Clean-In-Place (CIP) and Steam-In-Place (SIP)

Although CIP and SIP are in use throughout the pharmaceutical industry, it is usually more common, more extensive and sometimes more complex in a biopharmaceutical facility. This is because of the need to clean and sanitise large vessels and bioreactors and the process requirements for sterility as discussed above. CIP systems should be designed to minimise the potential for causing contamination. That is, adequate consideration should be given to the materials of constructions, the internal surface finishes, potential areas where water may be retained and the use of sealed storage vessels with hydrophobic vent filters. When in operation, CIP systems are expected to provide total

Biopharmaceutical Validation: an overview 333

surface coverage, clean down to a defined level and reduce endotoxin and bioburden. SIP is usually taken to mean "steam-in-place", although the acronym is also used for "sterilise-in-place" or "sanitise-in-place". Sanitisation or sterilisation can be performed using a number of media including hot air, superheated water, chemicals and steam. Many biopharmaceutical systems are very flexible and capable of being configured in a number of different ways. Prior to validation, it is important to establish the preferred configuration and SIP methodology otherwise an unrealistic burden may be placed on the validation team. For instance, some bioreactors are capable of being sterilised empty and full, by steam in the jacket or steam in the vessel. In some cases, each port and filter can be sterilised separately or as part of a complete system sterilisation. Sometimes there are choices in the steam path. Sometimes the system may be able to be sterilised by both superheated water and steam. Such flexibility results in an enormous potential number of configurations and permutations. Therefore it is important to define and to validate the way the system will eventually operate.

5.6 Utilities

One of the most obvious characteristics of a facility manufacturing products of biological origin compared to a standard pharmaceutical secondary facility is the number of utilities that are required and the flexibility that is built in to allow connection of any process skid to any utility. Usually purified water, WFI, clean steam, compressed air and process gases (e.g. oxygen, nitrogen and carbon dioxide) are provided for process use. In addition, waste systems for contained waste, process waste and solvent waste are usually needed. In order to allow maximum flexibility, these services may be provided in manifold systems or utility panels. These panels (often automated) must be qualified. IQ can consider these in isolation, but during PQ these should be tested as part of the system they serve. The services themselves must be validated sequentially as discussed previously. Typically, the purified water system supplies the clean steam and the WFI generator. It therefore should be qualified at each stage prior to the systems it serves. Process gases should be tested with regard to the specifications for the process. In these cases, the compendial requirements may not be appropriate as these were formulated with administration to patients in mind rather than as a requirement for biological cell growth.

It is not a regulatory requirement to validate waste systems except insofar as to demonstrate that sufficient precautions have been taken to prevent backflow into the process stream. By definition, the waste stream is the discard from the product and cannot impact on product quality. Some companies, however, choose to qualify the waste systems for reasons of safety assurance rather than regulatory compliance.

5.7 Laboratory Support

It is a natural consequence of a facility that is highly serviced and has a requirement for sterile production, that considerable laboratory support will be required during validation. Microbiological and chemical analysis will be required at PQ for all clean utilities. Similar support will be required when cleaning validation is being performed. The microbiologists will be required to process environmental samples taken during PQ of classified areas and will also be expected to check surfaces and clothing during PV. Bioburden and endotoxin testing to demonstrate CIP effectiveness and the processing of biological indicators to demonstrate sterilisation also falls under the microbiologist's remit. When scheduling these activities, the required expert resource may turn out to be the limiting factor.

5.8 Automation

Facilities producing active bulk product and biopharmaceutical facilities often share the common characteristic that there is far more automation than in a typical secondary facility. Most modern facilities of any size tend to have automated control over the air handling systems. These vary in sophistication. In some cases, these systems also control or monitor other utilities. Typically these building management systems will monitor and control environmental temperature, humidity and room pressure differentials. They will monitor the alarm status of critical clean utilities and they will be able to store and report on historical data. Some facilities may also have process control systems. These are usually capable of communicating with systems that are already automated and are also able to directly control and monitor items that do not have their own control systems. Some systems can be configured to electronically produce batch records, recording data both on line and off line. Testing these systems involves simulation both in the supplier's factory and on site and functional tests when all software is fully installed and hardware is connected. Until the process control system is qualified, it will not be possible to complete the qualification of systems that it controls.

6. CONCLUSION

In addition to being a requirement of the regulatory authorities, validation provides a number of other tangible benefits. These result in a facility operating in compliance with its design and producing product of consistent quality. The key to successful validation is to clearly define the requirements at each qualification stage, to integrate the resources into the broader project scope and to carefully plan, control and monitor all activities. The ideas presented in this chapter provide a broad overview of the field of validation. Further details on specific validation issues can be found in chapters 14 and 15.

BIOGRAPHY

Stephen Slater is a Regional Validation Manager for Raytheon Engineers and Constructors(RE & C). He is a microbiologist by training and has worked extensively in the biopharmaceutical industry both within manufacturing and as a consultant. RE & C provide world-wide GMP compliance and validation services from offices located in the United Kingdom, the United States, Germany and Puerto Rico.

REFERENCES

1. The Rules Governing Medicinal Products in the European Community, Volume IV (1992). Good Manufacturing Practice for Medicinal Products. The European Commission, Brussels.
2. Food and Drugs Administration (May 1987). Guidelines on the General Principles of Process Validation. Food and Drugs Administration, USA.
3. Food and Drugs Administration 21 (1996). Code of Federal Regulations Parts 210 and 211. Food and Drugs Administration, USA.
4. United States Pharmacopoeia, 1995. 23rd Edn. U.S. Pharmacopeil Convention, USA.
5. European Pharmacopoeia, 1997. 3rd Edn. The Council of Europe.
6. The International Conference on Harmonisation (May 1997). Guideline on Validating Analytical Procedures. http://www.ich.org/ich1/html.
7. Good Automated Manufacturing Practice (GAMP) (July 1996)

Chapter 14

Validation of Biopharmaceutical Chromatography Systems

K.F. Williams (*) and C.J.A. Davis (**)
(*) Validation Technologies (Europe) Ltd., Sutton Place, 49 Stoney St., Nottingham, NG1 1LX, UK. (**) Tanvec Ltd., Alexandra Court, Carrs Road, Cheadle, SK8 2JY, UK

Key words: validation, chromatography, biopharmaceutical

Abstract: The importance of validation of biopharmaceutical systems is introduced in the previous chapter. The key elements of validation are summarised by Bala (1) whilst the general areas of validation are detailed by the PDA (2). More specifically the importance of validating the biopharmaceutical chromatographic system becomes apparent when one realises that the system is being used for product purification purposes. The quality and product failures of the past, combined with public protection issues, led the US Food and Drug Administration (FDA) to publish the current Good Manufacturing Practices (cGMPs) in the Code of Federal Regulations (CFR) in 1979. The principal of validation was introduced by the FDA almost immediately thereafter to demonstrate that a process truly functions as intended with supporting documentation.

This tenet applies equally to chromatographic systems, where the aim of the validation is to show that purification has functioned as intended during standard process conditions and to provide the supporting documentation to prove this.

The difficulty arises in that the FDA has never really identified the manner in which validation is to be carried out and so industry has led the way in defining methods to satisfy the validation requirement.

This chapter is an overview of those methods with specific reference to chromatographic systems.

1. INTRODUCTION

Chromatographic separation is dependant upon the differential partition of molecules between a stationary phase (the column matrix) and a mobile phase (the buffer solution). Large scale protein purification generally utilises five molecular properties:
1. Net charge and distribution of charged groups - ion exchange chromatography.
2. Biospecific affinities - affinity chromatography.
3. Size and shape - gel permeation.
4. Hydrophobicity - hydrophobic interaction chromatography.
5. Metal binding - immobilised metal ion affinity chromatography.

It is the utilisation of the different molecular properties that make chromatography a powerful tool in purification techniques. This power, and the complex purifications required to manufacture therapeutic proteins, make chromatography the common unit operation it is.

The aim of the next sections are to highlight the key elements of chromatographic systems and the method of validating them.

2. KEY ELEMENTS OF CHROMATOGRAPHY PROCESS AND SYSTEM.

2.1 Typical Processing Steps

A chromatography process commonly consists of the steps detailed below. Whilst this is true for positive adsorption/desorption chromatography negative chromatography and gel permeation chromatography differ slightly:

- Equilibration; where the column matrix is prepared for adding the product.
- Load; addition of the product.
- Wash; where the residual load material is flushed from the column and in some cases an element of purification occurs.
- Elution; where the product is released from the matrix.
- Clean; where residual contaminants, bound specifically and non-specifically, are removed from the matrix.
- Storage; where the matrix is left in a solution suitable for stable storage of the matrix until the next use.

These steps make up the process and must be combined with the appropriate hardware, support systems and support processes in order that therapeutic product can be produced.

2.2 Typical Hardware

The hardware consists of the buffer tanks required by the process, the feed vessel containing the product, the eluate vessel that will receive the product, the column, the pump and the instrumentation needed to control the process. A typical process flow diagram for a chromatography process is illustrated in Figure 30.

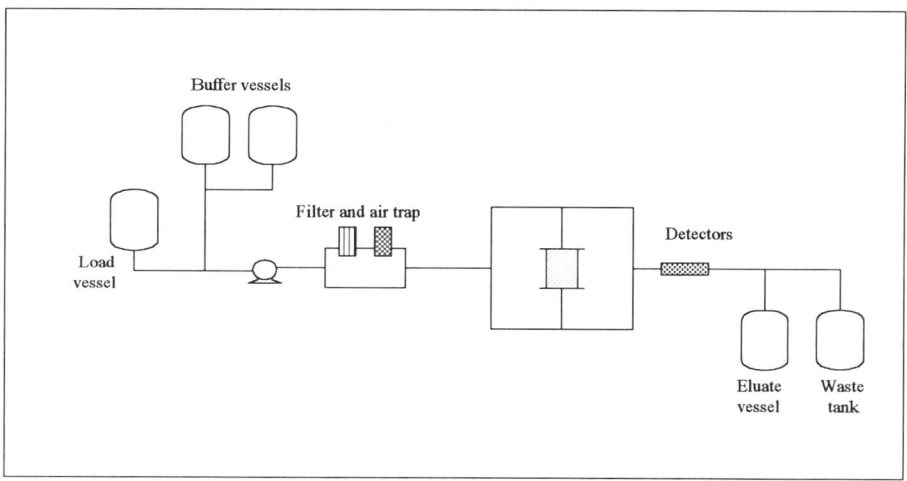

Figure 30. Process flow diagram for a typical chromatography process

2.3 Typical Control System

It is important when developing the process itself that the associated control strategies are kept as simple as possible. This will facilitate the process design, validation and subsequent operation and maintenance. Chromatography systems are generally automated to some degree and the scope and type of system will depend upon the plant, the process and the economics of production. The configuration for automation systems is summarised by Ransohoff (1). There are two main microprocessor based control systems, DCS (distributed control systems) and PLC/SCADA (programmable logic controller/supervisory control and data acquisition) systems.

In both systems the operator interface is at a terminal remote from the software and the input, output and control electronics. The issues surrounding the choice of automation systems are the same from industry to industry and they consist of the factors influencing the choice of hardware and software, such as cost, robustness, ease of configuration to name but a few. More specifically to the healthcare industries are considerations around validation and this led to the Good Automation Manufacturing Practice (GAMP) Guide (2) being written as guidance in specifying and developing automation systems for both the user and the supplier.

During the course of this chapter the automation validation aspects will follow the principles set out in the GAMP guide although the scope of the chapter allows only a low level of detail to be pursued.

2.4 Support Systems

The support systems primarily consist of the facility in which the process will be run. The facility includes the fabric of the room and the environmental control systems that ensures a controlled environment suitable for therapeutic processing. The environmental conditions should be specified within the project documentation and validated. The room fabric should include floors, doors, windows and walls that are non-shedding, inert to any potential spills and easily cleaned without pockets that will collect dirt or be difficult to clean. The environmental control system is the heating, ventilation and air conditioning system (HVAC) and air filtration that will maintain the environment within predetermined specifications and classifications. It should be noted that if the processing environment is high quality there may be financial savings in removing items such as buffer tanks and eluate vessels to a lower quality environment.

2.5 Support Operations

A chromatography process can not run in isolation of operations such as cleaning-in-place (CIP), steaming-in-place (SIP) and sanitisation. It is essential these support operations are considered during the process development stages, especially sanitisation where the chromatography matrix is chemically treated to minimise bioburden before processing starts. The CIP and SIP systems will tend to be designed separately but when it comes to testing those portions involved in the chromatographic process they should either be tested as part of the chromatography system or as components of the

Validation of Biopharmaceutical Chromatography Systems 341

CIP and SIP testing. The choice should be made at the Validation Master Plan (VMP) stage.

2.6 Design Considerations

The validation of chromatography process should be considered during process development. In addition to keeping the process as simple as possible and those factors mentioned in 2.4 and 2.5 above, the following items should be considered:

- Minimising leachates from the chromatography matrix during processing, sanitisation and storage. Any leachates should be identified and limits for their presence in the product stream set.
- Purification with respect to level of removal of contaminants from the feed stream.
- Product recovery levels required.
- Preventing or minimising ingress of additional contaminants from processing equipment.
- Identify key process steps and positions that require sample points.
- Maximising matrix stability - the effective useful lifetime of the matrix.

Whilst developing the process it is imperative that once the process has been fixed it will be possible to validate the removal of contaminants, nucleic acids, pyrogens and other process specific items. The development of the process must be thoroughly documented as this is a requirement to support subsequent regulatory submissions (3).

A key element of process development and the writing of the process description to describe the process is the identification of critical quality parameters. Critical quality parameters are those which are critical to producing product of the specified quality (4). Other parameters are then defined as courtesy parameters and the need for this differentiation will be detailed below.

The validation of the support processes must be considered as part of the development programme. A key part of this is the development of assays to support the process and process validation. The latter should include assays to support equipment cleaning and matrix leachate quantitation. In developing any assays the validation of the assays should be considered.

The process and equipment specifications should be developed so that the equipment parameters ranges are well within the specification range. This flexibility comes from the fact that the process parameter specification is set wider than the equipment specification. The validation parameter specification should in turn be wider than the process specification. Finally the regulatory requirement or standard needed to be achieved should be have a wider parameter range than the validation criteria chosen.

This approach may seem simplified but if the essence is followed then an equipment specification will never cause a chromatographic process to fail validation.

3. KEY ELEMENTS OF VALIDATION PROCESS.

Validation, unlike most of the developments in legislation seen in the last 30 years, is not a set of rules or national standards, nor is there a book on "how to", which if followed exactly, would guarantee that product, plant and production methods would be acceptable. In this light the following section is a distillation of the key elements that are commonly seen from project to project and that the authors have tried and tested with success.

3.1 Prospective and Retrospective Validation

Prospective validation should occur when a chromatographic process is being designed as part of an existing or new process build. The validation effort on the chromatographic part of the process starts with a clean sheet and as the process is designed the various validation activities and documentation requirements follow the project and validation flow detailed in Figure 31.

Retrospective validation occurs when a chromatographic process is already in place and the process needs to be tested and documented in order to, for example, obtain a product license for a new market. In many quarters retrospective validation is perceived to be more difficult, particularly where there is a lack of equipment documentation, but this is often overcome by the fact that the chromatography process has a large amount of historical data. The same sequence of events illustrated in figure 2 would occur, but some of the testing would refer to historical data.

For ease and clarity the next sections will consider and address prospective validation only. The following sections are structured to folllow the sections illustrated on the right hand side of Figure 31.

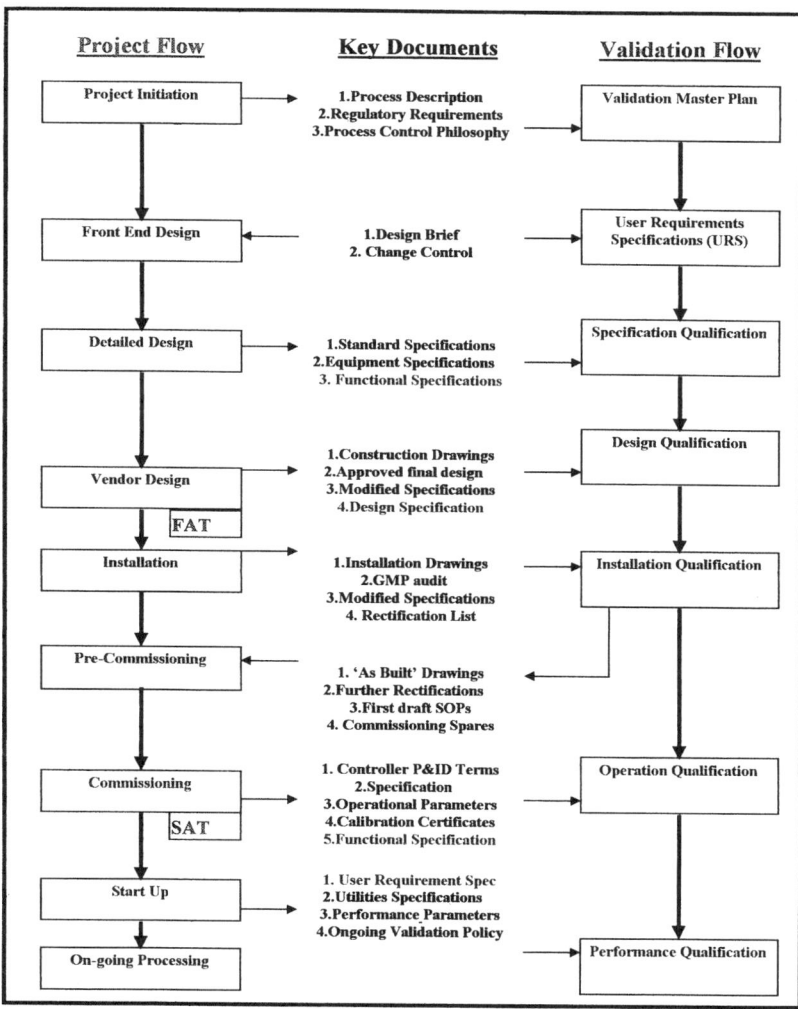

Figure 31. Project and validation flow and associated documentation

3.2 Validation Master Plan

The starting point for any validation effort is a Validation Master Plan (VMP). This is the step in the validation effort that is most important but is often the most poorly addressed due to either time pressures or a lack of vision. Time spent here at the planning stage will more than be repaid further down the line in the execution stages.

The validation of the chromatography step will often be part of a project covered by an all encompassing VMP or could be a stand alone project if a new column is being added to an existing process. In either case this document and the approach chosen is critical.

The purpose of the validation plan is as follows:

- *Provide a description of the systems* - this is a description of the chromatographic steps with the key process parameters for validation.
- *Address regulatory requirements* - inspecting authorities expect to see that there is a plan in place for validating the chromatographic process and that it was followed.
- *Outline a consistent approach to validation* - this sets out the ground rules for the validation such that if (and it almost always is) it is performed by different groups they will have followed a common methodology and reported their findings in a consistent manner.
- *Outline acceptance criteria* - the broad acceptance criteria for the chromatographic step or process are outlined here with the details being put in the individual protocols.
- *Outline responsibilities* - this defines who is responsible for what activity and ensures that there is no confusion and that all activities are assigned. This is best tackled by creating a matrix whereby each of the major activities has a responsible person assigned, even if they are not going to carry out the activity.
- *Create a project plan / milestones* - with any good project management effort the sequence of events and their timing are critical to achieve project end dates and allocate resource, and a validation project is no exception.
- The key documents required to write the VMP are:
- The process description, so that the systems can be determined.
- The process control philosophy so that the level of automation and hence the validation requirements of that automation can be determined.
- The regulatory requirements that the design, installation and operation have to adhere to.

The VMP should not be put on a shelf and filed away. Everyone involved in the project should have access to a copy and it should be modified as the need arises during the project. This modification requires that a change control system be instigated at the appropriate time.

3.3 Validation Management

Validation should be treated as a project and controlled as such with a project manager empowered to take the decisions required for the successful implementation of the validation programme. The validation programme should be planned and the plans fully documented in order that progress can be quantified. The use of milestone achievements is recommended to boost morale during a validation programme. The plans should be adequately resourced and all tasks must be scheduled with sufficient contingency time to accommodate the delays that will inevitably occur. One of the keys to successful validation programming is ensuring that all the prerequisite conditions for a task are completed before a task is started.

3.4 Change Control

At the VMP stage, the change control system for documenting changes and deviations should be agreed upon. This change control system should be suitable for both the equipment and the control system changes that will arise.

The question of when to implement change control in the project and validation flow is one of personal preference. Experience shows that implementing it too early is restrictive and too late results in uncontrolled changes. A formal change control system is best implemented once the detailed design has been finalised and the Specification Qualification (SQ) has been signed off. Any changes from here on in need to be recorded as they may contradict the agreed design and specifications. It is however recommended that a system for controlling and monitoring changes is implemented for the early stages of the project.

Despite a comprehensive process design and review programme, commissioning and validation will inevitably unearth errors in design and operation and process improvements will be identified. The appropriate changes, made after careful consideration of the need to make changes and the consequences of those changes to the project timelines, the process and the cost implications, should be made within the agreed change control framework.

It is important that contractors are aware of the change control process and it's importance within the project framework. This change control system should include representation of the key project personnel defined in the VMP and must be strictly adhered to in order that the validation efforts are not compromised.

3.5 User Requirements Specification (URS)

This step in the validation process is the primary time that the user gets to have their say as to how the chromatographic control system and processing equipment should function, and the broad operating parameters required of them. The deign brief is key to give the user ball park values to help create the URS for the process equipment and control system.

The URS is a key document in helping to develop the process equipment specifications and the hardware and software design specifications for the control system. It is also used (if written correctly) to test the chromatographic system during the Performance Qualification.

3.5.1 Process Equipment

At this stage the process description and design brief should provide enough information for the user to specify the key parameters of the process, such as the materials of construction, approximate capacities of eluate and feed vessels, the buffer tank number and approximate sizes, the flow rates and column sizes, and any sampling points that may be required.

This process equipment URS should be as comprehensive as possible and kept up to date as project requirements change. It is imperative that if the URS data changes the reasons for the change are clearly documented within the URS.

3.5.2 Control Systems

At this point in time the exact type of the control system does not have to be decided. The aim of this document is for the user to be able to specify the following key aspects of the chromatographic control system:

- Its main function and interfaces to both the chromatography system and the operators.

- The operational requirements of the control system such as data handling speed, security of access, operating environment (both physical and digital).

- Any system constraints such as timescales, maintenance and existing systems.

- How the control system development for the chromatographic system should be developed, and what will be tested.

The control system URS can be used directly to tender from chromatographic suppliers and they provide direct responses to each of the sections within it. The URS is well suited to act as the basis for PQ testing of the system.

3.6 Specification Qualification (SQ).

The two URS documents will enable the creation of equipment specifications for the chromatographic equipment and hardware and software design specifications for the control system. It is a good idea to ensure that the specifications are to the right standards and the correct documentation is in place for the next stages of the project. A matrix of documents, such as the one shown in Appendix 2, is recommended, even for a simple chromatographic project.

3.6.1 Process Equipment

The equipment specifications will either stipulate a skid mounted package, which will be designed and built away from site, or the individual components will be specified so that there is a specification for each of the equipment items, such as the chromatography vessels, the pump, the column . The specifications can be modified at this point and it is recommended that a list of the modifications and concessions be kept. It is a good practice to audit the potential vendors, particularly before the final contract is awarded.

3.6.2 Control System

The functional specification is normally written by the vendor and it describes in detail what the control system will do. A comparison should be made of the functional specification against the URS. Further revisions of this document can be developed with the user. Auditing of the software supplier is also good practice at this stage.

The functional specification can act as a good basis for operational qualification (OQ) of the control system.

3.7 Design Qualification (DQ)

Design Qualification is carried out to ensure that the vendor's design fully meets the requirements of the qualified specifications. There is a documentation requirement the minimum of which is listed in Appendix 2.

The detailed design and the vendor design are sometimes the same stage but in large contracts the engineering contractor would do a large amount of the design leading to a detailed specification for the vendor to build from. In most chromatographic purification projects, however, the vendor or the user is in the best position to design the system for the required process. Whichever is the case the important documents here are the construction issue drawings and the amended specifications to include all changes up to that point.

Design qualification is carried out on the approved designs and rectifications to the design are made. Once the design has been finalised, normally by sign off on the drawings and the specifications, formal change control should be implemented.

It is imperative that operating procedures are developed during detailed design in order to allow the design to be qualified and ensure it is operable. This work, commonly carried out using marked up piping and instrumentation diagrams, allows the development of validation protocols for operational qualification and process qualification. Additionally it allows the development of commissioning plans which will facilitate the progression from mechanical completion through to a validated facility minimising project delays. Operating procedures must include processing procedures, sampling procedures, and equipment preparation procedures such as cleaning-in-place and steaming-in-place.

3.7.1 Process Equipment

The final sizes, layout and orientation of the chromatographic equipment will have been determined at this point. It will still be predominantly on paper and so changes now will be less onerous than when the design is committed to fabrication. This is why a good practice is to carry out a design GMP audit as part of the design qualification.

Once the design qualification is complete the buffer tanks, vessels, pumps, column or skid can be manufactured.

3.7.2 Control System

The control system will be designed in two parts: he hardware and the software. The software design is normally based on functionality modules, for example 'SIP' module and 'CIP' module, and so the nature and content of the software modules needs to be designed. The hardware consists of the hardware design specification, the usual mechanical and electrical specifications (as per process equipment) and, where applicable, the network design specification. For testing purposes all of the above documents can come under the umbrella document known as the Design Specification. All of the separate design specifications should be checked against the functional specification as part of the Design Qualification. The Design Specification is a good document basis for writing the Installation Qualification of the control system.

Instruments become an issue at this stage of the project and the parameters they are measuring define their type. The definition of critical quality parameters during process development allows the instruments used to control the process to split into two categories: critical and courtesy instruments. The critical instruments will need to be validated, calibrated and maintained to a higher standard than the courtesy instruments and the impact of this upon subsequent activities should not be underestimated.

3.8 Installation Qualification

The aim of installation qualification is to ensure that the item or system has been fabricated, installed and calibrated to specification and the qualified design.

As mentioned above the fabrication of chromatography systems can either take place on-site or off-site at the manufacturer's premises. The choice of location is dependant upon the size of the individual systems, the scope of the complete system and the preferences of the those involved. What is important during construction is that the manufacturer's quality systems are followed throughout the construction period. These should include items such as materials control, standard operating procedures and appropriately trained staff.

3.8.1 Process Equipment

If the system is constructed on site it will be possible, and prudent, to conduct ongoing quality control (QC) of the fabrication process and installation. This QC should check that the pipework and valves are installed in accordance with the approved general arrangement drawings (GAs) and piping and instrumentation diagrams (P&IDs). Additionally the quality of the work should be monitored to ensure it is in-line with the quality manuals and systems. Any deviations should be recorded on construction deviation forms which should allow for recording corrective actions taken. This should all be documented and the documents maintained in order that they can form part of the IQ process.(see Section 3 of the Document Matrix in Appendix 2)

Factory acceptance tests (FAT) are not a substitute for the validation steps that follow, IQ and OQ. FAT should cover the whole system and demonstrate that the equipment has been constructed and functions in accordance to its specifications.

If the chromatography system is constructed off site it is vital that the subsequent connections to existing plant systems are made in a controlled manner. These may include connections to services and other process pipework, by welds or hygienic connections, or may consist of simple electrical connections. Validation should consider all aspects of these connections.

The majority of the documentation should be delivered to site when the equipment is delivered. On a large project the volume of this documentation means that its reception at site should not be overlooked.

3.8.2 Control system

The equivalent to FAT for the control system would be the hardware testing, software module testing and software integration testing. The hardware should be tested against the hardware design. The software module testing should reference the software module design specification and should test each of the functional modules. The software integration testing shows that the software once 'loaded' onto the hardware, still functions as intended.

The installation qualification begins by hardware acceptance testing. This ensures that all instrument loops are checked and that the connections from the field instruments to the control modules and data acquisition modules are correct.

Where necessary the network installation should be tested also.

3.9 Commissioning

When construction is complete the chromatography system should be commissioned to ensure it fulfils the design intent before undergoing formal Operational Qualification. This commissioning involves wet testing the system and if it occurs before passivation, should not use solutions that may be corrosive, such as low ionic strength water and high ionic strength buffers. It is important that commissioning is an ordered activity with plans, timelines and an appropriate documentation system to record what was done, the results of activities, any issues that rise and the actions taken to resolve those issues. It is useful to use the draft OQ documentation as commissioning procedures as this will serve as a check on the OQ procedures and can increase confidence that OQ will not uncover any nasty surprises.

In order for the commissioning to be meaningful it is necessary that calibration has occurred prior to commencement. If this is done then the data collected at the commissioning stage can be used to support the Operational Qualification.

3.10 Operational Qualification.

Operational qualification ensures that the system, and here this includes process equipment together with the control system, operates to the parameters defined in the specification.

Validation of Biopharmaceutical Chromatography Systems 351

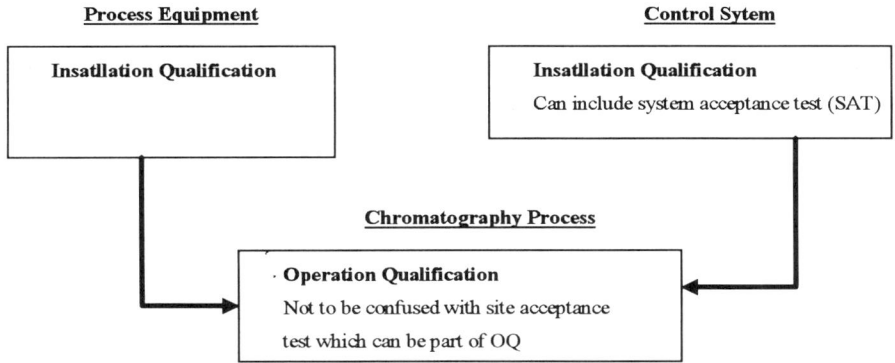

Figure 32. Combining the process equipment and the control system to perform OQ

Fundamentally operational qualification of the equipment should check that fluids flow where they should when they should and at the flow rates required by the process. When writing the procedures reference should be made to the user requirement statements and the equipment specifications drawn up during the design process. However the testing should focus on the requirements of the process and not the capabilities of the equipment.

OQ should include validation of the support processes such as CIP and SIP. CIP validation should check coverage of vessels and tanks, flow rates within pipes, maintenance of desired cleaning agent concentrations throughout cleaning cycles and the removal of residual cleaning agents to predetermined specifications. SIP validation should ensure the target temperature/time profiles are reached throughout the vessel. This is achieved by identifying likely "cold spots" within the system and probing these with temperature probes or biological indicators.

The validation of column sanitisation procedures is essential to reliable and reproducible production of hygienic product. It is important to minimise bioburden loading onto the column by filtering all solutions on the column and using suitable bioburden reduction processes on the associated equipment, CIP and/or SIP. The sanitisation process itself should be validated in laboratory studies where the sanitisation procedure is run on small scale columns inoculated with a range of microorganisms considered typical by the regulatory authorities. These studies will be intended to show the sanitisation procedure reduces the microbial count within the process matrix by an appropriate amount. The validation studies will need to be supported by studies carried out by the column manufacturer that show the column design is inherently hygienic. These studies need to be supported by an ongoing column monitoring programme that demonstrates the column is maintained in a hygienic state.

The OQ programme will include the column packing process and subsequent column qualification tests. This will be the first time the column will be packed and it is important that it is carried out correctly, due to the expense of column matrices and the criticality of correctly packed columns to the purification process. The qualification tests, height equivalent to a theoretical plate (HETP) measurements, asymmetry factor (A_f) measurements and pressure flow characteristics will be carried out after column packing to demonstrate successful column packing. These tests are summarised in Appendix 1.

The control system will be tested integrally as part of all of these operational checks. The sequences, alarms and similar items. will have already been tested at the vendors fabrication site. All that is required here is to test that physically the control system has been commissioned properly and that the various control terms have been established and optimised.

3.10.1 Support Operations

It is at this point that the support operations such as SIP, CIP and sanitisation are qualified. They will have gone through there own validation sequence up until now, but they are an integral part of the chromatographic step. They have to be validated to the specifications laid out in the process description.

Performance Qualification.

This is normally the simplest validation step if all the previous ones have been carried out correctly. In essence, the aim of performance qualification of the chromatography process is to run product through the column and show that the purification has occured to the desired limits, usually set out in the URS.

CIP and sanitisation is proven within limits after processing using the validated assay methods

4. SUMMARY

This paper summarises the validation process for biopharmaceutical chromatography processes. It details the key elements of the chromatographic process before examining the key elements of the validation process, with reference to chromatography systems. In recognition of the importance of examining the process in totality, reference is made throughout the paper to the support activities, such as CIP and SIP and automation. The validation process covers the project life cycle from user requirement specifications through all the qualification stages, including commissioning.

BIOGRAPHY

Keith Williams has been a senior consultant with Validation Technologies (Europe) Ltd for over three years. He has a degree in Genetics and Microbiology and a masters in Biochemical Engineering with some 10 years experience in the Biopharmaceutical and Validation fields.

VTEL are a process engineering and validation consultancy group, to most of the blue chip companies in the pharmaceutical sector, with offices in Nottingham and Dublin and sister offices in Montreal, Toronto, Newark, and Boston.

Chris Davis is a Process Engineering Consultant for Tanvec Ltd where he inputs to a wide variety of pharmaceutical projects. As a chartered chemical engineer he is involved in all aspects of process plant design from consultancy work and feasibility studies through to detailed design and commissioning. He has extensive experience in the large scale design and operation of biopharmaceutical plants in general and of chromatography systems in particular.

Tanvec are an international design and consultancy company specialising in the pharmaceutical and fine chemicals industries. They can provide a full range of services from consultancy through to the provision of turn key packages for facility design and build. In addition they can provide comprehensive service in the area of regulatory compliance.

REFERENCES.

1. Bala, G. (1994). Pharmaceutical Engineering, 57 - 64.
2. Parenteral Drug Association. (1992). Journal of Parenteral Science and Technology, 46, (3), 87-97.
3. Ransohoff, T. C. *et al.* (1990). Biopharm., 20-26.
4. Dannapel, B. *et al.* (1997). Pharmaceutical Engineering, 76-92.
5. Steiner, J. (1997). Clinical Development Strategic, Pre-Clinical, Clinical and Regulatory Issues. Interpharm Press.
6. PMA QC Section, Pharmaceutical Technology Europe: January 1994, 37 - 42.
8. Guideline on General Principles of Process Validation
9. Guide to Inspection of Computerised Systems in Drug Processing (02/83)
10. Guideline for Submitting Documentation for Sterilisation Process Validation in Applications for Human and Veterinary Drug Products

11. Guideline for submitting Documentation for the stability of Human Drugs and Biologics (02/87)
12. Guideline for Submitting Documentation for the Manufacture of and Controls for Drug Products (02/87)

APPENDIX 1: METHODS FOR THE DETERMINATION OF COLUMN QUALITY PARAMETERS

SAMPLE APPLICATION

1. Equilibrate the column according to standard procedures.
2. Make up a solution of sample in the column equilibrant. The sample should be a substance that can be detected by the downstream detectors but is inert to the column matrix. It is common to use a 2% (v/v) solution of acetone as this generally fufils the criteria described. It is important that the substance used is approved for use according to standard QC procedures.
3. Add the sample to the column in a manner that minimises the dilution of the sample before it is disperesed within the matrix. If acetone is used it is common to add 2% of the column volume.
4. If acetone is used it is important to ensure the downstream UV detector is set at suitable range to detect the sample. This may require reducing the sensitivity of the detector.
5. Elute the sample at a flow rate which minimises turbulent flow, which would adversely affect the results. This is often the operational flow rate for small columns with narrow bore tubing. However larger columns with larger diameter pipe may need to be run at lower flow rates.
6. When sample peak has eluted calculate the HETP and A_f as detailed below.

HETP AND A_F CALCULATIONS.

1. Obtain two plots of the detector against time. One should cover the period from when the sample elution began (plot 1), whilst the second (plot 2) should include an enlarged print of the sample peak.
2. An example of the measurements to made from the plot is shown below in figure X. On plot 1 draw a vertical line from the top of the peak down to the x-axis and note the time. Subtract the start time from this value to obtain the elution time, T_e.
3. On plot 2 draw two lines parallel with the straight rising and falling sections of the peak ensuring they cross the x-axis. Draw a horizontal line across the peak base at the same absorbance value as the baseline reading. Calculate the time encompassed in the region of the horizontal line by the lines parallel to the rising and falling peak sections. This the peak width time, T_w.

4. On plot 2 calculate the peak height, h, the difference between the top of the peak and the baseline.
5. At 10% of the peak height draw a horizontal line across the peak and calculate the width of the peak on the leading and tailing edges, the bisector being the vertical line. The leading width is T_a and the tailing width is T_b.
6. Calculate the HETP and A_f as detailed below.
   N = number of theoretical plates = $16(T_e / T_w)^2$.
   HETP = L / N (cm) where L is the bed depth in centimetres.
   $A_f = T_b / T_a$

Diagram 1: HETP and A_f plot measurements

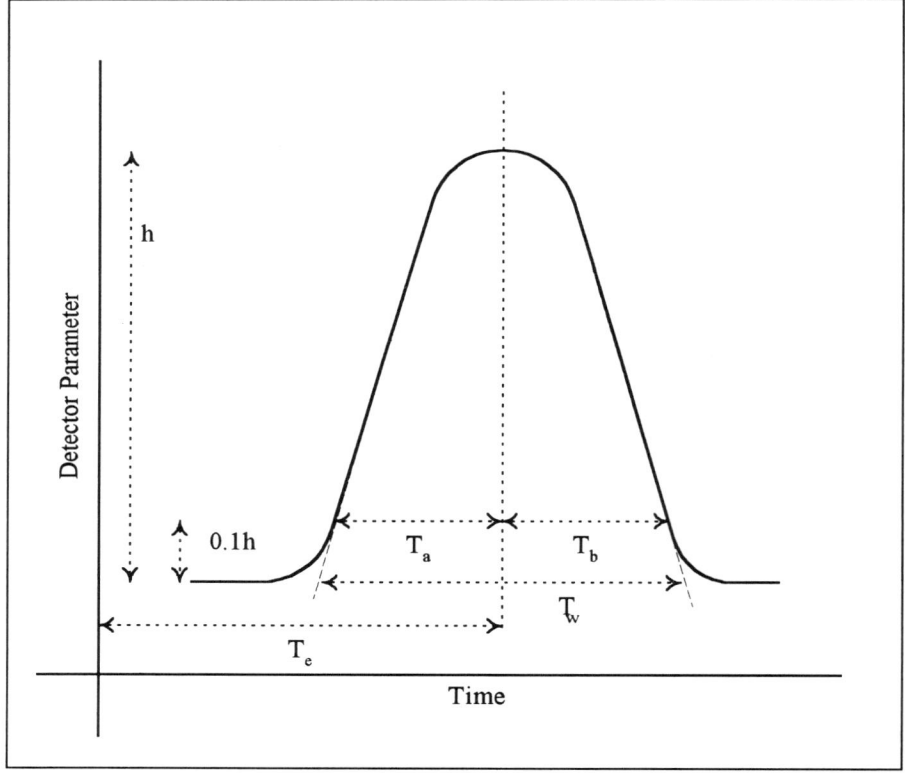

GENERIC PRESSURE FLOW CURVE PROTOCOL.

1. Equilibrate column with the buffer that will be used to carry out the measurements. It is common to use the buffer that will generate the highest pressure drop across the column when it is in use.
2. Run the buffer through the column at a low flow rate, wait for system to stabilise and then record the pre- and post-column pressures. Increase the flow rate and repeat the above.
3. Continue until the operating flow rates are exceeded.
4. Repeat in decreasing increments.
5. Plot the data.

APPENDIX 2: A TYPICAL DOCUMENT MATRIX

Section	Description
1.0 Specification Qualification	
1.1 User Requirement Specification	Definition of User Requirements
1.2 Standard Specifications	List of Standard User Specifications User Standards
1.3 List of System Specifications	Specifications used Tender Issue of Specifications and Documentation (must includes both URS)
1.4 Specification Amendments	List the amendments to Tender Issue Specification Construction Issue of Specification and Documents Supporting Documentation
1.5 Purchase Orders	List of the Amendments to Orders Copy of Purchase Orders Copy of Purchase Order Amendments Minutes of Meetings with Vendors
1.6 Specification Qualification Certificate	Approval of Specification Qualification

2.0 Design Qualification		
2.1 Vendor Drawings	List of All Vendor Design Drawings	
	General Arrangement Drawings	
	Plans and Elevations	
	Detail Drawings	
	Cross Sectional Drawings	
	P & I Drawings	
	Weld Details	
	Nozzle Orientation/Details	
	Electrical Schematics	
	Electrical Wiring Diagrams	
	Loop Diagrams	
	Control Panel Layouts	
	Pneumatic/Hydraulic Control Diagrams	
2.2 Vendor Materials and Equipment	List of Vendor Materials	
	Valve List	
	Lines/Pipes List	
	Instrument List	
	Pump List	
	Filter List	
	Pressure Vessel List	
	Lubrication List	
	Proposed Materials of Construction	
	Equipment Weights	
	Service Consumptions	
	Motor List	
	Electrical Load List	
	Cable List	
	Termination List	
	Control Panel List	
	Control Hardware List	
	I/O List	
	Parts Lists	
	Recommended Spare Parts List	
2.3 Vendor Technical Specifications	List of the Vendor's Technical Specifications	
2.4 Vendor Documentation	Vendor QA System	
	Vendor Quality Plan for System	
	Software Test/Fault Logs	
2.5 Design Review	Test Protocol	
	Design GMP audit	
2.6 List of Exclusions/Changes/to	Revision Pages to Specifications	

Specification	
2.7 Design Qualification Certificate	Approval of Design Qualification

3.0 Installation Qualification	
3.1 Vendor Drawings	As Built Drawings
	Amendments/Changes to DQ Drawings
3.2 Materials used in Installation	Materials Certificates (inc. surface finishes)
	Amendments/Changes to DQ Materials
3.3 Inspections	Agreed Inspection Protocol
	Inspection Reports
	Installation GMP Audit
	Third Party or Statutory Inspection Reports
	Copies of User Audit Reports
3.4 Factory Acceptance Test (FAT)	Agreed FAT Procedures & Instrument Calibrations
	FAT Reports
	Concessions/Requests/Technical Queries
3.5 Weld	Materials Certificates
	Welders Qualification Documentation
	Weld Location/Identification List
	Weld Inspection Reports
3.6 Cleaning and Passivation List	Agreed Procedure For Cleaning and Passivation
	Cleaning/Passivation Acceptance Reports
3.7 Installation and Operating/ Maintenance Manuals	List of Manuals
	Supplied Spare Parts List
	Operating and Maintenance Manuals
	Installation Checklist
	Installation Instructions
	Instrument Calibration Certificates
3.8 Final Installation Inspection	List of Inspections
	Agreed Inspection Protocol
	Final Installation Inspection Report
3.9 Installation Qualification Certificate	Approval of Installation Qualification

4.0 Operational Qualification

4.1 Site Acceptance Tests (SAT)	List of Site Acceptance Tests
4.2 Documentation	Calibration Methods
	List of Instruments to be Calibrated
	Instrument Calibration Certificates
	Agreed SAT Procedures
	Site Acceptance Tests Record Sheets
	SAT Report
4.3 System Acceptance Certificate	Certificate to Vendor for System
4.4 OQ Documentation	Agreed List of OQ Tests
	Approved OQ Procedures
	OQ Test Record Sheets
	OQ Test Report
4.5 System Hand-Over Certificate	Report to User for System
4.6 Operational Qualification Certificate	Approval of OQ

Chapter 15

Validation of Water for Injections (WFI) for Biopharmaceutical Manufacture

Paschal Baker and Wael Allan
Raytheon Engineers & Constructors UK, Validation & GMP Compliance Group

Keywords: Design, qualification, sampling philosophies and regimes

Abstract: This chapter discusses design and validation of Water for Injection generation and distribution systems. It covers not only the validation activities required for new systems, but also the approach taken for existing systems.

1. INTRODUCTION

One of the main problems encountered in industry is water systems, despite the fact that virtually all pharmaceutical and biopharmaceutical facilities possess a pharmaceutical grade water system. The major reason for this is that many water systems were designed many years ago, were not designed in accordance with Good Manufacturing Practices (GMP) and sterile manufacturing requirements, or the design was inconsistent with GMP requirements of industry (1).

Before discussing the validation of biopharmaceutical water systems, it is beneficial to view the types of water systems available for the manufacture of pharmaceutical and biopharmaceutical products. A look at design considerations is vital as this has an impact on the approach to validation.

2. TYPES OF WATER SYSTEMS

For the manufacture of pharmaceutical and biopharmaceutical products, it is essential that water does not contain chemicals which may react with active ingredients in the product. In addition, it must be free from pathogenic organisms, toxic substances and any other contaminants which may adversely affect the product, its appearance, shelf-life and other similar properties. If the water is being utilised in the manufacture of a parenteral product, the water must also have a bioburden less than 10 cfu/100ml (colony forming units per ml) and be free from pyrogens.

The two principal types are Purified Water (PW) and Water for Injection (WFI). Occasionally a lower grade of water is used in bulk pharmaceutical facilities, i.e. water produced by deionisation or demineralisation. This type of water has no pharmacopoeial specifications but it would certainly be expected to meet the specifications for potable water (2) as a minimum. Whatever grade of water is used, it is a requirement that the quality is controlled.

2.1 Water Specification

2.1.1 Purified Water

Purified Water (PW) is defined in the United States Pharmacopoeia 23 (USP 23) (3) as water obtained by distillation, ion exchange treatment, reverse osmosis or other suitable process.

It is defined in the current European Pharmacopoeia (4) as water prepared by distillation, by means of ion exchange or by any other appropriate method. Purified water is not to be used in direct contact with the preparation of parenteral solutions, however it can be used for example, in the washing of product contact equipment, providing that the final rinse is with WFI.

2.1.2 Water for Injection

Water for Injection is defined in the USP 23, 1995 as water obtained by distillation or reverse osmosis. The only method of preparation stated in the current British Pharmacopoeia (5) and Ph. Eur. (4) is distillation of potable or purified water.

A comparison of PW and WFI is given in Table 36.

Validation of Water for Injections (WFI) for Biopharmaceutical Manufacture

Table 36. Comparison of Water Quality Requirements for USP 23

	USP Purified Water	WFI Water
pH	5.0-7.0	5.0-7.0
Chloride	≤ 0.5 mg/l	≤ 0.5 mg/l
Sulphate	≤ 1.0 mg/l	≤ 1.0 mg/l
Ammonia	≤ 0.1 mg/l	≤ 0.1 mg/l
Calcium	≤ 1.0 mg/l	≤ 1.0 mg/l
Carbon dioxide	≤ 5.0 mg/l	≤ 5.0 mg/l
Heavy metals	≤ 0.1 mg/l as Cu^{2+}	≤ 0.1 mg/l as Cu^{2+}
Oxidisable substances	meets specified test	meets specified test
Total solids	≤ 10 mg/l	≤ 10 mg/l
Bacteria	≤ 100 cfu/ml	≤ 10 cfu/100 ml
Endotoxins (Endotoxin Unit, EU)	-	≤ 0.25 EU/ml

As can be seen from Table 36, the requirements for chemical content are the same for PW and WFI. The differences between the two types of water are microbiological limits and the specification for endotoxins. The specification for bacterial content is much tighter for WFI, and there is no requirement to test for endotoxins in PW. In the fifth supplement (6) to the USP 23 the testing requirements have been modified to remove the necessity for extensive chemical testing and substitute with tests for conductivity and Total Organic Carbon (TOC), which are easier to perform, more reliable and quantitative. These changes will be discussed later in this chapter.

Common to both PW and WFI is the pre-treatment of feed water. Pharmacopoeia usually refer to drinking water or town water, however in many cases, the starting point could be artesian well water.

3. PRE-TREATMENT

Pre-treatment is normally characterized by two sequential activities, filtration followed by primary water treatment, and these are described below.

3.1 Deep Bed Filtration

Deep Bed Filtration uses either sand or multimedia filters to remove colloidal materials and extraneous particles (7). Sand filters use silica sand, whereas multimedia filters consist of a variety of sands, coarse anthracite over graded sand and quartz. Multimedia filters provide a more uniform distribution of particulate solids throughout the depth of the bed than sand filters. The filtration process through a filter bed will continue until the filter bed becomes blocked with particulates, restricting the flow of water through the bed. To remove the particles blocking the bed, the filter is backwashed, rinsed and can then be re-used. The different materials utilised in a multimedia bed are of varying densities and form layers which remain in place even during backwashing. Therefore, this bed achieves better distribution of the particulate solids. Generally, backwashing in a sand filter causes the fine sand to collect at the top. Therefore, a sand filter will tend to become blocked with colloidal material nearer the top of the bed during backwashing, with the lower area of the bed remaining relatively unused.

3.2 Activated Carbon Filtration

Activated carbon filters are used to remove chlorine and soluble organics in order to prolong the life of the anion bed in the deioniser. As filters lose their adsorptive capacity, chemical testing of the downstream water should be conducted in order to determine when the filter needs to be replaced (8). As organic materials are trapped by carbon filters, this coupled with chlorine removal provides an ideal growth medium for bacteria. High microbial counts may compromise the ability of the distillation process, as there is a potential for endotoxin carryover, therefore it is vital to maintain the filters. Bacterial growth can be eliminated by sterilisation with steam or by caustic soda.

3.3 Water Softening and Deionisation

Softening is an ion exchange process in which calcium and magnesium ions are removed and exchanged with sodium ions. It is performed to prevent the formation of calcium compounds (scale) in the Reverse Osmosis units. As with activated carbon filtration, high microbial counts may challenge the distillation process and compromise the ability of the distillation process to effectively remove microbial contaminants and endotoxins. There is also an increased likelihood of carry over of viable cells and chemical impurities by entrainment. Therefore, it is necessary to monitor conductivity of the water and microbial counts, in order that an increase in either or both of these characteristics is an indication that regeneration is required. The softener is regenerated using concentrated sodium chloride solution. However, as it becomes necessary to regenerate the softener more frequently, the resin beds are nearing exhaustion and should be replaced.

3.4 Decarbonation

This treatment step is dependant upon the quality of feed water. Generally, if the level of bicarbonate is in the order of 15-20mg/l, then decarbonation is performed. The presence of high levels of bicarbonate and carbon dioxide can decrease the life of the ion exchange resins.

4. PURIFICATION

Following pre-treatment of the feed water, there are several generation and purification steps needed to produce water of the required specification.

There are a number of methods for the production of WFI, however the generation methods usually use either reverse osmosis as the final process, or most commonly, distillation. Both distillation and reverse osmosis (RO) are listed as acceptable methods of WFI generation in the USP, but RO is not included as production method in the British and European Pharmacopoeia. However, if in the licence application for a particular product pharmacopoeial WFI is not specified, there are other methods of generation available to the manufacturer. WFI is a requirement for parenterals, dialysis fluids and irrigation solutions. It is also a requirement for water for nebulisation in the European Pharmacopoeia.

4.1 Reverse Osmosis

Reverse Osmosis is a physical process in which the pre-treated water is passed across a semi-permeable membrane under high pressure. This high pressure, in excess of the osmotic pressure, causes the water to pass through the membrane, leaving a high concentration of dissolved salts upstream of the filter. RO is thus capable of reducing the total dissolved solid (TDS) content of a raw feed water to 5% of its initial value, providing a high quality treated water which is suitable for a wide range of applications.

With two-stage RO the permeate, or treated water stream, is passed through a second array of membranes in order to further reduce the TDS. At the present time, RO is the most common process used for water pre-treatment, but since the previous treatment steps do not completely rid the system of bacteria, there will always be bacteria on the upstream side of the RO, which require control and monitoring. Therefore, there is always a potential risk of production of large quantities of endotoxins, which may pass through the membrane should a leak develop. In this instance, there is a significant possibility that biological impurities will be carried over to the downstream side of the membrane and result in contamination of the distribution system. WFI generation systems utilising RO as the method of purification require careful consideration during validation, due to the potential risks associated with contamination of the distribution system.

4.2 Distillation

Distillation is a method of obtaining chemically and microbiologically pure water. Stills purify the feed water by vaporisation, separation of droplets from the vapour by circular velocity or entrainment separators and condensation of the purified vapour in a heat exchanger. There are three main methods of distillation, namely single effect, multiple effect and vapour compression stills, all of which operate using the same general principles.

Most of the problems which occur with distillation are related to poor feed water quality, as discussed in the earlier sections. Poor quality feed water can overchallenge the ability of the distillation system to effectively remove endotoxins. If the feed water is extremely contaminated, these contaminants may be entrained in the distillate. In addition, if the pre-treatment system is of poor design, or is not functioning correctly, there may be large quantities of hard water deposits in the feed water to the still. This will form scale on the heat exchanger which will eventually lead to clogging, or decreased throughput.

5. DESIGN GUIDELINES

When considering the requirements for validation of a WFI system, it is essential to first consider all critical steps in the process within the design of the system.

There are many guidelines detailing the requirements for design of a WFI system. One of the more recent guidelines is the ISPE (International Society of Pharmaceutical Engineers) & FDA Baseline Guide, Volume 4 for Water and Steam (1). The key areas of design are covered below.

5.1 Materials of Construction

Below is a list of recommended guidelines for WFI systems.
- Product contact materials of construction are usually 316L stainless steel
- Polished pipework (often electropolished)

- Gaskets and elastomers such as EPDM, PTFE and PVDF must be able to withstand steam sterilisation and be resistant to passivation solutions
- Butt weld and sanitary fittings, no threaded joints
- Tri-clamp fittings
- Orbital welding and weld logging including 100% borescope, videoed
- Zero Dead Leg valves
- Double tubesheet Shell & Tube heat exchangers
- Temperature sensing devices should be mounted in wells. Pressure sensing instruments should be flush mounted with sanitary seals to isolate the instrument workings from the WFI

5.2 Holding and Distribution Systems

Below is a list of recommendations for design and installation regarding the holding and distribution systems.
- Sterile hydrophobic vent filters of 0.2 micron rating should be fitted on storage tanks. In hot services, the filter must be steam jacketed or electrically heated to prevent condensation
- WFI should be kept between 65 and 80°C and should be drawn off immediately before use. It should not be allowed to stand for more than 8 hours before use. Ambient water loops must be sanitised frequently.
- Piping must be installed in accordance with current GMP regulations, which state that piping must be mounted in a continuous slope, in order that it is completely drainable
- Equipment tubing and instrumentation must be completely drainable, with no cracks or hard-to-clean areas
- Sanitary pumps, complete with flushed double mechanical seals
- No dead legs of > 6 times pipe diameter.
- 2 - 2.5 m/s circulating velocity
- Passivation upon installation
- Routine sanitisation with ozone or heat
- Routine maintenance
- Sampling points
- UV lamps

Recently, some of these design features have been questioned. An example of this is the ISPE Baseline Guide, section 8.3.1, which proposes that all of the design features mentioned may provide a level of security, but that they may not be necessary in every system. However, the key to an effective WFI system is the design and it is difficult to justify not including these design features, since it is the omission of one or more of these design features which often leads to validation failures. This observation is illustrated in a case study presented later in this chapter.

The ISPE guideline addresses many design features separately, for example:

1. Pipework polished finish. Many systems are now specified as 100% electropolished throughout. The major benefit of electropolishing is ease of cleaning. However, this does not appear to be a relevant issue in a WFI system. The ISPE states that the benefits for electropolish or finishes smoother than 0.8 micron RA (180 grit) is questionable. "Mill finish" tube has been recommended as a sound minimum specification.
2. Sloped pipework. Systems are normally specified as having sloped pipework to be fully drainable. In the absence of an SIP (steam in place) system the need for sloped pipework is for draining following passivation and chemical sanitisation, rather than a process need. If the system is not designed to be completely drainable this can cause problems. In this instance the mechanism to remove all traces of chemicals is to flush the system and sample the rinse water. This process is repeated until there are no residual chemicals in the rinse. The ISPE recommend that while there is no requirement for systems which do not use SIP to be fully drainable as long as water is not permitted to stagnate in the system, it is good engineering practice to allow for drainage of both pipework and equipment.
3. Welding. Orbital welding is generally recommended, although some manual welding will always be required. 100% inspection by borescope is becoming the industry standard, with a videotaped record. However, 100% photographic or Radiographic analysis is neither cost effective or infallible (1). Appropriate sampling is the preferred approach. Careful consideration should be paid to orbital welding of dead legs. 100% orbital welding can create dead leg problems due to types of equipment used.
4. Dead Legs. Zero Dead Leg Valves are becoming more common, but are increasing expensive. A good alternative may be the use of minimum distance diaphragm valves. The standards can be conflicting and difficult to interpret. For example, certain regulations state that the maximum dead leg should be six times the pipe diameter for pharmacopoeial grade water systems, and 2.5 times the pipe diameter for clean steam systems, where other regulations state 3 times the pipe diameter. The ISPE stand-point is as follows. "We propose avoiding a hard rule of thumb for maximum dead legs. Good engineering practice required minimising the length of the dead legs and there are many good designs available for doing do."
5. Sanitary Fittings and Pumps. The recognised industry standard is a tri-clover or similar fitting. However, there are other hygienic fittings which are perfectly acceptable. Ultra-high specification "WFI" pumps are very

expensive, and it is becoming recognised that many other standard hygienic design pumps are acceptable for both PW and WFI systems. A typical USP WFI system is shown in the following figure (Figure 33).

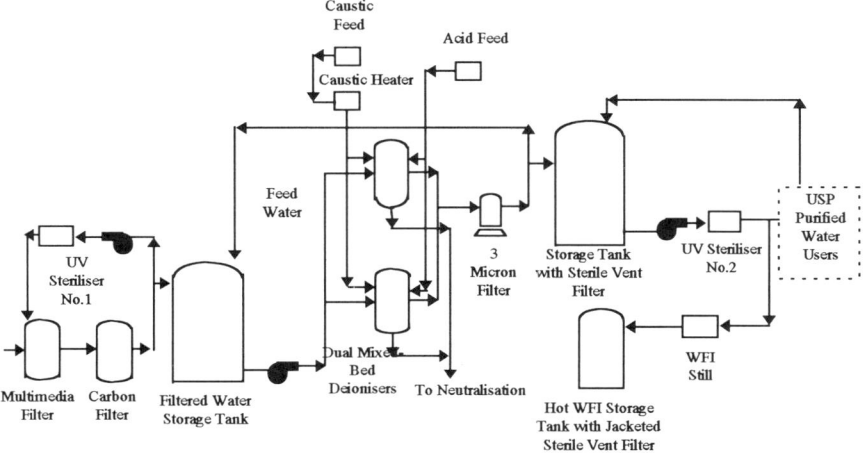

Figure 33. A Typical WFI System

The system flow diagram shows production of pharmacopoeial quality PW and WFI, based upon ion exchange and multi-effect distillation and consists of the following components:
– Multi-media filter
– Carbon filter
– Carbon filter circulation pumps
– Ultraviolet steriliser (pre-treatment)
– Filtered water storage tank
– Filtered water transfer pumps
– Mixed bed deioniser
– Acid feed system
– Caustic feed system (NaOH)
– Caustic heater
– 3 micron filter
– Ultraviolet steriliser (purification/generation)
– Storage tank
– Multiple effect still

This system includes ultraviolet disinfection. UV disinfection can be an effective method of reducing the numbers of micro-organisms in the water and may be necessary because removal of chlorine by the carbon filter leaves the water prior to de-ionisation steps susceptible to microbial build-up, which is not always totally removed by the deionisers and RO units. Filtration through sterilising grade filters will achieve the same objective, however, such filters would require frequent replacement to avoid bacterial overload and "grow-through". A combination of UV disinfection and filtration is probably most effective. However, UV can increase the amount of endotoxins in the water, as when the bacteria are disrupted, endotoxins are released (7). As the growth period of bacteria can be relatively rapid, unless UV can demonstrate effective reduction in bacteria on a per system basis its use can be of limited value (9).

The hot caustic is used to sanitise the carbon beds and also during regeneration of the anion exchanger. The 3 micron filter is used to remove any resin fines remaining in the water following the mixed bed deioniser process step.

6. VALIDATION

Having discussed the basic design principles for a USP WFI system, it is time to consider the requirements for validation of the system. Much of the design basis will have a direct effect on the types of sampling and testing required to validate the system.

6.1 Introduction

Validation of a WFI system is intended to assure that the water provided by the system will consistently and repeatedly achieve the quality attributes specified, when operated in accordance with the relevant operating procedures. The philosophies adopted for the validation of new and existing systems differ in specific ways which will be discussed later in this section.

It is essential to rely upon the system design and operation to ensure compliance with the appropriate regulations and validatability. The design and operation of the system will determine the quality of the water produced. Sampling will allow the quality of water being produced to be quantified.

6.2 Installation Qualification

The Installation Qualification (IQ) is a documentation process which verifies that the physical components of the system have been installed according to design specifications. The IQ will serve as the final major component and system quality audit prior to Operational Qualification (OQ).

Traditionally, when the installation and commissioning has been completed, IQ will take place. More recently, the Validation Turnover Package (VTOP) approach has become increasingly used for new systems and equipment. The VTOP consists of documentation (specifications, manuals, drawings etc) that fully characterises a system. The VTOP is compiled in a formal and organised package to serve as part of the Installation Qualification (IQ). This serves as a cost and time effective way of performing part of the IQ, and requires an integrated effort by engineering and construction departments, commissioning and validation.

The VTOP will cover all general engineering and technical documentation and drawings, electrical components and instrumentation calibration, PLCs and other automated systems, construction, installation and certification documentation, testing and commissioning documentation and a report listing any discrepancies. All testing to be conducted should be carried out independently.

The following is a description of typical sections of a VTOP for a WFI system, the sections may contain original documents, copies or references to the location in which they can be found.

1. General Documentation - to include all specifications, purchase orders, engineering documentation and Operation and Maintenance manuals.
2. Electrical Components and Instrumentation Calibration - to include reference to the electrical standards, hardware and software inventory and specifications, wiring diagrams, schematic diagrams, loop diagrams, interlock descriptions, critical instrument listing and calibration procedures and certificates.
3. PLCs and Automated Systems - to include Quality Programme / Technical Development Standards, software identification and associated documentation (i.e. functional specifications, annotated ladder logic, software functionality flow charts etc.).
4. Construction, Installation and Certification Documentation - to include quality standards / criteria, reference procedures and standards, construction and installation logs, materials certificates, welding procedure, inspection, map and certification, zero dead leg verification, slope verification, pressure test procedure and certification, cleaning and passivation reports.
5. Testing and Commissioning Documentation - to include Factory Acceptance Testing (FAT) and commissioning documentation.
6. Filter Integrity Testing of filters within the generation and distribution systems, as applicable.

Most of the points covered in "traditional" IQ activities can be covered in the form of a VTOP, which enables the documentation gathering process, as the vendor / manufacturer is directly involved in the provision of this documentation.

6.3 Operational Qualification

Operational Qualification (OQ) is a testing process which evaluates functionality by demonstrating that the system is operational and will perform within the required operating parameters. The OQ also serves as a final major component and systems operational audit prior to conducting the Performance Qualification. During the OQ, data is collected concerning critical processing parameters which could affect performance. For a new system, the data generated during the OQ execution will be verified against the equipment specification. For an existing system, often the vendor specification documents are not available, and therefore the data collected may be used as a baseline.

Generally, the OQ covers verification of operation of motors and pumps, monitoring of critical system parameters, i.e. temperature, pressure, flow etc. and extensive testing of the control system to assure correct functionality both for normal operation and in alarm conditions. In addition, it is necessary to verify that the sterilisation cycle is operating as intended, in accordance with the relevant specifications. This is achieved by conducting a minimum of three sterilisation cycles whilst temperature mapping the system using appropriately calibrated thermocouples. The temperature throughout the system should be monitored at one minute intervals or more often, and the data should be evaluated to determine the location of any cold spots and potential problematic areas. The data gathered from these tests will be utilised during the PQ activities.

6.4 Performance Qualification

Performance Qualification (PQ) is a testing process which evaluates the effectiveness and consistency of the system during production as defined by the manufacturing specification. The PQ verifies that the system will function in accordance with design and required specifications. During the PQ, data is collected concerning critical parameters with the purpose of evaluating the system performance.

There are two main sections to the PQ testing, verification of performance of the sterilisation cycle, and testing of the generation and distribution systems, to verify the quality of water produced. The performance of the sterilisation cycle must first be verified using suitably calibrated thermocouples and performing F_o calculations based upon the D-value of the most resilient bacteria in the local environment. The thermocouples should be placed such that the entire system has been taken into account and covered, and any potential cold spots determined by testing conducted during OQ activities should be targeted, as these locations represent the worst case.

The USP test specifications were modified in 1997 in the fifth supplement to the USP 23. Previously, the testing called for a number of separate tests for chemical content. The fifth supplement has eliminated the majority of this testing in favour of conductivity measurements and Total Organic Carbon (TOC) - an indirect measure of organic molecules present in the water, more in line with standard equipment used in industry to monitor the performance of pharmaceutical water systems. The microbiological and endotoxin tests remain unchanged. Table 37 shows the comparison of the testing required for the USP 23, and the fifth supplement.

Table 37. Comparison of Testing Requirements for WFI

Test	USP 23	Fifth Supplement to USP 23
pH	5.0-7.0	5.0 - 7.0
Chloride	≤ 0.5 mg/l	deleted
Sulphate	≤ 1.0 mg/l	deleted
Ammonia	≤ 0.1 mg/l	deleted
Calcium	≤ 1.0 mg/l	deleted
Carbon dioxide	≤ 5.0 mg/l	deleted
Heavy metals	≤ 0.1 mg/l as Cu^{2+}	deleted
Oxidisable substances	meets specified test	alternatively perform the test specified for TOC
Total solids	≤ 10 mg/l	deleted
TOC	not listed in USP 23 - new requirement of fifth supplement	meets the requirements of the USP specified test, alternatively perform the test for oxidisable substances
Conductivity	not listed in USP 23 - new requirement of fifth supplement	meets the USP specified test
Bacteria	≤ 10 cfu/100 ml	≤ 10 cfu/100 ml
Endotoxins (Endotoxin Unit, EU)	≤ 0.25 EU/ml	≤ 0.25 EU/ml

Subsequently, in the USP 23 eighth supplement (1998) (10) the requirement for pH measurement has also been removed.

In order to verify the performance of a system with regards to water quality PQ there are three sampling phases, which are discussed below:

1. Phase 1 - Extensive monitoring for a period of at least 14 days. Each use point should be monitored daily and tested for microbial contaminants, endotoxin content and particulate content. Conductivity and TOC measurements should be taken daily from a single sample point. This phase usually lasts for 14 days.
2. Phase 2 - Upon successful completion of phase 1, the second phase of monitoring will follow the same time period and sampling scheme as the first, but with a reduction in sampling points. This is conducted to prove that the system is able to consistently and repeatedly meet the requirements of the WFI specification.
3. Extended loop quality testing. This phase is designed to demonstrate that the correct operation of the WFI generation and distribution systems will consistently produce water of the required quality. The extended loop testing will be conducted over the course of a year (52 weeks), and results from phases 1 and 2 may contribute to this time period. This phase of testing is also intended to prove that seasonal variations do not affect the quality of water and to verify that all measurements taken and sterilisation frequencies are satisfactory.

6.4.1 Establishing a Sampling Plan

The FDA guidelines state that each usepoint on the WFI system should be monitored during the validation activities. However, this is not always possible, as some WFI systems have a great many sampling points, which would make it unrealistic to sample each and every one. One approach which is being used more commonly is to sample each use-point for 14 days and then to reduce the sampling during phase 2. The key to successfully achieving such a sampling scheme is to select points on the system which provide an overall picture of the entire system, including worst case points, such as at the end of a cold leg, or in difficult to clean areas. A sampling scheme is describe in section 6.6.3 for systems with many use points.

As discussed earlier in this chapter, the approach towards validating a new system will differ in several ways from the approach used when validating an existing system. The following sections detail two validation case studies, one for a new system and the other, an existing system.

6.5 Validation Approach for a New System

The following section suggests an approach for validating a new system which starts with the design of the system.

Validation of Water for Injections (WFI) for Biopharmaceutical Manufacture

6.5.1 System Description

For the purposes of this example, a typical WFI system is described, the major items are as follows; a USP Purified Water feed system, a multiple effect distillation unit, storage tank, circulation pump, delivery pipework and two point of use heat exchangers. The WFI system is PLC controlled, and constantly monitored for critical parameters, including feed water conductivity, column temperature and condensate conductivity. The PLC will also reject the distillate prior to feeding into the WFI tank if it does not meet the setpoint values.

The feed water (PW) enters the first heat exchanger of the pre-treatment system and is pre-heated (tube side) by the pure steam condensate and by the distillate coming from the last column. The heated feed water then passes into two pre-heaters, where the distillate is used as the heating medium. The water then feeds into the first column of the distillation unit. The distillate is piped into the pre-heaters of the pre-treatment system. The distillate from the first column is fed to the second column and in turn to the remaining columns comprising the multiple effect distillation unit. The pure steam produced in the last effect is condensed and cooled in the heat exchangers, then fed to the WFI tank awaiting distribution. A schematic of the system is shown in Figure 34.

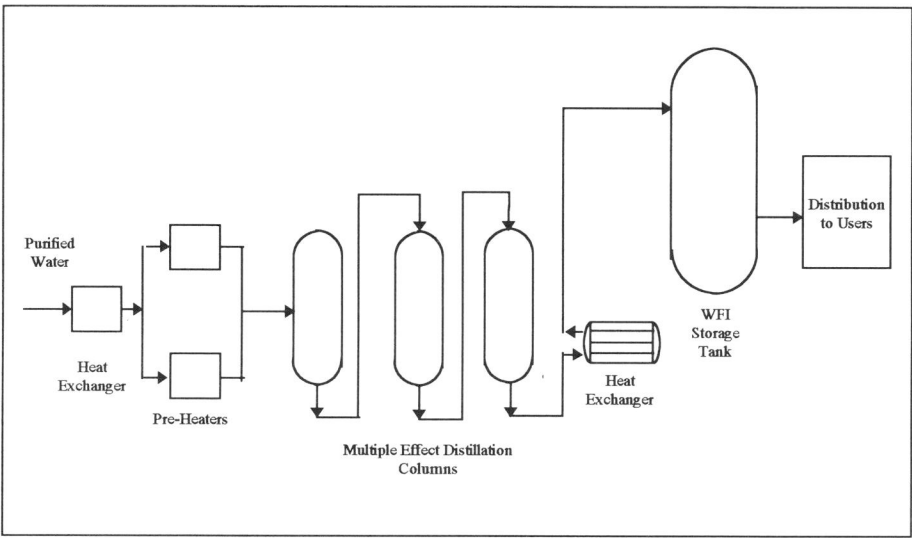

Figure 34. Schematic Diagram of New WFI System

In this particular example, the purified water validation will not be considered. However, it should be noted that the validation of the WFI system would be dependent upon the successful completion of the PW system.

6.5.2 Design Audits

When validating a new WFI system, it is important to ensure GMP compliance and sound design of the system prior to installation. GMP reviews of the system design are normally performed at 30% and 70% of completion of design. This review highlights any design characteristics which are non-compliant with GMP and they can then be corrected. Regular GMP audits play a key role in facilitating the validation exercise and provide an element of traceability, documenting all decisions made and modifications planned.

6.5.3 Installation Qualification

The VTOP method can be a very effective way of obtaining all the information required, which is normally collected during IQ. Information from vendors, manufacturers and sub-contractors, as applicable, is compiled in the VTOP to include certification for materials of construction, weld documentation, technical specifications, factory and site acceptance testing where performed and operational & maintenance manuals. The VTOP package is then reviewed during IQ activities, significantly reducing the amount of time required gathering information. If adopting this approach towards validation, it is key that all user requirements, including a list of documentation required, are clearly defined and are understood and agreed by all parties involved.

6.5.4 Operational Qualification

The OQ will consist of extensive operational testing of the system controls to ensure repeatability and consistency of operation. Some aspects of OQ are often performed during Site Acceptance Testing, and if documented correctly these can constitute a part of the OQ. Temperature mapping during sterilisation cycles should be performed to verify the operation of the cycle(s) and confirmation of sterilisation. Additionally, any cold spots in the system will be identified and subsequently be used during the PQ protocol execution.

Validation of Water for Injections (WFI) for Biopharmaceutical Manufacture

6.5.5 Performance Qualification

The system is again mapped to provide a geometric pattern of heat distribution during sanitisation. The potential cold spots identified during OQ testing will be targeted during these challenges.

In addition, a sampling plan will be established as discussed earlier in this chapter, for phases 1 and 2. Following this Phase 3, the extended loop testing will commence. All data from this sampling will be analysed and any out of specification results will be the subject of an investigation, as per normal production conditions. The sampling would involve on-line monitoring of TOC and conductivity in addition to microbial and endotoxin analysis. It is recommended that full chemical testing in accordance with the European Pharmacopoeia be conducted at the beginning and end of the loop and selected testing be performed at each stage in the generation system. The types of testing performed should reflect worst case conditions.

The approach described here, encompassing design, commissioning and installation, has been used successfully on many FDA approved systems.

6.6 Validating an Existing WFI System - Case Study

An established European biopharmaceutical facility requested a GMP Audit of their existing WFI system, in order to determine the status of the system with regards to current regulations and standards. Existing Piping and Instrumentation Diagrams (P&IDs) were reviewed and initial recommendations were made based upon the findings of the audit. These recommendations included addressing control issues, and several physical modifications to the system.

The following section describes the system, and components.

6.6.1 System Description

The system is a single loop system, with branch loops for each floor of the building supplying WFI at 80°C. Users take water at 80°C direct from the loop with single lines or through coolers at 20°C via single lines. The basic materials of construction are standard pipe hand welded stainless steel 1.4571 (equivalent to 316 Ti), vessels and coolers (also 316 Ti), sterile type filters, stainless steel ball valves, and 316 Ti stainless steel control valves. In recent months, the demand for WFI has grown, and the system is now stretched beyond its capacity. The system is frequently drained, which causes a system shut-down, in turn the pumps fail and the loop cools.

The system can be divided into three sections: pre-treatment, generation and storage & distribution.
1. Pre-Treatment - City water is filtered through a rough filter in order to remove particles, and then treated in mixed-bed deionisers. The water is then further treated via sterile filters.
2. Generation - the pre-treated water is fed into distillation columns, the resulting distillate is fed into a dedicated storage tank.

1. Storage and Distribution - There are two storage tanks for the WFI system, both of which are heated. The main loop is divided into three sub-loops, each serving different areas of the facility. The WFI is circulated around the loop at 80°C via a centrifugal pump. The return water from all three sub-loops is routed via a heat exchanger prior to being re-circulated, and does not circulated via the storage tanks.
A schematic of the system is shown in Figure 35.

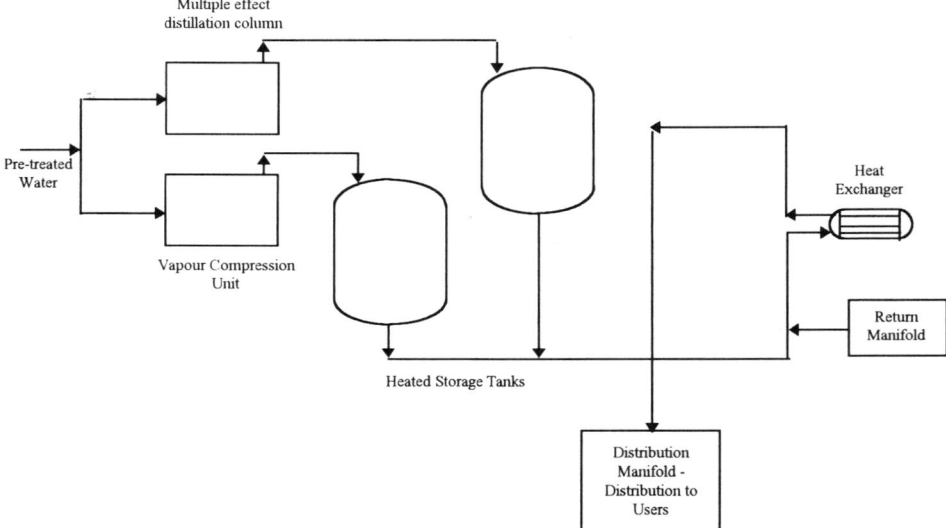

Figure 35. Schematic Diagram of Existing WFI System - Case Study No. 2

6.6.2 System Audit

The following list details the initial findings of the system audit.
- The materials used in the construction of the pipework do not meet current industry standards (i.e. SS 316Ti versus SS 316L). Most of the welding is manual, and it is unclear as to whether the system was passivated upon installation.

Validation of Water for Injections (WFI) for Biopharmaceutical Manufacture

- There are numerous dead legs downstream of the storage tanks In addition, many of the fittings and system components / instruments, including the circulation pump, are non-sanitary. Most of the valves installed are non-sanitary ball valves.
- The two storage tanks are fitted with internal heat exchangers, which are difficult to clean. The current practice is to use jacketed tanks, which eliminates the need to install hard to clean equipment within tanks.
- The return loop heat exchanger is of non-sanitary design, and coolers for each area served by the distribution system are mixed, some being of sanitary design, the rest being non-sanitary.
- The vent filters on the storage tanks are not heated which could cause blinding.
- Only part of the pipework is steamed in place (SIP). The tanks are not included in this activity. No temperature mapping studies of the sanitisation cycle have ever been performed.
- The current sampling scheme is inadequate in terms of identifying intermittent / recurring problems, and out of specification measurements are not fully investigated.

Based upon the findings and recommendations resulting from the initial audit, a further in-depth review of the system was performed and as the P&IDs did not reflect the "as-built" status of the system, current as-built drawings were prepared.

6.6.3   System Review and Sampling

It was decided to monitor the system for a period of three weeks in order to determine the performance of the system, prior to deciding to validate the system. A sampling plan was prepared, based upon the following philosophy, this case study refers to a project undertaken prior to issue of the fifth supplement to the USP, and therefore is based upon the previous sampling requirements of the USP. Full chemical testing is no longer a requirement of the USP, and this greatly reduces the need for periodic sampling for chemical testing, as TOC and conductivity monitoring can be conducted continuously on-line.

a) Particulate Sampling - the furthest warm use-point on the distribution loop on a per floor basis was taken as the worst case. It was considered that the hot use-point was more prone to particulate generation (i.e. corrosion) and that be sampling the end point on the distribution line prior to the filter, any potential particulate problems for that specific floor could be detected throughout the distribution ring. Samples were taken once per week.

b) Chemical Sampling - chemical sampling was performed on the generation line prior to the storage tank, in order to show the general efficiency of the water generation. Additionally, a sample was taken from the return line of the distribution loop to highlight any problems with the distribution system. It was considered sufficient to take samples once per week.

c) Microbial Sampling - The sample points were selected in order to provide an overall microbial analysis of the generation and distribution systems. The selection of these points was based upon isolating possible areas where microbial growth may proliferate (for instance, dead legs, no-flow areas, hard to sterilise piping and worst case distribution points). The location of filters also influenced the position at which the sample was taken. Where the in-line filter was fixed, the sample was taken after the filter, if it was removable, the sample was taken prior to the filter. Two/three samples were taken per week in order to provide an overall view of the generation and distribution systems.

d) Endotoxin Sampling - the sampling points were selected based upon the same criteria as for the microbial sampling plan, but samples were taken once per week only.

A small section of the sampling matrix for this system is shown in Figure 36.

Sample Date	\	Sample Points																			
		1				2				3				4				5			
		C	B	E	P	C	B	E	P	C	B	E	P	C	B	E	P	C	B	E	P
	1	X	X							X				X	X				X		
	2						X			X	X	X						X	X	X	X
	3		X												X						
	4						X			X	X	X									
	5		X			X	X							X	X				X		
	6																				
	7																				
	8	X	X								X				X			X	X	X	
	9						X		X	X	X	X							X		
	10		X											X	X						
	11						X			X	X										
	12		X			X	X							X	X			X	X		
	13																				
	14																				

Figure 36. Typical Sampling Plan for Existing System

The sampling matrix shows the sampling plan for Chemical (C), Microbial (B), Endotoxin (E) and Particulate (P) content, for a period of two weeks.

- Sample point 1 - Pre-treatment
- Sample point 2 - Distribution header
- Sample point 3 - Hot loop use point
- Sample point 4 - End of cold dead leg
- Sample point 5 - Use point

In addition to the sampling points identified in the matrix, several others were chosen, to generate an overall "map" of the distribution system. Due to the large size of the distribution loop, it was considered unnecessary and uneconomical to sample every use point during this initial system review. The results of the three week sampling plan showed erratic readings and indicated general lack of control of the system. All endotoxin sampling results were within the USP specification, hence no system modifications were recommended as a result of this sampling. Several of the samples taken for particulate analysis did not meet the required specification. This was attributed to the fact that some of the materials of construction, such as non-stainless steel valves were neither industry standard nor of hygienic design and additionally were prone to corrosion. It is also possible that particles may accumulate over a period of time, as the system had not been drained, cleaned and flushed to remove debris.

All samples taken for chemical analysis met the specified criteria. However, there were several failures of samples taken for microbial analysis, at pre-treatment, generation, storage and distribution sample points. The failures in the pre-treatment system were attributed to evaporation of chlorine from the feed water, allowing bacteria to grow. Another possible cause could be that the deionisation columns may harbour micro-organisms. An outlying sample taken for the storage area showed unusually high counts for micro-organisms, however it is unlikely that the contamination actually occurred in the tank, as all other readings remained well within the required limits. It is possible that this reading was obtained due to poor sampling technique, or more likely due to contamination resulting from the sampling valve. Generally, the counts taken in the generation and storage areas were all within the specified range, as were the distribution header and return manifold. This would indicate that, despite the high microbial counts in the pre-treatment section, they do not seem to be carried over into the storage and distribution loop.

Samples taken in the distribution loop, downstream of a non-sanitary stainless steel globe valve and associated flowmeter constructed of carbon steel showed high microbial counts. There were several other failures throughout the distribution system, especially in dead leg locations, and downstream of a non-sanitary heat exchanger.

Several modifications were recommended, some changes in the operation of the system and some mechanical modifications.

6.6.4 Operational Modifications

Inadequate operational controls had to be reviewed and rectified, including the lack of Standard Operating Procedures (SOPs), the lack of sterilisation logs and monitoring of the system. In addition, controls were put in place to prevent running the system below its low level switch in the storage tanks, as this could cause potential contamination of the main loop from sub-loops, as water may siphon back into the main loop. It was recommended that sterilisation logs be kept for each sub-loop of the system, and the system be monitored daily.

The low level switch alarm should be set to a higher position, to prevent the storage tanks from completely draining, and warning systems should be put in place which will monitor and warn the appropriate personnel when the water should not be used. The frequency of sterilisation had to be established, as current procedures stating sterilisation must take place once per week, were not followed, and possibly not sufficient for the elimination of microbial growth and maintenance of sterility. A guideline of sterilising three times per week was recommended, however should this not be feasible, point of use filters should be employed, which should be sterilised before each use and connected in a sterile manner before the system is on-line.

6.6.5 Mechanical Modifications

The user was advised to replace all non-sanitary type fitting, connections, pumps, valves, flowmeters and heat exchangers with sanitary industry standard ones. Non-sanitary components were generally constructed of unsuitable materials which were prone to corrosion, and harbouring microbial growth.

Pipework downstream of the valves should be modified to reduce the length of piping where possible. All sub-loops should be made to be self-draining, which may include the re-orientation of valves and filters.

Overall, it was recommended that the whole system be drained, chemically cleaned, and flushed before use. If this course of action was agreed, careful consideration must be paid to the types of chemicals used, as they may react some of the construction materials. Long term this procedure should be repeated at least once a year, with the view of replacing the entire system in the future.

The user decided that the system should be replaced, as the costs associated with modifying the system in order to be in compliance with current GMPs and industry standards, in addition to the increased operational modifications that would be required far exceeded the benefits.

7. CONCLUSIONS AND RECOMMENDATIONS

This chapter has discussed the considerations for design of a WFI system for biopharmaceutical manufacture and has shown the inconsistencies between regulatory requirements and industry standards. Industry is becoming increasing aware that, while there are a great many design recommendations, some of which are essential to assure the quality of water, a number of them need to be considered on a case by case basis, as they may not offer significant improvements to the particular system. The benefits provided to a system by a single component may be outweighed by the costs incorporated in installing it.

When attempting to retrospectively validate an existing WFI system, several points must be considered. It is often prudent to perform a review of the system prior to continuing with the validation exercise, as described in sections 6.6.2 and 6.6.3 of this chapter. This provides a basis for either continuing with the validation, or to determine if the system is validatable in its current state or requires modifications / replacement. Modifications to an existing system can prove costly and the benefits of making the modifications should be carefully considered and compared with alternative solutions, i.e. replacement of part or the entire system. Another deciding factor in the determination of whether to attempt to validate such a system may be the availability of necessary documentation, i.e. materials certificates and technical specifications. In some cases, such as the system described in section 6.6, the only practical solution is to replace the entire system.

The approach recommended for validation of a new WFI system is a truly integrated one, requiring input from all departments concerned from the outset (i.e. Engineering, Construction, Procurement, Commissioning and Validation). The VTOP approach to validation provides a sound alternative to "traditional" IQ activities, which invariably involve often laborious document hunting and information gathering. Using the VTOP approach, the IQ activities are reduced significantly, as the vendor / contractor compiles all necessary documentation. The VTOP package can then simply be reviewed for compliance with GMP, user requirements and as-built status, before the OQ activities can begin. It is worth noting here, that if this approach is used and the contractor / vendor is required to produce the information required, the provision of these documents should be written into the User Requirement

Specification in the initial stages of tender. It is essential that all parties involved understand who is responsible for providing documentation.

BIOGRAPHY

Raytheon offers the largest dedicated validation and GMP Compliance organisation in the industry. Raytheon experience spans from qualification of individual equipment through to GMP compliance of critical systems and facilities, in addition to computer and process validation. Raytheon's approach to validation goes beyond guaranteeing regulatory compliance by providing innovative and quality solutions.

Paschal Baker is a Validation Project Leader with a degree in Process Biotechnology. Her experience covers a working knowledge of laboratory techniques, analytical methods and fermentation. She has extensive experience in validation, including on-site protocol development and execution, review of existing systems and documentation, and the preparation of summary reports, Standard Operating Procedures and Validation Master Plans. Her particular fields of validation expertise include utilities, aseptic filling, isolation technology and filter integrity testing. Paschal is a member of the PDA, PS and ISPE.

Wael Allan has extensive experience in the pharmaceutical industry, primarily in design, commissioning and validation. As Regional Director for the UK Validation and GMP Compliance Group, he is responsible for a group of over forty consultants providing validation and GMP compliance support services to the healthcare industry. Wael has been active in arranging FDA meetings and presentations for a number of client companies and has taken a major part in the in-house training of client companies in validation and GMP. Wael is a chartered engineer and a member of the PDA and ISPE.

REFERENCES

1. ISPE & FDA Baseline Pharmaceutical Engineering Guide, Volume 4: Water and Steam Guide, Draft Working Document (Revision B), Revised 30 October 1997.
2. Code of Federal Regulation, 40CFR, Part 141.
3. United States Pharmacopoeia (1995). 23, National Formulary 18.
4. European Pharmacopoeia (1997). 0008, p.1724-1725.
5. British Pharmacopoeia (1993). (HMSO).
6. Fifth Supplement to the USPXXIII-NF.
7. Brown, J. et al. (1991). Water Systems for Pharmaceutical. J. Pharmaceutical Engineering, 11(4), 15-23.

8. Design Concepts For the Validation of a Water for Injection System (1983). Parenteral Drug Assoc., Technical Report No. 4.
9. Artiss, D. (1983). Water Systems Validation. In: Validation of Aseptic Pharmaceutical Processes. Carleton & Agalloco. pp. 207-251.
10. Eighth Supplement to the USPXXIII-NF.

Chapter 16

Information retrieval and the biopharmaceutical industry: an introductory overview

Patricia O'Donnell
Librarian (Engineering and Science), University of Limerick, Ireland

Key words: bibliographic databases, on-line searching, search strategy, information retrieval

Abstract: This chapter reviews bibliographic sources of information relevant to the biopharmaceutical industry and describes their evolution from printed indexes to electronic services deliverable to the desktop.

1. INTRODUCTION

Everyone who works in science and technology is well aware of the increase in available data and literature. Expressions such as information explosion and information overload are now commonly used to describe the bewildering array of information sources available. It is certainly true that there has been a very large increase in both scientific information and the modes of access to that information and scientists seeking information of either a specific or general nature now have a wide range of search and reference tools available to them. The efficient use of these tools is an essential part of their research training and it is the purpose of this chapter to review some of the essentials of bibliographic databases as we now find them.

In this chapter, the term 'database' is used in the bibliographic sense, that is with reference to databases of records describing the scientific literature. Biopharmaceutical researchers are well accustomed to using databases or databanks of DNA/protein/gene sequences to search and manipulate data to synthesise new information. The information contained in such sequence databases is considered to be in the public domain and freely available.

Searching bibliographic databases is therefore not difficult for such users. There are however some difficulties that do arise, one example being how to choose which is the most appropriate database to search for a particular need or in a particular context. Unfortunately there are no simple guidelines and the choice very often depends on experience, familiarity or just intuition.

It is important to distinguish between the increase in the amount of genuine new information available and the increase in the number of ways to access the information. Until recently information was stored on and retrieved from paper copy. Indexes such as Chemical Abstracts and Biological Abstracts were established and published by learned societies, essentially as a service to science in general. The practicalities and economics of printing imposed a discipline which resulted in tightly defined records and rules of presentation. Publishing and printing costs tended to limit the diversity of such indexes, and also perhaps more importantly limited the number of libraries which could afford to keep them. Increasing computer power, and the growth of rapid and reliable international telecommunications systems, has allowed the development of a variety of new ways of storing and accessing information. Additionally the effect of computerisation on publishing has been to make the assembling and dissemination of information a much more economically worthwhile operation. Textbooks, journals and indexes are now relatively easy to produce and can be quickly distributed throughout the world.

In just about five years the 'information landscape' of science and technology has been transformed. There is little information which is not now available or has the potential to be made available via the internet. The concept of the 'invisible college' of a group of researchers who communicated informally has become a 'virtual college' as they now communicate by email, take part in discussion groups on the internet, and take part in computer conferences, as well as collaborate on publications. Journal articles appear in both paper and digital forms, the full text of patents is available to view on the computer screen complete with drawings, conference abstracts, papers and even the accompanying audio visual support materials can be viewed via the world wide web. Although it is possible to make almost any information available on the web, the existing general search engines cannot deal with a multitude of unstructured, unindexed documents - retrieval is still a problem.

Although methods of information storage and retrieval have changed the rules established for paper copy still exert influence. The indexing and abstracting tasks performed by the producers of traditional bibliographic databases continue to be important in organising and retrieving information. The web is unlikely to be 'standardised' as that would defeat its idea and purpose, and the intelligent agents and robots who will wander the net retrieving information to suit a particular brief have yet to be developed.

2. BIBLIOGRAPHIC DATABASES

Printed indexes to the scientific literature such as Index Medicus, Chemical Abstracts, Science Citation Index, etc. have been around for a long time, and are familiar to those who started their research career twenty or more years ago, but it is likely that most people currently working in biopharmaceuticals have never used them in their printed format. The limitations of print are obvious, it takes a long time to work through many heavy volumes of indexes and many sheets of notes are prepared as details are transcribed in the library. Search strategies tended to be simple because there are limited access points. You can search for an author or a key word, but not combinations of terms. Printed indexes are on the way out although the time scale of the demise is not clear, and publishers have not yet confirmed their intentions. However, it is not likely that computer-based indexes will be extensively back dated, and anyone who needs to look at what was being published prior to about 1970 will need to search the literature using paper copy.

2.1 Development

In order to understand the structures that now exist it is useful to review briefly the development path from paper indexes. With the advent of computerised type setting, direct computer storage and retrieval of the bibliographic information was an obvious step forward. Index Medicus was the first database to be computerised in 1963 and was searchable only in batch mode. This meant that searchers sent their search requests to a central computer and the results were mailed back. This saved the researchers' time, but was expensive and did not allow for any interaction with the database or immediate modification of a search strategy.

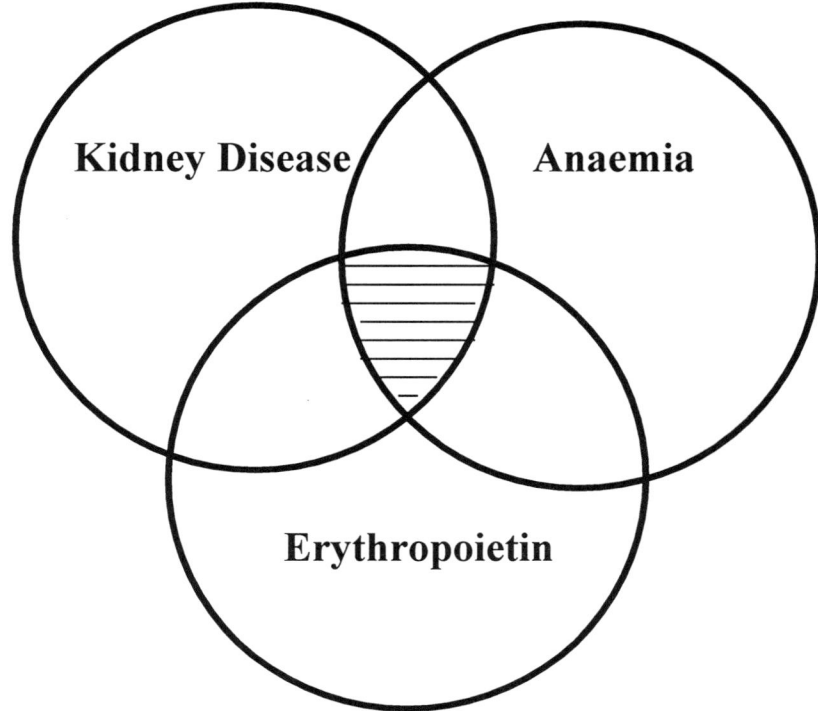

Figure 37. Logical search statement

In the early 1970s advances in communications provided online connection that allowed the searcher at a computer terminal to interact directly with a remote computer, using a low speed dial-up telephone line, and interrogate databases loaded there. The computerised databases of bibliographic information used to produce the printed publications could now be accessed directly by the searcher. The computerised databases were also renamed with more 'technical' sounding names, for example, Index Medicus became Medline and Science Citation Index became SciSearch.

Major commercial hosts or vendors such as Dialog, STN, Questel-Orbit, Data Star, DIMDI, date from this time and now provide access to hundreds of databases. They are not databases in themselves, nor do they produce databases, a situation that can lead to confusion because the same database may be available from more than one host service and in addition may be called something different by each host. For example, the Chemical Abstracts file is called CAS, CA, CHEM or Chemabs by different hosts even though the content of each file is still based on the American Chemical Society publication Chemical Abstracts. Additionally, although the basic search methods are the same the various vendors all use different search systems and command languages. Another option for large organisations was to lease databases on magnetic tape and load on a local mainframe to facilitate in-house researchers.

The breakthrough in end user or user friendly literature searching came with optical disc technology and the publication of the same databases on CD-ROM. Now a relatively small library could allow the end user access to selected databases without having to worry about connect charges. However, as with on-line services, a problem with CD-ROM systems is that several different producers can publish the same database or different subsets of that database, with different search software, and so again choices have to be made by the subscriber.

2.2 Online

There are a number of advantages in using online files rather than paper indexes to search the literature. The depth of indexing provides several access points which makes precise and complex searches possible. Complex searches can be performed in minutes and, on most host services, it is possible to search a 'cluster' or group of several databases simultaneously. This is especially useful for multidisciplinary topics, and advanced graphics systems can facilitate molecular structure searching. This is likely to become more important in the future

There are of course a number of disadvantages to online searching. The cost of a search is made up of a charge for the time spent connected to the database - the rate per hour varies with each database - and a charge for each reference viewed or downloaded. The command languages are different for each host and are generally not user friendly. It is difficult for the infrequent searcher to become familiar with more than one host's command language because each host uses its own search software. Browsing, essential for researchers, is not simple and it takes time and training to make full use of the features provided by the host system

2.3 CD-ROM

CD-ROM is a medium ideally suited to a research centre or university library where access by a wide range of users, of various levels of experience, is required. The data can be made available over local area networks allowing the individual at a desktop computer to search without online time costs. The early CD-ROM products each had their own search interface or on-screen display and this resulted in problems similar to those posed by the different command languages used by on-line hosts. However, these are gradually being standardised into Windows type interfaces which look familiar to most end users even though they may function differently. Browsing is easy and searchers can spend more time learning how to take advantage of all the features of the database.

Databases on CD-ROM are not entirely up to date because they are usually updated less frequently than the online file and so for a comprehensive literature review an additional search in the online file may be needed. For large databases there may be several discs for a range of years. This necessitates repeating the search strategy for each year or range of years. Most search software now includes a 'save search' or 'repeat search' feature which simplifies this process.

2.4 World Wide Web

In order to produce a product which combines the advantages of online and cd-rom media the host services are developing user friendly interfaces for online searching such as STN Easy and Dialog Web. In a sense, technology has come full circle in that large organisations can now lease or buy databases for networking as before, but now they are being implemented on intranets and using web interfaces for searching. Web technology has also enabled the online hosts to take advantage of its hypertext nature to provide new features for linking references and connecting literatures in new ways, sometimes described as 'data mining'.

3. INFORMATION NEEDS OF THE BIOPHARMACEUTICAL INDUSTRY

The biopharmaceutical industry is relatively new with its beginnings in the early 1980s. The information needs of biopharmaceutical researchers can range through many subject areas and documentation types. Because biopharmaceutical science and biotechnology are interdisciplinary subjects this presents some difficulties for information searching and identifying appropriate sources. The information needs of biopharmaceutical researchers could be grouped into four broad areas: (a) Scientific and technical literature; researchers are both producers and consumers. (b) Patent information which can be used in a number of contexts, as a source of scientific and technical information, as an alerting mechanism for licencing opportunities and as data for competitor analysis. (c) Business and company information is needed for market research, industry news, economic forecasts, financial and company reports. (d) Regulatory information is vitally important and there are databases of regulatory information available on many of the database host services. However, the regulatory bodies have now developed their own web sites. This allows more immediate updating and is probably a medium more suited to the delivery of this kind of information.

The large bibliographic databases contain information relevant to biopharmaceutical research but there are also some small specialised databases, focusing specifically on biotechnology. They may contain information from newsletters or obscure conferences which might never appear in the larger sources and can be indexed in greater depth with the needs of these particular users in mind. Within biopharmaceutical research the discipline boundaries are fluid and scientific from areas such as clinical medicine, chemistry, biochemistry, biology, medicine, drugs, pharmaceuticals, microbiology veterinary medicine, etc. can be relevant.

Some major bibliographic databases which cover these subject areas are Medline, Biosis, IPA, Embase, Science Citation Index, Current Biotechnology Abstracts and Chemical Abstracts. Brief details of these databases are given at the end of the chapter. Patent databases and several specialist drug and pharmaceutical databases are published by various publishers and these are hosted on a number of database hosts. Derwent is an important publisher for patents and drug information. There are hundreds of business databases available but two which focus specifically on biotechnology are BioBusiness and Chemical Business NewsBase.

3.1 What is contained in a bibliographic database?

Bibliographic databases consist of large numbers of records each of which describes a single publication such as a book, journal article, patent or conference paper. The publications are described by means of the title, the author, the abstract and the indexing terms. This data has to be sufficient to communicate to the searcher what the content of the paper is and also be sufficiently detailed to allow it to be retrieved even though it is not known to the searcher.

The database fields common to bibliographic databases are author, title, corporate source or author affiliation, source, indexing terms, and accession number to identify the record in the database. Other fields which may not be included in all bibliographic databases include abstract, language, document type and classification codes.

For example, a brief database record for a recent paper could look like this:

> Title: Nervous excitement over neurotrophic factors.
> Author: Gary Walsh
> Corporate Source: University of Limerick
> Source: Bio/Technology, 1995, Volume 13, Number 11, Pages 1167-1171

Each database producer has its own standardised policies about the format of the information entered in each of the fields but these policies vary from one producer to the next which has implications for searchers. One very simple example will show some of the variations which arise and of which searchers need to be aware.

A search for the paper mentioned above in three online databases gave the following details:

	Embase	**Scisearch**	**CBA**
AU	Walsh G.	Walsh G	Walsh, G
CS	University of Limerick	Univ Limerick	Univ. Limerick
SO	Bio/Technology (USA)	Bio-Technology	Bio/Technol. (Bio/Technology)
DT	field not included	Article	Journal
LA	English	English	English
AB	no abstract	no abstract	50 word abstract

This illustrates a number of points.

- The same information is input in a different way in each of the three fields AU, CS, and SO for all three databases. It is important to check the database and host search guides or help files to make sure you are using the search term in the correct format; this is particularly important when searching for authors' names.

- The same fields are not included in all the records. If the document type is not included in the record then a search cannot be limited to a particular document type.

- The author did not include an abstract and it was the producers of the small specialist database CBA who wrote the abstract. The database producers have different policies about including abstracts. An author abstract may be included or amended. If the author does not include an abstract the service may or may not provide one.

The searcher does not need to remember these specifics for each database. The producers and host services provide training programmes and plenty of support in the form of search guides, database descriptions, help sheets, quick search guides and even training files. These training files are usually a subset of a real file and can be accessed for a low connect charge in order practice the command language and try out new search features. Training programmes are now also being delivered via the internet. More details of these developments can be found on the web pages of the host services listed at the end of this chapter.

3.2 The Search Process

The same information search process applies whether looking for information in reference books, paper indexes, CD-ROM, online or www databases. The following key steps are essential for a satisfactory search result.

- Define the search or information need
- Identify the concepts in natural language
- Select a file
- Choose search terms and define a search strategy or profile
- Run the search
- Review the results and modify the search strategy if necessary
- Print or download the results

3.2.1 Define the search or information need

What is the scope of the search, a comprehensive literature review or just a few recent references? What is the time scale, is there a date before which the topic did not exist? What is known already, authors, relevant papers which could be used for a citation search? What type of documents are needed, reviews, patents? What languages are acceptable, English only, all languages?

3.2.2 Identify the concepts in natural language

Describe the search topic in natural language and then list the concepts. For a subject search two or three concepts would be sufficient for a preliminary search. For example, in a search for review articles (in any language) on the therapeutic use of recombinant human erythropoietin in the treatment of anaemia in kidney disease, the concepts are:

Concept 1	Concept 2	Concept 2
recombinant human erythropoietin	kidney disease	anaemia

The concept 'therapeutic use' could be included but, in this example, if the other concepts are identified in a review it could be assumed that 'therapeutic use' is implied.

3.2.3 Select a file

Medline and Embase are obvious sources for this search but the other major databases already mentioned will also contain some references. Most host services provide a directory type index to all the databases which allows a search statement to be run in a number of databases simultaneously in order to identify those files with the greatest number of records. The records cannot be viewed, just a report of how many are in each database.

3.2.4 Choose search terms and define a search strategy or profile

The search terms are those terms which may be used to describe the concepts. Depending on the file chosen there may be descriptors or thesaurus terms to describe the concepts. In addition the list of search terms must include synonyms, spelling and phrase variations, abbreviations etc.

	Concept 1	Concept 2	Concept 3
	recombinant human erythropoietin	kidney disease	anaemia
Search Terms	rhErythropoietin rhEPO recombinant human erythropoietin	kidney kidneys renal	anaemia anemia

Information retrieval and the biopharmaceutical industry: an introductory overview

| | human recombinant erythropoietin | | |

The search terms are combined in a logical search statement (Figure 37) and limited to review articles. There is no need to search earlier than 1989 when rhEPO was first used therapeutically.

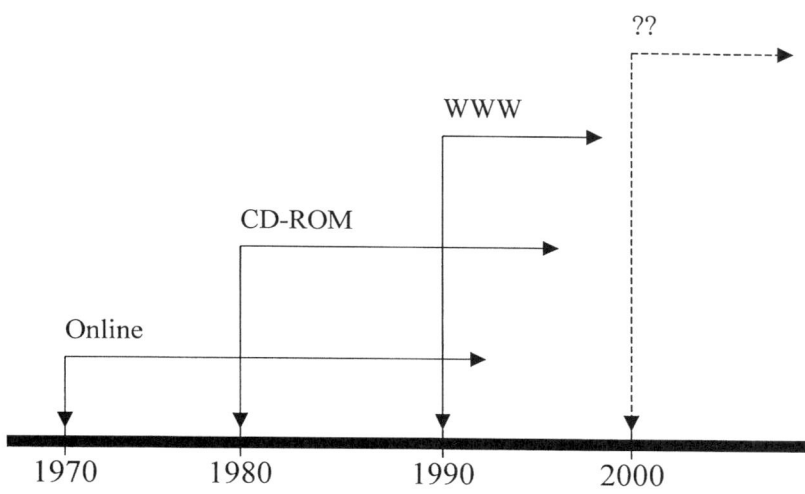

Figure 38. Timeline depicting development of information technologies

3.2.5 Review the results and modify the search strategy if necessary

The titles and indexing of references can usually be viewed free to check relevance. If the results set of references is too big, it can it be reduced by limiting by language, time span or more specific search terms. If the set of results is too small, any limitations such as document type in the example above can be removed and, there may be more search terms that describe the concepts which could be included.

3.2.6 Print or download the results

First it is necessary to decide on a format. A bibliographic format includes the minimum details needed to locate a publication and a full record format usually includes an abstract and the indexing terms. Print, or preferably download to a file, from where the references can be brought into a word processing or personal bibliographic software package.

Personal bibliographic software packages such as EndNote or Reference Manager support the researcher in the management of references, in developing personal databases of references and in the production of correctly formatted bibliographies ready for publication.

3.3 Examples of database services that index information of interest to biopharmaceutical research:

Biobusiness (Biosis) provides information about business applications of biological and biomedical research scanned from approximately 600 publications.

Biosis Previews (Biosis) is the database of the biological and life sciences literature but it also includes some clinical medicine, pharmacology and biochemistry. More than 6000 journals are abstracted.

Chemical Abstracts (American Chemical Society) divides its subject coverage into 80 sections including pharmaceuticals, pharmaceutical analysis, biochemical genetics. More than 12,000 journals are monitored and since 1995 electronic documents have been added.

Chemical Business NewsBase (Royal Society of Chemistry) provides information which affects chemical and pharmaceutical markets and products relating to legislation, environmental aspects, sales, markets and company results. It is updated twice a week.

Current Biotechnology Abstracts (Royal Society of Chemistry) is a small database focussing on biotechnology. Besides scientific and technical publications it includes news, patents and general information about forthcoming events and new books.

Embase (Elsevier Science) is the online version of Excerpta Medica which also covers the biomedical literature but excludes nursing, veterinary medicine, psychology and dentistry, which are included in Medline. Embase does place particular emphasis on the literature of drugs and pharmacology. This database is similar in size to Medline, indexing approximately 3,500 journals, and includes a high proportion of European titles.

International Pharmaceutical Abstracts (IPA) (American Society of Health-System Pharmacists) is abstracted from more than 800 pharmaceutical and medical journals covering the world literature of pharmaceutical research, development and technology.

Medline (U.S. National Library of Medicine), the online version of Index Medicus, runs from 1966 and includes the whole field of biomedicine; approximately 3000 journals are indexed. Medical Subject Headings (MeSH) is the list of descriptors used to index the articles. The depth of indexing varies depending on the type of journal. An article from a major clinical journal will be indexed with 10-12 terms whereas articles from paramedical journals are indexed with four to five index terms. Medline is inexpensive to search and on several web sites it is offered free of charge. Depending on the range of years offered and the search options available this may suit a particular search but the speed of web access can be variable. Medline does have a North American bias in its choice of literature.

Science Citation Index (Institute for Scientific Information) is a multidisciplinary index to the literature of science and technology. It indexes about 4,500 journals and its unique feature is citation indexing. This allows the searcher, using authors' references to prior articles as search terms, to find more recent literature which has cited the same articles and thus make new knowledge connections.

3.4   Some Pertinent Web Addresses

Database producers:

American Chemical Society
http://www.acs.org/

American Society of Health-System Pharmacists
http://www.ashp.org/public/about.html

Biosis
http://www.biosis.org

Chemical Abstracts Service
http://www.cas.org/

Derwent
http://www.derwent.com/index.html

Elsevier Science
http://www.elsevier.nl/

Institute for Scientific Information
http://www.isinet.com/

Royal Society of Chemistry
http://www.rsc.org

U.S. National Library of Medicine
http://www.nlm.nih.gov/
Database host services:

Datastar
http://products.dialog.com/products/datastar/datastar_about.html

Dialog
http://www.dialog.com/

DIMDI
http://www.dimdi.de/homeeng.htm

Questel-Orbit
http://www.questel.orbit.com/

STN
http://www.fiz-karlsruhe.de/stn.html

Database services which provide web access:

DialogWeb
http://www.dialogweb.com/

Profound
http://www.profound.co.uk/

STN Easy
http://stneasy.cas.org/

Personal Bibliographic Software Products:

Research Information Systems
http://sun.risinc.com/

EndNote
http://www.cherwell.com/ProdHome/endnotehome.html

Reference Manager
http://sun.risinc.com/rm/rmprod.html

ProCite
http://sun.risinc.com/pc/pcprod.html

Gateways to internet resources:

Organising Medical Networked Information (OMNI)
http://omni.ac.uk

Biomednet
http://www.biomednet.com

4. THE FUTURE

Developments in telecommunications and information technology have changed and will continue to change the way in which researchers publish and disseminate new scientific knowledge. These developments make it possible for faster publication of research results in electronic journals, on personal web pages or on departmental, university or company web sites. Retrieving information from these disparate and non standardised sources can be difficult and time consuming. Hundreds of search engines are available which use various search strategies and criteria for relevance ranking of results. For those who are interested the Search Engine Watch site (http://www.searchenginewatch.com) provides details about search engines, how they work and how they compare with one another. The site also includes tutorials to help searchers better use search engines.

The scientific researcher however, may not have the time or inclination to study the intricacies of even one, let alone several, of these search engines.

Gateway services have been developed which concentrate on particular subject areas. Two examples of interest to the biomedical researcher are OMNI (Organising Medical Networked Information) and BioMedNet. These gateway services do the work of retrieving, evaluating and organising information to suit a particular community of users and are becoming in a sense the electronic equivalent of the traditional bibliographic databases with the huge added value of being a single jumping off point for retrieving a variety of different kinds of information.

BIOGRAPHY

Patricia O'Donnell currently works in the information services department of the library at the University of Limerick. She has also worked as a technical information officer and teacher. She is interested in information skills teaching and was a member of the UL team which was involved with other European libraries in the early stages of the Educate project (http://educate.lib.chalmers.se/).

Chapter 17

Information technology and the internet as a resource of biopharmaceutical information

J.P. Jenuth, D. Fieldhouse, J.C.-M. Yu
Base4, Bioinformatics Inc.

Key words: Information technology, groupware, high-throughput screening, drug discovery, internet, databases, software

Abstract: The vast quantities of scientific data modern technologies and computer driven analytical equipment now generated renders data storage and analysis by traditional means almost impossible. This chapter explains how such information may be stored and analyzed electronically and overviews major sources of bioinformation on the internet.

1. INTRODUCTION

The amount of information being gathered, stored, searched, and analysed by the biopharmaceutical (biotechnology-derived drugs or pharmaceutical products) industry has been rapidly increasing over the last few decades. This growth has been due primarily to the emergence of new technologies that have enabled researchers to gather large volumes of information rapidly through the use of robots and from specialised machines such as DNA sequencers. Figure 39, for example, illustrates the explosion of information collected within the EMBL database, which is presently gathered primarily by automated DNA sequencers.

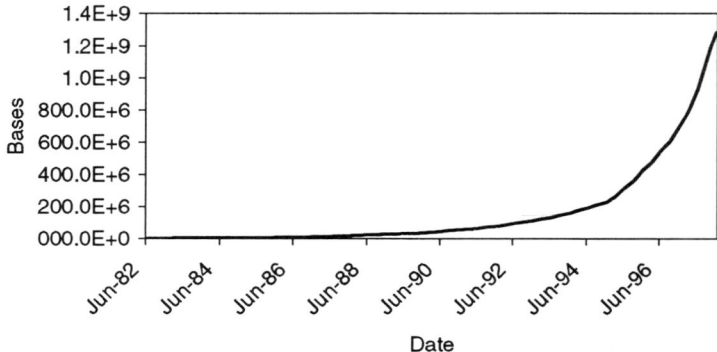

Figure 39. Number of bases sequenced in the EMBL nucleotide database

Computer technology has been used to analyse and store this data because of the sheer volume and the computationally intensive nature of most data analysis. At the same time, the way by which this data has come to be shared amongst people and organisations has evolved from sharing results by magnetic based media and regular mail to the use of private networks and the internet. In this chapter we will review the types of information biopharmaceutical companies gather; how this information is stored, analysed, and displayed; and how the biopharmaceutical industry has embraced the use of computer networks and the internet.

2. INFORMATION IN THE CORPORATE ENVIRONMENT

Over the last couple of decades biopharmaceutical companies have been deluged with an ever-increasing information load. This information has historically been being gathered in the form of data from low-throughput experiments being gathered in notebooks and individual machines. For any given individual researcher, this method of keeping track of day-to-day information works rather well. The system, however, starts to break down as the amount of data generated increases and, as the variety of sources of information that must be stored and analysed increases. We will discuss how biopharmaceutical companies and the software industry have tried to deal with data in lab notebooks, biodata from multiple sources, as well as storing and sharing corporate data through intranets.

2.1 Lab notebooks - What to do with them?

Scientists are espousing the concept of replacing today's conventional paper notebooks with an electronic version. Acting as a "cyberlink" replacement, collaborative electronic notebook systems (CENS) would facilitate research among scientists, and accelerate the capture, use, transfer, and archiving of critical and proprietary research data and records. The issue of data storage space is also addressed by ELNselectronic notebooks (ELNs). Today's paper notebooks cannot handle terabytes of data generated by high-throughput screening (as discussed in the next section), especially as 2-D and 3-D NMR and mass spectrometry, molecular modelling and visualisation are the methods which are beginning to dominate drug and biotech research. Furthermore, once these data are captured in ELNs electronic notebooks, there is a need for full-text indexing and retrieval for all information held within the notebooks, including chemical structures, spectra, documents, DNA sequences and protein structures. The ultimate benefit of ELNselectronic notebooks would be to expedite laboratory and R&D projects, particularly in situations where large amounts of data must be captured as records, over the protracted life of a drug discovery project.

Offerings of electronic notebook software programs include the 'Electronic Laboratory Notebook' (ELN) from ChemSW (wwwhttp://www.chemsw.com) which claims to be a powerful database program for chemists incorporating Windows OLE 2.0 technology to link all of one's information from other Windows programs to the ELN. To accommodate a broader base of users, however, the Collaborative Electronic Notebook Systems (CENS) Consortium™ has been established, which is hosted by the CENS Association (wwwhttp://www.censa.org). The CENS Consortium™ has issued a special membership call to computer, communications systems, and software vendors to jointly drive the development of products using emerging standards such as XML (Extensible Markup Language), CORBA™ (Common Object Request Broker Architecture), Java™, JavaBeans™ and others. The CENS Consortium is also seeking companies involved in amongst other things, groupware, document workflow, image and records management systems; the World Wide Web and Web publishing systems, to assist in the management and dissemination of information gathered into the ELNselectronic notebooks. To this end, participation in the Consortium™ will benefit scientists as well as the staff involved in IT/automation staff, regulatory affairs, quality assurance and patent attorneys.

2.2 Managing multiple sources of information

One of the biggest hurdles encountered by the biotechnology and pharmaceutical industry today, is the integration and analysis of information from disparate machines and by researchers located in different research centres around the world. Subsequent integration of this information with external data from collaborators, or with data or from the public domain (such as the Human Genome Project), further complicates the problem. As an example of managing multiple sources of information in a drug discovery context, let us consider high-throughput screening for which major advances have occurred. Multi-axis robotic arms, stackers, multi-channel washers and liquid handlers for plate manipulation speed up the screening process so that hundreds of thousands of assays can be carried out in a matter of days, while where it used to take six months to do 50,000 assays. However, each assay plate contains up to 96 or 384 wells in which one or more compounds can be screened. Quantification of a reaction between the compound(s) and target(s) in each well will yields raw data for each well. This data is obtained from various pieces of equipment including plate readers and gamma counters. Validation and analysis of raw data will be performed on individual computers, while the log files for the robotic arm activity may reside on a different computer, dedicated solely for controlling this arm. Once a hit is identified, deconvolution of compressed wells must be performed, and eventually followed by secondary screening. Plates associated with a barcode and compounds with identification numbers in a corporate database, facilitating the identification and location of the compound responsible for a successful hit. Clearly, high-throughput screening yields a deluge of data and information, but how does one 'tie' these pieces of information together for meaningful and effective analysis?

From the late 80's and early 90's the need for software to manage protocols, compounds, plates, robotic arms, data capture, validation, analysis, storage and presentation, became apparent. In some instances, collaborative efforts between large pharmaceutical companies and software vendors were initiated to develop high-throughput screening software. Currently available screening software includes ActivityBase from ID Business Solutions (http://www.idbs.co.uk), MDLScreen from MDL Information Systems (http://www.mdli.com/wel.html), and RS^3 Discovery HTS from Oxford Molecular Group (http://www.oxmol.com), all of which address HTS quite well but with emphasise on different aspects of the screening process. An HTS module from Chemical Design, is also available as part of their Chem-X system, however the announcement at time of writing that Oxford Molecular Group is to acquire Chemical Design may possibly bring a halt to its further development.

2.3 Groupware for Intranets and Extranets

A survey of groupware on the internet today yields multiple sites and services devoted to groupware. The term groupware is apparently distinct from CSCW or "Computer-Supported Cooperative Work" which is the study of how people work together using computer technology. Whereas "groupware" is often used to specifically denote the technology that people use to work together, "CSCW" refers to the field that studies the use of that technology. Intranets, which are, private networks that use the global connectivity and open standards of internet technology, are one of most important corporate computing platforms.

In 1989, Lotus' release of the first version of Notes put groupware, a new class of software, on the map (www.lotus.com). However, driven by the popularity of the internet, Lotus Notes 4.6 is now joined by Lotus Domino, which is web-browser based. Domino is self-described as "an applications and messaging server with an integrated set of services that enable you to easily create secure, interactive business solutions for the internet and corporate intranets".

Further propelled by the shift from closed, proprietary systems to open systems, Oracle InterOffice combines an open, scalable platform with the familiar Wweb browser interface for universal access, to deliver a full range of groupware, document management and web publishing capabilities (wwwhttp://www.oracle.com). OpenText Corporation has garnered its fair share of the groupware market since it introduced Livelink Intranet. Like InterOffice, (wwwhttp://www.opentext.com). Livelink is fully Web-based and open-architected to ensure rapid deployment, requiring only a standard Web browser on users' desktops to access its full functionality. The strengths of Livelink include management of documents, knowledge and projects, as well as graphical workflow maps and a powerful search engine (http://www.opentext.com).

Although Lotus Notes is used in many pharmaceutical and biotechnology companies, it is not specific for to these industries. At Base4 Bioinformatics, we have recognised the need for groupware aimed specifically at expediting drug discovery and development. PharMatrix, powered by Livelink, is a collaborative knowledge management tool that encompasses drug discovery and/ development projects, tasks and personnel, communication via email, discussion threads and bulletin boards, as well as document management. Enhancements allow sharing of sequences and alignments, protein structures and data, as well as automated retrieval of sequences, business intelligence, patent information and scientific literature from public databases. Moreover, PharMatrix allows users to make wWorkflows; these are particularly useful for tracking the status of a project, or a particular compound or procedure. Integration of PharMatrix with laboratory equipment, other software or corporate databases, yields an intelligent wWorkflow. For example, completion of an unattended procedure such as gamma counting can automatically 'updatetell' the Workflow initiated in PharMatrix so that this particular step is finished and thus moves on to the next step. The person responsible for this next step will be automatically notified via the In Box (visit wwwhttp://www.Base4.com for more information about PharMatrix). Like ELNs electronic notebooks and CENSA, the aim of PharMatrix is to capture records electronically to facilitate research among scientists, to be able to search these records and also to easily bring new staff up to speed on all aspects of a project. But on a higher level, PharMatrix integrates horizontally in a matrix-fashion, the disparate groups involved in a drug discovery or development project, such as biochemistry, molecular biology, pharmacology, marketing, regulatory affairs, clinical research, patent attorneys and manufacturing.

3. BIOINFORMATION ON THE INTERNET

3.1 Brief history of bioinformation on the Internet

The Internet was conceived in the late 1960's under a contract from the United States Department of Defence. It connected four universities in the US and was designed to provide communications in the event that one of the sites was destroyed in a nuclear attack. Throughout the 70's and 80's networking between sites over shared digital lines remained within the realm of academics and the military; little or no connectivity to these networks was established by biotechnology companies until 1995. Most large pharmaceutical and biotechnology companies shared data either using proprietary dedicated networks, phone lines or by physically shipping magnetic media. Little or no corporate presence was found on the Internet, primarily due to the fact that the backbone of the Internet was owned by a US government agency (NSFNET) that did not allow commercial traffic over the network. With the emergence of the world wide web (WWW) and NCSA's Mosaic web browser in 1993 the Internet was transformed from a network that shared primarily e-mail, files via ftp, and text based information through gophers to a network that was able to share a much richer set of information.

The WWW was adopted very quickly having less than 130 sites in June of 1993 to over 23,000 sites in June of 1995. From 1995 (when the Internet became a commercial network) to Mid 1998 the number of WWW sites has grown to over 2,000,000 (Figure 40). (See http://www.isoc.org for more information). Users can easily find information using sophisticated search engines that were originally developed in the early days of the WWW as well as newer ones as and many companies and organisations post information specific to their industries or interests.

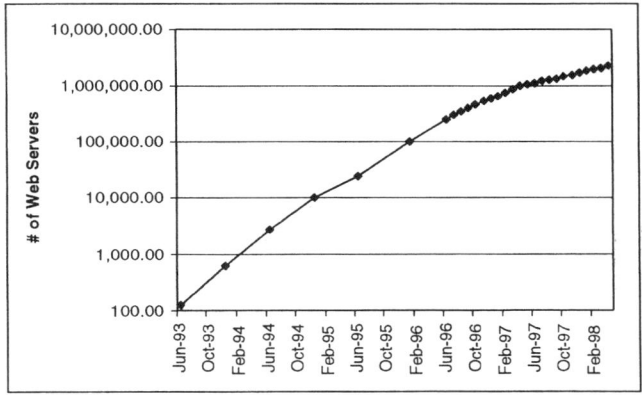

Figure 40. Increase in the number of web servers on the Internet, 1993-1998

3.2 Biopharmaceutical information available on the Internet

3.2.1 Public sites

Bioinformation from publicly funded research has been available on the Internet for many years in the form of files that could be uploaded by anonymous FTP, e-mail list servers, UUCP newsgroups, gophers and most recently through the World Wide Web. Descriptions for each of these services can be found on the Internet easily using one of the available search engines. (See http://www.delphi.com/navnet/faq/history.htm for more information about Internet protocols). We will concentrate on sites information that are is available over the World Wide Web, as they are the best-supported sites with much richer presentation of information. We will highlight sites to perform searches of databases that contain sequence specific information and review some of the tools available on the Internet to analyse sequence information.

3.2.1.1 Public Databases

Since this topic is covered in the previous chapter we will not discuss databases in detail. Many web sites have been developed and are maintained for researchers to perform a variety of searches against the many publicly available databases. Public databases have been catalogued in a searchable database that can be found on the Internet. This list is available at Infobiogen in France at http://www.infobiogen.fr/services/dbcat/ and can be searched, using the Sequence Retrieval System (SRS) at Infobiogen, as well as the Genome Mine at Base4 Bioinformatics in Canada - http://telomere.base4.com/srs5/ and many other SRS sites.

3.2.1.2 Nucleic acid and protein sequence searching

The number and sophistication of sites available on the Internet to perform analysis of sequence information has been increasing. The first sites that were posted in the early 1990's consisted of simple text based web pages. Users were able to enter either a query sequence or search term, and a text based page (or possibly a graphic embedded in a file which could either be printed or viewed with a separate program) was returned. Since these early beginnings sophisticated Java based applets have been developed that allows users to interactively perform operations on their query sequences and/or view results in ways that were not previously possible. We will outline some of the more useful sites below and give examples of sites that have incorporated special graphical capabilities into the analysis.

We will first take a brief look at tools to perform sequence searches followed by annotation searches. A small number of algorithms have been developed that will perform searches on protein or nucleotide databases to find sequences with similarities to a query sequence. Web interfaces have been developed for users to access these search tools over the Internet. Searches using the Smith-Waterman algorithm (1) are the most rigorous and resource intensive. This method does not use any first-pass approximation methods employing 'word matching' (k-tuples) to reduce the data thoroughness of the search. Public servers for Smith-Waterman searches can be found at http://www.timelogic.com/expsw.htm, http://www.dna.affrc.go.jp/htdocs/MPsrch/index.html and http://www2.ebi.ac.uk/blitz.html. Pearson and Lipman (2, 3), to primarily increase the speed at which searches could be achieved, developed Fasta in the late 1980's. Fasta is not guaranteed to find the best alignment between a query sequence and a database; it may miss matches. This is because it uses a strategy (an approximation method) that is expected to find most matches, but sacrifices complete sensitivity in order to gain speed. Fasta servers can be found at http://www.ebi.ac.uk and the BCM Launcher at http://kiwi.imgen.bcm.tmc.edu:8088/search-launcher/launcher.html. Blast (4, 5) was also developed to increase the speed at which database searches could be achieved by sacrificing sensitivity in similar ways to Fasta. The most recent version of Blast uses a new algorithm that speeds searches even further over the version one release and produces gapped alignments (which was not available in the first version). A variant of Blast called position specific iterative Blast (psi-Blast) uses information from any significant alignments from a blast search to construct a position-specific score matrix, which replaces the query sequence for a new round of database searching. The most widely used web interface to Blast is located at the NCBI: http://ncbi.nlm.nih.gov.

While the availability of tools and databases to search the majority of publicly available sequence information is well established, few good tools have been able to integrate the many databases which contain textual information. The major problem has come from the distributed nature of the annotations for sequence data and the many different formats this data is stored in. The EBI, NCBI and DDBJ, however, have been co-operating (see http://www.ebi.ac.uk/ebi_docs/embl_db/ft/collab.html) by sharing sequence information which has resulted in a database found at the NCBI and mirrored at other sites that contains nearly all the publicly available sequence information less annotations in Fasta format. These same three organisations have also mounted a concerted effort to standardise the feature table annotations produced from the nucleic acid databases from EMBL, DDBJ, and Genbank. Although the features tables have been standardised, each organisation keeps their own database format and distributes flat files in different formats. Descriptions of these flat files can be found by using FTP to retrieve the user manuals from the directories in which the databases reside at the NCBI, EMBL and DDBJ. Annotations for these databases can be searches using search engines at each respective site using tools over the WWW. Entrez, available at NCBI, is a tool that enables one to not only search the annotations of Genbank, but will also present one with records that are related to each result. As well, one can easily navigate to databases such as OMIM, Medline, Swissprot and others for which cross-reference data is available in the Genbank records. More information about Entrez searches can be found at the NCBI (http://ncbi.nlm.nih.gov). There are, however, over 400 biological databases that contain biological information (see http://www.infobiogen.fr/services/dbcat/), of which few are linked to Genbank records. To help solve this problem the European equivalent to the Entrez browser called the Sequence Retrieval System (SRS) was created. It is a system that can index the annotations of sequences from any database that is available as a flat file. Using the query tools provided, SRS users can easily uncover associations between entries from one database to another even though direct references do not exist between the databases. As well, one can derive new sequence databases based on annotation searches that can be then searched using the sequence search tools mentioned earlier. SRS has being implemented at over 30 sites. Links to all the SRS sites and the databases that are currently indexed can be found at http://srs.ebi.ac.uk:5000/srs5list.html. Two of the larger SRS sites can be found at http://srs.ebi.ac.uk:5000 and, http://www.infobiogen.fr/srs5/.

3.2.1.3 Public Domain software

An abundance of software is available for researchers from all scientific disciplines. This software has been traditionally shared by either magnetic media through the mail or by FTP over the Internet by researchers whose institutions had network connections. Most biological software quickly became accessible as the Internet was being adopted by most academic institutions in the early 90's. With the large number of programs and sites hosting these programs an effort to catalogue this software was established at the EBI. The result of this work is The Biocatalogue, which can be found at http://www.ebi.ac.uk/biocat/biocat.html. It is a software directory of general interest in molecular biology and genetics.

With the introduction of the WWW, not only could software be shared over the Internet, it could reside on a server that could execute the analysis and return the results using HTML based graphicsdisplays. A good searchable catalogue of sites can be found using the Biotoolkit, which is hosted by BioSupplyNet at http://www.biosupplynet.com/cfdocs/btk/btk.cfm. The Biotoolkit is a companion to the popular book "Bioinformation on the World Wide Web 1997" which contains a thorough list of web sites for molecular biologists.

The newest type of software being offered over the Internet is Java based. Java based software can be imbedded in web pages and has the ability to perform calculations and produce graphically superior output of results from data analysis interactively. Two types of Java applications exist, applets are applications that are uploaded from the server which are executed on the local machine, servlets are Java applications that are executed on the server and the local machine renders the results using a small Java applet. Servlets are typically only found in environments such as local Intranets where the bandwidth of the data connection is sufficient to pass the online display information. We expect that the number and complexity of Java applications will undoubtedly increase over the next few years as Java matures. A number of web based sites have been assembled that allow users to quickly find most of the available Java enabled sites and tools. Java based software sites can be easily found by taking a look at the bioWidgets consortium home page at http://goodman.jax.org/projects/biowidgets/consortium/ or at the Java-based Molecular Biology Workbench home page http://www.embl-heidelberg.de/%7Etoldo/JaMBW.html.

3.2.1.4 Journals and Patents

The IBM Patent Server (wwwhttp://www.patents.ibm.com/ibm.html) gives access to U.S. Patent & Trademark Office (USPTO) patent descriptions dating back to January 5, 1971. Scanned, multipage patent images are also available for patents issued from 1974 to present. One can retrieve and search over two million patents. Other Internet tools for patents include Patscan from University of British Columbia (http://www.library.ubc.ca/patscan/), PatentWeb and TrademarkWeb from MicroPatent (http://www.micropat.com/index.html), general information from the US Patent and Trademark Office (http://www.uspto.gov/), as well as biotechnology-specific patent information from the US Department of Agriculture (http://www.nal.usda.gov/bic/Biotech_Patents). In contrast to expensive patent searches, a non-profit group recently began posting U.S. DNA patents on a free online database (http://geneticmedicine.org) (6). The DNA Patent Database (DPD), a joint project of the Georgetown University's Kennedy Institute of Ethics and the Foundation for Genetic Medicine, allows free public access to the full text and analysis of all DNA patents issued by the United States Patent and Trademark Office (PTO). Apparently, the patents would have cost $27,000 if purchased from the PTO. The DPD contains more than 9000 nonplant, DNA-based patents issued from 1980 through December 1997 and will be updated quarterly. Its purpose is "to provide information on some of the most fundamental policy questions in biotechnology," including the patenting of human DNA. The patents can be searched by key word, sequence, or inventor.

Presented in its three official languages, English, German and French, the European Patent Office (EPO) can be accessed at wwwhttp://www.european-patent-office.org. An abundance of information is available here, including a 'toolbox for applicants' with downloadable application forms, prices and costs information, the European Patent Convention, and a European Patent Attorney database. 'Patent info products' offers a link to PATLIB, EPIDOS-INPADOC databases and the European Patent Register on-line which provides detailed information on all European and Euro-PCT patent applications. The information includes bibliographic data such as title of the invention, classification, publication dates, name and address of the applicant, inventor, attorney and the latest information about the status of the granting procedure of the patent application. Also at the EPO site, 'Patent information on the net' (http://www.european-patent-office.org/online/index.htm) contains valuable links to patent databases and patent information providers, some of which are specific for biotechnology, such as the Biotech Law Web Server.

Free online access to Medline is available at a few sites. The National Library of Medicine in the USA hosts The Grateful Med (http://igm.nlm.nih.gov/). In addition to MEDLINE, HealthSTAR, PREMEDLINE, and AIDSLINE the Internet version of the Grateful Med offers free access to AIDSDRUGS, AIDSTRIALS, DIRLINE, HISTLINE, HSRPROJ, OLDMEDLINE, and SDILINE. PubMed is available at the NCBI (http://ncbi.nlm.nih.gov). The PubMed search system is drawn primarily from MEDLINE and PREMEDLINE. The PREMEDLINE database provides basic citation information and abstracts before the full records are prepared and added to MEDLINE. PubMed also provides access to the molecular biology databases included in NCBI's Entrez retrieval system. Finally Medline is available at the BioMedNet which is also the gateway to HMSBeagle, another excellent site (http://biomednet.com) site. One can register for free, which will allow and haveone to access to the Medline database. One advantage of using this site is that search histories can be saved.

Pharmaceutical and business information can also be searched using commercial services. One of the most comprehensive and user friendly sites services available today is called Profound from Dialog Corporation (http://www.profound.com/). This service allows one to search titles and abstracts from over 500 data sources. As well as performing one off searches, users can perform automated searches when the databases are updated. The results from these searches are either e-mailed to the user or they can be automatically deposited into a GroupWare product such as Base4's Pharmatrix Intranet (http://www.base4.com/pharmatrix.html).

As the popularity and availability of the Internet has increased, many publishers of scientific journals have made available on line versions of articles containing full text including figures. Current online publications are either distributed as PDF files, which can be viewed by Adobe Acrobat, postscript files, or can be viewed on-line using a web browser. Most online journals can be accessed for a fee. These fees can be paid directly by individuals or by libraries in institutions or companies in which agreements have been made with the publishers. Check with your librarian for more information as to which online journals may be available to you. We will review some of the more popular on-line journals followed by sites (maintained by publishers) that point to the more specialized technical journals. In the area of general biology, Science and Nature have online access to their full text journal articles. Both have online web browser based access as well as the availability of PDF files. One advantage of the online version is the availability of hot links to the listed citations in each article through either Medline or links to the actual full text article. Thousands of journals are available on-line directly from publishers. These include Elsevier (http://www.elsevier.com), Oxford Press (http://www.oup.co.uk/jnls/), Harcourt Brace (http://www.hbuk.co.uk/), cell press (http://www.cellpress.com/) and many others. Other services are available such as the BioMedNet library (http://biomednet.com/library/) for registered users of BioMedNet (registration is free) which gives one access to on-line journal articles, usually for a fee. Check with your librarian for more information as to which online journals may be available to you

3.2.2 Commercial Sites

In contrast to publicly funded research organisations, most biopharmaceutical companies generate large amounts of information but do not want to make it accessible to the public. Due to the possibility that data security could be compromised, many organisations have been reluctant to connect to the Internet. However, with the appropriate security firewall arrangement, companies can have full access to the Internet and its resources with little risk of security violations through the public Internet. Most biotechnology and pharmaceutical companies have web sites on the Internet that can be found by using any of the more popular general-purpose search engines.

Employees of biopharmaceutical companies often use the publicly available sites to routinely launch searches of sequences and proteins against the many publicly available databases (See chapter 29 for details of relevant databases). Although the public Internet tools work well for many simple analyses, no guarantee of the confidentiality for any query exists and these tools cannot be easily used for high throughput analysis nor can they be modified for specific tasks. Because of these factors biopharmaceutical companies typically create copies of the public databases on internal computers which have publicly available tools for analysis on them and/or commercially available on in-house developed software packages.

3.3 Search engines

There are numerous web sites offering valuable information for the biopharmaceutical industry. such as Bbioinform (www.bioinform.com) and Bioinformer (http://bioinformer.ebi.ac.uk/). Bioinform (http://www.bioinform.com) is a bioinformatics news service, which offers behind-the-scenes reports and forward-looking analysis on the deployment of information technology in support of genomics and biotechnology worldwide. The site also contains a collection of Web-based bioinformatics resources. Similarly, Bioinformer (http://bioinformer.ebi.ac.uk/) is a quarterly newsletter which focuses on bioinformatics research, developments, and services at the European Bioinformatics Institute (EBI) and elsewhere. The newsletter features on news from the EBI, interviews, issues in bioinformatics, news focusing on novel technologies and research, updates and announcements of software, databases and services, and a compiled list of worthwhile conferences for bioinformaticians.

BioSpace (http://www.biospace.com/) is a business-to-business Internet "hub" site launched on March 15, 1995 to serve the global biopharmaceutical industry. The overriding goal of BioSpace is to facilitate informed and timely decision-making by providing critical and organised bioscience information in one location on the World Wide Web. This site contains the latest news, company information, industry events and career opportunities. Other news service providers on the Internet include Bio Online, which contains information and services related to biotechnology and pharmaceutical research, development, and manufacturing. It combines resources of biotechnology companies, biotechnology centres, research and academic institutions, industry suppliers, government agencies, and non-profit special interest groups. Yahoo http://biz.yahoo.com/news/biotechnology.html) and Infoseek (see http://www.worldpharmaweb.com/) both have biotechnology and pharmaceutical news sections containing articles that companies release over the various newswires.

Biopharma (wwwhttp://www.biopharma.com) offers access to a database with basic information about biopharmaceuticals in the U.S. marketplace, concentrating on product identity, composition, intellectual property, manufacture, marketing, companies and approvals. The Biotechnology Information Institute provides the information for those in the biotechnology and pharmaceutical industries, as part of their ongoing Biopharm Registry and Reference Project.

Representing the Pharmaceutical Research and Manufacturers of America, at PhRMA (wwwhttp://www.phrma.org/), one finds sections on Drugs in Development, industry Facts and Figures, consumer Health Guides series and Current News covering issues dealing with pharmaceuticals. The

Drugs in Development portion contains a publicly available database that can be searched by disease, indication or drug names. The most comprehensive and accordingly, the most costly, of sites for information pertaining to biopharmaceuticals is Current Drugs (http://current-drugs.com), which offers a range of products designed to address the competitor intelligence needs of the pharmaceutical and biotechnology sectors. Foremost among these is the Investigational Drugs database (IDdb), a searchable system available over the Internet for researching, tracking and analysing the R&D activities of thousands of companies and research groups involved in drug discovery and development. The ID Patent fast-alert allows researchers to review weekly published patent summaries on the Internet via BioMedNet (http://BioMedNet.com/cd/pfa) and ChemWeb.

BIOGRAPHY

Base4 Bioinformatics Inc. (www.Base4.com) provides integrated systems solutions for the biotechnology and pharmaceutical industry to expedite the drug discovery process. Base4 provides products and services for target selection, groupware and laboratory data flow. Headquartered in Mississauga, Canada, Base4 is a distributed company with nodes in Montreal and Toronto with additional nodes planned for Vancouver and San Diego. The authors are research scientists at Base4 Bioinformatics.

REFERENCES

1. Smith, T.F. and Waterman, M. (1981). Identification of common molecular subsequences. Journal of Molecular Biology, 147, 195-197.
2. Pearson, W.R. and Lipman, D.J. (1988). Improved tools for biological sequence analysis. Proceedings of the National Academy of Sciences USA, 85, 2444-2448.
3. Pearson, W.R. (1990). Rapid and sensitive sequence comparison with FASTP and FASTA. Methods in Enzymology, 183, 63-98.
4. Altschul, S.F. et al. (1997). Gapped BLAST and PSI-BLAST: a new generation of protein database search programs. Nucleic Acids Research, 25, 3389-3402.
5. SITE VISIT: DNA Patent Free-for-All (1998). Science, 280, Number 5365.

Chapter 18

Marketing Issues for the (Bio)pharmaceutical sector

Scott Spinka
President, CareMerica Inc., 16508 Kingspointe, Lake Lane, Chesterfield, Mo., USA

Key words: Marketing plan, differentiation, product launch, roll-out

Abstract: The Marketing Plan is the nucleus of the company's entire marketing and sales effort. It should be a document which the uninitiated can understand and use. It is often fruitless and costly to begin developing marketing programs and spending money in advance of making careful and global strategic assessment. This chapter focusses on this and related issues central to the successful marketing of (bio)pharmaceutical products.

1. INTRODUCTION

(Bio)pharmaceutical marketing is driven by a number of important elements. As with all successful marketing plans, it is fundamental to formulate programs to satisfy customer needs.

The starting point for developing a comprehensive marketing plan is thoughtful strategic assessment. Setting a baseline for where you wish to go is the beginning. It is often fruitless and costly to begin developing marketing programs and spending money in advance of making careful and global strategic assessment.

Where you wish to go doesn't sound too complex does it? Experience provides that the difficulty in this journey is not usually where you want to go; but, where you are.

A strategic baseline review will provide the information you seek regarding market targets, competition mapping, product/corporate differentiation, short and long term goals, opportunities, exposures, and more.

Establishing a strategic foundation will provide definitive insight for strategy and program development. A novelist wouldn't write a book without first deciding its general plot and characters. Similarly, a successful (bio)pharmaceutical marketer needs to embrace a long term understanding of the company's corporate and product goals and then construct a marketing plan which is consistent with these goals.

Understanding the landscape is critical. Taking an overview of competitive positions, customer needs, business offerings, and intended marketing strategies can pay handsomely to those diligent enough to map these various elements.

2. STRATEGIC BASELINE REVIEW

A strategic baseline review is a "snap shot" assessment of the business; what is working; what is not. It is an examination of the business; how products are sold, distributed, priced, packaged, marketed, differentiated, and how they perform.

The review should embrace company strengths and weaknesses as well as opportunities and exposures. A good place to begin is to visit customers; after all, successful (bio)pharmaceutical marketing is all about customers; understanding their needs; meeting their needs. It's amazing how many medical marketers think they can devine the needs of customers without meeting or speaking with any. The key is customer interface; of course, this axiom is generally associated with "market driven" companies. Building a product to satisfy identified customer needs stands a far greater chance for success than creating products which may fulfil no customer need.

"Technology driven" companies are a different animal; they develop products based upon a technological base. Technology driven companies must create their markets; customer driven companies on the other hand have an established market.

So, we should make a short list of the customer needs your product satisfies. This information is critical to your marketing success. You may wish to expand the list; but, experience has shown the short list will be most valuable. The essence of this methodology is to determine what elements should be used to promote the sale of your product/service.

3. MARKETING PLAN

As you develop your marketing plan the document should begin with objectives.

Make a list of the objectives you wish to fulfil. These should be important corporate, product, financial, and non-financial, targets. These objectives will become the nucleus of your plan. The Marketing Plan should flow from "objectives" to "strategies" to "programs". The strategies should be constructed to accomplish the objectives; the programs should be developed to support the strategies. Think of it this way:

- Objectives - what you wish to accomplish.
- Strategies - tactics you will use to accomplish the goal.
- Programs - activities you will design and implement to support the strategies.

The objectives you construct should be measurable. In this way you will be able to take benchmarks of your progress.

There are many important components of a comprehensive Marketing Plan. It is a compendium of salient information about the company, its customers, and products. The plan should incorporate a situation analysis, customer identification, customer needs, what the company will do for the customer, corporate marketing objectives and resulting strategies to achieve these objectives, product positioning analysis, competitive analysis and profiles, activities which support the marketing strategies.

A Marketing Plan should clearly state the company's "unique selling proposition" and product/service differentiation. The plan should incorporate a breakdown of the proposed marketing mix along with sales forecasts and expense estimates.

Marketing Plans, although comprehensive, should be constructed for ease of understanding and concise presentation regarding the company's proposed methods for achieving its sales goals. The plan should be periodised to reflect the company's realistic chronological expectations and should provide a good information template for other company disciplines e.g. operations/ manufacturing. The Marketing Plan should demonstrate a contingency for unanticipated results and highlight specific expected opportunities and exposures. The Plan should specify product pricing strategies as well as specific promotions which may alter these strategies. Marketing Plans are living documents which should incorporate on-going reassessment.

The Marketing Plan should define the company's business purpose, expectations of customers, plus, desired and perceived market image. A communication plan should be incorporated either by adjunct or integration into the Marketing Plan and detail how the company will present its important messages. The Marketing Plan is the nucleus of the company's entire marketing and sales effort. It should be a document which the uninitiated can understand and use.

3.1 Marketing Mix

As you build the Marketing Plan for your product/service, it is important to assess the "marketing mix" best suited for your market and/or circumstances.

The marketing mix is an assembly of elements which will impact the sales success of your product/service:

1. Product
2. Pricing
3. Promotions to Customers
4. Promotions to Sales
5. Distribution
6. Packaging
7. Selling
8. Advertising to Customers (end user)
9. Trade Advertising
10. Public Relations
11. Market Research/Information
12. Conventions/Societies/Organisations

Within this set of elements are subsets which provide nearly unlimited dimension which can affect product/service success.

- Product elements:
    - Size
    - Colour
    - Shape
    - Quality
    - Safety
    - Warranties/Guarantees
    - Design Features
    - Ease of Use
    - Other
- Price elements:
    - Brand (Private or Generic)
    - Impulse vs. Planned Purchase
    - Costs
    - Desired Margins
    - Service Requirements
    - Seasonality

- Life Cycle
- Market Sensitivity
- Purchase/Rent/Lease (options)
- Legal Restrictions
- Promotional Sensitivity
- Competition
- Packaging

- Promotions to customer elements:
  - Trials
  - Displays
  - Cost
  - Incentives
  - "Free" Goods
  - Bundling To Other Products
  - Education
  - Special Event Tie Ins (e.g. Sports; Annual Society Meeting)
  - Samples
  - Contracts
  - Joint Marketing

- Promotions to sales elements:
  - Trade Shows
  - Displays
  - Contests
  - Incentives
  - Sales Literature
  - Bundling To Other Products

- Distribution elements:
  - Mixing of Channels
  - Retail
  - Wholesale
  - Distributors (Stocking/Non Stocking)
  - Manufacturer Representatives
  - Agents
  - Telemarketing
  - Mail
  - Direct

- Packaging elements:
  - Cost
  - Shape
  - Other Uses

- Disposability
- Graphics
- Protection
- Visibility
- Style
- Reusability
- Promotional Uses
- Other

- Selling elements:
  - Organisation
  - Training/Education
  - Expenses
  - Compensation
  - Order Mechanics
  - Recruiting
  - References
  - Key Accounts
  - Managers
  - Method of Customer Contact

- Advertising (customer) elements:
  - Print Media
  - Video
  - Ad Placements
  - Monitoring/Measurements
  - Direct Mail
  - Product Messages
  - Samples
  - Television/Radio
  - Conventions
  - National/Regional Meetings
  - Societies
  - Frequency
  - Image

- Trade advertising elements:
  - Copy
  - Print Media
  - Direct Mail
  - Video
  - Ad Placements
  - Monitoring/Measurements
  - Product Messages

- Samples
- Television/Radio
- Conventions
- National/Regional Meetings
- Frequency
- Image

- Public relations elements:
  - "Steady Drip" Theory
  - National/International/Local Markets
  - Free and Cost
  - Image
  - Frequency
  - Monitoring/Measurements

- Market research/information elements:
  - Sales Organisation
  - Customers
  - Non Customers
  - Competitive Information
  - Purchased Data
  - Focus Groups
  - Trade Meetings (Conventions)
  - BRC (Buyer Response Cards)
  - Existing Literature

Constructing the optimum marketing mix is dependent on these factors. One must determine the elements which will bring the best differentiation to the product/service and provide the best return on investment.

3.2 Differentiation And Product Positioning

Differentiation and product positioning play crucial roles in the successful marketing of (bio)pharmaceutical products. Proper differentiation is needed to provide the product its appropriate competitive distinction and assist customer understanding of its features and benefits. There is a general marketing axiom that should never be forgotten: products that are not differentiated, are by default, price differentiated.

What is differentiation? Differentiation is distinguishing one's offering in favour of all others. Differentiation can be accomplished in many ways; product attributes; packaging; price; capacity; deliver ability; service; knowledge; training; warranty/guarantees; geography, etc.

Value satisfaction is what differentiation is meant to create in the mind of the customer. Theodore Levitt (1), suggests each product has a Total Product Concept. At the core of this concept is the "generic product"; next, the "expected product"; further differentiated is the "augmented product"; and finally, the "potential product". These product levels are dependent upon the depth and skill of product differentiation.

Most successful companies and products have a unique selling proposition (USP) associated with their real and implied offerings. The unique selling proposition is a theme and/or image which the company uses to drive market perception. It creates a unique perception for customers regarding companies and products. Consider Coca Cola an example. The USP for Coca Cola is "It's The Real Thing". Coca Cola has spent many years and millions of dollars promoting this USP; its advertising constantly focuses upon the idea that, although there maybe many colas on the market, there is only one COKE..."the real thing"!

Coca Cola made a serious violation of its USP several years ago when it came out with "NEW" Coke. This message made the public feel Coca Cola was no longer the real thing. Coke's error cost the company millions of dollars and eventually led to the return of Coca Cola Classic; "the real thing". USP's are important mechanisms for creating market perceptions and differentiation.

Differentiation which creates product sales is achieved by successfully communicating and imparting value satisfaction to the customer. Although (bio)pharmaceutical products are quite different from consumer products, there are many similarities in the successful marketing of each. Differentiation may be accomplished by making the product larger than any other; faster than any other; the least expensive, etc. The customer seeks value and evaluates products and companies on the basis of how much value satisfaction he/she can obtain through said company's product. The challenge is to determine what your customers perceive to be most valuable to them and then give it to them.

It's important to review and re-review how customers perceive your product/service. Understanding customer interests and needs is critical to successful marketing programs. One of the classic problems of failed marketing is "sheltered validation"; wherein, marketers mistakenly believe they know what customers want without actually asking customers. Getting into the field, meeting customers, observing their environment, understanding customer problems, is vital to good marketing.

(Bio)pharmaceutical products often take ten or more years of research and hundreds of millions of dollars of investment to obtain regulatory market approval.

Marketing Issues for the (Bio)pharmaceutical sector 429

This level of commitment demands professional marketing expertise. (Bio)pharmaceutical product marketing is no place for on-the-job training.

3.3 Product Distribution

Qualified distribution is critical to successful marketing.

Understanding the distribution channels of one's product market and securing appropriate contractual linkage to proven motivated distributors is job one.

Of course, good distributors are predicated by many things:

- Territorial coverage (geography)
- Number of sales representatives
- Dedication to clinical or market specialisation
- Profitability
- Marketing resources
- Number of product lines carried
- Competitive products carried
- Sales management
- Strategic interests
- Stocking inventories
- Margins

To select your optimum distributor requires significant evaluation. Like most things related to marketing, starting with a customer is a good beginning.

Customers know which distributors meet or exceed their expectations. They can reel off a list of perspective companies you should evaluate. Launching a (bio)pharmaceutical product is an extremely costly event. It is important to test or "pilot" all tactics including product positioning, proposed product names/trademarks, collaterals, and strategies. Marketing managers should develop a "pre launch plan" to support launch analysis.

3.4 Pre-Launch Planning

"Pre launch" planning should incorporate in-depth examination of launch options and issues:

- National Launch
- Regional Roll Out
- Inventory Requirements
- Launch "Windows"
- Strategic and Operational Objectives
- Marketing Mix Decisions
- Training Requirements
- Compensation Programs
- Promotional Strategies

Each of the elements listed above are critical to your ultimate sales and marketing success.

3.4.1 Regional Rollouts

If a regional rollout of your product is possible, it can be an excellent training ground for subsequent national launch. In general, I am biased to initiate regional roll out programs which can provide opportunities to limit exposures, test strategies and prices, and "go to school" regarding customer response and potential launch problems. If your preparation has provided the support mechanics for a national launch, you can quickly accelerate the roll out if necessary.

Roll outs are useful because they provide limited geography to control and experience. The information learned from a roll out can be beneficial when moving to national application. Pricing and product positioning can be tested. Roll outs usually provide valuable training information. Issues central to customer acceptance will likely surface and strategic changes which may be necessary to capture total market acceptance are identified early. Sometimes roll outs can not be done. Competitive actions and market timing may suggest an aggressive full scale product launch is needed.

This approach, although sometimes necessary, is least desirable. Launch exposures are very high. Customer information is not fully known. Issues to consider include:

- Price
- Product Availability

- Readiness Of Support Materials/Programs
- Sales Training and Customer Education
- Distribution Effectiveness
- Viability
- Image
- Likely Competitive Action
- Impact On Current Products
- Product Positioning
- Bundling Opportunities
- Product Performance
- Other

During regional roll outs and national full scale launches, communication is paramount. Sales and marketing must be a cohesive team. A good way to begin a regional roll out is to use your general Marketing Plan foundation as an overlay for customising a regional plan. Discussions with field representatives should highlight the unique characteristics of the region and particular nuances which may require adaptation. It is important to use the primary strategies and programs intended for national application as the roll out venue is a testing ground. Each of the primary Marketing Plan strategies should be tested and performance graded.

4. PRIMARY MARKET MESSAGES

Primary market messages should be tested and evaluated. Of course, all market messages, images, and graphical presentations should be tested in advance of going to any part of the market. This is usually done with focus groups. These presentations are critical to marketing success.

As you evaluate the product's primary market message, consider the product's differentiation and the most compelling customer need it successfully satisfies. It is important to avoid throw away terms as part of any primary message.

Throw away terms are phrases or words which have been overused by the industry and as such lose their impact and customer meaning; an example is: high quality. The term "high quality" has lost its impact in health care related marketing. High quality is considered "entry level" to all health care markets;

certainly no successful health care manufacturer would suggest to the market that his/her product is "low quality".

In the above example, it is acknowledged that health care customers really do expect and require "high quality"; but, using this term in today's markets brings little or no return on investment. It is important to develop market messages that capture the interest and imagination of customers. These messages are best developed as short and simple statements. Often one or two words can be pivotal in the product's positioning and success.

The product name is also an important component to successful marketing. The name should be unique and memorable. If possible the name should power the product's differentiation and features/benefits.

Naming a product capable of developing several hundred million dollars in annual sales is a science. Experts should be consulted and utilised to exact the best presentation. Registered trademarks will be extremely important in this context.

5. ECONOMIC CONSIDERATIONS

Since the evolution of DRG's in the 1970's and more recently health care reform of the 1990's, the consideration of product economic impact on health care costs has been paramount. In the case of higher market priced prescription medications and biotechnological/genetic products, it is critical to demonstrate cost effective viability to gain market acceptance. Of course, these mechanisms are best considered during the early stages of clinical validation versus late stage pre marketing.

Most successful economic considerations are taken into an analysis known as "outcomes". It is usually most cost effective to set up economic outcome studies to run parallel with clinical trial programs. In this way, economic proof can be obtained with clinical validation.

These studies should be developed within the context of marketing requirements. Marketers should anticipate study endpoints needed to achieve needed market interest and positioning. These studies should be routinely monitored to insure needed data is accumulating and meaningful.

Cost effect comparisons of competitive products can be very effective if data is economically favourable and coupled to clinical equivalence. Hand in hand with economic assessment is product reimbursement. In the case of formulary acceptance, reimbursement is often a fulcrum to product success. Over the counter products to the contrary, reimbursement is critical to many medical market applications. Reimbursement is a complex issue. In US markets there are hundred's of intermediaries/payers which are a backdrop to successful reimbursement. Each is capable of assigning independent and somewhat arbitrary compensation for your product. In each case, a reimbursement code is needed to orchestrate payment. In many cases, payment amounts may vary.

Standardisation of payment amounts and codes is somewhat elusive; however, the Federal government via Medicare and Medicaid provide guidance to most insurers and payers. Health maintenance organisations (being both provider and payer) are significantly motivated to use products which demonstrate conclusive economic benefit along with clinical superiority or equivalence. Economic models are often effective in presenting cost effect scenarios. It is important to note products may be significantly higher priced as long as the overall economic impact is lower.

Another factor particular to (bio)pharmaceuticals and economics is delivery mechanics. How will the product be delivered to the patient? Will it require sophisticated auxiliary equipment for delivery? Will it require special monitoring and what will be the costs for these adjuncts? Do competitive products require similar support or is the new product eliminating the need for these support systems? It will be important to make assessments of competitive comprehensive cost positions.

Outcome assessment is enhanced by third party validation. If you can orchestrate a credible competent organisation to conduct this research and you are confident of the outcome, this approach can provide valuable dividends.

Another important element for consideration is cannibalisation. Is the new product you are introducing to the market going to cannibalise existing products? If so, many factors should be considered. Inventories, pricing, trade-ins, product life cycles (patents), competitive strategies, and forecasting are but some of the issues you should review.

Product forecasting is a science; but not an exact science. In the case of (bio)pharmaceutical marketing, there are a number of independent organisations which chart the performance of pharmaceutical products/markets. This information is useful in assisting your forecast efforts. It should be understood these assessments are best considered relative data.

Determining the difference between the various sources of market data estimates of your own company's annual sales/unit performance and what you know to be the actual historical information will provide a "forecast constant" (fudge factor) which can assist your efforts establishing the relative values of this data.

Using these relative data will allow you to forecast with reasonable accuracy various market criteria. Generally, one can put reasonable faith in the relative basis of market growth percentages and market share. Actual unit performance and sales must be adjusted according to the "forecast constant" you have determined.

Entering a market arena not previously developed is a different matter. In this case, forecasting is best defined by careful examination of focus group data and demographic analysis. Working closely with medical societies and examining data developed via government statistics will assist your development of this information.

A critical issue for all product launches is "forecast exposure". It is important to realise many factors will contribute to or detract from a successful product launch, e.g. timing of regulatory approval.

As a myriad of things can "fall out of bed" during a new product launch, performance factoring should be considered to exact the profit exposure the company is willing to take in years one and two.

Generally, I believe first year profits should be taken into the financials at half value in order to protect the corporation/company from debilitating exposure. Of course, this is a politically charged matter; many will wish to push for full value numbers. In product launches, "Murphy's Law" usually applies. By taking half value profits year one, the company can limit it's downside and enjoy the possibility of substantial upside performance.

All product launches should be benchmarked to provide management the opportunity to make business adjustments based upon performance. Usually, these benchmarks are taken quarterly or in some cases monthly. Of course, management must realise marketing launch expense is often sloped in inverse relation to first year quarterly sales; in other words, resources applied at the beginning of the launch will be substantially larger than later (relative to return per dollar sales). Most product launches follow a bell curve development pattern.

In early stages substantial resources may provide only minor returns (the missionary period). As the launch moves forward and product acceptance begins to takes hold, larger returns will be possible for the same or smaller resource dedication (growth phase). Late in the product life cycle (mature products), one should consider milking revenues from a product which will not return greater sales regardless of revenue dedication ("cash cows"); and finally, managers should be diligent about product termination (resources exceed returns).

In some cases, launching a new product can result in cannibalisation of existing products. This scenario can be planned; unfortunately in some cases it is not. An example of planned cannibalisation is when a new proprietary product is introduced to replace an older product coming off patent. Protecting one's market share and profit position is as important as developing incremental sales; sometimes more.

Cannibalisation requires exceptional forecasting skills. Marketing managers must determine how to drive new product integration, pricing of both products, and promotions that will result in manageable inventory control of both lines. Sharp competitors will do their best to unseat loyal customers during this period.

5.1 Product Pricing

Product pricing is important to any successful product launch. Setting the product price is a science. Many books have been written on the subject and various strategies abound. It is important to assess competitive positions and price. Product differentiation will play a critical role in setting product price. As I have said before, products not differentiated, will by default, be price differentiated. Pricing is best first considered in a broad context. Mapping the competitive price environment will help you determine a likely entry point for your analysis.

Many factors will contribute to the ultimate price determination: product benefits, competitive products, market stratification, image development, reimbursement, historical pricing of similar products, potential market volume, proprietary position, margin goals/requirements, unique selling proposition, market demand, and other.

Understanding customer needs and how your product satisfies those needs will be important. Customers are willing to pay in direct relation to their "value satisfaction". After you have done a paper "price" analysis, you should then consider doing customer surveys, focus groups, and test marketing. These activities will assist your efforts to narrow in on the appropriate product price. It is important to realise the old axiom, it is easier to start high and then come down. On the other hand, a product priced too high will never get the launch momentum established.

An important element of the price is the gross margin. Taking a margin discount during launch can be accomplished in many ways; some provide an opportunity to retrieve the discount; others do not.

Rebates may also provide incentives which instil an advantageous customer price perception. Step pricing is a mechanism to lock customers into meeting their performance commitments. This is often an issue with large purchasers who might forecast big potential orders. The key is not simply large orders. Pull through is the necessary ingredient. Big stocking orders are great; but, market demand is more important.

(Bio)pharmaceuticals can set up "bundling" opportunities. In the situation wherein several products may be used by the same customer, it is possible to orchestrate this type of pricing. This is an excellent strategy for locking out competitors. Using this approach, several products are discounted as they are purchased together. Any singular offering may not be able to compete one on one with its similar competitive counterpart.

6. PERCEPTION

In (bio)pharmaceutical marketing, like other types of marketing, perception is very important. How the company, its products, services, customer programs are perceived by customers is pivotal to ultimate marketing success. Defining product image and perceived value is a creative process. Once again, making it successful depends on how well you understand customer needs and desires.

Many pharmaceutical products are marketed by perception. Over the counter medications are an example. This is not to say a strong data position is not valuable. For certain type of products, clinical data can be the most meaningful part of (bio)pharmaceutical marketing.

Marketing clinical data requires good message mechanics. Galvanising a clear, concise, compelling, and brief message is the challenge. Physicians enjoy in-depth statistical clinical data; but, complexity often results in confusion; an axiom to always remember is: when customers are confused...they do not buy.

Data related to certain criteria such as accuracy or survival rates are critical. The key to the presentation of such elements is distillation accompanied by in-depth backup. Consider an HIV diagnostic test. If you were the person to be tested, accuracy of the test would be crucial to you; still, statistical data can be interpreted many ways. Considering such a test, specificity and sensitivity and false positives and turnaround times and other factors can create differing positions.

7. BRANDING

Many pharmaceuticals become branded products. Establishing brand identification is a process which should begin with deliberate strategic assessment. Where do you wish to go and why? What objectives are to be achieved? What is the corporate message we wish to reinforce? What are the customer needs and how can this product bring satisfaction? Is it a value added product? Is there specialised technology which may be used as part its differentiation? What is the central theme that should penetrate all product presentations?

Brand development is a function of message mechanics, presentation, frequency of delivering the message, and differentiation. Message mechanics is the science of communicating easy to understand compelling messages.

Name development is a critical part of product branding. Developing a product name requires creativity and continuous cross examination of characteristics important to the products differentiation and customer needs.

The integration of all elements of strategic importance during this distillation will result in a short list of criteria which can assist in product name development. Creating market demand is a marketing job. Distributors, for all their value, can not be counted on to create market demand. This is an important axiom. Many manufacturers must learn this rule each year.

7.1 Creating market demand

Having the cure for death really means little if you are the only one who knows it. An aspect of creating market demand is market awareness and visibility. Visibility is critical to demand; people must know your product exists before they can be interested in it, or buy it. This sounds like a simple rule; yet, many health care companies don't seem to understand it.

Visibility can be created through a broad range of marketing mix elements. The mix that will work best for you largely depends on the type of product you are selling and your customers.

Market demand also depends on image development and product positioning.

Image development is also a creative process. It is significantly impacted by graphic presentation. Capturing customer attention and creating a credible message that is compelling and memorable is the objective. In today's market, this is most effectively accomplished with graphic presentation. Today's consumers and clinician's are constantly bombarded with data and images of all types. Physicians must read voluminous amounts of information just to keep up with the changing medical literature and they are further taxed by continuing education requirements.

Getting your message seen, understood, and acted upon, requires strong image development. Graphics can be developed which will deliver your important messages without the need for long and cumbersome verbiage. Customers do not want to read long and protracted presentations about anything unsolicited.

Consider that your customer doesn't know your product exists; and further, doesn't care that they do not know. This is a good position to begin image development. Can a graphic be developed that will communicate the product's unique selling proposition and will it stimulate customer interest? The answer is yes! Make your presentations as visual as possible. They will perform better and they will be remembered.

Graphics should be developed which can serve as a nucleus to complete campaigns. If the graphic is correctly developed, it can be used in all types media advertisements (with the obvious exception of radio). Providing customers memorable images of your product and company is the foundation for creating market demand. Consider how powerful the market image of, e.g., Tylenol is.

Physicians when polled about market presentations may offer they would prefer to see cold hard data. Under certain circumstances this is probably true. On the other hand, these physicians are also people affected by good marketing. In the environment of today's health care industry physicians possibly more than anyone can be successfully marketed. The question isn't what can you provide them once they are interested; it is what can you provide to get them interested.

After you have succeeded in capturing physician interest, then and only then, will all that data have a chance to influence clinical minds. The first order of business is capturing the interest.

7.2 Product Positioning

Product positioning is another area for careful development. To position your product correctly within a market requires understanding the competitive landscape of the market and the product's performance capabilities.

An important consideration in positioning is the opportunity to gain market share. Does your product have the opportunity to gain market share? When placed into a competitive environment how will it be received? If it has multi applications, which ones stand the greatest opportunity for success; will one application's chances far exceed the others? Does one application have the opportunity for significantly reduced competition? Should the product be positioned as a middle of the road beta blocker or a breakthrough offering for rheumatoid arthritis?

Criteria for positioning include; but, are not limited to: pricing, distribution, packaging, competition, life cycle, patents, marketing resources, regulatory requirements, technology.

Market mapping will help you establish product position. Mapping is a mechanism for sorting the various positions of all competitive products in your market. It is an overlay of information about each company/product. The information will define competitive price structures, differentiation, distribution agreements, geography coverage, potential strengths and weaknesses.

This exercise should provide a good perspective of niche potentials. It will also assist in setting pricing and distribution strategies. If you position a product to be the low cost market solution - this somewhat defines its strategies and marketing requirements. On the other hand, if the product is positioned to be in direct competition with the number one market leader, this sets a different set of requirements.

Good market positioning seeks the greatest market share capture while securing the highest margin with the least costs. Regarding positioning, I favour corporate positioning to be in concert with product positioning (if possible). Generally, customers are comforted and positively influenced by companies that demonstrate strategic commitment to clinical interests. Corporations are usually well served to demonstrate this commitment and market slogans which communicate it.

Corporate slogans are strategically very important. Look about the industry and you will see many of the most successful companies driving carefully conceived corporate messages. A well designed corporate slogan can greatly enhance the differentiation and power of a (bio)pharmaceutical product launch. The two, if properly constructed, can work in tandem to communicate strategic commitment to the customer. Companies which do not use corporate slogans are missing out on a great opportunity to capture customer imagination. Looking upon the broad (bio)pharmaceutical market, customers are influenced by companies that demonstrate strategic commitment to clinical areas and/or areas of technological research.

8. PRODUCT SCALE-UP

Prior to product launch there are other things to consider and carefully control. One of the most demanding times for any medical manufacturer is "scale up". Many a product has fallen out of bed as a result of scaling up problems.

It is one thing to control a product when it is micro managed within the confines of a research laboratory; all of the variables are at arms length. It is quite another to scale up production and suddenly go from making a few samples to making literally thousands or millions of doses. In the midst of scale-up, many product factors can significantly change.

It is critical that marketing managers keep a careful vigil with manufacturing engineering as scale-up unfolds. Assessments should be bench marked to assure all basic launch tenets remain in tact.

9. IMPORTANCE OF TRAINING

Upon product launch there are other factors which can make or break product acceptance; one is training. Training encompasses many theatres; sales representatives, distributors, end users, and perhaps many others.

Training may require several different modalities depending on the target audience. One of the primary requirements of all training is building competence and confidence of those participating. Did the programs effectively teach the trainee the intended material. Is the trainee capable of performing the desired tasks? If the program involves product use, can the trainee now properly and effectively use the product? Separately, how will the training be performed? Will it be face to face or through some type of correspondence media, e.g. video? Objectives should be developed and tested to insure training requirements are achieved.

One of the prime requirement of training is good communication. Separately, clinicians require "real world" training scenarios versus hypothetical in-services. They will quickly know whether or not your program is credible.

The cost of training during initial product launch may be significantly higher than later in the launch cycle. It is critical that initial training be sufficient and professional. Getting off to a good start is very important. If the launch is thwarted by poor utilisation due to training problems, chances of success will be greatly diminished. Also it is important to develop "reference clinicians" and/or "reference centres" wherein knowledgeable clinical people with product experience may assist new potential users with their experience and insights. This credibility and third party reference is very valuable in advancing the product launch. As the launch progresses, training methods can be modified and designed with less labour intensive programs.

Many (bio)pharmaceutical products may require significant technical service to successfully support product usage. Consider the example of bone marrow transplantation technology. Obviously, technical support may play an important role in the successful use and clinical performance of the product, e.g. cell separation, immobilisation, etc. Many of these support elements, although independent of your product, may link their performance to your success. It is incumbent on product managers to understand the complete technology chain in product dependent procedures.

10. PRE-LAUNCH MARKET PLANNING

As we assess all of the variables in (bio)pharmaceutical product marketing, we understand pre launch market planning is very important. Some companies give little preparation to these things. I strongly recommend pre launch market planning to begin no less than one year prior to the objective product launch date. You will recall, some aspects of product marketing need to be planned as far back as pre clinical schedules, e.g. outcomes analyses.

Determining proper positioning, image development, pricing, training, distribution and other elements may take many many months of assessment. Waiting until a few months prior to launch is a most serious mistake.

BIOGRAPHY

CareMerica Inc. is a highly specialized international health care consulting and marketing firm. Scott Spinka is the Founder and President of CareMerica and a past Director of the Medical Marketing Association; the company provides out source "medical marketing", strategic assessment, and full media creative services (www.caremerica.com).

REFERENCES

1. Levitt, T. (1983). The marketing imagination. The Free Press.

Chapter 19

Viral mediated gene therapy

Brendan Murphy
Department of Applied Science, Limerick Institute of Technology, Moylish Park, Limerick

Key words: Gene therapy. Viral Vectors; Retroviruses; Adenoviruses; Vector targeting; Genetic disease; Cancer; Infectious disease; HIV.

Abstract Gene therapy has recently received attention as a novel strategy for the correction of human disease at the genetic level. Several methods of transferring genes into somatic mammalian cells have been developed and have led to a range of gene therapy clinical trials for both inherited and acquired disorders. The majority of clinical trials have used viral vectors of animal origin to mediate gene transfer, with retroviruses, adenoviruses and adeno-associated viruses constituting the vehicles of choice. While the principle of gene therapy has been proven, with appropriate therapeutic responses being noted, its potential has not yet been realised in large scale clinical trials. Critical areas for future development remain in the design of more efficient, safer, vectors for gene delivery. Viral vectors are currently undergoing extensive modifications to produce vectors custom-made for particular clinical applications.

1. INTRODUCTION.

The possibility of using the gene as a drug for the correction of human disease, at the genetic level, has been transformed from a concept to a clinical reality over the past decade. This approach to disease intervention has, commonly, become known as gene therapy. While previously the majority of all diseases have been treated symptomatically, the development of gene-based therapeutics is moving intervention towards a more causative approach and, as such, represents a new paradigm in the treatment of disease. The techniques are founded on the remarkable progress that has been made in molecular biology over the last twenty years. These include powerful recombinant DNA techniques for the analysis and manipulation of nucleic acids and sophisticated strategies for their transfer to somatic mammalian cells. Combined with the tracing of the origin of numerous disease states at the molecular level, these advances have contributed to the increasing viability of gene therapy.

The potential indications for gene therapy are widespread and include the correction of monogenic inherited disorders, multifactorial disorders, and a broad range of cancers and infectious diseases. Monogenic inherited disorders (mutation of a single gene) were the first disease states for which gene therapy was applied. Examples include inherited immune deficiency disorders (such as adenosine deaminase (ADA) deficiency), sickle cell anaemia, haemophilia and cystic fibrosis. The identification of the defective genetic locus and isolation of its normal sequence are pre-requisites for the design of gene therapy protocols targeting such disease states. Multifactorial disorders, including coronary and arterial diseases (that have a pathology involving the confluence of a number of genetic and environmental components) are amenable to gene therapy through the introduction of genes encoding products that function at the cellular level to inhibit the progression of the disease. Gene therapy is applicable to numerous cancers through the targeting of somatic mutations of cellular genes involved in their initiation and progression. Other strategies involve the delivery of genes encoding specific anticancer agents or immunomodulatory products. Gene therapy is also applicable to chronic infectious diseases, such as herpes virus, hepatitis and HIV infections, through enhancement of immune responses by post-exposure vaccination. Nucleic acids designed to, directly, target the infectious and replicative cycle of the infectious agent can also be delivered. Other gene therapy protocols that do not involve the transfer of genes with direct therapeutic benefit have also been designed. These involve the transfer of marker genes that function by facilitating the tracking of the fate of cells in the body.

Effective gene therapy involves the transfer of genes to a broad range of target tissues in anatomically diverse locations. This can involve either *ex-vivo* or *in-vivo* strategies depending on the clinical target. The *ex-vivo* approach involves explanted autologous cells to which the therapeutic gene is transferred to the somatic tissue *in-vitro*, prior to re-implantation of the modified cells in the recipient for clinical effect. The *in-vivo* approach involves direct introduction of the therapeutic genes into the target tissue of the recipient in applications where tissue explantation is not as efficient or feasible.

Whether a gene therapy strategy involves an *ex-vivo* or *in-vivo* approach, the delivery of nucleic acid to augment, replace or inhibit altered or inappropriately functioning genes, requires vector systems to aid in the delivery of the therapeutic gene to the appropriate intracellular site. Data from clinical trials is available from vector systems of both viral and non-viral origin. The non-viral vectors have largely consisted of plasmid-based systems, involving physio-chemical methods of gene delivery, and are reviewed in the subsequent chapter. This chapter focuses on the viral-based vector systems for gene delivery and overviews the applications and potential of viral-mediated gene therapy.

2. CATEGORIES OF VIRAL VECTORS

There are four categories of viral vectors currently undergoing clinical trials in gene therapy protocols. These include the retroviruses, adenoviruses, adeno-associated viruses and herpes-simplex viruses. Each category has its own inherent set of advantages and limitations and the choice of viral vector developed and tested depends on an array of factors. These include the type of target tissue; the size of the exogenous DNA; the nature of the gene products; and the required duration of expression of the gene product. An ideal vector would be target cell specific and capable of delivering the therapeutic gene and regulatory elements of sufficient size for the particular clinical application. It would be capable of regulated gene expression for the appropriate time duration to achieve the desired clinical response. It would also be capable of avoiding the host defence system and would not induce immunogenic or inflammatory responses, or generate replicative competent viruses. As the indications for gene therapy are widespread each clinical application has its own optimal vector type and many investigators are now creating designer vectors tailor-made for particular clinical targets.

Viral vectors of animal origin were an obvious choice in therapeutic gene transfers as they have evolved sophisticated mechanisms to introduce their nucleic acid into specific recipient cells while simultaneously being adept at evading host defence mechanisms. Of the categories of viral vectors currently undergoing clinical trials, the retroviruses and the adeno-associated viruses transfer their DNA directly to the chromosomal DNA of the target cell in a stably integrated form, while the adenoviruses and the herpes-simplex viruses introduce their DNA into the nucleus in an unintegrated (and therefore transient) form (Table 38).

Table 38. Features of viral vectors used for gene delivery in gene therapy trials

Vector	Genome Size	Insert Capacity	Location of Transduced Vector	Expression of Viral Genes
Retrovirus (MoMuLV)	10 kb	8 kb	Integrated	No
Adenovirus	8 kb	7-8 kb	Extrachromosomal	Yes
Adeno-associated virus	4.7 kb	4-5 kb	Integrated	Yes
Herpes simplex virus	152 kb	50 kb	Extrachromosomal	Yes
Vaccinia virus	187 kb	25 kb	Extrachromosomal	Yes
Avipox virus	260 kb	4 kb	Extrachromosomal	Yes

2.1 Retroviral Vectors

Retroviruses are a large class of enveloped viruses and contain their genetic information in double-stranded RNA genomes. This is reverse-transcribed by the viral-encoded reverse-transcriptase enzyme to a proviral DNA molecule which is then stably integrated into the genome of infected cells. The integrated DNA is expressed over extended periods of time (1).

Since the first gene therapy protocol involving a retrovirus vector for the treatment of adenosine deaminase (ADA) deficiency (2) the retrovirus vectors have become the vehicles of choice for therapeutic gene transfer. They constitute the majority of vectors used in clinical trials thus far (60% of clinical trials have used retrovirus vectors). The popularity of retroviral vectors has stemmed, in part, from the fact that huge advances have been made in the understanding of the molecular biology of retroviruses over the last two decades. They also transduce many cell types at very high efficiencies; some retroviral vectors transduce up to 100% of target cells. Retroviral entry into target cells is dependent on the presence of specific viral receptors on the target cell. Of the retroviral vectors, the class that have received most attention are recombinant vectors derived from the Murine Retrovirus, Moloney Murine Leukaemia (MoMuLV) (3, 4).

Wild-type retroviral genomes comprise of approximately 10 kb of RNA and have a relatively simple prototypical organisation of genes (Figure 41(a)). The termini of the genome are bounded by 5' and 3' long terminal repeats

(LTR's) that include sequences critical for the integration of the proviral genome. LTR sequences also confer promoter and enhancer activities. A sequence that provides a viral packaging signal (Ψ) required for correct packaging of viral RNA is also present downstream from the 5' LTR. The endogenous viral RNA bounded by the LTR's encodes the genes *gag, pol, and env* (Figure 41(a)). These specify group specific antigen core proteins, reverse transcriptase/replication functions and viral envelope proteins, respectively.

In the production of retroviral vectors only those endogenous DNA sequences required to function in *cis* are retained while the viral vector itself is rendered replication incompetent in recipient cells by deleting the essential viral functions *gag, pol* and *env*. The deletion of these endogenous sequences provides space (within the limits of the packaging capacity) for the insertion of exogenous sequences specifying the therapeutic function and its regulatory elements. This produces a recombinant, replication-defective vector, which is then replicated and produced to high titres in packaging or helper cell lines (Figure 41(a) and (b)).

Packaging or helper cell lines express the viral proteins Gag Pol and Env, which are able to function in *trans*. Their expression and *trans*-functionality supports the production and packaging of recombinant vector which itself contains all the necessary *cis* acting functions to produce progeny virions. Packaging cell lines that yield up to 10^7 viral particles per ml have been developed (5). A prime concern in the large-scale production of vector in packaging cell lines is to eliminate the breakout of replication competent retroviruses (RCR's). To achieve this the viral RNA in the packaging cell is split into two transcription units, which are integrated into different regions of the host chromosome. One transconjugant unit encodes the *gag* and *pol* genes while the other encodes the *env* gene. Additional measures include the deletion of the packaging sequence and the deletion of the 3'LTR and its replacement with a polyadenylation sequence.

Retrovirus vectors can accommodate up to 9 kb of exogenous DNA and their primary advantage is the stable introduction of pro-viral DNA into the target cell by integrating into host DNA in the absence of the expression of viral proteins (Figure 41 (b)). They are also highly efficient at gene transfer and can transduce up to 100% of target cells. The exogenous DNA introduced to the target cell has the capacity to survive for the life span of the host cell and be passed on the progeny cells upon mitotic division. Limitations of retroviral vectors include the highly labile nature of the viral particle and associated difficulties in purifying high virion titres, and their inability to infect non-proliferating cells in differentiated tissue. The random nature of the integration of the proviral genome in the transduced target cells carries the risk of interfering with host cellular functions potentially causing insertional mutagenesis or the activation of oncogenes (Table 39).

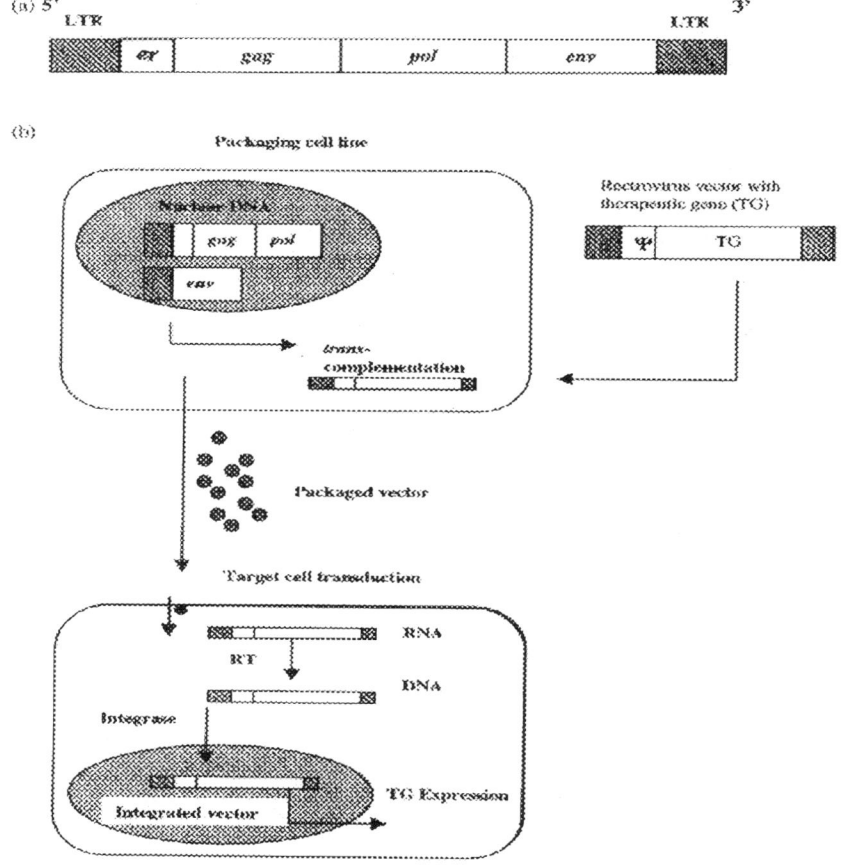

Figure 41. (a) Prototypical organisation of the wild-type rectrovirus genome depicting the long terminal repeat (LTR) sequences at its termini, the packaging signal (Ψ), and the three protein coding regions *gag, pol* and *env* (not to scale). (b) Method of producing replication-incompetent retrovirus vectors and transduction of target tissue. In the vector the endogenous *gag, pol* and *env* sequences are deleted and replaced with exogenous DNA encoding the therapeutic gene products. The vector is transfected to a packaging cell line for the replication and production of the vector by trans-comlemetation from chromosommolly integrated *gag, pol,* and *env* sequences. Packaged vectors are then used to transduce the appropriate target cell. Vector RNA is converted to proviral DNA by the enzyme reverse transcriptase (RT). The proviral genome is randomly integrated into the target cell genome were its therapeutic gene (TG) is expressed.

Viral mediated gene therapy

Table 39. Advantages and limitations of retrovirus vectors for gene therapy applications.

Advantages	Limitations
• Their molecular biology is very well understood.	• They do not infect non-proliferative cells in differentiated tissue.
• Integration of the vector means transferred genes are maintained for the life of the host cell.	• They risk activating oncogenes or inactivating normal cellular functions by causing insertional mutagenesis.
• High efficiencies of transduction of target cells (up to 100% for some retroviral vectors).	• Problems encountered with transducing specific cell types such as hematopoietic stem cells.
• Immunogenic viral proteins are not expressed.	• Rates of transfer and gene expression can vary dramatically between different cells and patient.
• Vector production is relatively straight forward.	• Retroviral particles are relatively labile and difficult to purify and concentrate in high enough titres.

2.2 Lentiviral-based retroviral vectors

The inability of MoMuLV-based retroviral vectors to transduce non-proliferative cells is a significant limiting factor and precludes their use for *in vivo* gene therapy protocols to non-cycling target tissue such as hepatocytes, neurones, myofibres and hematopoietic stem cells (6). The lentiviruses, which include the human immunodeficiency viruses (HIV) are unique among retroviruses in being able to infect quiescent cells due to their possession of specific regulatory genes.

HIV, which specifically targets CD4-positive T-cells has been developed as a lentivirus vector, in part, as its host range can be expanded beyond CD4-positive cells by pseudotyping other envelope glycoproteins (7) and for their applications in AIDS gene therapy protocols through the direct delivery of genes encoding antiviral activity to CD4 and macrophage and monocyte infected HIV cells.

A key concern in HIV-based vector development and production is the avoidance of the generation of replication-competent viruses. The wild-type HIV genome encodes *gag*, *pol* and *env* sequences flanked by LTR sequences as characteristic of retroviruses. The LTR's encode *cis* acting sequences that function in integration, transcription and polyadenylation (8). As with MoMuLV-based vectors the HIV vector is rendered devoid of viral protein coding sequences and only retains the *cis*-acting functions. Vectors are produced in packaging cell lines that express *trans*-acting proteins required for vector replication and production (9). One plasmid contains *gag* and *pol*, the other *env*, while both lack HIV packaging sequences and contain the cytomegalovirus immediate early promoter and simian virus 40 polyadenylation signals (8). These precautions are necessary to prevent the generation of replication-competent viruses.

Lentivirus-based vectors combine the advantage of long-term gene expression because of stable integration of vector into chromosomal DNA of target tissue with the ability to transduce non-proliferative tissue. As the molecular biology of HIV has been extensively studied they are very suitable viruses for vector development. HIV-1 derived vectors have been employed to successfully transduce non-proliferative neurons (10) and hematopoietic cells (11). These studies have involved pseudotyping HIV-1 vectors with vesicular stomatitis virus glycoprotein-G envelope protein.

2.3   Adenoviral Vectors

The adenoviruses comprise non-enveloped double-stranded DNA viruses that are responsible for respiratory and eye infections in humans. Though adenoviruses have a natural tropism for the epithelium of the respiratory tract they can also transduce most other cell types giving a broad target cell range. Adenoviral vectors have been the second most popular vehicles of choice for gene delivery after retroviral vectors and, to date, have been employed in 15% of gene therapy clinical trials.

In contrast to retroviruses, adenoviruses can efficiently infect non-proliferative cells (infection of proliferative cells often results in cell lysis) and integration of adenoviral sequences into host chromosomal DNA is not an integral part of the its life-cycle. However inefficient integration can occur in non-dividing cells at high multiplicities of infection. While adenoviruses display efficient gene delivery to a broad host cell range their principle drawbacks have stemmed from the direction of cytotoxic and humoral immune responses of the host cell against viral antigens. This results in the destruction of transduced cells and the build-up of immunity to vector re-administration.

While there are more than 50 known human adenovirus serotypes, the majority of vectors have been constructed from serotype Ad5, an adenovirus implicated in respiratory infections in infants. The adenovirus genome consists of a linear double-stranded DNA molecule of approximately 35 kb, with a prototypical molecular organisation more complicated than retroviruses. The genome is bounded by inverted terminal repeat (ITR) sequences that function as replication origins and, also contains a sequence (Ψ) required for packaging of the viral genome (Figure 42(a)). The infective cycle of the virus is composed of two phases, early and late, whose transcriptional units involve early (E) genes and late (L) genes. The early phase involves a number of regulatory steps that occur prior to viral DNA synthesis. The early genes E1, E2, E3 and E4 encode proteins that regulate viral and cellular gene expression, viral replication, and inhibition of cellular apoptosis. Expression of the early viral gene E1 is required for the initiation of the viral replicative and lytic cycle by *trans*-activation of the other early viral genes. It renders cells capable of intensive viral DNA and protein synthesis using host-encoded functions. The late phase, under the control of the major late promoter (MPL), involves the synthesis of structural proteins and viral DNA and the concommittant assembly of virions.

Adenovirus vectors for gene therapy have the E1 coding region deleted and therefore are replication incompetent. In many vectors, an additional early phase gene, the E3 region is also deleted. The E3 region encodes functions that inhibit the host immune response and has been found to be non-essential for viral replication in cultured cells (12). The total capacity of adenoviral vectors containing these deletions for exogenous DNA is approximately 8 kb. In E1 deleted vectors production of the replication-incompetent vector for administration to target cells is carried out in an E1 complementing human embryonic kidney cell line designated 293. The 293-cell line contains a copy of the wild-type Ad5 E1 region integrated in its genome. Its gene product functions in *trans* to complement the E1 deleted vectors facilitating their replication and production (Figure 42(b)). Recently, a novel E1 complementing cell line has been developed that eliminates the danger of producing replication-competent adenoviruses by homologous recombination. This encodes the minimum E1 sequence necessary for complementation but lacking any overlapping sequence with the vector (13). Unlike the retroviruses, the adenoviruses are easy to manufacture to high titres (10^{13} virions per ml) and, their isolation is amenable to ion exchange chromatography for large-scale production of quantities required for therapeutic purposes in clinical trials (14). Adenoviral particles remain stable when subjected to extensive purification (Table 40).

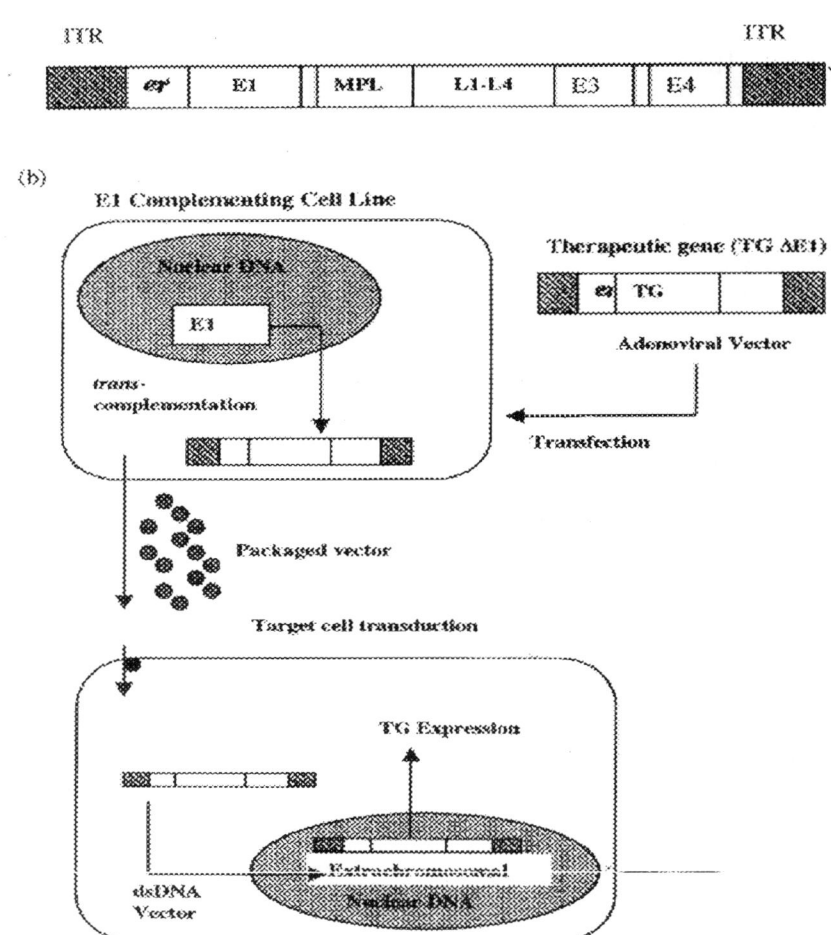

Figure 42. (a) Organisation of the wild-type adenovirus Ad5 genome indicating the inverted terminal repeat (LTR) sequence, the packaging signal (Ψ) and the relevant transcriptional units; the early phase genes E1 and E4, the major late promoter (MPL) and late phase genes L1 to L4 (not to scale). (b) Method of producing replication incompetent adenovirus vectors and transduction of target cells. In the vector the E1 region is deleted (ΔE1) and replaced with the therapeutic gene of interest (the E3 region can also be deleted if packaging space is required). The vector is transfected to an E1 complementing cell line and packaged vectors are then used to transduce the target tissue. After entry into the target cell the double-stranded DNA (dsDNA) vector migrates to the nucleus where its therapeutic gene (TG) is expressed from an extrachromosomal position.

Table 40. Advantages and limitations of adenovirus vectors for gene therapy applications

Advantages	Limitations
• They exhibit a broad target cell range.	• They evoke non-specific immunogenic and inflammatory responses.
• Transduction is independent of the replicative status of the host cell.	• A limited expression period (weeks to months) necessitates periodic vector re-administration.
• The epichromosomal nature of the transduced particle eliminates the possibility of insertional mutagenesis.	• T-cell responses reduces efficacy of vector re-administration.
• They can be easily produced to high titre (10^{10} - 10^{13} virions per ml) and the viral particle exhibits high stability during extensive purification.	• Infection of proliferative cells can result in cell lysis.

Adenovirus vectors infect target cells by attaching to specific cellular receptors via a fibre and penton and form a cytoplasmic endosome After lysis of the endosome the linear double stranded genome is delivered to the nucleus where it exists extrachromosomally. The unintegrated nature of the genome results in only transient expression of the therapeutic genes delivered to the nucleus. The duration of gene expression is further reduced due to CD4 and CD8-dependent immune responses initiated as a result of 'leaky' expression of viral proteins. These T-cell responses result in the elimination of transduced cells and repeated administration of the vector is inhibited due to their elevated levels. An additional, potentially, harmful consequence of adenovirus administration is the induction of inflammatory responses thought to be a downstream effect of the activation of the transcription factor NF_KB by the adenovirus (15). To overcome these difficulties second-generation adenoviral vectors that are less immunogenic have been developed as they contain additional viral sequences deleted. These include deletions in the E4 region, a transcriptionally complex region that regulates the change from early to late phase gene expression. However, while E4 defective vector-transduced cells have been found to contain the transduced DNA for longer periods of time, its long-term expression has not been enhanced possibly due to the absence of viral promoters in the E4 deleted region (12). Other recent developments have included the design of adenoviral vectors that contain only the inverted terminal repeat regions required for replication. These so-called gutless vectors require helper viruses to complete their replicative cycle but production problems particularly associated with the separation of the vector from the helper virus have yet to be overcome (5).

2.4 Adeno-associated Virus Vectors

Adeno-associated viruses (AAV) comprise a non-pathogenic parovirus family that are indigenous to humans. They are not autonomous in that they require a helper virus (the adeonoviruses or certain other viruses such as herpes virus) to complete their replicative cycle. The AAV genome consists of a small (4.7 kb) single strand of DNA with an extremely simple prototypical organisation comprising just two genes, designated *rep* and *cap* (Figure 43(a)). The *rep* gene encodes a family of four overlapping proteins involved in replication and integration while the *cap* gene encodes three proteins involved in encapsidation. These are flanked by inverted terminal repeat (ITR's) which contain palindromic sequences that function in *cis* during viral replication.

As both the Rep and Cap proteins can be supplied in *trans*, AAV vectors are constructed by deleting their coding regions and replacing them with exogenous DNA containing the appropriate therapeutic gene and transcriptional elements (Figure 43 (b)). Their capacity for exogenous DNA is not large as it can only slightly exceed the 4.7 kb length of the wild-type genome. AAV vectors lack all coding sequences and during vector production these are provided in *trans* in packaging cell lines. The standard packaging cell line has been the 293 cell line (as with adenovirus vectors) which has been transfected with a plasmid that provides the *rep* and *cap* functions in *trans*. As adeno-associated viruses require a helper virus for their replication, a 293 cell line transfected with the vector and plasmid is subsequently infected with an adenovirus (helper virus) to facilitate the replicative cycle. Both the AAV vector and adenovirus particles are then assembled and packaged in the cell line. The AAV vector is subsequently purified from total cell lysates through CsCL gradients (16).

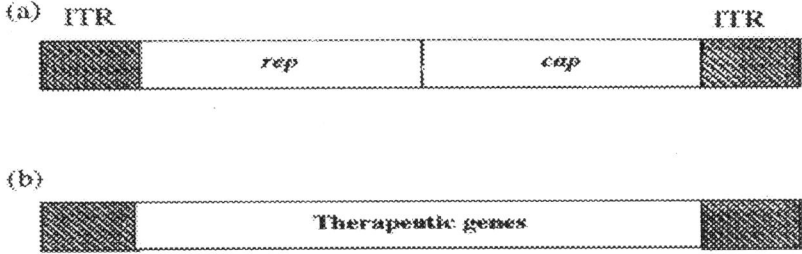

Figure 43. (a) Prototypical organisation of the wild-type adeno-associated genome consisting simply of two genes rep and cap bounded by inverted terminal repeats (ITRs) (b) AAV vectors are constructed by replacing the rep and cap genes with the therapeutic sequences of interest.

The AAV vectors have a number of advantages over the adenovirus and retrovirus vectors. As they are structurally very simple and lack viral protein coding regions and they do not provoke immunogenic and inflammatory responses like the adenoviruses. Long term gene expression has also been noted, and this has been attributed to integration of the vector in the target cell. Although integration is less efficient than with retrovirus vectors their infection is independent of the replicative state of the host cells. Though wild-type adenovirus integrates preferentially into a specific site on chromosome 19, the recombinant vectors lack this specificity possibly due to the absence of Rep protein (5, 17). An additional limiting factor has been the impaired ability of the integrated single-stranded vector genome to synthesise its complementary strand. In the infectious cycle of wild-type AAV, second-strand synthesis is stimulated by the E1 and E4 advenoviral proteins but with AAV vectors this is dependent on cellular factors (5).

The main limitations of the AAV vectors have been difficulties in generating large quantities of the vector for clinical purposes. Purification procedures are complex and result in low viral titres (10^4 virions per ml) and contamination of vector with helper adenovirus has also been a problem. However, major technical improvements in vector development, that involve replacing the helper virion with adenoviral plasmids, have recently been reported [18]. The adenoviral plasmid harbours the minimal adenoviral functions necessary for vector replication and AAV vector stocks with undetectable adenoviral contamination have been reported. These are important developments as AAV vectors have been found to transduce a wide range of cell types at high frequency, with stable transduction and gene expression occurring independently of the helper virus [5].

2.5 Additional virus vectors

A number of additional viruses have been developed as gene therapy vectors. These include herpes simplex virus, vaccinia virus, and poxvirus vectors. The use of these viruses as gene therapy vectors expands the scope and efficiency of gene delivery to potentially facilitate therapeutic gene transfers to a wider range of clinical conditions. However, common problems associated with the development of these viruses as vectors have been the increased difficulty in generating replication incompetent viruses and safety issues associated with high levels of immunogenicity and cytotoxicity, stemming from the large size and complex replicative cycles of the wild-type genomes (Table 38).

Herpes simplex viruses (HSV) have received a lot of attention as potential gene therapy vectors. HSV can establish a latent infection in neurons of the peripheral and central nervous system by existing as an episome with tight control of viral lytic genes. Though it has a natural tropism for the central nervous system, HSV can infect a wide variety of other cell types yielding broad applications in gene therapy.

The genome of the virion SHV-1 is a linear double-stranded DNA molecule of 152 kb. This large genome is composed of two distinct segments designated unique long (UL) and unique short (US) and each is bounded by inverted repeat sequences. This large genome encodes over 75 gene products that are classified as immediate early (IE), early (E) or late (L) depending on their sequence of expression during the replicative cycle. The genome encodes at least 20 essential (replicative and structural) genes but also has many non-essential regions that are not necessary for viral replication in cell culture and represent some 50% of the wild-type genome. This allows for the deletion of large segments of the viral genome and replacement with exogenous DNA, without effecting the replicative ability of the virus. HSV-derived vectors, therefore, have a very large capacity for exogenous DNA (up to 50 kb) and can deliver multiple therapeutic sequences to target tissue.

For application as a vector, the HSV virion has to be rendered replication-incompetent. This is achieved by deleting or inactivating essential IE genes that regulate E and L genes involved in viral replication and encapsidation. Viruses defective in IE genes involved in the early stage of the regulation of the lytic cycle can enter a state of latency (19) in which the double-stranded DNA genome remains extrachromosomal. Inactivated functions essential for replication are provided in *trans* in suitable complementing cell lines during vector production. HSV vectors are considered safe in that they can be produced to high titres in the absence of replication competent viruses.

However, a major limiting factor in the use of HSV vectors has been cytotoxicity associated with low-level expression of undeleted viral genes.

Elimination of this toxicity has proved difficult, requiring the manipulation of several gene products (20). As a result of the large genome size and complex biology of HSV, many of the characteristics associated with this toxicity have yet to be elucidated. A recent approach to the elimination of toxicity has involved the generation and use of amplicons. Amplicons are HSV-derived virus vectors that contain a plasmid encoding the HSV packaging sequence and origin of replication. They are generated in a packaging cell line that is co-transfected with the plasmid and a cosmid containing the entire HSV genome but with a deleted packaging sequence. Only the plasmid sequences are packaged with all critical viral functions provided in *trans* from the packaging-defective viral sequences on the cosmid (5).

Vaccinia virus and Pox virus vectors have been developed for the delivery of therapeutic genes for specific clinical purposes. Vaccinia virus vectors can accept large inserts of exogenous DNA (up to 25 kb) and have been used to deliver IL-1 sequences to tumour cells (21). Pox virus vectors can accept approximately 4 kb of exogenous DNA and have been used at the interface between gene therapy and vaccination due to their ability to infect human cells in the absence of viral replication (22).

2.5.1 Limited Replication Competent Vectors

As replication-defective viral vectors are limited in their ability to diffuse through tissue their efficiency in particular clinical applications, such as the *in vivo* treatment of tumours, is limited. A number of innovations have been designed to tackle this limitation by permitting limited rounds of replication, with consequent infection and transduction of neighbouring cells without introducing additional safety concerns.

One strategy has involved the design of, so-called, plasmoviruses that combine the advantages of both viral and non-viral methods of gene transfer. Plasmoviruses combine the simplicity of plasmids in clinical grade vector production with the efficiency of viruses for transduction and gene delivery to specific cell types. One plasmovirus vector has been constructed from the MoMuLV retrovirus by replacing the *env* gene with exogenous DNA that generated a plasmid (including therapeutic sequences) and inserting a transcription unit coding for *env* (23). The *env* sequence was designed to have minimum homology with proviral sequences eliminating the generation of replication competent viruses. After transfection to tumour cells by physical means (subcutaneous injection) the plasmoviruses generated replication incompetent recombinant retroviruses that were capable of carrying out a single round of transduction of neighbouring tumour cells (23).

A simpler approach using adenoviral vectors has recently been reported (24). This involved transcomplementation of an E1 deleted adenovirus with a co-delivered exogenous plasmid (pE1) encoding the E1 sequence. One round of viral replication was facilitated. The design of these strategies represents an effective method for therapeutic gene targeting of tumour cells and may represent a prototypical approach for the treatment of cancers involving superficial tumours in cavities where vector administration is feasible.

3. THERAPEUTIC GENE TARGETING USING VIRAL VECTORS.

An emerging need in gene therapy is the targeted delivery and expression of therapeutic sequences in specific cell types. The majority of vectors developed thus far have little or no cell specificity except for specific tissue tropisms such as the adenovirus tropism for epithelial tissue or HSV tropism for neurons of the peripheral and central nervous system. Retroviral vectors exhibit no specific tissue tropisms, with the one exception of HIV-derived vectors, which offer specificity for CD4-positive cells. As both adenviral and retroviral vectors can deliver therapeutic genes to a very broad range of tissue types the development of targeting systems has been recommended for enhancing the safety and efficiency of gene therapy protocols (25). Targeted delivery is recommended if the transferred gene encodes a toxic gene product designed for the elimination of specific cells or to enhance the therapeutic gene dose in specific cells by preventing dilution of the vector in collateral tissue. It also reduces the dependency on *ex-vivo* protocols which are more complex and expensive (25). Dramatic improvements in trageting efficiency are required to yield protocols that are practical for *in vivo* gene therapy.

Cell-specific targeting of viral vectors has been achieved by modifying adenoviral and retroviral vectors to bear specific antibodies, ligands of receptors to bind specific cell types. Antibodies directed against the viral envelope proteins can be attached to specific tissue cell surfaces using streptavidin and primary antibodies that recognise cell-specific receptors or antigens. Hematopoietic cells have been targeted with retroviral vectors using the stem cell factor receptor (SCF) using DNA-polylysine-streptavidin complexed with biotin-SCF (26). Gastrointestinal cancer cells have been targeted with MoMuLV retroviruses displaying a single chain variable fragment antibody to the carcinembryonic antigen (CEA) predominately found in cancers of the colon and stomach (27). High efficiency targeting of adenoviral vectors using similar strategies have also been reported (28, 29).

Mutation of viral envelope protein coding sequences to generate modified envelopes capable of recognising specific cellular receptors or antigens provides an alternative strategy. Retroviral envelope proteins have been altered to express chimeric protein with a cell-specific ligand (30). Chemical modification of viral envelope proteins has also been used to direct infection to specific cellular receptors. Asialation of retroviral glycoprotein receptors has been used to direct vectors to hepatocyte-specific asialoglycoprotein receptors (31). Other developments in cell targeting have involved the design of chimeric vectors that contain features of two or more viruses. Such pseudotyping can lead to constructs carrying the core and nucleic acid of one virus and the viral envelope proteins of another virus thus potentially changing the spectrum of cell-type transduction (5, 31).

An additional level of specificity is conferred on viral vectors by targeting the integration of the exogenous DNA of the vector to specific regions of the target cell genome. While the production of such defined alterations in eukaryotic genomes has, particularly, been recommended for monogenetic defects, the strategy has widespread applications in gene therapy. Its potential advantages include the correction of both recessive and dominant defects at precise target loci, the delivery of only defective portions of genes rather than the entire coding region, and, the elimination of risks of insertional mutagenesis and oncogene activation associated with random integration (32).

Vectors designed for positional targeting within the genome contain a segment of DNA homologous to the target region. Homologous recombination can direct the insertion of the vector at the target sequence by a single reciprocal exchange that duplicates the region of homology. Alternatively, it can involve the replacement of the region of homology by the vector in homologous recombinaton events involving two reciprocal exchanges (32). DNA-RecA protein complexes can be included in the vector to promote strand exchange and pairing. While viral vectors have not received extensive study for gene targeting (due to limitations in the size of homologous DNA the can deliver) initial studies with adenoviral vectors targeting the *FGR* locus of mouse ES cells have yielded encouraging results (32). The ability of adenoviruses to deliver double-stranded DNA segments to nuclei at high efficiencies does render them suitable for developments as targeting vectors by homologous recombination.

Targeting retroviral integration by homologous recombination is complicated by interference generated by the random integration mediated by their integrase enzymes. However, novel approaches are being developed that involve tethering the integrase enzyme to choose a specific target sequence. The HIV integrase has been fused to the sequence-specific DNA binding domain of the phage lambdha repressor. The resulting integrase conferred a high degree of preferential integration to lambdha repressor DNA binding sites (33). Fusion of sequence-specific DNA binding domains to the integrase enzymes of retroviruses may offer a powerful approach to positional targeting of retroviral vectors and the concommittant elimination of disadvantages such as insertional mutagenesis. An additional innovation in retroviral vector development has been the development of MoMuLV vectors that excise the viral component of the vector after the delivery of the exogenous DNA to the chromosome. Excision is mediated by the phage PI site-specific excisionase enzyme, Cre, which acts at specific (LoxP) sites accommodated on the vector. These novel developments may create new generation vectors with enhanced safety and efficiency of gene delivery.

4. CLINICAL APPLICATIONS OF GENE THERAPY: AN OVERVIEW.

The majority of the worlds gene therapy clinical trials (72%) have been initiated in the United States. Europe accounts for 20% of trials, while the remaining 8% are spread world-wide between Eastern and Australia. Although gene therapy started with the treatment of genetic disorders, these indications represent only a minority of gene therapy targets (14%). The majority of trials (62.5%) target cancer therapy. Other significant targets include infectious diseases (8%) and gene marking protocols (11%) (34). Approximately 80% of trials use viral vectors to mediate gene transfer. While the majority of trials are in phase I some have progressed to phase II (34).

4.1 Gene therapy and genetic disorders.

The low number of trials involving genetic disorders in part stems from the rarity of many of the disorders and the associated low number of patients available for clinical trials. In excess of 70% of clinical trials for genetic disorders have been for the treatment of cystic fibrosis. The gene for this monogenetic autosomal recessive disorder was cloned in 1989. The gene was identified as the Cystic Fibrosis Transmembrane Regulator (*cftr*) and the CFTR protein functions as a multi-ion, chloride channel in nasal and lung epithelial cells. It facilitates cAMP-mediated chloride ion transport across the epithelial cells. The absence of the epithelial chloride ion channels leads to chronic lung disease in CF sufferers. CF is the most common lethal inherited disorder affecting individuals of Northern European descent, with an incidence of approximately 1 in 2500 live births, and this frequency has contributed to the interest in developing gene therapeutic approaches.

Gene therapy for CF aims to transfer the *cftr* gene to nasal and lung epithelial cells. The clinical studies that have been initiated have used three classes of gene transfer vectors. These include adenoviruses, adeno-associated viruses, and cationic lipids (see next chapter). While most studies have used nasal epitheluim as a target, lung tissue has also received some attention (35). The majority of published trials have involved the use of adenovirus-mediated gene delivery (35) and some trials have reported inflammatory responses as a result of adenovirus administration. Adenovirus vectors remain extrachromosomal giving a limited expression period that necessitates vector re-administration. Repeat administration of *cftr*-containing vector has been found to be less effective, most likely as a result of immune responses directed against the vector. Preliminary results from a phase 1 clinical trial using an adeno-associated virus have also indicated functional correction of chloride ion transport deficiency (35). Adeno-associated vectors can become integrated and, long-term persistence of the vector has been detected in the absence of toxicity (AAV vectors lack all viral coding regions; Figure 43). While functional correction of the CF genetic defect has been demonstrated in principle, many issues remain to be addressed. These include improved transduction efficiencies to yield greater numbers of transduced cells (given the large cell mass in lung tissue), and improvements in vector design to eliminate cytotoxicity and yield long-term expression of the therapeutic gene.

The liver has been an important target for gene therapy protocols involving genetic diseases. Liver cells have been targeted for the correction of hypercholesteremia, phenylketonuria, haemophilia, anaemia and a number of enzymatic defects (36). Transduced hepatocytes are capable of secreting the therapeutic gene product directly into the blood. Many protocols have involved the use of adenoviral vectors as adenoviruses localise in the liver following intravenous injection (36). Transduction of hepatocytes occurs at high efficiencies (over 90%) at relatively safe doses (37). Complete phenotypic correction of phenylketonuria has been achieved in mice using adenovirus vectors thus, underlining the potential of this approach. However, as with CF gene therapy, the most significant limiting factors have been the short-term expression of therapeutic genes due to immunogenic elimination of transduced hepatocytes and the build-up of immunity to vector re-administration.

Gene therapy has also been applied to correct genetic mechanisms associated with non-mendelian disorders. Neurodegenerative diseases represent an important class of such target disorders for gene therapy. These include Alzheimers, Parkinsons and Huntingtons disease. The targets for their therapy includes the underlying causative genes, neurotrophic factors that affect neuronal function and survival, and secondary metobolic and neurotransmitter functions (38). Vectors derived from the Herpes Simplex viruses have mainly been used in protocols for these diseases because of their natural tropism for post-mitotic neuronal cells. However, retroviral vectors have been used to deliver cytotoxic agents to tumors of the CNS because their inability to transduce non-mitotic cells gives a mechanism for targeting of replicating tumor cells (38, 39).

4.2 Gene Therapy and Cancer

The majority of gene therapy protocols approved for clinical trials involve cancer therapies, despite the fact that many cancers arise from the alteration of a number of genetic loci by multi-step processes (40). A wide variety of approaches have proven to affect tumour cell growth both *in-vitro* and *in-vivo*. The most widely applied approaches have involved tumour-suppressor gene therapy, immunomodulatory gene therapy and suicide gene therapy. Additional strategies have involved the protection of bone marrow during chemotherapy by their transduction with drug resistant markers and combination therapies aimed at eliciting cumulative therapeutic effects (40,41).

The loss of tumour supressor genes such as p53 and overexpression of oncogenes such as K-*RAS* have been implicated in a number of malignant cancers. Wild-type p53 protein regulates the expression of genes that contribute to uncontrolled proliferation of cells and its inactivity contributes to malignancy in a number of cancers. Transduction of cancer cells with p53 coding sequences has the potential to restore p53 expression and inhibit malignant cell growth. Viral vectors have been extensively used to deliver tumour suppressor genes to cancer cells. It has been noted from a large number of trials that gene transfer to cancer cells with viral vectors is more efficient than to other tissue types, with vectors spreading through cancer cell matrices (40, 42). Vector transduced cells have also been found to mediate bystander killing of non-transduced cells and in several models have been found to induce apoptosis of p53 cells (41-43). Adenovirus vectors encoding the p53 gene have been shown to inhibit human head, neck and lung cancers, human colon cancers in mouse models, and liver cancers (40). A phase 1 clinical trial involving the injection of lung tissue with a retrovirus carrying p53 was reported to suppress tumour growth in 6 of 9 patients in the absence of any associated toxicity (40). A number of studies have combined viral-mediated p53 gene therapy with chemotherapy and have reported superior tumour control than with the use of either strategy independently (44, 45).

Immunomodulatory cancer gene therapy involves the production of autologous cellular vaccines by *ex-vivo* transduction of immunostimulatory molecules and their subsequent implantation to induce a systemic immune response aimed at tumour cell destruction, and vaccination against tumour recurrence (39). Immunological destruction of tumour cells *in vivo* has been reported after vaccination with cells that produce immune activating cytokines. Several phase I clinical trials involving autologous retrovirally modified tumour cells transduced with IL-2, TNFC and GM-CSF into melanoma, colorectal, renal cell carcinoma, neuroblastoma and breast cancer cells are underway (39). Another approach involves rendering tumour cells more immunogenic through the transduction of co-stimulatory molecules such as B7 which enhances T cell activation. While transfection with B7 inhibits tumour growth, covaccination with the cytokine IL-2 has been found to significantly enhance antitumour activity (46, 47). In attempts to circumvent the high cost and times required for *ex vivo* production of vaccines a number of *in vivo* strategies have also been designed. Direct injection of adenovirus vectors encoding IL-2 and IL-12 into established murine breast cancer tumours (48) and metastic colon cancers (49) caused significant suppression of tumour growth.

Suicide gene therapy involves the transduction of a gene whose product converts a non-toxic pro-drug into a toxic substance. The pro-drug is administered systemically and activated in transduced cells. The most widely used suicide gene therapy involves the use of the Herpes Simplex virus thymidine kinase (HSV-*tk*) gene which converts gancyclovir into cytotoxic gancyclovir triphosphates that function as nucleoside analogues inhibiting DNA synthesis. The first such protocols approved involved retroviral vector-mediated transduction of HSV-*tk* to brain tumours [40] and the strategy is now being applied to a range of cancer with significant tumor suppression being noted (41). This is, in part, the result of a significant bystander killing effect of non-transduced cells due to diffusion of the toxins. Adenoviruses have been used to transduce liver tumour cells with the *Escherischia coli* cytosine deaminase (*CD*) gene which converts 5-fluorcytosine to the chemotherapeutic agent 5-fluorouracil. The *CD* gene has also been used for gastrointestinal tumours and diffusion of toxic 5-fluorouracil also mediates a significant bystander killing effect (40).

The use of one or more gene therapy strategies to augment traditional cancer therapies is also under investigation. The transduction of tumours with p53, HSV-*tk* and *CD* genes have been reported to sensitise cancer cells to radation, suggesting that combination therapy may provide cumulative effects and aid in the control of advanced tumours in particular (50, 41). Additional protocols have been developed to enhance chemotherapeutic approaches to cancer gene therapy through the protection of hematopoietic stem cells from the toxic effect of chemotherapy. The multiple drug resistant gene (MDR1) pumps chemotherapeutic drugs such as doxorubicin, taxol and actinomyin D from within the cell, and transduction to bone marrow stem cells may allow higher doses of chemotherapy with reduced toxicity (40, 39). Protocols for the treatment of breast and ovarian cancer have been approved for patients receiving taxol (40).

While a wide range of cancer gene therapy strategies and protocols have been reported, and have been proven to cause tumour regression both *in-vitro* and *in-vivo*, complete cancer remissions have not been achieved. One limitation is the restriction of therapies to regional tumours, as viral vectors for systemic administration have not yet been developed. Despite significant bystander killing effects of a number of protocols, improved transduction efficiencies will also be necessary to mediate complete remissions. This can, in part, be achieved through improvements and innovations in vector design such as the limited replication-competent vectors described. The limitations of single gene therapy strategies have also led investigators to combine therapies to produce synergistic effects. These approaches combine cytokine genes with suicide genes, in various combinations, and with radiotherapeutic and chemotherapeutic treatments and have enhanced the rate of tumour regression (36). One recent report of very significant tumour regression, involved a combination of four therapies (51). It involved standard radiotherapy, suicide gene therapy mediated by an adenovirus construct containing both the *CD* and HSV-*tk* genes, and adenovirus-mediated cytopathic effects. The authors demonstrated 2-3-fold synergistic effects (51).

4.3 Gene therapy for infectious disease: The Aids Model

The majority of gene therapy clinical trials for infectious diseases have been for HIV infections. Most protocols have involved intracellular immunisation strategies which involve *ex-vivo* transduction of antiviral genes to mature CD4 positive lymphocytes or haematopoietic CD34-positive stem cells using retroviral vectors. Their reinfusion allows repopulation of HIV infected individuals with HIV resistant cells (52). Other approaches involve immunotherapy protocols aimed at enhancing the host immune response by *in-vivo* transfer of vectors expressing HIV genes, or their *ex-vivo* transfer to cytotoxic T lymphocytes (CTL's) prior to their reinfusion (52).

Targets for anti-HIV gene therapy have included the regulatory protein Rev and Tat, and the structural proteins Gag, Env, and Vpx. The Rev and Tat proteins are essential viral proteins required for regulation of viral gene expression. Mutagenesis of *tat* and *rev* sequences has yielded proteins with *trans*-dominant negative phenotypes that inhibit, in *trans*, the activity of the wild-type proteins. The expression of *trans* dominant negative proteins renders CD4 positive T lymphocytes resistant to HIV infection (52, 53). A phase I clinical trial with a Rev *trans*- dominant inhibitor has yielded encouraging results with reinfused *ex-vivo* modified autologous T-lymphocytes having a selective advantage over unmodified cells *in-vivo* (52, 54). In preclinical studies, transdominant inhibitors have also been used to interfere with viral assembly by inhibiting the Gag protein. While encouraging results have been observed from a number of protocols using *trans*-dominant negative proteins, the anti-viral properties have been of low potency and their intracellular nature has meant activity only in directly transduced CD4+ cells (52, 55). Recently, anti HIV activity has been demonstrated in CD34 positive stem and progenitor cells using a Rev transdominant inhibitor transduced *ex-vivo* at high efficiencies with a MoMuLV retrovirus vector (56). These developments are of significance as in addition to CD4 positive cells, monocytes and macrophages represent a significant target for HIV infection.

Other anti-HIV gene therapy strategies have involved interferons (IFNs), antisense RNAs, ribozymes, and intrabodies. Exogenous addition of IFN to HIV infected cells efficiently inhibits viral replication (52). Gene therapy approaches have involved the introduction of exogenous HIV-inducible IFN genes into target cells. Strong anti-viral properties have been reported for CD4 positive cells transduced with IFNα, β and γ genes using retoviral vectors (52, 57). Viral replication has also been inhibited through the direction of antisense RNA sequences against the *tat, rev gag* and *env* genes, and with ribozymes directed against the HIV *tat, gag* and *pol* genes (52, 58). A recent innovation in intracellular immunisation strategies has been the use of so-called intrabodies, single chain antibodies directed against viral proteins. They cause HIV protein to be sequestered in the endoplasmic reticulum. Tat intrabodies have been found to strongly inhibit replication (59).

Immunotherapy strategies involve immunisation of target cells by *in-vivo* expression of viral proteins stimulating induction of cellular and humoral immune responses. Preclinical data have been encouraging, with one recent study involving intramuscular injection of a retrovirus vector expressing the Env protein in mice and non-human primates (60, 61). HIV-specific immune responses, as a result of activation of cytotoxic T-lymphocytes, were noted. However, the compromised status of the immune system of HIV infected patients may be a limiting factor for immunotherapy approaches.

5. SUMMARY

While the initial promise of gene therapy has not yet been realised in the clinic, the principal of gene therapy for the correction of human disease at the genetic level has been clearly established. In pre-clinical and clinical studies gene therapy has been shown to mediate biological responses relevant to the correction of both inherited and acquired disorders. To realise the full potential for gene therapy a critical area for future development is the design of more efficient, safer vectors tailor-made for particular clinical applications.

Viral vectors have been shown to hold particular promise for gene delivery but require further modifications to enhance transduction efficiencies and to eliminate cytotoxicity and immunogeneticity. Targeting viral vectors to specific tissue types and directing vector delivery to specific intracellular sites is also of particular importance. Innovations such as the design of chimeric viral vectors, plasmoviruses, and the tethering of integrase enzymes with site-specific recombinases may represent prototypical approaches to facilitate such targeting. As the principle rate-limiting step to gene therapy is the safe delivery of therapeutic genes to target tissue, the development of these second generation viral vectors will be of critical importance to realising the clinical potential of gene therapy. As the human genome project will provide in excess of 100,000 human genes, each with a potential therapeutic application, the logic and the potential of gene therapy is inescapable.

BIOGRAPHY

Brendan Murphy lectures in Biochemistry at Limerick Institute of Technology, Moylish Park, Limerick, Republic of Ireland. He obtained his Ph.D. in Molecular Genetics from the University of Limerick in 1997 and has undertaken post-doctoral research developing postgraduate curricula for education and training for the biopharmaceutical sector. His principle research interests are in the area of genetic recombination and virology.

REFERENCES

1. Varmus, H. (1988). Retroviruses. Science, 240, 1427 - 1435.
2. Anderson, W.F. (1992). Human gene therapy. Science, 256, 808 - 813.
3. Eglitis, M.A. and Anderson, W.F. (1988). Retroviral vectors for introduction of genes into mammalian cells. Biotechniques, 6, 608 - 614.
4. Tolstoshew, P. and Anderson, W.F. (1990). Gene expression using retroviral vectors. Current Opinion in Biotechnology, 1, 55 - 61.
5. Robbins, P.D. et al. (1998). Viral vectors for gene therapy. Trends in Biotechnology, 16, 35 - 40.
6. Miller, D. G. et al. (1990). Gene transfer by retrovirus vectors occurrs only in cells that are actively replicating at the time of infection. Molecular and Cellular Biology, 10, 4239 - 4242.
7. Londau, N. et al. (1991). Pseudotyping with human T-cell leukemia virus type 1 broadens the HIV-1 host range. Journal of Virology, 65, 162 - 169.
8. Parolin, C. and Palu, G. (1997). HIV-1 vectors for gene therapy. Minerva Biotechnology, 9, 139 - 147.
9. Yu, H. et al. (1996). Inducible human immunedeficiency virus type 1 packaging cell lines. Journal of Virology, 70, 4530 - 4537.

10. Zufferey, R. et al. (1997). Multiply attenuated lentiviral vector achieves efficient gene delivery in-vivo. Nature Biotechnology, 15, 871 - 875.
11. Akkina, R.K. et al. (1996). High efficiency gene transfer into CD34+ cells with a human immunodeficiency virus type 1-based retroviral vector pseudotyped with vesicular stomatitis virus envelope glycoprotein. Journal of General Virology, 70, 2581 - 2585.
12. Leppard, K.- N. (1997). E4 gene function in adenovirus, adenovirus vector and adeno-associated virus infections. Journal of General Virology, 78, 2131 - 2138.
13. Bout, A (1997). PER. C6: a novel packaging cell line for RCA-free production of EI-deleted recombinant adenoviral vectors. Cancer Gene Therapy, 4, 324.
14. Shabram, P.W. (1997). Analytical anion - exchange HPLC of recombinent type - 5 adenoviral particles. Human Gene Therapy, 8, 453 - 465.
15. Clesham, G.J. (1998). High adenoviral loads stimulate NFKB-dependent gene expression in human vascular smooth muscle cells. Gene therapy, 5, 174 - 180.
16. Snyder, R. et al. (1996). Production of recombinant adeno-associated viral vectors. In: Dracopoli, N., Haines, J., Kref, B., Muir, D., (Eds) Current protocols in Human Genetics, pp 12.1.1 - 12.1.23. John Wiley and Sons Publisher, New York.
17. Halbert, C.L. et al. (1995). Adeno - associated virus vectors transduce primary cells much less efficiently than immortalised cells. Journal of Virology, 69, 1473 - 1478.
18. Salvetti, A. et al. (1998). Factors influencing recombinant Adeno-Associated virus production. Human Gene Therapy, 9, 695 - 706.
19. Dobson, A.T. et al. (1990). A latent non-pathogenic HSV - 1 - derived vector stably expresses β-galactosidase in mouse neurons. Neuron, 5, 353 - 360.
20. Glorioso, J.C. et al. (1995). Development and application of herpes simplex virus vectors for human gene therapy. Annual review of microbiology, 49, 675 - 710.
21. Peplinski, G. R. et al. (1995). Construction and expression in tumuor cells of a recombinant vaccinia virus encoding human interlukin-1 beta. Annual Surgical Oncology, 2, 151-159.
22. Wilkinson, G. W. and Borysiewicz, L. K., (1995). Gene Therapy and Viral Vaccination: The interface. British Medical Bulletin, 51, 205-216.
23. Noguiez-Hellin, P. et al. (1996). Plasmoviruses: Nonviral/viral vectors for gene therapy. Proceedings of the National Academy of Sciences, USA., 93, 4175 - 4180.
24. Han, J.S. et al. (1998). A method of limited replication for the efficient in vivo delivery of adenovirus to cancer cells. Gene Therapy, 9, 1209 - 1216.
25. Liebert, M.A. (1998). Cell-specific targeting with retroviral vectors. Human Gene Therapy, 9, 767 - 770.
26. Schwarzenberger, P. et al. (1996). Targeted gene transfer to human hematopoietic progenitor cell lines through the C-kit receptor. Blood, 87, 472 - 478.
27. Konishi, H. et al. (1998). Targeted strategy for gene delivery to carcinoembryonic antigen-producing cancer cells by retrovirus displaying a single-chain variable fragment antibody. Human Gene Therapy, 9, 235 - 248.
28. Cristiano, R.J. et al. (1993). Hepatic gene therapy: efficient gene delivery and expression in primary hepatocytes utilising a conjugated adenovirus-DNA complex. Proceedings of the National Academy of Sciences, USA., 90, 11548 - 11552.
29. Curiel, D.T. et al. (1992). High-efficiency gene transfer mediated by adenovirus coupled to DNA-polylysine complexes. Human Gene Therapy, 3, 147 - 154.
30. Kosahara, N. et al. (1994). Tissue-specific targeting of rectroviral vectors through ligand-receptor interactions. Science, 266, 1373 - 1376.

31. Walther, W. and Stein, U. (1996). Targeted vectors for gene therapy of cancer and retroviral infections. Molecular Biotechnology, 6, 267 - 286.
32. Yanez, R.J. and Porter, A.C.G. (1998). Therapeutic gene targeting. Gene Therapy, 5, 149 - 159.
33. Bushman, F. (1995). Targeting retroviral integration.
34. www.wiley.co.uk/genetherapy.
35. Alton, E.W.F.W. et al. (1998). Towards gene therapy for cystic fibrosis: a clinical progress report. Gene Therapy, 5, 291 – 292.
36. Paillard, F. (1998). Cancer cells under the fire of combined therapies. Human Gene Therapy, 9, 1259 - 1260.
37. Eisensmith, F.C. and Woo, S.L.C. (1997): Viral Vector-Mediated gene therapy for Hemophilia B Thrombosis and Haemostasis, 78, 24 – 30.
38. Friedman, T. (1994). Gene Therapy for neurological disorders. Trends in Genetics, 10, 210-214.
39. Culver, U.W. and Blaese, R.M. (1994).
40. Roth, J. A., and Christiano, R. J., (1997). Gene Therapy for Cancer: What have we done and where are we going? Journal of the National Cancer Institute, 89, 21-29.
41. Hall, S. J. et al. (1997). Gene therapy 97. The promise and reality of cancer gene therapy. American Journal of Human Genetics, 61, 785 - 789.
42. Cai, D.W. et al. (1993). Stable expression of the wild-type p53 gene in human lung cancer cells after retrovirus-mediated gene transfer. Human Gene Therapy, 4, 617 - 624.
43. Fujiwara, T. et al. (1993). A retroviral wild-type p53 expression vector penetrates human lung cancer spheroids and inhibits growth by inducing apoptosis. Cancer Research, 53, 4129 - 4133.
44. Roth, J. A., (1996). Modification of tumour suppressor gene expression in non-small cell lung cancer (NSCLC) with a retroviral vector expressing wild (normal) p53. Human Gene Therapy, 7, 861-874.
44. Gjerset, R.A. et al. (1995). Use of wild-type p53 to achieve complete treatment sensitisation of tumour cells expressing endogenous mutant p53 molecular carcinogens. 14, 275-285.
45. Nguyen, D. M. et al. (1996). Gene therapy for lung cancer: enhancement of tumour suppersion by a combination of sequential systmeic cisplation and adenovirus-mediated p53 gene transfer. Journal of Thoracic Cardiovascular Surgery, 112, 1372-1377.
46. Salvadori, S. et al. (1995). B7-1 amplifies the response to interlukin- 2 secreting tumor vaccines *in-vivo*, but fails to induce response by naive cells in-vivo . Human Gene Therapy, 6, 1299 - 1306.
47. Gaken, J. A. et al. (1997). Irradiated NC adenocarcinoma cells transduced with both B7.1 and interlukin- 2 induce CD4+ - mediated rejection of established tumours. Human Gene Therapy, 8, 477 - 488.
48. Addison, C.L. et al. (1995). Intramural injection of an adenovirus expressing interlukin -2 induces regression and immunity in a murine breast cancer model. Proceedings of the National Academy of Sciences, USA., 92, 8522 - 8526.
49. Caruso, M. et al. (1998). Adenovirus-mediated interlukin-12 gene therapy for metastatic colon cancer. Proceedings of the National Academy of Sciences, USA., 93, 11302 - 11306.

50. Khil, M. S. et al. (1996). Radiosensitisation by 5 - fluorocytosine of human colorectal carcinoma cells in culture transduced with cytosine deaminase genes. Clinical Cancer Research, 2, 53 - 57.
51. Freytag, S. O. et al. (1998). A novel three-pronged approach to kill cancer cells selectively: concomitant viral, double suicide gene, and radiotherapy. Human Gene Therapy, 9, 1323 - 1333.
52. Sorg, T. and Methali, M. (1997). Gene therapy for AIDS. Transfusion Science, 18, 277 – 289.
53. Vandendriessche, T. et al. (1995). Inhibition of clinical Human Immunedeficiency Virus (HIV) type 1 isolates in primary CD4+ lymphocytes by retroviral vectors expressing anti-HIV genes. Journal of Virology, 69, 4045 - 4052.
54. Woffendin, C. et al. (1996). Expression of a protective gene prolongs survival of T-cells in HIV 1 infected patients. Proceeding of the National Academy of Sciences, U.S.A., 93, 1889 - 1894.
55. Malim, M. H. et al. (1992). Stable expression of *trans*-dominant Rev protein in human T - cells inhibits human immunedeficiencey virus replication. Journal of Experimental Medicine, 176, 1197 - 1201.
56. Davis, B. R. et al. (1998). Targeted transduction of CD34+ cells by transdominant negative rev-expressing retrovirus yields partial anti-HIV protection of progeny macrophages. Human Gene Therapy, 9, 1197 - 1207.
57. Vieillard, V. et al. (1995). Autocrine interferon-β synthesis for gene therapy of HIV infection: increased resistance to HIV 1 in lymphocytes from healthy and HIV-infected individuals. AIDS, 9, 1221 - 1228.
58. Sczakiel, G. et al. (1992). Tat- and Rev-directed antisense RNA expression inhibits and abolishes replication of Human Immune Deficiency Virus type 1: a temporal analysis. Journal of Virology, 66, 5576 – 5581.
59. Mheshilkar, A. M. et al. (1995). Inhibition of HIV-1 TAT-mediated LTR transactivation and HIV-1 infection by anti-TAT single chain intrabodies. EMBO Journal, 14, 1542-1551.
60. Sajjadi, N. et al. (1994). Recombinant retroviral vector delivered intramuscularly localises to the site of injection in mice. Human Gene Therapy, 5, 693 - 699.
61. Irwin, M. J. et al. (1994). Direct injection of a recombinant retroviral vector induces human immunedeficiency virus-specific immune responses in mice and nonhumans primates. Journal of Virology, 68, 5036 - 5044.

Chapter 20

Pharmaceutical gene medicines for non-viral gene therapy

A. Rolland, S. Sullivan, K. Petrak
GENE MEDICINE, INC., 8301 New Trails Drive, The Woodlands, Texas, YSA

Key words: Gene therapy, DNA, liposomes, genetic disease, cancer, clinical trial, plasmid, delivery systems.

Abstract: Gene therapy entails the introduction of a selected gene into a specific somatic cell such that subsequent expression of the gene achieves a therapeutic goal. One of the technical challenges to developing successful gene therapy protocols remains the development of safe and effective gene delivery systems. This chapter focuses upon the many non-viral approaches to achieving target-specific gene delivery. After discussing plasmid-based gene medicines and their manufacture, lipid-, polymer- and polypeptide-based gene delivery systems are presented in detail. A selected review of clinical trials undertaken using non-viral delivery systems is then presented.

1. SOMATIC GENE THERAPY: MYTH OR REALITY?

Imagine... replacing a defective or missing gene in a patient's body to prevent or treat diseases - supplementing the body with new proteins that will induce a therapeutic benefit - orchestrating a cellular and humoral immune response - controlling intracellular events to create a new quality of pharmacological response! All of these scientific dreams from several decades ago could become reality with the advent of somatic gene therapy. Recent advances in genomics, with the discovery of the structure and function of thousands of genes in the human genome, associated with the discovery of novel methods to control the transfer and the expression of these genes in specific biological targets, have radically changed the potential of this 'genetic revolution' and enhanced the promise of somatic gene therapy (1-7).

Since the first clinical trial in gene therapy in 1990 (8) - the *ex vivo* introduction of the adenosine deaminase (ADA) gene with a retroviral vector into the lymphocytes of patients suffering from the inherited defect, severe combined immunodeficiency (SCID) - more than 300 additional gene therapy clinical trials have been initiated worldwide for more than 2,000 patients using one of the following strategies (9-15): i) an *ex vivo* approach whereby cells of the patient are removed from the body, transduced with viral vectors or other non-viral methods described in this chapter, and re-introduced as genetically-modified cells in the patient's body, ii) an *in vivo* direct transduction of the target cells using viral vectors such as replication-defective retroviruses, adenoviruses, adeno-associated viruses (as described in the previous chapter) and iii) a direct *in vivo* administration of non-viral plasmid-based systems. This latter approach is being presented in this chapter that describes the various elements that can be incorporated into a non-viral, or synthetic, gene delivery system as well as the genetic sequences that can be engineered into a plasmid expression system.

Gene therapy using non-viral methods is designed to provide specific cells of a patient with the *genetic software* necessary to produce therapeutic proteins for the prevention, correction or modulation of a disease. Such methods are intended to overcome the limitations associated with the clinical use of protein drugs, including their low bioavailability, poor pharmacokinetics, chemical instability and relative high cost. In addition, gene therapy has the unique ability to effect the intracellular distribution of the expressed protein in defined compartments (e.g., mitochondria or cell membranes) and to direct antigens to a specific pathway (MHC-class I or class II) in order to modulate a preferred immune response. These potentials are unique to a gene therapy approach which can control the location and the function of an administered gene while it may be difficult, if not impossible, to achieve the same effects by the administration of the corresponding proteins.

By realizing the tremendous potential of gene therapy, which can combat a disease at the molecular level, the field has evolved over the recent years from being limited to the replacement of genetically defective genes in inherited disorders (e.g., cystic fibrosis, muscular dystrophy, ADA deficiency, familial hypercholesterolemia, phenylketonuria) to finding applications for acquired diseases. Even for acquired diseases, the thought process has evolved from applying gene therapy to end stage or untreatable diseases by conventional approaches - the best example being cancer - to treating, in a prophylactic or therapeutic mode, other acquired diseases such as inflammation, cardiovascular diseases and infectious diseases. For gene therapy to become a valid therapeutic approach, besides achieving success in clinical trials, products will need to be developed as safe, convenient and cost-effective pharmaceuticals that are administered by conventional routes (15-16).

1.1 Plasmid-based Gene Medicines

Turning genes into medicines for target-specific therapy requires the ability to control both the location and the functioning of an administered gene in the patient's body. Therefore, plasmid-based gene medicines comprise a number of different elements, as described below, to control the various events from their administration site to the nucleus of the target cell. These key events will vary according to the biology and (patho)physiology of the biological target. As a consequence, the premise is that each target associated with a defined route of administration will require specific gene delivery and plasmid-based expression systems. The development of target-specific, non-viral gene therapies requires the combination of a synthetic gene delivery system that can control the biodistribution and access of the expression plasmid to the desired target cell and a gene expression system that regulates the amount, fidelity and duration of expression of the gene product. The spatial and temporal modulation of gene function *in vivo* is thus critical to enable safe and effective gene therapy.

A gene medicine is composed of three major elements: *i)* a gene encoding a therapeutic protein, *ii)* a plasmid-based expression system, and *iii)* a synthetic gene delivery system. These products are intended to have low toxicity due to the use of synthetic components for gene delivery (minimizing for instance the risks of immunogenicity generally associated with viral vectors) and non-integrating plasmids for gene expression. Since no integration of plasmid sequences into host chromosomes has been reported *in vivo* to date, they should neither activate oncogenes nor inactivate tumor suppressor genes. This built-in safety with non-viral systems contrasts with the risks associated with the use of most viral vectors. As episomal systems residing outside the chromosomes, plasmids have defined pharmacokinetics and elimination profiles, leading to a finite duration of gene expression in target tissues (18, 12). Such properties of gene medicines should enable a physician to control gene-dosing regimens according to therapeutic needs.

One of the major challenges in effective synthetic gene delivery resides in the ability to circumvent the numerous barriers that a plasmid will encounter from its administration site to the nucleus of the target cell. The premise that each biological target will require a unique combination of delivery elements to overcome key-limiting steps in the overall gene transfer process *in vivo* still holds true. It is now widely recognized that there will not be a *magic bullet* in non-viral gene therapy. According to the route of administration and the tissue intended to be transfected, synthetic gene delivery systems will have to be adjusted to control the *D*istribution (e.g., dispersion in a solid tissue such as muscle following intramuscular administration; distribution to a specific tissue after intravenous injection), *A*ccess to the target cell (e.g., extravasation through liver sinusoids to access hepatocytes), *R*ecognition (including uptake by the target cell via either passive adsorptive mechanisms or receptor-mediated endocytosis) and *T*rafficking within the cell (e.g., release from the endosomal compartment following uptake, decomplexation from the carrier/plasmid system, and translocation to the nucleus). These steps can be described as the *DART* concept of target-specific gene delivery.

On their quest for the 'holy grail' in gene therapy, scientists will have to decipher the complex pathways and limiting events that prevent efficient *in vivo* gene transfer. Over the last years, the number of mechanistic studies to define the potential rate-limiting steps in plasmid delivery have increased very significantly with the concurrent development of new analytical techniques. They have enabled the design of more adequate synthetic delivery elements to overcome some of these biological barriers. As the field of gene therapy evolves, it is becoming more apparent that the early applications of DNA plasmids in isotonic saline (so-called 'naked' DNA) to *in vivo* gene transfer might turn out to be relatively limited. Although 'naked' DNA was shown several years ago to transfect a variety of cells *in vivo*, the lack of stability of such unprotected plasmids to enzymatic degradation associated with irreproducibility in uptake (by still undefined but inefficient processes) has led to highly variable expression and biological responses in animal models (19-28). The very low bioavailability of 'naked' plasmids in most tissues also requires high doses of plasmids to be administered to generate a pharmacological response.

The field of non-viral gene delivery has therefore evolved into the development of more advanced synthetic delivery systems. Such systems, obtained by the assembly of plasmids with various delivery elements, are designed to affect the steps described above, by for instance, protecting plasmids from premature degradation in biological milieu using effective condensation or modification of plasmid surface properties. Condensing carriers have been designed to interact with plasmids by ionic interactions to compact plasmids into particulates of defined hydrodynamic size and surface properties (7). Additional strategies that include the modulation of the plasmid surface charge and hydrophobicity by interaction with protective, interactive non-condensing systems (e.g., PINC^{TM} polymers) have shown advantages over the use of 'naked' DNA for direct administration to solid tissues (30-33). A 'naked' plasmid is indeed a negatively charged, relatively hydrophilic, large colloid. Such physicochemical characteristics prevent plasmids from i) effectively dispersing through extracellular matrices (e.g., muscle, solid tumors), ii) crossing biological barriers such as most endothelia as well as the blood-brain barrier, and iii) being taken up efficiently by cells because of the charge-charge repulsion at the surface of the highly negatively charged plasma membranes (16, 7). It is therefore essential to modulate the stability (enzymatic, physical, chemical), as well as the colloidal and surface properties of plasmids to enable their effective and reproducible delivery to the desired target cells. As described below, recent studies have focused on designing novel delivery elements that will address the protection, condensation, uptake, targeting, endosomal release, intracellular decomplexation and nuclear localization of plasmid expression systems. The complexity of such novel systems also requires analytical methods that permit the quality control of the formulated plasmids. The preparation of well-characterized systems is a critical step in the development of gene medicines as pharmaceutical products. A number of these different gene delivery systems and their potential applications to human gene therapy will be presented below.

Once the plasmid has accessed the nucleus of the target cells, efficient transcription and translation processes need to occur to generate sufficient gene product to provide a pharmacological response. Plasmids can be designed to contain specific genetic elements that control the levels of therapeutic protein being produced, the fidelity or accuracy or gene expression, as well as the duration and timing of expression. Theoretically, single or multiple genes can be inserted in a plasmid without size limitations. The accuracy of protein production is important not only from a therapeutic perspective but also for regulatory reasons. Only the intended protein should be expressed with no other products that may result from alterations in transcription, RNA processing or translation. Duration of expression *in vivo* is highly variable depending on the cells expressing the transgene and the genetic elements incorporated in the expression plasmid. As plasmids do not integrate into the host genome, their nuclear half-life will control the duration of gene expression. In non-dividing or slowly dividing cells, such as skeletal muscle, the residence of intact plasmids in myocytes can extend for several weeks to months. In rapidly proliferating cells, such as cancer cells, the residence time does not exceed a few days without manipulating genetic elements that may, for instance, anchor the plasmid to the nuclear matrix or provoke plasmid episomal replication.

The enhancer/promoter region of an expression plasmid will determine the levels of expression. Most of the gene expression systems designed for high levels of expression contain the intact human cytomegalovirus (CMV) immediate early enhancer/promoter sequence (34-37). However, down-regulation of the CMV promoter over time has been reported in tissues. The hypermethylation of the CMV promoter, as observed when incorporated into retroviral vectors (38-39), has not been observed for episomal plasmids *in vivo* (40). Nevertheless, the CMV promoter silencing could be linked to its sensitivity to reduced levels of the transcription factor NF-κB (41). The activity of the CMV promoter has also been shown to be attenuated by various cytokines including interferons (α and γ), and tumor necrosis factor (TNF-α) (42-45). In order to prolong expression *in vivo* and ensure specificity of expression in desired tissues, tissue-specific enhancer/promoters have been incorporated in expression plasmids (42). The chicken skeletal α-actin enhancer/promoter has been shown to provide high levels of expression (equivalent to the ones achieved with a CMV-driven construct) for several weeks in non-avian striated muscles (46). With such tissue-specific plasmids, negligible or no expression has been reported in non-striated muscle tissues such as lung and liver (47). Other tissue-specific expression systems (e.g., liver-, lung-, tumor-specific) have been tested with variable success, including some which have displayed specific and persistent expression *in vivo* as compared to viral promoters (48-51, 37, 52).

Additional genetic sequences in the expression plasmids can be added to influence the stability of the messenger RNA (mRNA) and the efficiency of translation. The 5' untranslated region (5' UTR) is known to effect translation and it is located between the cap site and the initiation codon. The 5' UTR should ideally be relatively short, devoid of strong secondary structure and upstream initiation codons, and should have an initiation codon AUG within an optimal local context (53-54). The 5' UTR can also influence RNA stability, RNA processing and transcription. In order to maximize gene expression by ensuring effective and accurate RNA splicing, one or more introns can be included in the expression plasmids at specific locations. The possibility of inefficient and/or inaccurate splicing can be minimized by using synthetic introns that have idealized splice junction and branch point sequences that match the consensus sequence. Another important sequence within a gene expression system is the 3' untranslated region (3' UTR), a sequence in the mRNA that extends from the stop codon to the poly(A) addition site. The 3' UTR can influence mRNA stability, translation and intracellular localization (55). The skeletal muscle α-actin 3' UTR has been shown to stabilize mRNA in muscle tissues thus leading to higher levels of expression as compared to other 3' UTR (46, 56, 47). This 3' UTR appears to induce a different intracellular compartmentalization of the produced proteins, preventing the effective trafficking of the proteins to the secretory pathway and favoring their perinuclear localization.

One of the attractive features of plasmid expression systems is the possibility to express multiple genes from a single construct. These multivalent systems may find applications in the expression of heterodimeric proteins, such as antibody fragments, or in the *in vivo* production of multiple antigens to generate a potent immune response for genetic vaccination (57-60). In cancer immunotherapy, the co-expression of co-stimulatory molecules with a variety of cytokines may also lead to potent therapeutic effects.

Plasmids reside within the transfected cells as episomes and therefore will degrade, as pre-drugs, over time. The pharmacokinetic profile of plasmids mainly depends on the type of transfected cell and the sequences included in the gene expression system. Plasmids injected in skeletal muscle will reside in differentiated myocytes for several weeks and lead to persistent expression (61, 56). In other tissues (e.g., lung, tumor, liver), the duration of expression with similar expression systems (e.g., CMV-driven constructs) will be transient with low levels of expressed transgene after only a few days. These differences may be attributed to the intrinsic stability of plasmids in different cells as well as inactivation of the exogenous promoters. Strategies to improve the persistence of expression *in vivo* include the incorporation of replication elements in the expression system. The most widely investigated are derived from viral sources, such as SV40, bovine papilloma virus, Epstein-Barr virus and human papovavirus (62-64). These viral origin sequences require the co-expression of specific viral proteins (e.g., SV40 T antigen, bovine papilloma virus E1 and E2 proteins) to be functional. Some of these viral proteins potentially introduce safety concerns due to their immunogenic, toxic or oncogenic properties. Sequences associated with origins of replication and the proteins that interact with these sites may also increase nuclear retention, possibly by plasmid attachment to the nuclear matrix (65).

Some proteins are naturally produced by the body according to circadian rhythms. Therefore, it may be beneficial for certain gene therapy approaches to enable the production of therapeutic proteins in a pulsatile manner. In addition, providing the physician with the ability to control expression of therapeutic genes according to the needs of the patient, by turning genes on or off in response to the administration of low molecular weight drug molecules, brings gene medicines to a higher level of safety. Several systems are currently under development to regulate gene expression with low molecular weight drugs (66-69). The most advanced ones include the tetracycline-, the rapacycin- and the antiprogestin-regulated systems (GeneSwitch^{TM}) (70-71). Each system is based on highly active chimeric protein(s) that can bind, in a drug-dependent manner, to a specific DNA binding site built in the promoter region of the transgene. Upon administration of an antiprogestin molecule acting as an agonist at low doses, the drug would, for instance, bind to an expressed mutated human progesterone receptor (GeneSwitch^{TM} protein) that has lost the ability to bind to progesterone and other endogenous steroids. Upon drug-GeneSwitch protein interaction, the mutant receptor becomes activated and forms homodimers that bind GAL4 DNA binding sites that have been engineered in the expression plasmid upstream of a minimal promoter for the transgene. The binding of the activated GeneSwitch protein to the plasmid triggers the expression of the therapeutic gene. The expression of the

GeneSwitch has been shown to be even controllable by a tissue-specific promoter to enable tissue-specific, drug-controlled transgene expression (72).

1.2 Quality of DNA and Manufacture Scale

Development of manufacturing processes for pharmaceutical grade DNA plasmid must meet several criteria. The first is the manufacturing process itself, which must adhere to the Food and Drug Administration (FDA) guidelines for Good Manufacturing Procedures (GMPs). Secondly, the process must be cost-effective. Finally, the procedure should yield a final product that meets a series of specifications relating to purity, sterility and absence of pyrogens. Purity is a performance criterion in that contaminants, such as bacterial RNA and chromosomal DNA, proteins and lipopolysaccharides may interfere with plasmid formulation or in the case of "naked" DNA may interfere with plasmid bioavailability. An additional specification is the plasmid conformation. Bacterial plasmids can exist in several forms: linear, open circle, closed circle and supercoiled. The performances of several gene delivery systems have been shown to be sensitive to plasmid conformation. Hence, depending upon the performance criteria, the DNA conformation may also be a release specification. The manufacturing process for bacterial DNA plasmid generally consists of the following steps:

1. Fermentation of bacteria transduced with expression plasmid
2. Bacteria harvest
3. Bacteria lysis and processing
4. Purification by chromatographic techniques
5. Concentration by ultrafiltration
6. Sterile filtration

The recombinant protein field has yielded technology for propagating large batches of bacteria and separating it from growth media. High throughput procedures have been developed for bacterial lysis and precipitation. Several chromatography procedures, such as anion exchange (73), gel filtration (74) or affinity chromatography (75) have been developed to purify the DNA. The present cost for DNA manufacture on a gram scale approaches 10 to 20 dollars/mg. As the scale of the manufacturing process is increased, this cost should decrease. Process development is focused on increasing the scale of the lysis step, improving processing efficiency after lysis and reducing or completely eliminating the chromatography steps. With regard to plasmid conformation, 4 companies claim to have developed manufacturing processes that yield predominantly supercoiled plasmid. At least 4 companies also have had their manufacturing process approved by the FDA for pharmaceutical manufacture. In summary, DNA plasmid can be manufactured under FDA approved GMP guidelines. It can be manufactured on a pharmaceutical scale with a high yield of product, a high level of purity and an acceptable level of potency.

The next phase in product development is to increase transfection efficiency, thus minimizing the therapeutic dose. Gene delivery technology has been and is being developed to achieve this goal. The present gene therapy product candidates in the clinics arose from the initial technology development. As these clinical trials progress, deficiencies in the existing technology are identified. Basic and preclinical research is focused on developing technology to eliminate these limitations. The resultant technology should increase transgene expression levels, increase the duration of the expression and minimize transfection of non-target organs. The following section will focus on three areas of gene delivery technology: lipid-, polymer-, and polypeptide-based systems.

1.2.1 Lipid-Mediated Gene Transfer

In the initial stages of lipid-based gene therapy, the focus was on DNA uptake. However, results have shown that uptake may not be the only rate limiting step and steps post cell entry may also be rate limiting. These steps are release from endocytic vacuoles, dissociation of the DNA from the delivery carrier in the cytoplasm or nucleus (dependent upon delivery system) and nuclear uptake. Increased cytoplasmic access of plasmid can be approached in two ways: plasma membrane fusion and endocytic vacuole membrane fusion. The first scenario was developed by Dr. Papahadjopoulos and colleagues by packaging plasmid into liposomes with a composition that fused with cell membranes upon addition of extracellular calcium (76) which triggered the fusion event. The other method was to package the plasmid into liposomes that were composed of lipid capable of fusing with the endocytic vacuole membrane following endocytosis. The bilayers of these liposomes were stable at pH 7 but would become fusogenic at pH 5, the pH of endosomes and lysosomes. Antibody targeting of these liposomes to cancer cells grown i.p. was demonstrated and an inducible promoter was able to specifically induce transgene expression (77). An earlier study showed that, using standard liposome formulation and incorporating a galactose glycolipid into the lipid bilayer, decreased Kupffer cell uptake in the liver and increased hepatocyte uptake could be achieved (78). A plasmid encoding the preproinsulin gene was packaged into these liposomes. Increase in insulin levels was observed along with a decrease in blood glucose. There were several problems with the standard liposome approach, such as poor DNA trapping efficiency, inefficient release of plasmid, and serum stability. In lieu of these problems, isolated reports using conventional liposomes have yielded in vivo gene transfer to hepatocytes (79). In this case, the transgene was α1-antitrypsin as opposed to a marker gene such as luciferase, chloramphenicol acetyl transferase or beta-galactosidase. Unfortunately, no animal was used to determine if the levels of α1-antitrypsin were significant to produce a biological effect. Conventional liposomes have advantages in that the same technology used for small molecular weight compound delivery can be taken advantage of for gene delivery. However, no technology has been reported to date to overcome the main obstacles of low trapping efficiency.

An alternative approach that does overcome trapping efficiency and transfection efficiency are lipids or amphipathic detergents with a cationic head group to bind DNA. Some examples of the cationic lipids and helper lipids are shown in Figure 44.

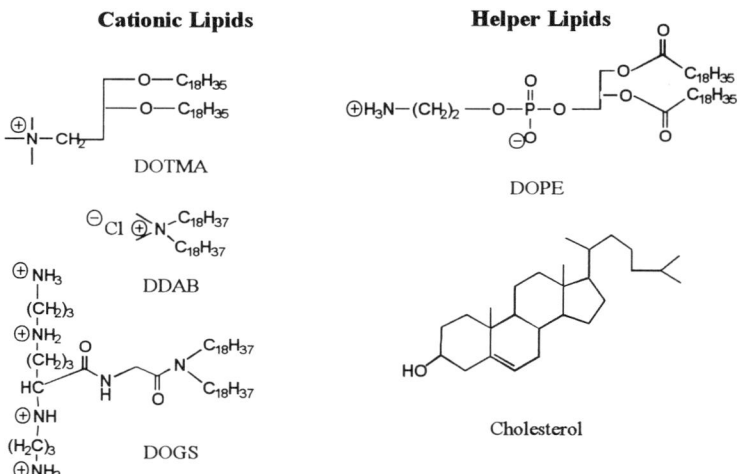

Figure 44. Examples of Cationic Lipids and Helper Lipids

Dr. Felgner developed the first cationic lipid for gene transfer, DOTMA, an ether linked diacyl glycerol with a choline head group. This lipid in combination with dioleoylphosphatidylethanolamine (DOPE) formed liposomes that bound DNA and the resultant complex transfected cells (80). Dr. Behr developed another lipid, termed DOGS, composed of a dialkylamine linked to spermine. This lipid by itself formed complexes with plasmids and yielded cell transfection (81). This new class of transfection reagents sparked the development of non-viral gene delivery systems for gene therapy.

Parameters that were optimized for maximal *in vitro* transfection efficiency were the charge ratio of cationic lipid and DNA, the requirement of a helper lipid (DOPE), the ratio between helper lipid and cationic lipid and the amount of DNA. Several general conclusion were generated from this research. These were:

1. Different cell types responded better to one lipid than another
2. Different optimal charge ratios were effective for different cell types
3. Optimization of charge ratio was a bell shaped curve for all cell types. Increase in ratio yielded increased transfection until a peak was observed. The decrease in expression as the ratio increased was due to cytotoxicity
4. Most transfection complexes were inactivated by serum. Hence, most transfection protocols required an incubation period of transfection complexes with cells in media containing no serum
5. The transfection activity was transient in that there was a finite window between addition of lipid to DNA and addition to cells, i.e., the transfection activity was transient

These last two points were major obstacles in progressing from *in vitro* to *in vivo* experiments. Short-term solutions consisted of mixing liposomes and plasmid just prior to administration. Using administration routes where serum components were not encountered circumvented serum instability. Hence, intrapulmonary, intramuscular and intratumoral administration were the initial focus for *in vivo* gene transfer (Table 41).

This led to a combination of these administration routes along with the first generation of cationic lipids to proceed to the initial clinical trials for treatment of cystic fibrosis and the development of genetic vaccines for cancer and infectious diseases. Expansion of the number of disease applications is realized in the ability to administer transfection complexes intravenously yielding gene transfer to specific organs and more importantly, specific cell types within that organ that are responsible for the disease state.

Table 41. Cationic Lipid-based Gene Delivery Systems

Cationic Lipid	Administration Route	Transgene	Species	Results	Reference
DOTMA	Intravenous and intratracheal	CAT	Mouse	Transfection of lungs by both administration routes	Brigham (82)
DOSPA	Intravenous	Alkaline phosphatase	Mouse	Transfection of lung, heart, muscle, spleen and liver	Hofland (83)
DOGS	Intravenous	Luciferase	Mouse	Long term expression in multiple organs	Thierry (63)
Guanidinium cholesterol	Intrapulmonary	Lac-Z[a]	Mouse	Transfection of lung epithelium	Oudrhiri (84)
Spermine cholesterol	Intrapulmonary	Lac-Z	Mouse	Transfection of lung epithelium	Eastman (85)
Liposomes with HVJ[b]	Intrapulmonary	Luciferase	Mouse	Transfection of multiple areas of lung epithelium	Yonemitsu (86)
DOTAP	Intrapulmonary	Lac-Z	Mouse	Transfection of lung epithelium	McLachlan (87)

[a] Lac-Z is β-Galactosidase
[b] HVJ is Hemagglutinin Virus of Japan

The first significant report of *in vivo* lipid-mediated gene transfer was by Brigham (82), who compared intravenous to intratracheal administration of cationic lipid/DNA plasmid complexes for *in vivo* gene transfer. The results showed that both administration routes yielded transgene expression in lungs. Peak expression was observed 3 days after administration and expression persisted for 7 days. A more controversial study by Zhu (88), using the same formulation showed transgene expression in the lung, spleen, heart, liver, kidney and lymph node in a dose-dependent fashion. Many laboratories tried to repeat these results but were unsuccessful. However, within the past two years there have been a number of reports using different cationic lipids and formulating them with DNA in various manners yielding similar, or better transfection results compared to this initial report (83, 89-93). In general, the findings are as follows:

1. The lung was the major transfected organ
2. The major transfected cell type was endothelium and transfection was restricted to microvasculature
3. Use of large liposomes yielded higher levels of expression than smaller liposomes in lungs
4. Expression was transient, decaying at least an order of magnitude 5 to 7 days after administration
5. Exceeding a ratio of lipid to DNA or a DNA dose or a combination of the two can result in side effects and animal death

There have been recent developments to reduce non-specific transfection and focus on targeting expression. Polyethylene glycol has been shown to prevent opsonization of liposomes for delivery of small molecular weight chemotherapeutics. This technology has been adapted for lipid-based gene delivery. The research team at Inex in Vancouver has shown that upon incorporation of PEG-lipid into lipid-based gene delivery systems, the circulation half-life is increased from minutes to hours. There have been no published reports showing transgene expression. This is not surprising due to PEG's ability to not only reduce particle opsonization but also inhibit non-specific liposome uptake. In a step toward *in vivo* targeted gene delivery, Dr. Papahadjopoulos, along with Dr. Hong, developed a five component system composed of spermidine condensed DNA, DDAB, Cholesterol, PEG-phosphatidylethanolamine and a PEG-PE with a single chain antibody to HER2. *In vitro* results showed that incorporation of PEG into the formulation completely inhibited transgene expression. However, attachment of a single chain antibody against HER2 receptors not only restored transgene activity but increased the expression level by 70-fold. These results identify technology that can be used to yield targeted *in vivo* gene transfer. This should greatly open up the field for novel therapeutic applications.

There has been more focus on the assembly of the transfection complexes rather than identifying new transfection reagents. For example, the initial protocols yielded complexes with transient activity. These were even employed for the early clinical trials where the lipid and DNA were mixed at the bedside just prior to use. The transient nature of the transfection activity was primarily due to aggregation where transfection complex size grew as a function of time such that the aggregates were too large to be functional. One hypothesis is that not all the cationic lipid-binding sites on the DNA were saturated making sites available for binding of cationic lipid associated with another DNA plasmid thus resulting in aggregation. Simply increasing the cationic lipid such that all the sites would be saturated was not sufficient. Formulation processes have been developed that allow the cationic lipid to bind and condense the DNA in the absence of liposome structure. The liposomes composed of cationic lipid/helper lipid are dissolved in detergent and added to plasmid. Once the lipid has bound to the plasmid, the detergent is removed by extensive dialysis. A detergent with a high critical micelle concentration, such as octylglucoside, is used to facilitate complete removal. The resultant transfection complexes are stable for several months when stored as a suspension at 4°C. Modifications to this procedure was required to increase the DNA concentration. In one case, the plasmid and detergent solubilized lipid are added in 3M NaCl. The detergent is removed by extensive dialysis against the same high salt concentration. The complex is then dialyzed against saline to yield an isotonic solution. The complexes can be formed at 1 mg of DNA/ml and upon i.v. administration yield transfection of lung, heart, spleen, muscle and liver. The primary transfected cells in each organ are in the endothelium. In another case, a polyethlyeneglycol-lipid derivative is solubilized in the detergent along with cationic lipid and helper lipid. The detergent is then removed by dialysis yielding the transfection complex. Inclusion of the PEG allows DNA concentrations of 0.5 mg/ml to be obtained.

Another method was to take a polyamine cationic lipid, DOGS, and dissolve it in organic solvent. The organic solvent was removed by evaporation yielding a film. The film was hydrated with a DNA suspension resulting in spontaneous formation of transfection complexes. Intravenous administration yielded transfection of the lung, liver, spleen and heart. Using a human papovavirus derived episomal system increased the duration of expression. Plasmids replicated extrachromosomally in lung 2 weeks post injection. This protocol not only generated active *in vivo* transfection complexes but stabilized the plasmid through episomal replication resulting in an increase in duration of expression.

1.2.2 Polypeptide-based Gene Delivery

The earliest use of protein-based transfection agents was in polio viral RNA research where methylated serum albumin and polyornithine were used to introduce viral RNA into HeLa cells yielding mature virions (94). This tool for introducing of foreign gene into cells has developed into a delivery technology for gene therapy. Table 42 shows examples of different polypeptide systems for gene delivery.

Table 42. Poly-L-Lysine-based Gene Delivery Systems

Ligand	Protein	Route	Target	Gene	Reference
AsOR[a]	Poly-L-lysine	Intrahepatic portal vein	Hepatocytes	CAT[b]	Wu (95)
Anti-secretory component Fab	Poly-L-lysine	Intravenous	Airway epithelium	Luciferase and Lac-Z	Ferkol (96)

[a]AsOR is asialoorosomucoid
[b]CAT is bacterial chloramphenicol acetyl transferase

Dr. Wu (95) first pioneered the concept of cationic polyaminoacid transfection agents for gene therapy. Poly-L-lysine was conjugated to asialoorosomucoid protein, a desialated glycoprotein with terminal galactose residues. Plasmids were complexed to the conjugate by electrostatic interactions and transfected into hepatocytes expressing the asialoglycoprotein receptor (97-98). This system was shown to be functional upon intravenous administration, yielding transfection of hepatocytes (99). The amount of administered DNA was large and transgene expression was transient with loss of activity within 1 week of administration. The other aspect was that transfection had to be completed by first pass due to complement activation by the poly-L-lysine. The positive transfection results showed the utility of poly-L-lysine as a standard transfection agent yielding further development with other targeting ligands. These targeting ligands included carbohydrates (100), vitamins (101), proteins (96, 102) and peptides (103). All displayed receptor-mediated targeting to cells expressing the appropriate receptor. Transfection could be competed with an excess of targeting ligand. However, transgene expression was low. Coformulation of poly-L-lysine/ligand conjugate with other agents have been shown to increase transgene expression and still maintain target specificity *in vitro*. For example, a combination of cationic liposomes with antibody conjugated poly-L-lysine yielded a synergistic increase in transgene expression in mouse endothelial cells (104). A combination of poly-L-lysine derivatized with transferrin, complexed with plasmid and packaged in negatively charged liposomes yielded receptor-mediated delivery to myogenic cells (105). The poly-L-lysine served to bind and condense the DNA along with associating a targeting ligand to the complex. The lipid, either cationic or anionic, served to facilitate interactions of the complex with membranes, i.e., fusion, resulting in release of the plasmid into the cytoplasm. A protein-based approach designed to improve poly-l-lysine transfection combined a multifunctional fusion protein with poly-L-lysine condensed plasmid. The fusion protein was a composite of TGF-α, the natural ligand for EGF receptor; the translocation domain of

Pseudomonas exotoxin A, and the DNA binding domain of yeast GAL4 transcription factor. This fusion protein was shown to increase transfection of EGF receptor expressing cells 150-fold compared to poly-L-lysine alone. Transfection of cells not expressing EGF receptor showed no increase in transfection with the fusion protein. These results are promising, but the utility of the technology needs to be demonstrated *in vivo*. Poly-L-lysine itself has been shown to activate complement. This would need to be overcome for i.v. administration but may not be as much of a concern for local administration. The other aspect is the immunogenicity. One purpose of non-viral gene therapy is to yield products that can potentially be administered chronically. An immunogenic complex could be neutralized by antibodies. The degree of immunogenicity will also be dependent on the administration route (i.m. and i.v. being less desirable).

1.2.3 Polymer-based Gene Delivery

Polymer-based gene delivery systems have only recently been developed for gene therapy applications. Two polymer systems have been shown to yield *in vivo* gene transfer. Table 43 summarizes these studies.

Table 43. Polymer-based Gene Delivery Systems

Polymer	Administration Route	Transgene	Results	Reference
PEI[a]	Intraneural to cerebral cortex, hippocampus, and hypothalamus	Luciferase and bcl-2	Luciferase positive neurons and glial cells in cortex, hippocampus and hypothalamus	Abdallah (106)
PEI	Intraarterially into renal artery	Luciferase	Luciferase positive proximal kidney cells	Boletta (107)
PVP[b]	Intramuscular	IGF-I	IGF-I detection in muscle 28 days post injection	Alila (56)
PVP	Intramuscular	Human Growth Hormone, Human Factor IX	hGH detected in muscle 28 days post injection	Anwer (47)

[a] PEI is polyethylenimine
[b] PVP is polyvinylpyrrolidone

PEI is a cationic polymer and PVP is a neutral polymer. PEI, first reported by Dr. Behr (108), was selected based on earlier research using low molecular weight polyamines that bind to DNA, i.e., spermine and spermidine. Formulation of this polymer with DNA yields transfection of many cell types with transgene levels higher than that obtained by commercially available lipids (108-109). The PEI used in the initial studies had a high polydispersity and the degree of branching was not controlled. A polymer with a more narrow polydispersity and controlled branching has been shown to improve on the originally reported levels of transfection (110). An explanation for the high level of gene expression is due to extensive endosome swelling and rupture as a result of the high buffering capacity of the polymer thus providing an escape mechanism for the polycation/DNA particles. This polymer can be derivatized with targeting ligands-yielding receptor mediated transfection (111). The polymer/plasmid complex can be further modified with polyethylene glycol, as described previously for the cationic lipids, to prevent opsonization and also have a terminal ligand, such as transferrin, to mediate cell uptake (112). The use of this polymer for gene transfer is fairly recent and has not been studied as extensively as the lipid systems. The levels of transgene expression can be comparable to, or even exceed in some cases, the expression from lipid-based systems. However, judging from the chemical structure and composition, a method by which this polymer could be metabolized *in vivo* is not obvious. For this to become a pharmaceutically applicable gene delivery system, the mechanism for how this polymer is voided from the body needs to be better understood.

The second polymer (IGF-I), PVP, is a protective, interactive, non-condensing system (PINC™). Intramuscular administration of the PVP/DNA plasmid encoding for human growth hormone (hGH) (47) or insulin-like growth factor-I (56) yielded long term expression of transgene, 28 days post injection, that was biologically active. The PVP gene delivery system has also been applied successfully to cancer immunotherapy. DNA plasmids encoding different cytokines was formulated with PVP and administered intratumorally. Tumors were completely irradicated and a second tumor challenge yielded no tumors, indicative of immunization of the mice against the tumor cells. Formulation of plasmid with PVP represents a significant improvement over the administration of plasmid alone. Intramuscular administration of DNA plasmids has been shown to yield transfection of muscle cells. Coformulation of DNA plasmids with PVP increases the transfection efficiency (extent and levels of expression) and maintains a simple two component system, thus improving performance and facilitating manufacture.

Other polymer systems are in development that improve on the described systems. One class of polymer are the Starburst™ dendrimers. These are highly structured polymers composed of primary and secondary amines. Plasmids bind to the surface of the polymer and the resulting complex interacts with cells to yield transfection (113). Modification of these polyamidoamine dendrimers to yield fractured dendrimers improved the transfection activity (114). A proposed reason for improved transfection activity is due to increased flexibility of the polymer. Another polymer system combines drug delivery technology with gene delivery technology in which a copolymer of poly (D,L-lactic acid) and poly-L-lysine grafted polysaccharide is being developed for i.v. administration (115). The poly-L-lysine increases the hydrophilicity of the PLGA polymer thus facilitating DNA loading and the oligosaccharide serves as a targeting ligand to bind cell surface lectins for increased uptake. Transfection activity remains to be demonstrated for this system.

Polymer-based gene delivery is the newest entry to the field. It offers the attractive aspect of a chemically defined vehicle gene transfer agent. The polymer can be chemically modified to yield desired physical properties, such as alterations in DNA binding affinity, ligand targeting and controlled release. Movement of the polymer-based systems away from tissue culture into animal models for performance evaluation should more clearly defined development paths for future research.

2. CLINICAL PERSPECTIVE OF GENE THERAPY

The logical application of the fundamental biomedical science to meet the needs of medicine opens up numerous opportunities to gene therapy. So far only relatively few have been tested in a clinical setting.

As of June 1995, a total of 597 subjects had undergone gene transfer experiments involving more than a dozen diseases. By May 1998, over 2,557 patients have been enrolled in gene therapy clinical trials.[1] The same source gives the total number of clinical trials in different phases as 329. Some 242 of these human gene transfer clinical evaluation are taking place in the USA[2]. Most of the US studies (146) are directed towards gene therapy of cancer, 23 focus on the treatment of AIDS and 16 on cystic fibrosis. Some 18% of the studies use non-viral means of transfection. Since 1990, the majority of requests for trial approval have been in the area of oncology. The data adapted from the above NIH reference are summarized in Table 44.

---

[1] www.wiley.co.uk/genetherapy
[2] http://www.nih.gov/od/orda/protocol.htm

Pharmaceutical gene medicines for non-viral gene therapy

Table 44. Gene Therapy Clinical Trials

Category	Disease/Disorder	Total # of Protocols	Non-Viral Protocols
Monogenic Diseases	α-1- Antitrypsin Deficiency	1	1
	Chronic Granulomatous Disease	3	-
	Cystic Fibrosis	16	4
	Familial Hypercholesterolemia	1	-
	Fanconi Anemia	1	-
	Gaucher Disease	3	-
	Hunter Syndrome	1	-
	Ornithine Transcarbamylase deficiency	1	-
	Purine Nucleoside Phosphorylase Deficiency	1	-
	SCID-ADA	1	-
	X-linked SCID	1	-
	Leukocyte Adherence Deficiency	1	-
	Canavan Disease	2	2
Cancer	All therapeutic approaches	146	27
Infectious Diseases	Human Immunodeficiency Virus	23	-
Other Diseases/Disorders	Peripheral Artery Disease	1	1
	Rheumatoid Arthritis	2	1
	Arterial Restenosis	1	1
	Cubital Tunnel Syndrome	1	1
	Coronary Artery Disease	3	1
	Total	210	39

Clinical investigations that use non-viral means of transfection, (namely "naked" plasmid or cationic lipid formulations) are tabulated below (Table 45).

Most of the human gene transfer protocols have so far reached only the early stages of clinical investigations (Phases I and II) (with one exception being the Phase III trial by Novartis, entered August 1996, using retroviral transfection with HSV-tk gene to treat glioblastoma patients). These have usually been small-scale clinical experiments intended to test the feasibility and safety of administering particular gene medicines and to evaluate the effects of expressing specific gene products. Most of the ongoing trials have not to date reported any clinical results. The ability of gene medicines to produce the desired therapeutic effect has been indicated in some studies (116-117) but not yet fully demonstrated in large clinical studies for any gene therapy protocol. Similarly, because clinical experience is still very limited, it is not possible to predict or to exclude short-term and long-term adverse effects of gene transfer therapy. Evaluation of the results of gene therapy protocols has been much hindered by the low frequency of gene delivery to target cells, and often by the lack of meaningful biochemical and clinical endpoints.

Table 45. Summary of main gene therapy clinical investigations underway which utilize non-viral approaches to gene delivery

Company	Product Indication	Gene	Clinical Phase (date of entry)	Lipid
GENEMEDICINE, INC./ Boehringer Mannheim	Head & neck cancer	IL-2	I (2/97)	DOTMA
Vical	Head & neck cancer	Allovectin-7 (HLA-B7)	II (9/97)	DMRIE
Genzyme	Cystic fibrosis	CFTR	I (4/95)	#67
Vical	Melanoma	Allovectin-7	II (9/95)	DMRIE
Megabios	Cystic fibrosis	CFTR	I/II	EDMPC
Targeted Genetics	Cancer	TgDCC-E1A	I	DC-chol
Boehringer Mannheim	Cystic fibrosis	CFTR	I	DOTAP
Vanderbilt University/ GENEMEDICINE, INC.	AAT deficiency	AAT	I	DOTMA
University of Kuopio/ GENEMEDICINE, INC.	Coronary disease	VEGF	II	DOTMA
University of Kuopio/ GENEMEDICINE, INC.	Vascular disease	VEGF	II	DOTMA
Vical	Metastatic renal cell carcinoma	IL-2	II	-
Vical/Pasteur Merieux Connaught	Malaria vaccine	PfCSP	I	-

2.1   Gene therapy and genetic disease

Perhaps the most logical and obvious application of gene therapy is to treat monogenic diseases in which an absence or malfunction of a single gene is directly responsible for the development of disease condition. Cystic fibrosis is a life-shortening autosomal recessive condition caused by mutation of the cystic fibrosis transmembrane conductance regulator (CFTR) gene. Although the disorder affects all secretory epithelia to varying degrees, the major cause of morbidity and mortality is inflammatory lung disease. Mutations within the CFTR gene result in defective AMP-mediated chloride ion transport. Also, absorption of sodium ion is increased in the airways. The overall outcome is the dehydration of airway secretions and impairment of mucociliary clearance, creating a fertile ground for opportunistic infections. This results in inflammation and lung damage with progressive loss of lung function.

The progress in this field has been extraordinary. The discovery of the CFTR gene was announced in September, 1989 (118-120), and less than 4 years later, the first human clinical trials (using viral vectors) were initiated (121). A number of reported studies have shown that liposome-mediated CFTR cDNA transfer can correct the cAMP-mediated Cl⁻ transport defect in CF transgenic mice (122-124).

Porteous *et al*. (125), as a prelude to CF clinical trials, provided evidence for safety and efficacy of DOTAP cationic liposome-mediated CFTR gene transfer to the nasal epithelium of patients with cystic fibrosis. A single dose of 400 µg pCMV-CFTR:2.4 mg DOTAP was administered in a randomized, double-blinded fashion to the nasal epithelium of eight CF patients, with a further eight receiving buffer only. Patients were monitored for signs and symptoms for 2 weeks before treatment and 4 weeks after treatment. Inflammatory cells were quantified in a nasal biopsy taken 3 days after treatment. There was no evidence for excess nasal inflammation, circulating inflammatory markers or other adverse events ascribable to active treatment. Transgene DNA was detected in seven of the eight treated patients up to 28 days after treatment and plasmid derived CFTR mRNA in two of the seven patients at +3 and +7 days. Transepithelial ion transport was assayed before and after treatment by nasal potential difference during drug perfusion and by SPQ fluorescence halide ion conductance. Partial, sustained correction of CFTR-related functional changes toward normal values were detected in two treated patients. The level of gene transfer and functional correction were comparable to those reported previously using adenoviral vectors or another DNA-liposome complex.

Caplen *et al*. (126) performed a double-blind placebo-controlled study on liposome-mediated CFTR cDNA gene transfer to the nasal epithelium in 15 delta 508 homozygous CF subjects (nine CFTR cDNA, six placebo). CFTR mRNA was detected in the nasal biopsies from five of eight treated patients. Sodium transport-related measurements were reported to be significantly reduced by about 20% towards the values found in normal subjects. As pointed out by the authors, it is important to note that the observed changes fall within the coefficient of variation of these measurements. Similar studies using pCMV-CFTR/DOTAP system are under way [G McLachlan *et al*., as above].

The ideal gene medicine for CF should be completely safe, highly efficient for entering specifically airway epithelial cells and expressing cystic fibrosis transmembrane conductance regulator (CFTR). It should be capable of transfecting nondividing cells of the lung airway surface, and should be of low immunogenicity. It has been estimated (127) that to be successful, the expression system will have to transfer CFTR cDNA to some 5-10 % of the airway epithelial cells and give rise to persistent expression of the gene. The current technology cannot meet such requirements. While the clinical studies on CF gene therapy have not demonstrated efficacy they at least refined the understanding of the barriers preventing effective CFTR transfer (128). Gene therapy for cystic fibrosis is not expected to be available in the clinic for a number of years yet. Based on the published and presented clinical results, however, it is reasonable to conclude that the principle has been proven – transfer and expression of CFTR cDNA *in vivo* in the human respiratory tract (at least as represented by the nasal epithelium) is feasible.

A similar clinical study, this time using AAT, was reported by Brigham *et al.* (117). Using DOTMA:DOPE/pCMV-AAT complex administered to the nose, it was shown that increased levels of human AAT transgene could be detected in the treated nasal epithelium for some 5-7 days after administration. It was also reported that the treatment with the AAT gene medicine reduced inflammation in the nasal epithelium (as judged by measuring the levels of IL-8 in the nasal washings), as compared to saline-treated control, and to an extent far superior to the administered purified AAT protein.

In the context of CFTR and AAT measurements of physiological effects it might be very significant to note further that plasmids in themselves, and especially when in a complex with cationic lipids, elicit responses other than just those attributable to the expression of the encoded protein. Freimark *et al.* (129) reported that administration of plasmid/lipid complexes to the lung airways leads to the induction of the Th1-associated cytokines IFN-γ and IL-12, and of TNF-α. The presence of cytokines may well interfere with the determination of parameters selected as biochemical and biophysical endpoints.

2.2 Gene therapy and cancer

Cancer gene therapy development represents the bulk of the current clinical studies. The complexity of cancer may appear to make this disease a bad choice for gene therapy. However, perhaps driven by the wide prevalence of cancer and the lack of viable drug-based therapy alternatives, several companies have made cancer the focus of their product leads. While the molecular biology of cancer is still far from being completely understood, the mutation of DNA by chemicals, radiation and viruses is a factor contributing to a loss of body's control over cell growth. The complexity of the disease is reflected in the existence of numerous strategies that are being employed to control such aberrant cell growth through the application of gene therapy. Those most widely used in the existing clinical trials are based on the action of:

- Tumor suppressor genes that make proteins capable of stopping cell growth. For example, p53 protein can detect defects in cellular DNA, and can either stop the process of cell division or it can cause the cell to undergo apoptosis (cell death).

- Suicide gene therapies. A gene such as thymidine kinase encodes an enzyme which converts a relatively innocuous substance gancyclovir to a cytotoxic metabolite. In this way not only the transfected cells but to some extent also the surrounding cancer cells can be killed.

- Immunotherapy based on expressing a gene to produce a protein capable of activating the immune system. Common approaches utilize expression of cytokines (e.g., IL-2, alpha interferon) or HLA-B7 (Allovectin-7).

- Anti-angiogenesis based on blocking the growth of blood vessels in the tumor area and thus "starving" the tumor (for example through the action of angiostatin and endostatin). This could then lead to a decrease in the rate of tumor growth, tumor size reduction or even tumor irradication.

Although several studies in cancer gene therapy have now been completed, clinical data has not yet been fully reported. A prerequisite to developing effective gene medicines is the demonstration of their safety. Nabel *et al.* (130) introduced the gene encoding a foreign major histocompatibility complex protein, HLA-B7, into HLA-B7-negative patients with advanced melanoma. Melanoma nodules were injected with a plasmid/cationic lipid (DMRIE:DOPE) complex. The transferred gene was expressed and was localized to the site of injection, and no apparent toxicity or anti-DNA antibodies were associated with this treatment. Regression of the treated and distal lesions was observed in one patient. This early work demonstrated that non-viral gene medicines provide an important and potentially safer alternative to viral approaches. Plasmids containing appropriate regulatory sequences can be made relatively easily and utilized to express a variety of different gene products. Both plasmid and liposomes remain stable for months upon storage.

A subsequent study (131) showed that T cells migrated into treated lesions of 6 out of 7 patients. Two of the two patients analyzed showed an enhanced tumor-infiltrating lymphocyte reactivity. Local inhibition of tumor growth was detected after gene transfer in two patients, one of whom showed partial remission. This patient was subsequently treated with tumor-infiltrating lymphocytes derived from gene-modified tumor; this resulted in a complete regression of residual disease. Stopeck *et al.* (132) using the same gene transfer procedure (employing Allovectin-7 cationic lipid system, Vical Inc.) treated 17 HLA-B7-negative, metastatic melanoma patients. In all, twelve patients received a single intra-lesional injection of the following dose of plasmid: 10 µg (4 patients), 50 µg (5 patients) or 250 µg (3 patients). Five patients received two or three injections of 10 µg of plasmid to a single tumor site at 2-week intervals. Tumor biopsies were obtained before therapy, and 2 and 4 weeks after gene injection, and analyzed for transgene expression by PCR, RT-PCR, flow cytometry, and immunohistochemistry. Ninety three percent of all biopsy samples obtained after gene therapy contained HLA-B7 plasmid DNA, mRNA, or the protein. In seven patients the injected nodule decreased in size by at least 25% as measured by radiological and physical examination. One patient with a single site of disease showed complete remission. The results so far on the safety profile and biological activity of this therapy warrants further studies to define its antitumor efficacy. Expanded Phase II and Phase III trials are planned. Phase II trial will be open to patients with metastatic, refractory, Stage III or IV disease that has not spread to other organs. Up to 70 advanced melanoma patients will be enrolled. The objective of the trial is a partial or complete response in at least 15% of the evaluable patients, persisting with a median duration of at least four months. In Phase III, the efficacy of the Allovectin-7 gene therapy when combined with standard chemotherapy in patients with unresectable, metastatic melanoma not previously treated with chemotherapy will be determined. According to the statement from Vical, Inc. who sponsor these trials, other tumor types under investigation include lymphoma, colorectal and breast cancer.

Rubin *et al.* (133) employed the same cationic lipid-based system (DMRIE:DOPE) containing the combination of the HLA-B7 gene and β-microglobulin (Allovectin-7), for the immunotherapy of hepatic metastases of colorectal carcinoma. Two administration schedules were used. Patients on the first schedule received an injection on day 1 and the injected lesion was biopsied to determine transfection every 2 weeks for 8 weeks. Doses were escalated from 10 µg to 50 µg to 250 µg with three patients treated at each level. The second schedule included multiple injections of 10 µg. Three patients received injections on days 1 and 15, and three patients on days 1, 15 and 29. The HLA-B7 protein was detected in five of eight patients by immunohistochemistry and in seven of 14 patients by fluorescence-activated cell sorting analysis. It was concluded that liposomal gene transfer by direct injection was feasible and non-toxic.

Safety and toxicity of catheter gene delivery to the pulmonary vasculature was examined by Nabel *et al.* (134) in a patient with metastatic melanoma. The patient received two treatments of HLA-B7 plasmid complexed to cationic liposomes into a right posterior basal pulmonary artery associated with a mass lesion. The treatments were well tolerated.

Murray *et al.* (135) reported the results of a Phase I trial of intra-tumoral liposomal E1A gene therapy in patients with recurrent and refractory breast cancer and head and neck cancer. Previous laboratory and animal studies showed that E1A, a tumor inhibitor gene, can inhibit tumor growth and suppress metastases, induce apoptosis and reverse the overexpression of oncogene HER-2/neu. In patients with cancer, overexpression of HER-2/neu is correlated with increased metastases and resistance to chemotherapeutic agents and hence poor prognosis. In the first study, three of three patients with breast cancer and ovarian cancer, treated with E1A, showed a downregulation of HER-2/neu. In a subsequent escalation Phase I study, nine patients with recurrent and unresectable cancer, and nine patients with head and neck cancer were treated. In 16 patients that could be evaluated for response, nine had stable disease, five had progressive disease and two had minor responses in treated tumors despite tumor progression at other untreated sites. In one of six patients who had repeated biopsies of treated tumor, no pathologic evidence of tumor was found. In four of seven patients evaluated to date, evidence of down regulation of HER-2/neu was reported. There are plans to initiate Phase II studies for E1A in head and neck cancer in the second half of 1998.

Marchand et al. (136) reported tumor regression responses in melanoma patients treated with a peptide encoded by gene MAGE-3. The MAGE genes are expressed in a significant proportion of various tumors but no expression has been observed in normal tissues (except testis). MAGE-3 gene codes for an antigenic nanopeptide which is recognized by autologous cytolytic T cells on major histocompatibility complex molecule HLA-A1, and has been found to be expressed in some two thirds of a large number of melanoma samples. Metastatic melanomas showed a higher proportion (76%) of positive samples as compared to primary melanomas (36%). Twelve tumor-bearing melanoma patients were included in the study. Six were withdrawn after 1 or 2 injections (of the scheduled 3 s.c. immunization injections of the synthetic MAGE-3.A1 peptide at monthly intervals) because of rapid progression of the disease necessitating other forms of treatment. Out of the remaining six patients that received all three immunizations, three showed tumor regression responses.

2.3 An additional application

A promising emerging application for gene therapy appears to be the generation *in vivo* of endothelial cell mitogens to promote angiogenesis in patients with limb ischaemia. Pre-clinical findings suggest that intra-arterial expression of vascular endothelial growth factor (VEGF) transgene can improve blood supply to the ischaemic limb (137-138). Clinical application was reported by Isner et al. (139) who administered 2 mg of plasmid encoding for human VEGF (phVEGF165) applied to the hydrogel polymer coating of an angioplasty balloon. By inflating the balloon, plasmid was transferred to the distal popliteal artery. Based on the treatment of a single patient the authors report that digital subtraction angiography 4 weeks after gene therapy showed an increase in collateral vessels at the knee, mid-tibial and ankle levels. This effect persisted for at least 12 weeks. Intra-arterial Doppler-flow measurements showed that the resting and maximum blood flows increased by 82% and 72%, respectively. A subsequent expanded clinical investigation (116) of this so called "therapeutic angiogenesis" aimed to show both the safety and feasibility of intramuscular gene transfer employing "naked" plasmid encoding VEGF165, and the potential therapeutic benefits in patients with critical limb ischemia. Gene transfer was performed in 10 limbs of 9 patients with non-healing ischemic ulcers or rest pain, or both, due to peripheral arterial disease. A total dose of 4 mg of plasmid encoding human VEGF165 was injected directly into the muscles of ischemic limb. Gene expression was evidenced by a transient increase in serum levels of VEGF. The formation of new collateral blood vessels was documented by contrast angiography in 7 limbs. Magnetic resonance angiography showed a qualitative evidence of improved distal flow in 8 limbs. Ischemic ulcers healed or markedly improved in 4 of 7 limbs, resulting in successful limb salvage in 3 patients recommended for below-knee amputation.

3. CONCLUSION

Gene therapy is still clearly in need of proving its worth in the clinic. Critical questions need to be answered through clinical testing: which diseases can be treated by gene therapy, which methods of administration are most effective and safe, which routes and schedules of administration are effective, etc. The frequently-voiced perceived "failure" of gene therapy clinical investigations has been linked to early unreasonable expectations, overselling of the results of pre-clinical and clinical studies, the lack of fundamental knowledge about the diseases relevant to the development and applications of gene therapy, the lack of sufficiently advanced technology, and to the lack of rigor in defining biochemical and disease endpoints (141). It has been argued (142) "that much of the most informative data in the near future will actually come from the clinical trials in which *in vivo* reactions to vaccine, delivery vectors and treatments can be assessed". This is seen as the only way in which it can be determined how gene medicines, which have been developed based on results obtained in imperfect model systems in animals, behave in human patients. Rigorously generated pre-clinical data should, however, continue to be the basis of well-designed clinical studies.

BIOGRAPHY

Alain Rolland (Fax: 9281) 364-0858; Tel: (281) 364-1150; e-mail: rollaa@GeneMedicine.com.) is Vice President, research, Sean Sullivan and Karel Petrak are directors and program heads of GENE MEDICINE INC.

GENE MEDICINE is a leader in the development of non-viral gene therapy products designed for the treatment or prevention of serious diseases. Gene medicines deliver the 'genetic software' to targeted cells in the body to produce therapeutic proteins or desired immune responses. The Company's core technology includes lipid-, polymer-, and peptide-based gene delivery systems, each able to be applied to specific clinical targets, and gene expression systems to regulate the production of multiple genes. GENE MEDICINE technology is potentially applicable to the treatment of a wide variety of diseases and disorders because of its capability to deliver therapeutic genes to several tissues and cell types and to control the expression of desired proteins. The Company's initial focus is on the development of gene medicines for treating certain cancers, neuromuscular disorders, cardiovascular diseases, and pulmonary diseases, as well as the development of genetic vaccines for treatment or prevention of infectious diseases. Gene medicines deliver instructions under controlled conditions to specific cells in the body to produce therapeutic results, effectively working with the body's own systems to fight and defend against disease.

REFERENCES

1. Anderson, W.F. (1992). Human Gene Therapy. Science, 256, 808-813.
2. Ledley, F.D. (1993a). Are contempory methods for somatic gene therapy suitable for clinical applications? Clinical Investigative Medicine, 16, 78-88.
3. Ledley, F.D. (1993b). Hepatic gene therapy: Present and future. Hepatology, 18, 263-273.
4. Ledley, F.D. (1994a). Development in somatic gene therapy. Exp. Op. Invest. Drugs, 3, 913-921.
5. Ledley, F.D. (1994b). Non-viral gene therapies. Current Opinion in Biotechnology, 5, 626-636.
6. Anderson, W.F. (1995). Gene Therapy. Scientific American, 124-128.
7. Rolland, A.P. (1998). From genes to gene medicines: Recent advances in nonviral gene delivery. Critical Reviews in Therapeutic Drug Carrier Systems, 15(2), 143-198.
8. Blaese, R.M. *et al.* (1995). T lymphocyte-directed gene therapy for ADA-SCID: Initial trial results after 4 years, Science, 270, 475.
9. Miller, A.D. (1992). Human gene therapy comes of age. Nature, 357, 455-460.
10. Kay, M.A. *et al.* (1993). *In vivo* gene therapy of hemophilia B: Sustained partial correction in factor IX-deficient dogs. Science, 262, 117-119.

11. Culver, K.W. and Blaese, R.M. (1994). Gene therapy for adenosine deaminase deficiency and malignant solid tumors. In: Wolff, J.A., Ed., Birkhauser, Boston, Gene Therapeutics: Methods and Applications of Direct Gene Transfer, 263-280.
12. Ledley, F.D. (1995) Nonviral gene therapy: The promise of genes as pharmaceutical products. Human Gene Therapy, 6, 1129-1144.
13. Knowles, M.R. et al. (1995). A controlled study of adenoviral-vector-mediated gene transfer in the nasal epithelium of patients with cystic fibrosis. New England Journal of Medicine, 333, 823-831.
14. Mendell, J.R. et al. (1995). Myoblast transfer in the treatment of Duchenne's muscular dystrophy. New England Journal of Medicine 333, 832-838.
15. Rolland, A. and Tomlinson, E. (1996). Controllable gene therapy using non-viral systems, in gene therapy and artificial self-assembling systems for gene transfer, In: Felgner, P., Heller, M., Lehn, P., Behr, J-P., and Szoka, F.C., Jr. (eds.). ACS Books, Washington.
16. Tomlinson, E. and Rolland, A., (1996). Controllable gene therapy: Pharmaceutics of non-viral gene delivery systems. Journal of Controlled Release, 39, 357-372.
17. Coleman, M.E. et al. (1994). Regulatory elements of the chick a-skeletal actin gene direct high level and tissue specific Development. Journal of Cell Biochemistry W25.
18. Ledley, F.D. and Ledley, T.S. (1998). Pharmacokinetic considerations in somatic gene therapy. Advanced Drug Delivery Reviews, 30, 133-150.
19. Wolff, J.A. et al. (1990). Direct gene transfer into mouse muscle in vivo. Science, 247, 1465-1468.
20. Wolff, J.A. et al. (1992a). Expression of naked plasmids by cultured myotubes and entry of plasmids into T tubules and caveolae of mammalian skeletal muscle. Journal of Cell Science, 103, 1249-1269.
21. Wolff, J.A. et al. (1992b). Long-term persistence of plasmid DNA and foreign gene expression in mouse muscle. Human Molecular Genetics, 1, 363-369.
22. Lin, H. et al. (1990). Expression of recombinant genes in myocardium in vivo after direct injection of DNA, Circulation, 82, 2217-2221.
23. Hickman, M.A. et al. (1994). Gene expression following direct injection of DNA into liver. Human Gene Therapy, 5, 1477-1483.
24. Kawabata, K. et al. (1995). The fate of plasmid DNA after intravenous injection in mice: Involvement of scavenger receptors in its hepatic uptake. Pharmaceutical Research 12, 825-30.
25. Riessen, R. et al. (1993). Arterial gene transfer using pure DNA applied directly to a hydrogel-coated angioplasty balloon. Human Gene Therapy, 4, 749-758.
26. Ulmer, J.B. et al. (1993). Heterologous protection against influenza by injection of DNA encoding a viral protein. Science, 259, 1745-1749.
27. Sikes, M. et al. (1994). In vivo gene transfer into rabbit thyroid follicular cells by direct DNA injection, Human Gene Therapy, 5, 837-844.
28. Meyer, K.B. et al. (1995). Intertracheal gene delivery to the mouse airway: Characterization of plasmid DNA expression and pharmacokinetics. Gene Therapy, 2, 450.
29. Rolland, A.P. (1996). Controllable gene therapy: Recent advances in non-viral gene delivery. In: Targeting of Drugs 5: Strategies for Oligonucleotide and Gene Delivery in Therapy. Gregoriadis and McCormack (Ed.), Plenum Press, New York, 79-95.
30. Mumper, R.J. et al. (1995 b). Interactive polymeric gene delivery systems for enhanced muscle expression. Pharmaceutical Research, 12, 80.

31. Mumper, R.J. et al. (1996). Polyvinyl derivatives as novel interactive polymers for controlled gene delivery to muscle. Pharmaceutical Research, 13, 701-709.
32. Mumper, R.J. et al. (1998). Protective interactive noncondensing (PINC) polymers for enhanced plasmid distribution and expression in rat skeletal muscle. Journal of Controlled Release, 52, 191-203.
33. Mumper, R.J. and Rolland, A.P. (1998). Plasmid delivery to muscle: Recent advances in polymer delivery systems. Advanced Drug Delivery Reviews, 30, 151-172.
34. Tsan, M-F. et al. (1995). Lung-specific direct *in vivo* gene transfer with recombinant plasmid DNA. American Journal of Physiology, 268 (Lung Cell. Mol. Physiol. 12), L1052-L1056.
35. Hartikka, J. et al. (1996). An improved plasmid DNA expression vector for direct injection into skeletal muscle. VR1012 constuction. Human Gene Therapy, 7, 1205-1217.
36. Tanner, F.C. et al. (1997). Transfection of human endothelial cells. Cardiovascular Research, 35, 522-528.
37. Yew, N.S. et al. (1997). Optimization of plasmid vectors for high-level expression in lung epithelial cells. Human Gene Therapy, 8, 575-84.
38. Challita, P.M. and Kohn, D.B. (1994). Lack of expression from a retroviral vector after transduction of murine hematopoietic stem cells is associated with methylation *in vivo*. Proceedings of the National Academy of Science 91, 2567-2571.
39. Rettinger, S.D. et al. (1994). Liver-directed gene therapy: Quantitative evaluation of promoter elements by using *in vivo* retroviral transduction. Proceedings of the National Academy of Science, 91, 1460-1464.
40. Loser, P. et al. (1998). Reactivation of the previously silenced cytomegalovirus major immediate-early promoter in the mouse liver: Involvement of NFkappaB. Journal of Virology, 72, 180-190.
41. May, M.J. and Ghosh, S. (1997). Rel/NF-kappa B and I kappa B proteins: An overview. Seminars in Cancer Biology, 8, 63-73.
42. Harms, J.S. and Splitter, G.A. (1995). Interferon-(inhibits transgene expression driven by SV40 or CMV promoters but augments expression driven by the mammalian MHC I promoter. Human Gene Therapy, 6, 1291-1297.
43. Gribaudo, G. et al. (1995). Interferon-α inhibits the murine cytomegalovirus immediate-early gene expression by down-regulating NF-kB activity. Virology, 211, 251-260.
44. Qin, L. et al. (1997). Promoter attentuation in gene therapy: Interferon-γ and tumor necrosis factor-α inhibit transgene expression. Human Gene Therapy, 8, 2019-2029.
45. Freimark, B.D. et al. (1998). Cationic lipids enhance cytokine and cell influx levels in the lung following administration of plasmid. Cationic lipid complexes, 160, 4580-4586.
46. Coleman, M.E. et al. (1995). Myogenic vector expression of insulin-like growth factor-I stimulates muscle cell differentiation and myofiber hypertrophy in transgenic mice. Journal of Biological Chemistry, 270, 12109-12116.
47. Anwer, K. et al. (1998). Systemic effect of human growth hormone after intramuscular injection of a single dose of a muscle-specific gene medicine. Humane Gene Therapy, 9, 659-670.
48. Manthorpe, M. et al. (1993). Gene therapy by intramuscular injection of plasmid DNA; studies on firefly luciferase gene expression in mice. Human Gene Therapy, 4, 419-431.

49. Molkentin, J.D. and Olson, E.N. (1996). Combinatorial control of muscle development by basic helix-loop-helix and MADS-box transcription factors. Proceedings of the National Academy of Sciences, 93, 9366-9373.
50. Wu, G.Y. et al. (1991). Receptor-mediated gene delivery in vivo. Partial correction of genetic analbuminemia in Nagase rats. Journal of Biological Chemistry, 266, 14338-14342.
51. Ferkol, T. et al. (1993). Regulation of the phosphoenolpyruvate carboxykinase/human factor IX gene introduced into the livers of adult rats by receptor-mediated gene transfer. FASEB Journal, 7, 1081-1091.
52. Walther, W. and Stein, U. (1996). Cell type specific and inducible promoters for vectors in gene therapy as an approach for cell targeting. Journal of Molecular Medicine, 74, 379-92.
53. Kozak, M. (1997). Recognition of AUG and alternative initiator codons is augmented by G in position +4 but is not generally affected by the nucleotides in positions +5 and +6. EMBO Journal, 16, 2482-2492.
54. Jansen, M. et al. (1995). Translational control of gene expression. Pediatric. Research, 37, 681-686.
55. Wickens, M. et al. (1997). Life and death in the cytoplasm: Messages from the 3' end. Current Opinion in Genetic Development, 7, 220-232.
56. Alila, H.A. et al. (1997). Expression of a biologically active human insulin-like growth factor-I following intramuscular injection of a formulated plasmid in rats. Human Gene Therapy, 8, 1785-1795.
57. Donnelly, J.J. et al. (1993). The signal for translational readthrough of a UGA codon in Sindbis virus RNA involves a single cytidine residue immediately downstream of the termination codon. Journal of Virology, 67, 5062-5067.
58. Conry, R.M. et al. (1996). Selected strategies to augment polynucleotide immunization. Gene Therapy, 3, 67-74.
59. Lew, D. et al. (1995). Cancer gene therapy using plasmid DNA: Pharmacokinetic study of DNA following injection in mice. Human Gene Therapy, 6, 553-564.
60. Rubin, J. et a. (1997) Phase I study of immunotherapy of hepatic metastases of colorectal carcinoma by direct gene transfer of an allogeneic histocompatibility antigen, HLA-B7. Gene Therapy, 4, 419-425.
61. Doh, S.G. et al.. (1997). Spatial-temporal patterns of gene expression in mouse skeletal muscle after injection of lacZ plasmid DNA. Gene Therapy, 4, 648-663.
62. Chiang, C.M. et al. (1992). Viral E1 and E2 proteins support replication of homologous and heterologous papilloma viral origins. Proceedings of the National Academy of Sciences USA, 89, 5799-5803.
63. Thierry, A.R. et al. (1995). Systemic gene therapy: biodistribution and long-term expression of a transgene in mice. Proceedings of the National Academy of Sciences USA, 92, 9742-9746.
64. Cooper, M.J. et al. (1997). Safety-modified episomal vectors for human gene therapy. Proceedings of the National Academy of Sciences USA, 94, 6450-6455.
65. Calos, M.P. (1996). The potential of extrachromosomal replicating vectors for gene therapy. Trends Genetic, 12(11), 463-466.
66. Liang, X. et al. (1996). Novel, high expressing and antibiotic-controlled plasmid vectors designed for use in gene therapy. Gene Therapy, 3, 350-356.
67. Rivera, V.M. et al. (1996). A humanized system for pharmacologic control of gene expression. Nature Medicine, 2, 1028-1032.

68. Liberles, S.D. et al. (1997). Inducible gene expression and protein translocation using nontoxic ligands identified by a mammalian three-hybrid screen. Proceedings of the National Academy of Sciences, USA, 94, 7825-7830.
69. Vegeto, E. et al. (1992). The mechanism of RU486 antagonism is dependent on the conformation of the carboxy-terminal tail of the human progesterone receptor. Cell, 69, 703-713.
70. Wang, Y. et al. (1994). A regulatory system for use in gene transfer. Proceedings of the National Academy of Sciences, USA, 91, 8180-8184.
71. Wang, Y. et al. (1997b). Positive and negative regulation of gene expressin in eukaryotic cells with an inducible transcriptional regulator. Gene Therapy, 4, 432-441.
72. Wang, Y. et al. (1997a). Ligand-inducible and liver-specific target gene expression in transgenic mice. Nature Biotechnology, 15, 239-243.
73. Hines, R.N. et al. (1992). Large-scale purification of plasmid DNA by anion-exchange high-performance liquid chromatography. Biotechniques, 12, 430-434.
74. Horn, N.A. et al. (1995). Cancer gene therapy using plasmid DNA: purification of DNA for human clinical trials. Human Gene Therapy, 6, 565-573.
75. Wils, P. et al. (1997). Efficient purification of plasmid DNA for gene transfer using triple-helix affinity chromatography. Gene Therapy, 4, 323-330.
76. Fraley, R. et al. (1981). Liposome-mediated delivery of deoxyribonucleic acid to cells: enhanced efficiency of delivery related to lipid composition and incubation conditions. Biochemistry, 20, 6978-6987.
77. Wang, C.Y. and Huang, L. (1987). pH-sensitive immunoliposomes mediate target-cell-specific delivery and controlled expression of a foreign gene in mouse. Proceedings of the Naional Academy of Sciences, USA, 84, 7851-7855.
78. Soriano, P. et al. (1983). Targeted and nontargeted liposomes for *in vivo* transfer to rat liver cells of a plasmid containing the preproinsulin I gene. Proceedings of the National Academy of Sciences, USA, 80, 7128-7131.
79. Alino, S.F. et al. (Human α1-antitrypsin gene transfer to *in vivo* mouse hepatocytes. Human Gene Therapy, 7, 531-536.
80. Felgner, P.L. et al. (1987). Lipofedction: a highly efficient, lipid-mediated DNA-transfection procedure. Proceedings of the National Academy of Sciences, USA, 84, 7413-7417.
81. Behr, J.P. et al. (1989). Efficient gene transfer into mammalian primary endocrine cells with lipopolyamine-coated DNA. Proceedings of the National Academy of Sciences, USA, 86, 6982-6986.
82. Brigham, K.L. et al. (1989). Rapid communication: *in vivo* transfeciton of murine lungs with a functioning prokaryotic gene using a liposome vehicle. American Journal of Medical Science, 298, 278-281.
83. Hofland, H.E. et al. (1997). *In vivo* gene transfer by intravenous administration of stable cationic lipid/DNA complex. Pharmaceutical Research, 14 (6), 742-749.
84. Oudrhiri, N. et al. (1997). Gene transfer by guanidinium-cholesterol cationic lipids into airway epithelial cells *in vitro* and *in vivo*. Proceedings of the National Academy of Sciences, USA, 4, 94(5), 1651-1656.
85. Eastman, S.J. et al. (1997). Optimization of formulations and conditions for the aerosol delivery of functional cationic lipid:DNA complexes. Human Gene Therapy, 8(3), 313-322.
86. Yonemitsu, Y. et al. (1997) HVJ (Sendai virus)-cationic liposomes: a novel and potentially effective liposome-mediated technique for gene transfer to the airway epithelium. Gene Therapy, 4(7), 631-638.

87. McLachlan, G. et al. (1996). Laboratory and clinical studies in support of cystic fibrosis gene therapy using pCMV-CFTR-DOTAP. Gene Therapy, 3(12), 1113-1123.
88. Zhu, N. et al. (1993). Systemic gene experssion after intravenous DNA delivery into adult mice. Science, 261, 209-211.
89. Li, S. and Huang, L. (1997). In vivo gene transfer via intravenous administration of cationic lipid-protamine-DNA (LPD) complexes. Gene Therapy, 4 (9), 891-900.
90. Song, Y.K. et al. (1997). Characterization of cationic liposome-mediated gene transfer in vivo by intravenous administration. Human Gene Therapy, 1, 8(13), 1585-1594.
91. Hong, K. et al. (1997). Stabilization of cationic liposome-plasmid DNA complexes by polyamines and poly(ethylene glycol)-phospholipid conjugates for efficient in vivo gene delivery. FEBS Letters, 400(2), 233-237.
92. Templeton, N.S. et al. (1997). Improved DNA: liposome complexes for increased systemic delivery and gene expression. Nature Biotechnology, 15 (7), 647-652.
93. Liu, Y. et al. (1995). Cationic liposome-mediated intravenous gene delivery. Journal of Biological Chemistry, 270 (42), 24864-24870.
94. Koch G. and Bishop J.M. (1968). The effect of polycations on the interaction of viral RNA with mammalian cells: studies on the infectivity of single and double-stranded poliovirus RNA. Virology, 35, 9-17.
95. Wu, G.Y. et al. (1994). Incorporation of adenovirus into a ligand-based DNA carrier system results in retention of original receptor specificity and enhances targeted gene expression. Journal of Biological Chemistry, 269(15), 11542-11546.
96. Ferkol, T. et al. (1995). Gene transfer into the airway epithelium of animals by targeting the polymeric immunoglobulin receptor. Journal of Clinical Investigation, 95, 493-502.
97. Wu, G.Y. and Wu, C.H. (1987). Receptor-mediated in vitro gene transformation by a soluble DNA carrier system. Journal of Biological Chemistry, 262, 4429-4432.
98. Wu, G. Y. and Wu, C.H. (1988). Evidence for targeted gene delivery to Hep G2 hepatoma cells in vitro. Biochemistry, 27, 887-892.
99. Wu, G.Y. and Wu, C.H. (1988a). Receptor-mediated gene delivery and expression in vivo. Journal of Biological Chemistry, 263, 14621-14624.
100. Martinez-Fong, D. et al. (1994). Nonenzymatic glycosylation of poly-L-lysine: a new tool for targeted gene delivery. Hepatology, 20, 1602-1608.
101. Mislick, K.A. et al. (1995). Transfection of folate-polylysine DNA complexes: evidence for lysosomal delivery. Bioconjugate Chemistry, 6, 512-515.
102. Foster, B.J. and Kern, J.A. (1997). HER2-targeted gene transfer. Human Gene Therapy, 8, 719-727.
103. Harbottle, R.P. et al. (1998). An RGD-oligolysine peptide: a prototype construct for integrin-mediated gene delivery. Human Gene Therapy, 9, 1037-1047.
104. Trubetskoy, V.S. et al. (1992). Cationic liposomes enhance targeted delivery and expression of exogenous DNA mediated by N-terminal modified poly(L-lysine)-antibody conjugate in mouse lung endothelial cells. Biochimica et Biophysica Acta- Gene Structure and Expression, 15, 1131(3), 311-313.
105. Feero, W.G. et al. (1997). Selection and use of ligands for receptor-mediated gene delivery to myogenic cells. Gene Therapy, 4, 664-674.
106. Abdallah, B. et al. (1996). A powerful nonviral vector for in vivo gene transfer into the adult mammalian brain: polyethylenimine. Human Gene Therapy, 7, 1947-1954.
107. Boletta, A. et al. (1997). Human Gene Therapy, 8, 1243-1251.

108. Boussif, O. et al. (1995). A versatile vector for gene and oligonucleotide transfer into cells in culture and *in vivo*: polyethylenimine. Proceedings of the National Academy of Sciences, USA, 92, 7297-7301.
109. Baker, A. et al. (1997). Polyethylenimine (PEI) is a simple, inexpensive and effective reagent for condensing and linking plasmid DNA to adenovirus for gene delivery. Gene Therapy, 4, 773-782.
110. Ferrari, S. et al. (1997). ExGen 500 is an efficient vector for gene delivery to lung epithelial cells *in vitro* and *in vivo*. Gene Therapy, 4, 1100-1106.
111. Zanta, M.A. et al. (1997). *In vitro* gene delivery to hepatocytes with galactosylated polyethylenimine. Bioconjugate Chemistry, 8 (6), 839-844.
112. Kircheis, R. et al. (1997). Coupling of cell-binding ligands to polyethylenimine for targeted gene delivery. Gene Therapy, 4, 409-418.
113. Kukowska-Latallo, J.F. et al. (1996). Efficient transfer of genetic material into mammalian cells using Starburst polyamidoamine dendrimers. Proceedings of the National Academy of Sciences, 93, 4897-4902.
114. Tang, M.X. et al. (1996). *In vitro* gene delivery by degraded polyamidoamine dendrimers. Bioconjugate Chemistry, 7, 703-714.
115. Maruyama, A. et al. (1997). Poly(L-lysine)-graft-dextran copolymer is a novel stabilizer of triplex DNA (I): stabilization of poly(dA).2poly(dT)triplex. Nucleic Acids Symposium Service, 37, 225-226.
116. Baumgartner, I. et al. (1998). Constitutive expression of phVEGF165 after intramuscular gene transfer promotes collateral vessel development in patients with critical limb ischemia. Circulation, 97, 1114-1123.
117. Brigham, K. Abstract in Proceedings of the American Thoracic Society Annual Meeting, April 28, 1998, Chicago, Ill. (submitted to Nature Medicine).
118. Rommens, J.M. et al. (1989). Identification of the cystic fibrosis gene: Chromosome walking and jumping. Science, 245, 1059-65.
119. Riordan, J.R. et al. (1989). Identification of the cystic fibrosis gene: Cloning and characterization of complementary DNA. Science, 245, 1066-73.
120. Kerem, B-S. et al. (1989). Identification of the cystic fibrosis gene: Genetic analysis. Science, 245, 1073-80.
121. Rosenfeld, M.A. et al. (1992). *In vivo* transfer of the human cystic fibrosis transmembrane conductance regulator gene to the airway epithelium. Cell, 68, 143-155.
122. Alton, E.W.F.W. et al. (1993). Non-invasive liposome-mediated gene delivery can correct the ion transport defect in cystic fibrosis mutant mice. Nature Genetics, 5, 135-142.
123. Hyde, S.C. et al. (1993). Correction of the ion transport defect in cystic fibrosis transgenic mice by gene therapy. Nature, 362, 250-255.
124. McLachlan, G. et al. (1996). Laboratory and clinical studies in support of cystic fibrosis gene therapy using pCMV-CFTR-DOTAP. Gene Therapy, 3, 1113-1123.
125. Porteous, D.J. et al. (1997) Evidence for safety and efficacy of DOTAP cationic liposome mediated CFTR gene transfer to the nasal epithelium of patients with cystic fibrosis.Gene Therapy, 4(3), 210-218.
126. Caplen, N.J. et al.. (1995). Liposome-mediated CFTR gene transfer to the nasal epithelium of patients with cystic fibrosis. Nature Medicine, 1, 39-46.
127. Crystal, R.G. (1995) The gene as the drug. Nature Medicine, Volume 1, Number 1.
128. Wagner, J.A. and Gardner, P. (1997). Toward cystic fibrosis gene therapy. Annual Review of Medicine, 48, 203-16.

129. Freimark, B.D. *et al.* (1998). Cationic lipids enhance cytokine and cell influx levels in the lung following administration of plasmid: cationic lipid complexes. Journal of Immunology, 160, 4580-4586.
130. Nabel, G.J. *et al..* (1993). Direct gene transfer with DNA-liposome complexes in melanoma: expression, biologic activity, and lack of toxicity in humans. Proceedings of the National Academy of Sciences, USA, 90, 11307-11311.
131. Nabel, G.J. *et al.* (1996). Immune response in human melanoma after transfer of an allogeneic class I major histocompatibility complex gene with DNA-liposome complexes. Proceedings of the National Academy of Sciences, USA, 93, 15388-15393.
132. Stopeck, A.T. *et al.* (1997). Phase I study of direct gene transfer of an allogeneic histocompatibility antigen, HLA-B7, in patients with metastatic melanoma. Journal of Clinical Oncology, 15, (1), 341-349.
133. Rubin, J. *et al.* (1997). Phase I study of immunotherapy of hepatic metastases of colorectal carcinoma by direct gene transfer of an allogeneic histocompatibility antigen, HLA-B7. Gene Therapy, 4, 419-425.
134. Nabel, E.G. *et al.* (1994). Safety and toxicity of catheter gene delivery to the pulmonary vasculature in a patient with metastatic melanoma. Human Gene Therapy, 5, 1089-1094.
135. Murray, J.L. (1998). Proceedings of American Society of Clinical Oncology Meeting, Los Angeles.
136. Marchand, M. *et al.* (1995). Tumor regression responses in melanoma patients treated with a peptide encoded by gene MAGE-3. International Journal of Cancer, 63(6), 883-885.
137. Tabata, H. *et al.* (1997). Cardiovascular Research, 35(3), 470-479.
138. Tsurumi, Y. *et al.* (1997). Arterial gene transfer of acidic fibroblast growth factor for therapeutic angiogenesis *in vivo*: critical role of secretion signal in use of naked DNA. Circulation, 96(9 Suppl), II-II3828.
139. Isner, J.M. *et al.* (1996). Clinical evidence of angiogenesis after arterial gene transfer of phVEGF165 in patient with ischaemic limb. Lancet, 348(9024), 370-4.
140. Orkin, S.H. and Motulsky, A.G. (1995). Report and Recommendations of the Panel to Assess the NIH Investment in Research on Gene Therapy.
141. Vile, R.G. (1996). Gene therapy for cancer, the course ahead. Cancer and Metastasis Reviews, 15, 403-410.

Index

α-Interferon 6
Abciximab 35, 39
Actimmune 21, 22
Activated carbon filters 366
Acute myocardial infarction 185
Adeno-associated Virus Vectors 454
adeno-associated viruses 445
adenosine deaminase 444
adenoviral plasmid 455
Adenoviral Vectors 450
adenoviruses 445
Adsorption 222, 241
aggregation 90, 221
aggregration 144
AIDS 111, 488
albumin 89
Alferon LDO 21
Alferon N 21
alkaloid 2
alteplase 185, 186
Amgen 4
Amgen Inc 109, 251
amino acids 231
Amplicons 457
Ampligen 21
angioplasty 47
anistreplase 189
Antibacterial agents 235
Antibiotics 3
Antibodies 3
Anticoagulant 11
Anticoagulants 11
antisense RNAs 465
Antisense technology 30
antithrombin III 11
antithrombotic 35
aplastic anaemia 111
Arrhenius equation 219
asparaginase 13
Avonex 21, 22

Baby Hamster Kidney cells 127
Behringwerke 7
BeneFix® 73
Beta-interferons 173
Betaseron 21
Betasteron 22
bibliographic databases 389
bioinformatics 418
Biological Abstracts 390
Biological License Application 118
Biological Response Modifier 117
bioreactors 332
Biosis 395
BioTropin 16
BLA 118
blood clotting factors 3
blood glucose 150
Blood products 3
bone marrow transplantation 111
bovine papilloma virus 477
Bowes melanoma cell 196
Brevundimonas dimunuta 332
Bulk Drug Substance 227

Cancer 462
CAPTURE Trial 59
carbohydrate fingerprinting 86, 91
cardiovascular drugs 35
cationic lipid 480
cationic polymer 486
CD4 464
CD-ROM 393
CEA-scan 26
Centralized procedure for Marketing
    Authorization applications 299
Ceramic Hydroxyapatite 88
Ceredase 14
Cerezyme 14
Chemical Abstracts 390

Chemical Business NewsBase 400
Chemical Degradation 223
Chemical Sampling 382
chimaeric 25
chimeric vectors 459
Chinese hamster ovary 76
Chinese Hamster ovary (CHO) cell 6
Chiron Corporation 80
CHO cells 77
chorionic gonadotrophin 15
Christmas disease 75
chromatographic purification 8
chromatographic system 337
chromatography 337
chronic granulomatous disease 23
circular dichroism 92
Cleaning Validation 320
cleaning-in-place 340
clomiphene 140
clotting factors 74
Coagulation Factor IX 73
coagulation factors 55
collaborative electronic notebook systems 407
Colony stimulating factors 18
colony-stimulating factors 109
colorectal carcinoma 494
Computerised Systems Validation 321
Computer-Supported Cooperative Work 409
Concurrent Validation 314
COOH-terminal analysis 90
coronary artery disease 45
coronary thrombosis 210
co-stimulatory 463
CPMP 102, 299
Creutzfeldt-Jacob disease 4
Crohn's disease 263
cryogranulation 230
Current Biotechnology Abstracts 395, 400
CVMP 299

cyclodextrins 235
cystic fibrosis 444, 460
cytokines 3, 462
cytomegalovirus 475
cytosine deaminase 463
cytotoxicity 481

deamidation 144, 219, 223, 233
Decarbonation 367
decentralized' procedure 297
degradation profile 219
Deionisation 366
dendrimers 487
Design Audits 378
Design Qualification 347
diabetes 149
diabetes mellitus 149
dicoumarol 11
Digitalis 2
digitoxin 2
digoxin 2
DNase 13, 219
Downstream processing 8
drug delivery 238
drug discovery 405
drug registration 161
drug substance 218
dystentry 2

E. coli 6
Edman degradation 130
electrospray ionization 90
Eli Lilly 4, 150
Embase 398
EMBL 405
embryo transfer 140
EMEA 176, 289
ENABLING DISCLOSURE 258
endotoxin 97, 110, 320

Index 509

Endotoxin Sampling 382
endotoxins 365
env 447
enzyme immuno assays 131
Enzymes 3
EPIC Trial 49
EPO 18
EPOGEN 19
Epstein-Barr virus 477
Equilibrium dialysis 234
Erythrina trypsin inhibitor 197
erythropoiesis 18
Erythropoietin 6, 18, 251
Establishment License Application 118
estradiol 141
European Federation of Pharmaceutical Industries Association 309
European Medicines Evaluation Agency 289
European Patent Convention 415
European Patent Office 415
European pharmaceutical law 289
European Public Assessment Report 303
European Union 289
excipients 8, 89, 231
Extranets 409

factor IX 74
Factor VIII 6
Factory Acceptance Tests 325, 349
Fallopian tube 140
familial hypercholesterolemia 472
Fasta 412
FDA 103
Fermentation 227, 228
fibrin 75
Fibrinogen 36, 75
Fibrinolysis 64
Filgrastim 109
fluorescence spectroscopy 92
follicle stimulating hormone 15, 125

follitropin 125
Follitropin beta 125
Food and Drug Administration 118, 337
formulation 89, 217, 230, 242
freeze-drying 236
Freund's adjuvant 160
FSH 15, 125

gag 447
GAMP 321
gap analysis 328
Gaucher's disease 14
Genbank 413
gene therapy 28, 443
Genentech 4
Genetics Institute 77, 251
Genotropin 16
γ-interferon 175
glucocerebrosidase 14
glycerol 234
glycine 90
glycoprotein (GP) IIb/IIIa 35
glycosylation 7
GMP 311
gonadotrophins 15
gonadotropic cells 125
gonadotropins 126
Gonal F 16
Good Automation Manufacturing Practice 340
Good Manufacturing Practice 311
Good Manufacturing Practices 363
GP IIb/IIIa receptor 65
Granulocyte-colony stimulating factor 6
GRANULOKINE 111
(GUSTO)-III 63

haematopoiesis 109
Haematopoietic growth factors 109
haemophilia 444

haemophilia A 10
haemophilia B 10
Haeomopoietic growth factors 18
helper cell lines 447
hematopoietic stem cells 449
Hemophilia 73
hemophilia B 73
hemorrhagic disorder 74
heparin 11, 57
hepatitis B surface antigen 24
hepatitis C 173
hepatocytes 449, 461
herpes virus 444
herpes-simplex viruses 445
HETP 352
Hirudin 12
histidine 90
Humalog 15, 16, 149, 171
human chorionic gonadotrophin 125
Human Genome Project 408
Human growth hormone 6, 15
human immunodeficiency viruses 449
human papovavirus 477
human serum albumin 231, 235
humanized antibodies 25
Humatrope 16
Humulin 14, 15, 149
HVAC 340
hybridoma technology 4
hypoglycaemic 150

ICH 104, 319
IgG 37
Immunoassays 134, 157
immunoglobulin 127
immunoscintigraphy 25
Immunotherapy 493
in European Pharmacopoeia 97
In vitro bioassays 134
inclusion bodies 186, 197
IND 117

Index Medicus 391
Infanrix HepB 24
Infergen 21
infertility 125
Information technology 405
infringement 250
inhibin 137
insertional mutagenesis 453
Installation Qualification 316, 349, 372
Insulin 3, 6, 14, 149, 479
insulin analogues 149
insulin lispro 15, 149
insulin-like growth factor-I 152
insulin-like growth factor-II 152
Insuman 16
integrin 36, 44
interferon 20
Interferon Beta 173
interferons 3, 20, 465
interleukin 20
Interleukin 2 6
Interleukin-2 24, 220
interleukins 3, 20
International Conference for
    Harmonisation 218
International Pharmaceutical Abstracts
    401
internet 405
intrabodies 465
intracytoplasmic sperm injection 125
intranets 406, 409
Intron A 21, 22
inventiveness 249
inverted terminal repeat 450
Investigational New Drug 117
in-vitro fertilization 125
isoelectric focussing 126
IVF 125, 136

ketoacidosis 150
kidney transplant rejection 25

Index

Koseisho 103
Kupffer cell 479

lentiviruses 449
leukaemia 111
leukemia 173
Leukine 19
LH 125, 137
Lipid-Mediated Gene Transfer 479
liposomes 239, 471, 479
Liprolog 16
lispro 149
litigation 249
long terminal repeats 446
Lotus Notes 409
luteinizing hormone 15, 125
Lyosphere 145

macrophage 449
Maillard reaction 233
mannitol 112
marketing authorization 295
Marketing Issues 421
Marketing Mix 424
Marketing plan 421
marketing resources 438
mass spectrometry 90
master cell bank 81
Medicines Control Agency 103
Medline 392
Mercury 2
Methods Validation 319
Microbial Sampling 382
mitogens 495
monoclonal antibody 25
monocyte 449
multiple sclerosis 173
muscular dystrophy 472
mutual recognition 297, 301

myofibres 449
myogenic cells 485

N-terminal sequencing 93
NH_2-terminal sequencing 90
Nanofiltration 89
Neorecormon 19
NEUPOGEN 109
neurones 449
neutropenia 109
neutrophils 109
New Drug Application 118
non-Hodgkin's lymphoma 115
non-viral gene therapy 471
novelty 249
Novolin 15
Nutropin 16
NV Organon 126

OKT 3 Monoclonal antibody 6
OKT3 25
oncogenes 449, 473
OncoScint 26
oocytes 140
Operational Qualification 317, 350, 373, 374
orphan drug' 176
Orthoclone 26
ovarian hyperstimulation 136
ovarian stimulation 125
ovulation 141
oxidation 144, 219, 220

p53 462
PACE-SOL 81
Packaging cell lines 447
Paired basic amino-acid cleaving enzyme 78

Particulate Sampling 381
Patent 249
patenting 249
Pecacuanha 2
Peptide map analysis 95
peptide mapping 86, 90
Performance Qualification 318, 352, 374
pH stabilisation 233
pharmaceutical legislation 289
Pharmaceutical regulation 289
phase transition 228
phenylketonuria 472
pituitary 125
Pivotal trial 137
plasma 10
plasma products 73
plasma protein fraction 10
plasmids 474
Plasminogen activator inhibitor 194
plasmoviruses 457
Platelet 35, 116
pol 447
Polyclonal antibodies 25
polyethylene glycol 234
poly-L-lysine 485
Polypeptide-based Gene Delivery 484
polysorbate surfactants 231
polysorbate-80 90
postmenopausal 125
posttranslational modifications 91
posttranslational processing 77
poxvirus 456
precipitation 221
pregnancy 125, 140
process patent 250
Process Validation 319
Procrit 19
Product License Application 118
product of Nature 250
product patent 250
Product positioning 438
Product Pricing 435

product stability 228
product-by-process patent 250
progesterone 141, 477
proinsulin 156
Project Management 330
Proleukin 21
Prospective Validation 314, 342
Protein unfolding 220
prothrombin 75
Protocol execution 329
Protropin 15, 16
PTCA 48
Pulmozyme 13
Puregon 16, 125
Purified Water 364
pyrogens 341, 478

Q-Sepharose 88
Qualification 311
quality 218, 311
Quality Assurance 312
Quinine 2

rapid acting insulin 149
RAPPORT 62
Receptor-binding assays 134
recombinant plasminogen activator 185
recombinant tissue-type plasminogen activator 186
Recombinant vaccines 24
Recombivax HB 24
Refludan 7
reocclusion 185
reperfusion 200
replicative competent viruses 445
reteplase 185
Retrospective Validation 314, 342
retroviruses 445
Re-Validation 315

Index

Reverse Osmosis 366, 367
reverse-phase HPLC 93
reverse-transcriptase 446
rheumatoid arthritis 18
ribozymes 465
Roferon-A 21, 22

Saccharomyces cerevisiae 7
Saizen 16
sanitisation. 340
scale up 439
scanning calorimetry 228
Science Citation Index 391
Scripps Clinic and Research Foundation 251
SDS-PAGE 90, 93
Search engines 418
Sequence Retrieval System 412, 413
Serostim 16
sickle cell anaemia 444
size-exclusion high-performance liquid chromatography 93
Smith-Waterman algorithm 412
SOPs 324
sorbitol 112, 236
Specification Qualification 345
spermatogenesis 125
spermidine 486
spermine 486
stabilisation 217, 218
stability 217
steaming-in-place 340
steam-in-place 333
sterilise-in-place 333
Sterility 331
Storage conditions 228
strategic baseline review 422
streptokinase 185, 186
sucrose 90, 236
Suicide gene therapy 463
Sweet's syndrome 116

Swissprot 413

T lymphocytes 464
taxol 2
Teratology 159
The European Commission 290
The rules governing medicinal products in the European Union 294
The Validation Master Plan 312
Therapeutic antibodies 261
Therapeutic enzymes 13
thrombin 45, 75
thrombocytopenia 12, 44, 52
Thrombolysis 185
thrombolytic therapy 185
thrombosis 35, 43
thymidine kinase 463
Tissue plasminogen activator 6, 12, 254
Total Organic Carbon 375
Total Product Concept 428
tPA 12, 43
transfection 481
Tritanrix. 24
tumor associated antigens 25
tumor suppressor genes 473
tumour necrosis factor 219
Turnover Packages 324
Twinrix paediatric 24
type I interferons 173

U.S. Patent & Trademark Office 415
ultracentrifugation 92
Ultrafiltration/Diafiltration 87
ultraviolet disinfection 371
unique selling proposition 428
unstable angina 49
urokinase 13, 186
User Requirements Specification 346

Vaccines 3
vaccinia virus 456
Validation 311
Validation Management 345
Validation Master Plan 341, 343
Validation Protocols 326
Validation Turnover Package 373

warfarin 11
Water for Injection 364
Water for Injections 363
Water Softening 366
Water Specification 364
Wellcome Foundation 264
Western blot analysis 117
whole plasma 74
working cell bank 7, 81
World Wide Web 394, 411

zidovudine 115